Introductory Lectures on

Knot Theory

K&E Series on Knots and Everything — Vol. 46

Introductory Lectures on Knot Theory

Selected Lectures Presented at the Advanced School and Conference on Knot Theory and its Applications to Physics and Biology

ICTP, Trieste, Italy 11 – 29 May 2009

Editors

Louis H Kauffman
University of Illinois at Chicago, USA

Sofia Lambropoulou
National Technical University of Athens, Greece

Slavik Jablan
The Mathematical Institute, Serbia

Jozef H Przytycki
George Washington University, USA

The Abdus Salam
International Centre for Theoretical Physics

IAEA

World Scientifi

NEW JERSEY • LONDON • SINGAPORE • BEIJING • SHANGHAI • HONG KONG • TAIPEI • CHENNAI

Published by

World Scientific Publishing Co. Pte. Ltd.
5 Toh Tuck Link, Singapore 596224
USA office: 27 Warren Street, Suite 401-402, Hackensack, NJ 07601
UK office: 57 Shelton Street, Covent Garden, London WC2H 9HE

British Library Cataloguing-in-Publication Data
A catalogue record for this book is available from the British Library.

INTRODUCTORY LECTURES ON KNOT THEORY
Series on Knots and Everything — Vol. 46

ISBN-13 978-981-4307-99-4
ISBN-10 981-4307-99-8

Contents

Preface

This book is the fruit of a three-week summer school and conference held at the Abdus Salaam International Centre for Theoretical Physics in Trieste, Italy from May 11 to May 29 of 2009. The editors of this volume were invited to organize the School by Le Dung Trang, then Director of the Mathematics Institute at ICTP. In the three-week summer school, many short courses on knot theory and its applications were given. In this book, we have collected papers that are primarily related to the school. Each paper endeavors to be self-contained and to tell a story about knots from the beginning. In this preface, we shall say a few words about each paper and some words about how the papers relate to one another. We hope that this will provide a guide for the reader of this book.

The program was called the "Advanced School and Conference on Knot Theory and its Applications to Physics and Biology". There were 21 school lecturers. The conference took place from 25 May to 29 May 2009. There were 62 talks and poster sessions. There were in total 137 participants, most of them young researchers. Among them 41 were female researchers. The participants were from 43 countries. More information, abstracts, school notes and conference material can be found on the ICTP webpage: <http://cdsagenda5.ictp.trieste.it/full_display.php?smr=0&ida=a08157>. We now give capsule descriptions of the papers in this book.

The paper by Anna Beliakova and Thang Le gives a picture of the invariants of 3-manifolds that have emerger both from Witten's functional integral, generalized Vassiliev invariants and from the work of Reshetikhin and Turaev via quantum groups. All these constructions of invariants depend upon the use of the Kirby calculus to formulate problems about 3-manifolds in terms of problems about knots and links. This paper contains a quick introduction to the Kirby calculus as well as a survey of invariants of 3-manifolds.

The paper by Scott Carter is an introduction to quandles. Quandles are an algebraic generalization of the fundamental group of the complement of a knot. The algebraic axioms for a quandle parallel the structure of the Reidemeister moves (a set of moves on knot diagrams that generate isotopies of knots and so reformulate the theory of knots in terms of their diagrams). This paper explores the subject of quandles through to the use of state summations and quandle cocycles and their applications to surfaces embedded in 4-dimensional space.

The paper by Chmutov is an introduction to Vassiliev invariants. This approach to knot invariants was first proposed by Vassiliev in terms of the space of all knots and singular embeddings of knots, but quickly became related to skein theory via Birman and Lin and Lie algebras and physics via Bar-Natan.

The paper by Patrick Dehornoy describes his results on braid-ordering and the motivations of this work in the set theory of large cardinals and relationships with quandles and generalizations of quandles. The reader will see the structural remark that the self-distributivity axiom, placed on a mathematical system with a single binary operation, states that each element of the system acts as a homomorphism of that system to itself. Quandles are self-distributive, but have extra structure. One may ask about systems that are only self-distributive. In early set theoretic investigations, Dehornoy found that models for such single-axiom systems (magmas) could be found by using large transfinite cardinals. Later he found that these ideas could be used and applied in the theory of braids, and the large cardinals were not strictly necessary.

The paper by Roger Fenn gives a different introduction to quandles, starting with the notion of coloring of a knot diagram and continuing to formulations of racks and crossed modules and solutions to the set-theoretic Yang–Baxter equation. Applications are given to classical knots and to welded and virtual knots and links.

The paper by C. McA Gordon is a tight survey of surgery and Dehn fillings for hyperbolic 3-manifolds. This paper can be used as a source of ideas and references for the reader who wants to know more about current methods in the surgery of 3-manifolds.

The paper by Ilyutko and Manturov on graph links is the study of a generalization of knot theory beyond virtual knot theory and related to it via the properties of the intersection graphs of Gauss diagrams for virtual knots. The reader may want to read this paper along with the paper by L. H. Kauffman that introduces virtual knot theory. The graph link generalization of virtual knot theory is very natural, and sheds light on many problems in virtual knot theory. Ilyutko and Manturov take conscientious effort to show examples where their theory is strictly more general than virtual knot theory.

The paper by Slavik Jablan and Radmila Sazdanovic is a survey of open problems related to writhe, unknotting number and the behavior of numerous invariants including properties of Khovanov homology.

The paper by Louis H. Kauffman and Sofia Lambropoulou on unknots and tangles is, in fact, an elementary introduction to the theory of rational tangles, the Conway theorem classifying rational tangles in terms of their fractions, and applications to the structure of unknot diagrams, relationships with the tangle model for DNA recombination and relationships of continued fractions with other areas of mathematics. This paper contains many examples that occurred in the courses taught in Trieste by Kauffman and Lambropoulou.

The paper by Louis H. Kauffman introducing virtual knot theory is an introduction to this diagrammatic theory of knots and links in thickened surfaces and it contains new material about the arrow polynomial of Dye and Kauffman and new material about parity for virtual knots (see also the paper on parity in this volume by V. O. Manturov). The paper contains a quick introduction to Khovanov homology and an even quicker introduction to the Dye–Kauffman–Manturov generalizations of Khovanov homology for the arrow polynomial. Virtual knot theory is a natural extension of classical knot theory and is an active field of research at the time of this writing.

The paper by Louis H. Kauffman introducing Khovanov homology is a first pedagogical step into this large and significant subject. Khovanov homology generalizes the Jones polynomial to a homology theory and is part of a continuing breakthrough in the study of invariants of knots and links.

The paper by Sofia Lambropoulou on braid equivalences and the L-move is a survey of Lambropoulou's use of the L-move (see the paper for this key concept) and its applications to proving Markov theorems for braids in many contexts. The L-move, pioneered by Lambropoulou, provides new and conceptual approaches to the relationships between braids and their closures in many differenct categories.

The paper by Vassily O. Manturov on free knots and parity is about generalizations of knot theory motivated by virtual knot theory and ways to make new invariants from the inevitable parity considerations that arise in virtual knot theroy. In the Gauss diagram of a virtual knot, a given chord may be intersected by an even or by an odd number of other chords. The intersection is always even for classical knots, but in the virtual case and its generalizations, this access to parity leads to a host of new results, many of which are examined in this paper by Manturov.

The paper by Kenneth C. Millet on physical knot theory is a thorough survey of the use of knot theory in the study of polymer chains and of the many problems in the theory of piecewise linear knots that are inspired by the physical questions.

The paper by Hugh R. Morton on knots, satellites and quantum groups is a self-contained introduction to methods for using the Homflypt polynonmial to study satellite knots. In the course of this motivated introduction one meets methods that are related to the symmetric groups and their representations, the construction of link invariants from quantum groups and the use of Kirby calculus to construct 3-manifold invariants. This paper can be read along with the papers by Beliakova and Le and by Pichai in this volume.

The paper by Jozef Przytycki is an introductory survey of the theory of knots and its history. The reader new to knot theory can start with this paper and learn about colorings, tangles, quandles, constructions related to knot diagrams and many other matters fundamental to the theory of knots and links. The reader will also learn how to associate a symplectic structure to a tangle boundary so that tangles become Lagrangians (a proof of this result has not been published before). This paper, starting from history and elementary principles, leads to novel mathematics.

The paper by Ramadevi Pichai is devoted to the detection of chirality and mutations of knots and links via the use of the Chern–Simons–Witten approach to link invariants. This paper gives a quick introduction to the form of Witten's methods and will serve to whet the reader's appetite for more about this relationship of knot theory with physical ideas.

The paper by Eric J. Rawdon on physical knot theory should be read in tandem with the paper by Kenneth C. Millett. These two papers give a wide view of this field of application for piecewise linear knots.

The paper by Renzo L. Ricca is an introductory paper on the history and properties of the linking number as defined by C. F. Gauss by an integral related to electromagnetic theory. The paper makes a plausible reconstruction of how Gauss constructed his concept and definition of linking, providing an exciting point of view on the past and present of this central concept for linking and writhing of space curves.

The book also includes the list of participants (p. 503) in the Summer School, the Schedule of Talks and Course of the School (p. 507), a conference photo (p. xii) and a photo of the organizers (p. xiii). A link to the ICTP Advance School and Conference website is as follows: http://cdsagenda5.ictp.trieste.it/display_level.php?fid=1/70

We thank all the participants and lecturers in the Summer School for devoting their energy and time to this course, and we thank the authors of these papers for the extra energy and time that they have devoted to the composition of their work. Still, with all the work in this volume made explicit, this book is just a snapshot of a small part of the flow of information and ideas that went on in Trieste. A dramatic video on irrational Dehn surgery, created during the Trieste meeting, can be found at: http://www.youtube.com/results?search_query=Dehn%27s+dilemma&aq=f

To the reader: We wish you could have been there. If you were there, we hope this volume helps to recapture that time and carry it forward.

The illustration on the back cover of this book was composed by Leonardo DaVinci around 1496. DaVinci is celebrated as a keen observer of Nature, a theorist of powerful and inventive imagination and a very great artist. His depiction of a knot came from his speculations and observations on the nature of space and how space might be filled with a weave. We challenge the reader to compute the Jones polynomial for the DaVinci knot!

On the front cover the reader will find a more recent example of the art of knot depiction by Slavik Jablan. Jablan's work reaches outward into 4-dimensional space from a 3-dimensional weaving pattern. We feel that both illustrations capture the spirit of the book.

ADVANCED SCHOOL AND CONFERENCE ON KNOT THEORY AND ITS APPLICATIONS
TO PHYSICS AND BIOLOGY
11 - 29 May 2009
ICTP, Trieste - Italy

LinKnot
2009/05/26 11:15

ON THE UNIFICATION OF QUANTUM 3–MANIFOLD INVARIANTS

Anna Beliakova and Thang Le

ABSTRACT. In 2006 Habiro initiated a construction of generating functions for Witten–Reshetikhin–Turaev (WRT) invariants known as unified WRT invariants. In a series of papers together with Irmgard Bühler and Christian Blanchet we extended his construction to a larger class of 3–manifolds. The unified invariants provide a strong tool to study properties of the whole collection of WRT invariants, e.g. their integrality, and hence, their categorification.

In this paper we give a survey on ideas and techniques used in the construction of the unified invariants.

2000 Mathematics Subject Classification: 57N10 (primary), 57M25 (secondary)

INTRODUCTION

Background. In the 60s and 70s Rochlin, Lickorish and Kirby established a remarkable connection between links and 3–manifolds. They showed that every 3–manifold can be obtained by surgery (on S^3) along framed links, and surgeries along two links give the same 3–manifold if and only if the links are related by a sequence of Kirby moves. This allows us to think of 3–manifolds as equivalence classes of framed links modulo the relation generated by Kirby moves.

After the discovery of the Jones polynomial in 1984, knot theory experienced the transformation from an esoteric branch of pure mathematics to a dynamic research field with deep connections to mathematical physics, the theory of integrable and dynamic systems, von Neumann algebras, representation theory, homological algebra, algebraic geometry, etc. Among important developments were the constructions of *quantum link invariants*

Key words and phrases. 3–manifold, unified quantum invariant, Habiro ring, Andrews identity, Frobenius map.

(generalization of the Jones polynomial), and of *the Kontsevich integral* (universal finite type invariant).

The quantum link invariants were extended to 3–manifolds by Witten and Reshetikhin–Turaev (WRT), the lift of the Kontsevich integral to 3–manifolds was defined by Le–Murakami–Ohtsuki (LMO). However, while the relationship between the Kontsevich integral and quantum link invariants is simple, the relationship between the LMO invariant and the WRT invariants is much more complicated and remains mysterious in many cases.

Let us explain this in more details. The Kontsevich integral of a knot takes values in a certain algebra $\mathcal{A}(S^1)$ of chord diagrams. Any semi–simple Lie algebra and its module define a map $\mathcal{A}(S^1) \to \mathbb{Q}[[h]]$ called a weight system. One important result of Le–Murakami and Kassel is that the following diagram commutes, where the $\mathfrak{sl}_2$ weight system uses the 2–dimensional defining representation.

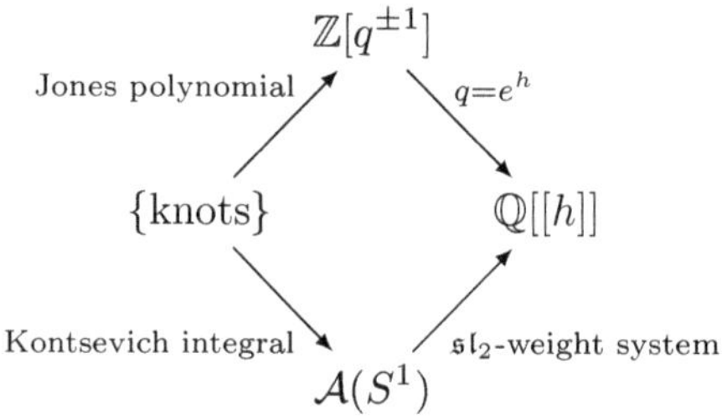

In particular, this proves that the Kontsevich integral dominates the Jones polynomial, and similarly all quantum link invariants coming from Lie algebras. Hence in addition to being universal for finite type, the Kontsevich integral is also universal for all quantum link invariants. It is conjectured that the Kontsevich integral separates knots.

Does there exist a similar commutative diagram for 3–manifolds? The quantum WRT invariant associates with a compact orientable 3–manifold M, a root of unity ξ and a semi–simple Lie algebra, say $\mathfrak{sl}_2$ for simplicity, a complex number $\tau_M(\xi)$. The LMO invariant takes values in a certain algebra $\mathcal{A}$ of Feynman diagrams. Every semi–simple Lie algebra defines a weight system map from $\mathcal{A}$ to $\mathbb{Q}[[h]]$. Hence, we have

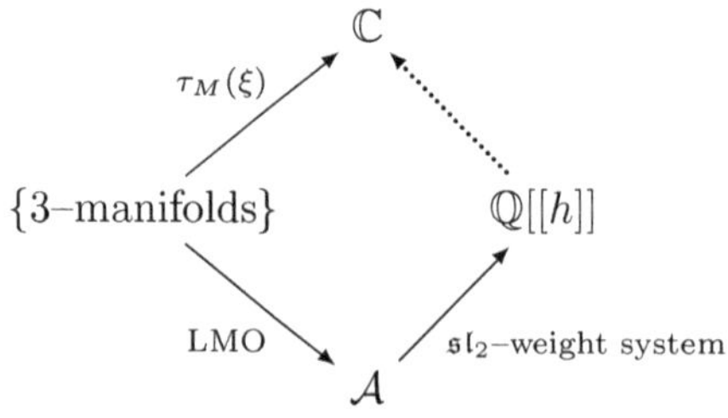

The image of the composition of the two bottom arrows is known as the Ohtsuki series [20]. Ohtsuki showed that for any rational homology 3–sphere M and a root of unity ξ of *prime* order p, the first $(p-1)/2$ coefficients, modulo p, of the Ohtsuki series are determined by $\tau_M(\xi)$. This is shown by the dotted arrow in the diagram.

This result of Ohtsuki raises many interesting questions: Does there exist any relationship between the LMO and WRT invariants at roots of unity of non–prime order? Is the whole set of WRT invariants determined by the LMO invariant? The discovery of the series prompted Ohtsuki to build the theory of finite type invariants of homology 3–spheres, which was further developed by Goussarov and Habiro.

In the 90s, Habegger, Garoufalidis and Beliakova showed that the LMO invariant is trivial if the first Betti number of a 3–manifold is bigger than 3, and for the Betti numbers 1,2 and 3, the LMO invariant is determined by the classical Casson–Walker–Lescop invariant. However, in the case of rational homology 3–spheres, the LMO invariant is more powerful, it is a universal finite type invariant in the Goussarov-Habiro sense.

The relationship between the LMO and WRT invariants at non–prime roots of unity remained open for quite a while. This is because all known techniques heavily rely on the fact that the order of the root is prime and can not be extended to other roots. However, recently Habiro's theory [7] of unified invariants provided a complete solution of this problem in the case of integral homology 3–spheres. In this case, also the question of integrality of the WRT invariants at non–prime roots of unity was solved simultaneously. Though intensively studied (see [17], [16], [6], [14] and the references there), the integrality of WRT invariant was previously known for prime roots of unity only. Note that a conceptual solution of the integrality problem is of primary importance for any attempt of categorification of the WRT invariants (compare [9]).

Unified invariants of integral homology 3–spheres. The unification of the WRT invariants was initiated in 2006 by Habiro. For any integral homology 3–sphere M, Habiro [7] constructed a *unified invariant* J_M whose evaluation at any root of unity coincides with the value of the WRT invariant at that root. Habiro's unified invariant J_M is an element of the following ring (Habiro's ring)

$$\widehat{\mathbb{Z}[q]} := \varprojlim_k \frac{\mathbb{Z}[q]}{((q;q)_k)}, \qquad \text{where} \quad (q;q)_k = \prod_{j=1}^{k}(1-q^j).$$

Every element $f(q) \in \widehat{\mathbb{Z}[q]}$ can be written (non–uniquely) as an infinite sum

$$f(q) = \sum_{k \geq 0} f_k(q)\,(1-q)(1-q^2)...(1-q^k),$$

with $f_k(q) \in \mathbb{Z}[q]$. When $q = \xi$, a root of unity, only a finite number of terms on the right hand side are not zero, hence the evaluation $\mathrm{ev}_\xi(f(q))$ is well–defined and is an algebraic integer. However, the fact that the unified invariant belongs to the Habiro ring is stronger than just integrality of $\tau_M(\xi)$.

The Habiro ring has beautiful arithmetic properties. Every element $f(q) \in \widehat{\mathbb{Z}[q]}$ can be considered as a function whose domain is the set of roots of unity. Moreover, there is a natural Taylor series for f at every root of unity. In [7] it is shown that two elements $f, g \in \widehat{\mathbb{Z}[q]}$ are the same if and only if their Taylor series at a root of unity coincide. In addition, each function $f(q) \in \widehat{\mathbb{Z}[q]}$ is totally determined by its values at, say, infinitely many roots of order 3^n, $n \in \mathbb{N}$. Due to these properties the Habiro ring is also called a ring of "analytic functions at roots of unity". Thus belonging to $\widehat{\mathbb{Z}[q]}$ means that the collection of the WRT invariants is far from a random collection of algebraic integers; together they form a nice function.

General properties of the Habiro ring imply that for any integral homology 3–sphere M, the Taylor expansion of the unified invariant J_M at $q = 1$ coincides with the Ohtsuki series and dominates WRT invariants of M at all roots of unity (not only of prime order). This is summarized in the following commutative diagram.

$$\begin{array}{ccccc}
 & \nearrow^{\tau_M(\xi)} & \mathbb{C} & & \\
 & & \uparrow{\scriptstyle q=\xi} & & \\
\{\mathbb{ZHS}\} & \xrightarrow{J_M(q)} & \widehat{\mathbb{Z}[q]} & \hookrightarrow & \mathbb{Z}[[1-q]] \\
 & \searrow_{\mathrm{LMO}} & & & \downarrow{\scriptstyle h=1-q} \\
 & & \mathcal{A} & \xrightarrow[\text{system}]{\mathfrak{sl}_2\text{–weight}} & \mathbb{Q}[[h]]
\end{array}$$

By $\mathbb{Z}$HS we denoted here the set of integral homology 3–spheres. In particular, this shows that Ohtsuki series has integral coefficients (which was conjectured by Lawrence and first proved by Rozansky).

Recently, Habiro ring found applications in algebraic geometry for constructing varieties over the non–existing field of one element [15].

Unified invariants of rational homology 3–spheres. In [2], we give a full generalization of the Habiro theory to rational homology 3–spheres. This requires completely new techniques coming from number theory, commutative algebra, quantum group and knot theory. Let us explain this in more details.

Assume M is a rational homology 3–sphere with $|H_1(M,\mathbb{Z})| = b$, where for a finite group G we denote by $|G|$ the number of its elements. Then our unified invariant I_M belongs to a modification $\mathcal{R}_b$ of the Habiro ring where b is inverted. Unlike the case $b = 1$, the modified Habiro ring is not an integral domain, but a direct product of integral domains, where each factor is determined by its own Taylor expansion at some root of unity. There is a decomposition $I_M = \prod_{c|b} I_{M,c}$, where $I_{M,c}$ dominates the set $\{\tau_M(\xi)|(\mathrm{ord}(\xi), b) = c\}$. If $b = 1$, then I_M coincides with Habiro's J_M. The invariant $I_{M,1}$ was first defined in [12].

Our results can be summarized in the following commutative diagram. Here we assume for simplicity that $b = p^k$ is a power of a prime and put $e_n := \exp(2\pi I/n)$ the primitive nth root of unity.

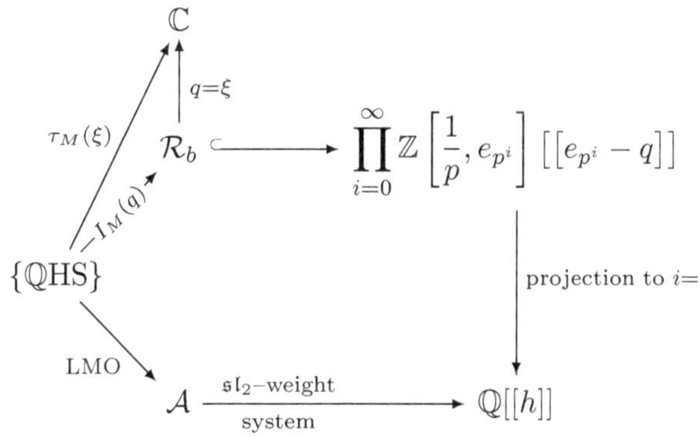

By $\mathbb{Q}$HS we denote the set of rational homology 3–spheres M, with $|H_1(M,\mathbb{Z})| = b$. In particular, for any $M \in \mathbb{Q}$HS, we generalize Ohtsuki series as follows. Let us fix a divisor $c < b$ of b, then the Taylor expansion of I_M at e_c is a power series in $(q - e_c)$ with coefficients in $\mathbb{Z}[1/b][e_c]$ which dominates the WRT invariants at roots of unity whose order has the greatest common divisor c with b. If $c = b = p^k$, then $I_{M,b}$ is a priory determined by a product of power series $\prod_{i\geq k} \mathbb{Z}\left[\frac{1}{p}, e_{p^i}\right][[e_{p^i} - q]]$, however, conjecturally, it is enough to consider the series in $q - e_b$.

The commutative diagram tells us that LMO invariant determines $I_{M,1}$, or the set of WRT invariants at roots of unity coprime with b. Since there is no direct way to obtain power series in $(q - e_c)$ from the LMO invariant, we conjecture the existence of the refined universal finite type invariant, dominating our power series. On the physical level of rigor, this means that the new refined invariant should capture more than just the contribution of flat connections from the Chern–Simons theory.

The methods used in the unification of the WRT invariants also led to the full solution of the integrality problem for quantum $SO(3)$ and $SU(2)$ WRT invariants. In [4, 3], we showed that $\tau_M^G(\xi)$ for any 3–manifold M and any root of unity is always an algebraic integer. Here $G = SO(3)$ or $SU(2)$. The integrality of the spin and cohomological refinements is work in progress.

Assume M is the Poincaré homology 3-sphere, obtained by surgery on a left-hand trefoil with framing -1. Then

$$I_M = \frac{1}{1-q}\sum_{k=0}^{\infty} q^k(1-q^{k+1})(1-q^{k+2})\dots(1-q^{2k+1}).$$

We expect that the categorification of the WRT invariants will lead to a homology theory with Euler characteristic given by I_M.

The paper is organized as follows. In Section 1 we recall the definitions of Kirby moves, WRT invariants and of the cyclotomic expansion for the colored Jones polynomial. In Section 2 we state our main results and outline the proofs. Section 3 is devoted to the discussion of the rings, where the unified invariants take their values. In addition, we construct generalized Ohtsuki series. The Laplace transform method, the Andrews identity, Frobenius maps and the integrality of WRT invariants are explained in the last Section.

1. Quantum (WRT) invariants

1.1. Notations and conventions. We will consider $q^{1/4}$ as a free parameter. Let

$$\{n\} = q^{n/2} - q^{-n/2}, \quad \{n\}! = \prod_{i=1}^{n}\{i\}, \quad [n] = \frac{\{n\}}{\{1\}}.$$

We denote the set $\{1, 2, 3, \dots\}$ by $\mathbb{N}$. We also use the following notation from q–calculus:

$$(x;q)_n := \prod_{j=1}^{n}(1 - xq^{j-1}).$$

Throughout this paper, ξ will be a primitive root of unity of *odd* order r and $e_n := \exp(2\pi I/n)$.

All 3–manifolds in this paper are supposed to be closed and oriented. Every link in S^3 is framed, oriented, and has components ordered. For a link L, let (L_{ij}) be its linking matrix, where for $i \neq j$, $L_{ij} := \mathrm{lk}(L_i, L_j)$ is the linking number of the ith and the jth components, and $L_{ii} = b_i$ is the framing of L_i, given by the linking number of L_i with its push off along the first vector in the framing.

Surgery along the framed link L consists of removing a tubular neighborhood of L from S^3 and then gluing it back with the diffeomorphism given by framing (i.e. the meridian of each removed solid torus, diffeomorhic to a neighborhood of L_i, is identified with the longitude of the complement twisted b_i times along the meridian). We denote by M the result of the surgery.

1.2. **The colored Jones polynomial.** Suppose L is a framed oriented link with m ordered components and $V_1, \ldots, V_m$ are finite-dimensional modules over a ribbon Hopf algebra. Then one can define the quantum invariant $J_L(V_1, \ldots, V_m)$ through the machinery of quantum link invariant theory, see [21]. The quantized enveloping algebra $U_h(sl_2)$ of sl_2 is a ribbon algebra, and for each positive integer n there is a unique $U_h(sl_2)$-module of dimension n. The quantum link invariant $J_L(n_1, \ldots, n_m)$, where n_j stands for the n_j-dimensional $U_h(sl_2)$-module, is usually called the colored Jones polynomial, with the n_j's being the colors. When all the colors are 2, $J_L(2, \ldots, 2)$ is the usual Jones polynomial, which can be defined using skein relation.

One can construct the colored Jones polynomial without the quantum group theory by first defining the Jones polynomial through the skein relation, and then defining the colored Jones polynomial by using cablings, see e.g. [10, 11].

Let us recall here a few well–known formulas. For the unknot U with 0 framing one has

$$J_U(n) = [n] \tag{1}$$

Moreover, J_L is multiplicative with respect to the disjoint union. If L_1 is obtained from L by increasing the framing of the ith component by 1, then

$$J_{L_1}(n_1, \ldots, n_m) = q^{(n_i^2-1)/4} J_L(n_1, \ldots, n_m). \tag{2}$$

If all the colors n_i are odd, then $J_L(n_1, \ldots, n_m) \in \mathbb{Z}[q^{\pm 1}]$.

1.3. **Kirby moves.** By Kirby theorem, any link invariant which does not change under Kirby moves is an invariant of a 3–manifold given by surgery on that link. Let us first recall what the Kirby moves are.

K1–Move (handle slide): For some $i \neq j$, replace the ith component L_i with L'_i, a band connected sum of L_i with a push off of L_j (defined by the framing), with $b'_i = b_i + b_j + 2\mathrm{lk}(L_i, L_j)$.

K2–Move (blow up): Add (or delete) a split unknotted component with framing ± 1.

These two moves are equivalent to the one Fenn–Rourke move defined as follows:

FR–Move Locally the following two pictures are interchangeable

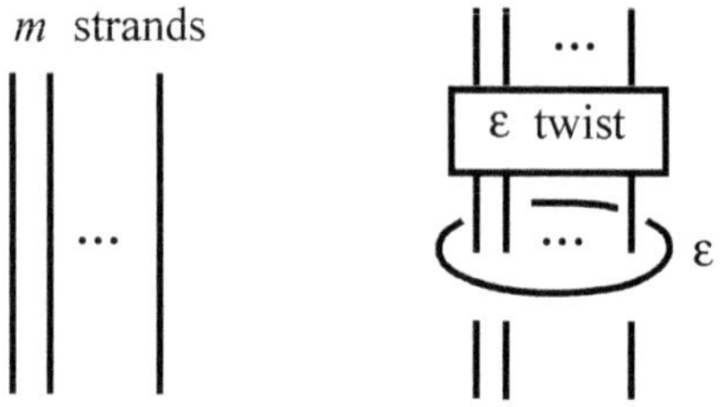

where $\varepsilon \in \{1, -1\}$ and the closed component has framing ε. Note that K2–Move corresponds to the case when $m = 0$.

The main idea of the construction of the WRT invariants is to make the colored Jones polynomial invariant under Kirby moves by averaging over all colors. To make this precise, we have to choose the quantum parameter q to be a root of unity, otherwise the sum would be infinite.

1.4. **Evaluation and Gauss sums.** For each root of unity ξ of odd order r, we define the evaluation map ev_ξ by replacing q with ξ.

Suppose $f(q; n_1, \ldots, n_m)$ is a function of variables $q^{\pm 1}$ and integers $n_1, \ldots, n_m$. In quantum topology, the following sum plays an important role

$$\sum_{n_i}^{\xi} f := \sum_{\substack{0<n_i<2r \\ n_i \text{ odd}}} \mathrm{ev}_\xi f(q; n_1, \ldots, n_m)$$

where in the sum all the n_i run over the set of *odd* numbers between 0 and $2r$.

In particular, the following variation of the Gauss sum

$$\gamma_b(\xi) := \sum_{n}^{\xi} q^{b\frac{n^2-1}{4}}$$

is well–defined, since for odd n, $4 \mid n^2 - 1$. It is known that, for odd r, $|\gamma_b(\xi)|$ is never 0.

1.5. **Definition of the WRT invariants.** Let

$$F_L(\xi) := \sum_{n_i}^{\xi} \prod_{i=1}^{m} [n_i]\, J_L(n_1, \ldots, n_m).$$

Theorem 1. [Reshetikhin–Turaev] *$F_L(\xi)$ is invariant under K1–Move.*

An important special case is when $L = U^b$, the unknot with framing $b \neq 0$. In that case $F_{U^b}(\xi)$ can be calculated using the Gauss sum and is nonzero.

Let σ_+ (respectively σ_-) be the number of positive (negative) eigenvalues of the linking matrix of L. Then we define

$$\tau_M(\xi) = \frac{F_L(\xi)}{(F_{U^{+1}}(\xi))^{\sigma_+}\,(F_{U^{-1}}(\xi))^{\sigma_-}}. \tag{3}$$

It is easy to see that $\tau_M(\xi)$ is invariant under K2–Move, and hence, by Theorem 1, it is a topological invariant of M called the $SO(3)$ WRT invariant. Moreover, $\tau_M(\xi)$ is multiplicative with respect to the connected sum.

Remark. If we drop the condition that the colors in the summation are odd, and sum over all (odd and even) colors, the result will be the $SU(2)$ WRT invariant $\tau_M^{SU(2)}(\xi)$. In this case, the order of ξ could also be even.

The $SO(3)$ WRT invariant extends naturally to the invariant of the pair (M, L'), where the manifold M contains a link L' inside. In this case we have to replace the surgery link L of M by $L \cup L'$ in all definitions, fix colors on L' and sum over all colorings of L only. We omit here the precise definition and refer to [2] for more details.

For example, the $SO(3)$ invariant of the lens space $L(b,1)$, obtained by surgery along U^b, is

$$\tau_{L(b,1)}(\xi) = \frac{F_{U^b}(\xi)}{F_{U^{\mathrm{sn}(b)}}(\xi)}, \tag{4}$$

were $\mathrm{sn}(b)$ is the sign of the integer b.

Suppose M is a rational homology 3–sphere. Then there is a unique decomposition $H_1(M,\mathbb{Z}) = \bigoplus_i \mathbb{Z}/b_i\mathbb{Z}$, where each b_i is a prime power. We renormalize the $SO(3)$ WRT invariant as follows:

$$\tau'_M(\xi) = \frac{\tau_M(\xi)}{\prod\limits_i \tau_{L(b_i,1)}(\xi)}, \tag{5}$$

where $L(b,a)$ denotes the (b,a) lens space. Note that $\tau_{L(b,1)}(\xi)$ is always nonzero.

Let us focus on the special case when the linking matrix of L is diagonal. Assume each $L_{ii} = b_i$ is a power of a prime or 1, up to sign. Then $H_1(M,\mathbb{Z}) = \oplus_{i=1}^m \mathbb{Z}/|b_i|$, and

$$\sigma_+ = \mathrm{card}\,\{i \mid b_i > 0\}, \quad \sigma_- = \mathrm{card}\,\{i \mid b_i < 0\}.$$

Thus from the definitions (3), (4) and (5) we have

$$\tau'_M(\xi) = \left(\prod_{i=1}^m \tau'_{L(b_i,1)}(\xi)\right) \frac{F_L(\xi)}{\prod_{i=1}^m F_{U^{b_i}}(\xi)}, \tag{6}$$

with

$$\tau'_{L(b_i,1)}(\xi) = \frac{\tau_{L(b_i,1)}(\xi)}{\tau_{L(|b_i|,1)}(\xi)}.$$

The collection of $SO(3)$ WRT invariants is difficult to study, since their definition heavily depends on the order of ξ. The following cyclotomic expansion will play an important role for the unification of the WRT invariants.

1.6. **Habiro's cyclotomic expansion of the colored Jones polynomial.** For non–negative integers n, k we define

$$A(n,k) := \frac{\prod_{i=0}^{k}\left(q^n + q^{-n} - q^i - q^{-i}\right)}{(1-q)\,(q^{k+1};q)_{k+1}}.$$

For $\mathbf{k} = (k_1, \ldots, k_m)$ let

$$A(\mathbf{n},\mathbf{k}) := \prod_{j=1}^{m} A(n_j, k_j).$$

Note that $A(\mathbf{n},\mathbf{k}) = 0$ if $k_j \geq n_j$ for some index j. Also $A(n,0) = q^{-1}J_U(n)^2$.

The colored Jones polynomial $J_L(\mathbf{n})$ can be repackaged into the invariant $C_L(\mathbf{k})$ as stated in the following theorem.

Theorem 2. [Habiro] *Suppose L is a link in S^3 having zero linking matrix. Then there are invariants*

$$C_L(\mathbf{k}) \in \frac{(q^{k+1};q)_{k+1}}{(1-q)}\,\mathbb{Z}[q^{\pm 1}], \quad \textit{where } k = \max\{k_1, \ldots, k_m\} \tag{7}$$

such that for every $\mathbf{n} = (n_1, \ldots, n_m)$

$$J_L(\mathbf{n}) \prod_{i=1}^{m} [n_i] = \sum_{0 \leq k_i \leq n_i - 1} C_L(\mathbf{k})\, A(\mathbf{n},\mathbf{k}). \tag{8}$$

Note that the existence of $C_L(\mathbf{k})$ as rational functions in q satisfying (8) is easy to establish. They correspond to the Jones polynomial colored by different elements of the Grothendieck ring of $U_q(\mathfrak{sl}_2)$, i.e. by linear combinations of representations. The difficulty here is to show the integrality of (7).

Since $A(\mathbf{n},\mathbf{k}) = 0$ unless $\mathbf{k} < \mathbf{n}$, in the sum on the right hand side of (8) one can assume that $\mathbf{k}$ runs over the set of all m–tuples $\mathbf{k}$ with non–negative integer components. We will use this fact later.

2. Main Results

Let us state our main results announced in the introduction.

For any positive integer b, we define the cyclotomic completion ring $\mathcal{R}_b$ to be

(9)
$$\mathcal{R}_b := \lim_{\overleftarrow{k}} \frac{\mathbb{Z}[1/b][q]}{((q;q^2)_k)}, \qquad \text{where} \quad (q;q^2)_k = (1-q)(1-q^3)\dots(1-q^{2k-1}).$$

For any $f(q) \in \mathcal{R}_b$ and a root of unity ξ of *odd* order, the evaluation $\mathrm{ev}_\xi(f(q)) := f(\xi)$ is well–defined. Similarly, we put

$$\mathcal{S}_b := \lim_{\overleftarrow{k}} \frac{\mathbb{Z}[1/b][q]}{((q;q)_k)} .$$

Here the evaluation at any root of unity is well–defined. For odd b, there is a natural embedding $\mathcal{S}_b \hookrightarrow \mathcal{R}_b$.

Let us denote by $\mathcal{M}_b$ the set of rational homology 3–spheres such that $|H_1(M,\mathbb{Z})|$ divides b^n for some n. Our main result is the following.

Theorem 3. [Beliakova–Bühler–Le] *Suppose the components of a framed oriented link $L \subset M$ have odd colors, and $M \in \mathcal{M}_b$. Then there exists an invariant $I_{M,L} \in \mathcal{R}_b$, such that for any root of unity ξ of odd order*

$$\mathrm{ev}_\xi(I_{M,L}) = \tau'_{M,L}(\xi) .$$

In addition, if b is odd, then $I_{M,L} \in \mathcal{S}_b$.

If $b = 1$ and L is the empty link, I_M coincides with Habiro's unified invariant J_M and $\mathcal{S}_1 = \widehat{\mathbb{Z}[q]}$.

One may ask what is the evaluation of I_M at an even root of unity in the case when b is odd. In her PhD thesis [5], Bühler shows that it coincides with $\tau_M^{\prime SU(2)}(\xi)$. Hence, for odd b, I_M dominates both $SO(3)$ and $SU(2)$ WRT invariants. An analogous result for b even is work in progress.

Compared to Habiro's case, the proof of Theorem 3 uses the following new techniques: 1) the Laplace transform method; 2) the difficult number theoretical identity of Andrews generalizing those of Roger–Ramanujan; 3) the Frobenius type isomorphism providing the existence of the b–th root of q in $\mathcal{R}_b$. In addition, we had to generalize the deep integrality result of Habiro (Theorem 2), to a union of an algebraically split link with any odd colored one.

The rings $\mathcal{R}_b$ and $\mathcal{S}_b$ have properties similar to those of the Habiro ring. An element $f(q) \in \mathcal{R}_b$ is totally determined by the values at many infinite sets of roots of unity (see Section 3), one special case is the following.

Proposition 4. [Beliakova–Bühler–Le] *Let p be an odd prime not dividing b and T the set of all integers of the form $p^k b'$ with $k \in \mathbb{N}$ and b' any odd* *divisor of b^n for some n. Any element $f(q) \in \mathcal{R}_b$, and hence also $\{\tau_M(\xi)\}$, is totally determined by the values at roots of unity with orders in T.*

The general properties of the ring $\mathcal{R}_b$ allow to introduce generalized Ohtsuki series as the Taylor expansions of I_M at roots of unity. In addition, we show that these Taylor expansions satisfy congruence relations similar to the original definition of the Ohtsuki series (see Section 3.5).

2.1. **Strategy of the proof.** Let us outline the proof of Theorem 3 and state the main technical results that will be explained later.

We restrict to the case $L = \emptyset$ for simplicity. We would like to define $I_M \in \mathcal{R}_b$, such that

$$\tau'_M(\xi) = \mathrm{ev}_\xi\,(I_M) \tag{10}$$

for any root of unity ξ of odd order. This unified invariant is multiplicative with respect to the connected sum.

The following observation is important. By Proposition 4, there is *at most one* element $f(q) \in \mathcal{R}_b$ such that for every root ξ of odd order one has

$$\tau'_M(\xi) = \mathrm{ev}_\xi\,(f(q))\,.$$

That is, if we can find such an element, it is unique, and we put $I_M := f(q)$.

2.2. **Laplace transform.** The following is the main technical result of [2]. A proof will be explained in the next Section.

Theorem 5. [Beliakova–Bühler–Le] *Suppose $b = \pm 1$ or $b = \pm p^l$ where p is a prime and l is positive. For any non-negative integer k, there exists an element $Q_{b,k} \in \mathcal{R}_b$ such that for every root ξ of odd order r one has*

$$\frac{\sum_n^\xi q^{b\frac{n^2-1}{4}} A(n,k)}{F_{U^b}(\xi)} = \mathrm{ev}_\xi(Q_{b,k}).$$

In addition, if b is odd, $Q_{b,k} \in \mathcal{S}_b$.

2.3. **Definition of the unified invariant: diagonal case.** Suppose that the linking number between any two components of L is 0, and the framing on components of L are $b_i = \pm p_i^{k_i}$ for $i = 1, \ldots, m$, where each p_i is prime or 1. Let us denote the link L with all framings switched to zero by L_0.

Using (8), taking into account the framings b_i's, we have

$$J_L(\mathbf{n}) \prod_{i=1}^m [n_i] = \sum_{\mathbf{k}\geq 0} C_{L_0}(\mathbf{k}) \prod_{i=1}^m q^{b_i \frac{n_i^2-1}{4}} A(n_i, k_i).$$

By the definition of F_L, we have

$$F_L(\xi) = \sum_{\mathbf{k}\geq 0} \mathrm{ev}_\xi(C_{L_0}(\mathbf{k})) \prod_{i=1}^{m} \sum_{n_i}^{\xi} q^{b_i \frac{n_i^2-1}{4}} A(n_i, k_i).$$

From (6) and Theorem 5, we get

$$\tau'_M(\xi) = \mathrm{ev}_\xi \left\{ \prod_{i=1}^{m} I_{L(b_i,1)} \sum_{\mathbf{k}} C_{L_0}(\mathbf{k}) \prod_{i=1}^{m} Q_{b_i,k_i} \right\},$$

where the existence of the unified invariant of the lens space $I_{L(b_i,1)} \in \mathcal{R}_b$, with $\mathrm{ev}_\xi(I_{L(b_i,1)}) = \tau'_{L(b_i,1)}(\xi)$ can be shown by a direct computation (we refer to [2] for more details).

Thus if we define

$$I_M := \prod_{i=1}^{m} I_{L(b_i,1)} \sum_{\mathbf{k}} C_{L_0}(\mathbf{k}) \prod_{i=1}^{m} Q_{b_i,k_i},$$

then (10) is satisfied. By Theorem 2, $C_{L_0}(\mathbf{k})$ is divisible by $(q^{k+1};q)_{k+1}/(1-q)$, which is divisible by $(q;q)_k$, where $k = \max k_i$. It follows that $I_M \in \mathcal{R}_b$. In addition, if b is odd, then $I_M \in \mathcal{S}_b$.

2.4. **Diagonalization using lens spaces.** The general case reduces to the diagonal case by the well–known trick of diagonalization using lens spaces. We say that M is *diagonal* if it can be obtained from S^3 by surgery along a framed link L with diagonal linking matrix, where the diagonal entries are of the form $\pm p^k$ with $p = 0, 1$ or a prime. The following lemma was proved in [12, Proposition 3.2 (a)].

Lemma 6. *For every rational homology sphere M, there are lens spaces $L(b_i, a_i)$ such that the connected sum of M and these lens spaces is diagonal. Moreover, each b_i is a prime power divisor of $|H_1(M,\mathbb{Z})|$.*

To define the unified invariant for a general rational homology sphere M, one first adds to M lens spaces to get a diagonal M', for which the unified invariant $I_{M'}$ had been defined in Subsection 2.3. Then I_M is the quotient of $I_{M'}$ by the unified invariants of the lens spaces. But unlike the simpler case of [12], the unified invariant of lens spaces are *not* invertible in general. To overcome this difficulty we insert knots in lens spaces and split the unified invariant into different components. This is also the reason why we need to generalize Habiro's integrality result to algebraically split links together with odd colored components.

3. Cyclotomic completions of polynomial rings

Since unified invariants belong to cyclotomic completions of polynomial rings, we outline their construction and the main properties. For simplicity, only the case b is a power of a prime is considered, the general case is treated in [2].

3.1. **On cyclotomic polynomials.** Recall that $e_n := \exp(2\pi I/n)$ and denote by $\Phi_n(q)$ the cyclotomic polynomial

$$\Phi_n(q) = \prod_{\substack{(j,n)=1\\ 0<j<n}} (q - e_n^j).$$

For example, $\Phi_1(q) = q - 1$ and $\Phi_2(q) = q + 1$. The degree of $\Phi_n(q) \in \mathbb{Z}[q]$ is given by the Euler function $\varphi(n)$. Suppose p is a prime and n an integer. Then (see e.g. [18])

$$\Phi_n(q^p) = \begin{cases} \Phi_{np}(q) & \text{if } p \mid n \\ \Phi_{np}(q)\Phi_n(q) & \text{if } p \nmid n. \end{cases} \tag{11}$$

It follows that $\Phi_n(q^p)$ is always divisible by $\Phi_{np}(q)$.

The ideal of $\mathbb{Z}[q]$ generated by $\Phi_n(q)$ and $\Phi_m(q)$ is well–known, see e.g. [12, Lemma 5.4]:

Lemma 7.

(a) *If $\frac{m}{n} \neq p^e$ for any prime p and any integer $e \neq 0$, then $(\Phi_n) + (\Phi_m) = (1)$ in $\mathbb{Z}[q]$.*
(b) *If $\frac{m}{n} = p^e$ for a prime p and some integer $e \neq 0$, then $(\Phi_n) + (\Phi_m) = (p)$ in $\mathbb{Z}[q]$.*

3.2. **Habiro's results.** Let us summarize some of Habiro's results on cyclotomic completions of polynomial rings [8]. Let R be a commutative integral domain of characteristic zero and $R[q]$ the polynomial ring over R. For any $S \subset \mathbb{N}$, Habiro defined the S–cyclotomic completion ring $R[q]^S$ as follows:

$$R[q]^S := \varprojlim_{f(q)\in\Phi_S^*} \frac{R[q]}{(f(q))} \tag{12}$$

where Φ_S^* denotes the multiplicative set in $\mathbb{Z}[q]$ generated by $\Phi_S = \{\Phi_n(q) \mid n \in S\}$ and directed with respect to the divisibility relation.

For example, since the sequence $(q;q)_n$, $n \in \mathbb{N}$, is cofinal to $\Phi_{\mathbb{N}}^*$, we have

$$\widehat{\mathbb{Z}[q]} \simeq \mathbb{Z}[q]^{\mathbb{N}}. \tag{13}$$

Note that if S is finite, then $R[q]^S$ is identified with the $(\prod \Phi_S)$–adic completion of $R[q]$. In particular,

$$R[q]^{\{1\}} \simeq R[[q-1]], \quad R[q]^{\{2\}} \simeq R[[q+1]].$$

Suppose $S' \subset S$, then $\Phi^*_{S'} \subset \Phi^*_S$, hence there is a natural map

$$\rho^R_{S,S'} : R[q]^S \to R[q]^{S'}.$$

Recall important results concerning $R[q]^S$ from [8]. Two positive integers n, n' are called *adjacent* if $n'/n = p^e$ with a nonzero $e \in \mathbb{Z}$ and a prime p, such that the ring R is p–adically separated, i.e. $\bigcap_{n=1}^{\infty}(p^n) = 0$ in R. A set of positive integers is R*–connected* if for any two distinct elements n, n' there is a sequence $n = n_1, n_2, \ldots, n_{k-1}, n_k = n'$ in the set, such that any two consecutive numbers of this sequence are adjacent. Theorem 4.1 of [8] says that if S is R–connected, then for any subset $S' \subset S$ the natural map $\rho^R_{S,S'} : R[q]^S \hookrightarrow R[q]^{S'}$ is an embedding.

Example: Assume $R = \mathbb{Z}$, $S = \mathbb{N}$ and $S' = \{1\}$, then we have that the map $\widehat{\mathbb{Z}[q]} \to \mathbb{Z}[[q-1]]$ is an embedding. This implies that, for any integral homology 3–sphere M, J_M is determined by the Ohtsuki series and this series has integral coefficients.

If ζ is a root of unity of order in S, then for every $f(q) \in R[q]^S$ the evaluation $\mathrm{ev}_\zeta(f(q)) \in R[\zeta]$ can be defined by sending $q \to \zeta$. For a set Ξ of roots of unity whose orders form a subset $\mathcal{T} \subset S$, one defines the evaluation

$$\mathrm{ev}_\Xi : R[q]^S \to \prod_{\zeta \in \Xi} R[\zeta].$$

Theorem 6.1 of [8] shows that if $R \subset \mathbb{Q}$, S is R–connected, and there exists $n \in S$ that is adjacent to infinitely many elements in $\mathcal{T}$, then ev_Ξ is injective.

Example: Consider again the case when $R = \mathbb{Z}$, $S = \mathbb{N}$ and put $\mathcal{T} = \{3^n | n \in \mathbb{N}\}$, then $3 \in S$ is adjacent to infinitely many elements of $\mathcal{T}$ and hence, for any integral homology 3–sphere M, the whole set of its WRT invariants is determined by the evaluations of J_M at roots of unity of order in $\mathcal{T}$.

3.3. **Taylor expansion.** Fix a natural number n, then we have

$$R[q]^{\{n\}} = \varprojlim_k \frac{R[q]}{(\Phi^k_n(q))}.$$

Suppose $\mathbb{Z} \subset R \subset \mathbb{Q}$, then the natural algebra homomorphism

$$h : \frac{R[q]}{(\Phi^k_n(q))} \to \frac{R[e_n][q]}{((q-e_n)^k)}$$

can be proved to be injective. Taking the inverse limit, we see that there is a natural injective algebra homomorphism

$$h : R[q]^{\{n\}} \to R[e_n][[q - e_n]].$$

Suppose $n \in S$. Combining h and $\rho_{S,\{n\}} : R[q]^S \to R[q]^{\{n\}}$, we get an algebra map

$$\mathfrak{t}_n : R[q]^S \to R[e_n][[q - e_n]].$$

If $f \in R[q]^S$, then $\mathfrak{t}_n(f)$ is called the Taylor expansion of f at e_n.

3.4. **Splitting of $\mathcal{S}_p$ and evaluation.** For every integer a, we put $\mathbb{N}_a := \{n \in \mathbb{N} \mid (a, n) = 1\}$.

Suppose p is a prime. Analogously to (13), we have

$$\mathcal{S}_p \simeq \mathbb{Z}[1/p][q]^{\mathbb{N}}.$$

Observe that $\mathbb{N}$ is not $\mathbb{Z}[1/p]$–connected. In fact one has $\mathbb{N} = \amalg_{j=0}^{\infty} p^j \mathbb{N}_p$, where each $p^j \mathbb{N}_p$ is $\mathbb{Z}[1/p]$–connected. Let us define

$$\mathcal{S}_{p,j} := \mathbb{Z}[1/p][q]^{p^j \mathbb{N}_p}.$$

Note that for every $f \in \mathcal{S}_p$, the evaluation $\mathrm{ev}_\xi(f)$ can be defined for every root ξ of unity. For $f \in \mathcal{S}_{p,j}$, the evaluation $\mathrm{ev}_\xi(f)$ can be defined when ξ is a root of unity of order in $p^j \mathbb{N}_p$. In [2] we proved the following.

Proposition 8. *For every prime p one has*

$$\mathcal{S}_p \simeq \prod_{j=0}^{\infty} \mathcal{S}_{p,j}. \tag{14}$$

Let $\pi_j : \mathcal{S}_p \to \mathcal{S}_{p,j}$ denote the projection onto the jth component in the above decomposition. Suppose ξ is a root of unity of order $r = p^j r'$, with $(r', p) = 1$. Then for any $x \in \mathcal{S}_p$, one has

$$\mathrm{ev}_\xi(x) = \mathrm{ev}_\xi(\pi_j(x)).$$

If $i \neq j$ then $\mathrm{ev}_\xi(\pi_i(x)) = 0$.

3.5. **On the Ohtsuki series at roots of unity.** Suppose M is a rational homology 3–sphere with $|H_1(M, \mathbb{Z})| = b$. By Theorem 3, for any root of unity ξ of order pr

$$\tau'_M(\xi) \in \mathbb{Z}[1/b][e_{pr}] \simeq \frac{\mathbb{Z}[1/b, e_r][x]}{(f_p(x + e_r))},$$

where

$$f_p(t) := \frac{t^p - e_r^p}{t - e_r}.$$

Hence we can write

$$\tau'_M(e_r e_p) = \sum_{n=0}^{p-2} a_{p,n} x^n \tag{15}$$

where $a_{p,n} \in \mathbb{Z}[1/b, e_r]$. The following proposition proven in [2] shows that the coefficients $a_{p,n}$ stabilize as $p \to \infty$.

Proposition 9. [Beliakova–Bühler–Le] *Suppose M is a rational homology 3-sphere with $|H_1(M, \mathbb{Z})| = b$, and r is an odd positive integer. For every non-negative integer n, there exists a unique invariant $a_n = a_n(M) \in \mathbb{Z}[1/b, e_r]$ such that for every prime $p > \max(b, r)$, we have*

$$a_n \equiv a_{p,n} \pmod p \quad \text{in } \mathbb{Z}[1/b, e_r] \text{ for } \; 0 \le n \le p-2. \tag{16}$$

Moreover, the formal series $\sum_n a_n (q - e_r)^n$ is equal to the Taylor expansion of the unified invariant I_M at e_r.

4. Laplace transform, Andrews identity and Frobenius maps

The aim of this section is to define the Laplace transform and to study its image.

4.1. **Laplace transform.** To define the unified invariant we have to compute

$$\sum_n{}^{\xi} q^{b\frac{n^2-1}{4}} A(n,k)$$

where terms depending on n in $A(n,k)$ look as follows

$$\prod_{i=0}^{k} (q^n + q^{-n} - q^i - q^{-i}) = (-1)^{k+1} \, (q^n; q)_{k+1} \, (q^{-n}; q)_{k+1} \, .$$

Formally, the last expression can be considered as a Laurent polynomial in q and q^n. Hence we only have to compute

$$\sum_n{}^{\xi} q^{b\frac{n^2-1}{4}} q^{na} \, ,$$

which can be easily done by the square completing argument. Let us state the result.

Suppose r is an odd number, b is positive integer and

$$c := (r, b), \quad b_1 := b/c, \quad r_1 := r/c.$$

Lemma 10. [Beliakova–Le] *One has*

$$\gamma_b(\xi) = c\, \gamma_{b_1}(\xi^c). \tag{17}$$

$$\sum_n{}^{\xi} q^{b\frac{n^2-1}{4}} q^{an} = \begin{cases} 0 & \text{if } c \nmid a; \\ (\xi^c)^{-a_1^2 b_1^*} \gamma_b(\xi) & \text{if } a = c a_1, \end{cases} \tag{18}$$

where b_1^* *is an integer satisfying* $b_1 b_1^* \equiv 1 \pmod{r_1}$.

This computation inspired us to introduce the following operator, called the Laplace transform. Remember that $\int e^{ax^k} f(x)\, dx$ is called the Laplace transform of f of order k.

Let $\mathcal{L}_{b,c;n} : \mathbb{Z}[q^{\pm n}, q^{\pm 1}] \to \mathbb{Z}[q^{\pm c/b}]$ be the $\mathbb{Z}[q^{\pm 1}]$–linear operator, called the (discrete) Laplace transform (of the second order), defined by

$$\mathcal{L}_{b,c;n}(q^{na}) := \begin{cases} 0 & \text{if } c \nmid a; \\ q^{-a^2/b} & \text{if } a = ca_1, \end{cases} \tag{19}$$

Lemma 11. [Beliakova–Le] *Suppose* $f \in \mathbb{Z}[q^{\pm n}, q^{\pm 1}]$. *Then*

$$\sum_n^{\xi} q^{b\frac{n^2-1}{4}} f = \gamma_b(\xi)\, \mathrm{ev}_\xi(\mathcal{L}_{b,c;n}(f)).$$

The point is that $\mathcal{L}_{b,c;n}(f)$, unlike the left hand side $\sum_n^{\xi} q^{b\frac{n^2-1}{4}} f$, does not depend on ξ. To prove Theorem 5 we need to show that $\mathcal{L}_{b,c;n}((q^n;q)_{k+1}$ $(q^{-n};q)_{k+1})$ is divisible by $(q^{k+1};q)_{k+1}$. For this we use the remarkable identity discovered by Andrews.

4.2. **Andrews identity.** To warm up we start with the identity from the Ramanujan "Lost" Notebook:

$$\prod_{k\geq 1} \frac{1}{(1-q^{5k-4})(1-q^{5k-1})} = \sum_{n\geq 0} \frac{q^{n^2}}{(1-q)(1-q^2)\dots(1-q^n)}$$

Only in the 60s, MacMahon gave a combinatorial interpretation of this identity as follows. Let $\lambda = (\lambda_1, \lambda_2, \dots, \lambda_t)$ with $\sum_i \lambda_i = n$ be a partition of n into non–increasing integers. Then the identity can derived from the fact that the number of partitions of n with all λ_i of the form $5k+1$ or $5k+4$ is equal to the number of partitions where $\lambda_i - \lambda_{i+1} \geq 2$ for all i.

This identity is a very special case of the Andrews identity we used in [2]: For any numbers $b_i, c_i, i = 1, \dots, k$ and positive integer N we have

$$1 + \sum_{n=1}^{N} q^{kn+Nn}(1+q^n)\frac{(q^{-N})_n}{(q^{N+1})_n} \prod_{i=1}^{k} \frac{(b_i)_n}{b_i^n} \frac{(c_i)_n}{c_i^n} \frac{1}{(\frac{q}{b_i})_n (\frac{q}{c_i})_n}$$
$$= \frac{(q)_N\,(\frac{q}{b_k c_k})_N}{(\frac{q}{b_k})_N\,(\frac{q}{c_k})_N} \sum_{n_k \geq n_{k-1} \geq \dots \geq n_2 \geq n_1 = 0} \frac{q^{n_k}(q^{-N})_{n_k}(b_k)_{n_k}(c_k)_{n_k}}{(q^{-N}b_k c_k)_{n_k}}$$
$$\times \prod_{i=1}^{k-1} \frac{q^{n_i} \frac{(b_i)_{n_i}}{b_i^{n_i}} \frac{(c_i)_{n_i}}{c_i^{n_i}} (\frac{q}{b_i c_i})_{n_{i+1}-n_i}}{(q)_{n_{i+1}-n_i} (\frac{q}{b_i})_{n_{i+1}} (\frac{q}{c_i})_{n_{i+1}}}$$

where $(a)_n = (a;q)_n$.

The point is that (for a special choice of parameters) the left hand side of the identity can be identified with $\mathcal{L}_{b,c;n}\left((q^n;q)_{k+1}\,(q^{-n};q)_{k+1}\right)$, where the right hand side is a sum with all summands divisible by $(q^{k+1};q)_{k+1}$.

In the case $b=\pm 1$, the computations are especially simple and we get that

$$\mathcal{L}_{-1,1;n}((q^n;q)_{k+1}\,(q^{-n};q)_{k+1}) = 2(q^{k+1};q)_{k+1}\,.$$

The same holds also for $\mathcal{L}_{1,1;n}$ up to units. This allows to write the explicit formula for I_M given in the introduction.

4.3. **Frobenius isomorphism.** It remains to show that the image of the Laplace transform belongs to $\mathcal{R}_b$, i.e. that certain roots of q exist in $\mathcal{R}_b$.

In [2] we proved the following.

Theorem 12. [Beliakova–Bühler–Le] *The Frobenius endomorphism F_b : $\mathbb{Z}[1/b][q]^{\mathbb{N}_b} \to \mathbb{Z}[1/b][q]^{\mathbb{N}_b}$, sending q to q^b, is an isomorphism.*

This implies the existence of the bth root of q in $\mathcal{S}_{b,0}$ defined by

$$q^{1/b} := F_b^{-1}(q) \in \mathcal{S}_{b,0}\,.$$

Let us mention that this result does not hold over $\mathbb{Q}$, i.e. for $y \in \mathbb{Q}^{\mathbb{N}_b}$ with $y^b = 1$ we have $y = \pm 1$.

Further, we introduce another Frobenius homomorphism

$$G_m : R[q]^{\mathbb{N}_b} \to R[q]^{m\mathbb{N}_b} \quad \text{by} \qquad G_m(q) = q^m.$$

Since $\Phi_{mr}(q)$ always divides $\Phi_r(q^m)$, G_m is well–defined.

This map allows us to transfer $q^{1/b}$ from $\mathcal{S}_{p,0}$ to $\mathcal{S}_{p,i}$ with $i > 0$, and hence to define a realization of $q^{a^2/b}$ in $\mathcal{S}_p$ with the correct evaluation.

4.4. **Integrality.** The proof of integrality of the SO(3) WRT invariants for all roots of unity and for all 3–manifolds is given in [4]. Note that even restricted to rational homology 3–spheres, this fact requires a separate proof, since the existence of $I_M \in \mathcal{R}_b$ does not imply integrality of the WRT invariants, unless $b = 1$. In this subsection we define a subring $\Gamma_b \subset \mathcal{R}_b$, such that for any $f \in \Gamma_b$, $\mathrm{ev}_\xi(f) \in \mathbb{Z}[\xi]$. The fact that $I_{M,1}$ belongs to this subring was proved in [4]. We conjecture that this holds in general.

For any divisor c of b, let us decompose $\mathbb{N} = \cup_{c|b}\, c\,\mathbb{N}_{b/c}$. Then

$$\mathbb{Z}[1/b][q]^{\mathbb{N}} = \prod_{c|b} \mathbb{Z}[1/b][q]^{c\mathbb{N}_{b/c}}\,.$$

Analogously, we have $\Gamma_b = \prod_{c|b} \Gamma_{b,c}$, where $\Gamma_{b,c} \subset \mathbb{Z}[1/b][q]^{c\mathbb{N}_{b/c}}$ is defined as follows.

Let $A_{b,c} = \mathbb{Z}[e_c][q^{\pm 1}, q^{\pm c/b}]$. Put $t = q^{c/b}$ and let $A^{(m)}_{b,c}$ be the algebra generated over $A_{b,c}$ by

$$\frac{(t;t)_m}{(q^c;q^c)_m} \; \frac{(e_c;e_c)_{(c-1)/2}}{\widetilde{(q;q)_m}}$$

where

$$\widetilde{(q;q)_m} = \prod_{i=1, c\nmid i}^{m} (1-q^i) .$$

Then every element $f \in \Gamma_{b,c}$ has a presentation

$$f = \sum_{m=0}^{\infty} f_m \frac{(q^{m+1};q)_{m+1}}{1-q}$$

with $f_m \in A^{(m)}_{b,c}$. For any root of unity ξ of odd order r with $(r,b) = c$ and $f \in \Gamma_{b,c}$, we have

$$\mathrm{ev}_\xi(f) = \mathrm{ev}_\xi \left(\sum_{m=0}^{(r-3)/2} f_m \frac{(q^{m+1};q)_{m+1}}{1-q} \right) .$$

Observe that for $m < (r-1)/2$,

$$\mathrm{ev}_\xi(\widetilde{(q;q)_m}) \mid (e_c;e_c)_{(c-1)/2} \text{ and } \mathrm{ev}_\xi \left(\frac{(t;t)_m}{(q^c;q^c)_m} \right) \in \mathbb{Z}[\xi] .$$

Conjecture. *For any $M \in \mathcal{M}_b$, there exists an invariant $I'_M \in \Gamma_b$, such that for any root of unity ξ of odd order $\mathrm{ev}_\xi(I'_M) = \tau_M(\xi)$.*

We also expect that $\Gamma_{b,b}$ is determined by its Taylor expansion at e_b.

References

[1] G. Andrews, *q–series: their development and applications in analysis, number theory, combinatorics, physics, and computer algebra*, regional conference series in mathematics, Amer. Math. Soc. **66** (1985)

[2] A. Beliakova, I. Bühler, T. T. Q. Le, *A unified quantum $SO(3)$ invariants for rational homology 3-spheres*, arXiv: 0801.3893

[3] A. Beliakova, Q. Chen, T. T. Q. Le, *On the integrality of Witten-Reshetikhin-Turaev 3-manifold invariants*, arXiv:1010.4750

[4] A. Beliakova, T. T. Q. Le, *Integrality of quantum 3–manifold invariants and rational surgery formula*, Compositio Math. **143** (2007) 1593–1612

[5] I. Bühler, *Unified quantum $SO(3)$ and $SU(2)$ invariants for rational homology 3–spheres*, PhD thesis, University of Zürich, 2010

[6] P. Gilmer, G. Masbaum, *Integral lattices in TQFT*, Ann. Scient. Ecole Normale Sup. **40** (2007) 815–844

[7] K. Habiro, *A unified Witten–Reshetikhin–Turaev invariant for integral homology spheres*, Invent. Math. **171** (2008) 1–81

[8] K. Habiro, *Cyclotomic completions of polynomial rings*, Publ. Res. Inst. Math. Sci. **40** (2004) 1127–1146
[9] M. Khovanov, *Hopfological algebra and categorification at a root of unity: the first steps*, math.QA/0509083 (2005), to appear in Commun. Contemp. Math.
[10] R. Kirby, P. Melvin, *The 3–manifold invariants of Witten and Reshetikhin–Turaev for* sl$(2, C)$, Invent. Math. **105** (1991) 473–545
[11] R. Lickorish, *An introduction to knot theory*, Springer 1997
[12] T. T. Q. Le, *Strong integrality of quantum invariants of 3–manifolds*, Trans. Amer. Math. Soc. **360** (2008) 2941–2963
[13] T. T. Q. Le, J. Murakami, T. Ohtsuki, *On a universal perturbative invariante of 3-manifolds*, Topology **37** (1998) 539–574
[14] T. T. Q. Le, *Quantum invariants of 3–manifolds: integrality, splitting, and perturbative expansion*, Topology Appl. **127** (2003) 125–152.
[15] Y. Manin, *Cyclotomy and analytic geometry over* $\mathcal{F}_1$, arXiv:0809.1564
[16] G. Masbaum, J. Roberts, *A simple proof of integrality of quantum invariants at prime roots of unity*, Math. Proc. Camb. Phil. Soc. **121** (1997), 443–454
[17] H. Murakami, *Quantum* $SO(3)$*–invariants dominate the* $SU(2)$*–invariant of Casson and Walker*, Math. Proc. Camb. Phil. Soc. **117** (1995) 237–249
[18] T. Nagell, *Introduction to Number Theory*, Almqvist & Wiksells Boktryckeri (1951)
[19] T. Ohtsuki, *Quantum invariants. A study of knots, 3-manifolds, and their sets*, Series on Knots and Everything, World Scientific **29** (2002)
[20] T. Ohtsuki, *A polynomial invariant of rational homology* 3*–spheres*, Invent. Math. **123** (1996) 241–257
[21] V. Turaev, *Quantum invariants of knots and 3–manifolds*, de Gruyter Studies in Math. 18 (1994)

Institut für Mathematik, Universität Zurich, Winterthurerstrasse 190, 8057 Zürich, Switzerland
E-mail address: `anna@math.uzh.ch, irmgard.buehler@math.uzh.ch`

Department of Mathematics, Georgia Institute of Technology, Atlanta, GA 30332–0160, USA
E-mail address: `letu@math.gatech.edu`

A SURVEY OF QUANDLE IDEAS

J. Scott Carter*
University of South Alabama
carter@jaguar1.usouthal.edu

This article surveys many aspects of the theory of quandles which algebraically encode the Reidemeister moves. In addition to knot theory, quandles have found applications in other areas which are only mentioned in passing here. The main purpose is to give a short introduction to the subject and a guide to the applications that have been found thus far for quandle cocycle invariants.

1. Introduction

A quandle is an algebraic system whose axioms are derived from the Reidemeister moves on classical knot diagrams. Any codimension-2, piecewise-linear embedding, whose image in Euclidean space is locally-flat, has a fundamental quandle. In particular, classical knots are completely classified, up to orientation, by their fundamental quandles. Of course, this means that the fundamental quandle is extremely hard to compute. In fact, I only know of one example (that of the trefoil) in which the fundamental quandle has been identified to a heretofore unrelated quandle (the Dehn quandle of simple closed curves on a torus in which the quandle operation is induced by Dehn twists[55]). The fundamental quandle of a classical knot describes the fundamental group and a peripheral subgroup, and it is in this respect that it classifies the knot. Section 2 contains this description.

Interesting and computable invariants come from representing the fundamental quandle into a finite quandle that is often, but not always, a dihedral quandle. Not only can we count the number of such quandle colorings, but we can separate different colorings into distinct subclasses that are indexed by the value of a quandle cocycle. The quandle cocycle invariants are described below.

The final section of the paper summarizes many of the applications that have been found for quandle cocycles. The list is certainly not

*Supported in part by NSF Grant DMS #0603926.

encompassing, but contains, to the best of my knowledge, a reasonable guide to the myriad results that have been obtained in the wake of the definition of the quandle cocycle invariants.[16]

First I turn to a synopsis of the history of the subject and some important definitions. My historical perspective will be tainted by my vision based upon contributions to the field that my collaborators and I have made.

History, Definitions, and Examples

Joyce's dissertation cites the 1929 work of Burstin and Mayer[10] as the earliest work on distributive groups. He points out that similar ideas underlie the core group of Moufang loops.[8,47] In 1942/3, Takasaki[61] introduced the notion of a *kei* — a set that possesses an involutory self-distributive operation. A 1959 correspondence between Gavin Wraith and John Conway re-introduced the idea in the context of a group that acts upon itself via conjugation. Therein the idea was named a wrack (after wrack and ruin of a group after the multiplication operation had been dismissed). According to Wikipedia and other sources, the name "wrack" was a bit of Conway word play upon Wraith's name. Modern authors have returned to the spelling "rack."

In the terminology that was later introduced by Joyce,[39] a kei is an involutory quandle. At approximately the same time of Joyce, Matveev[45] introduced self-distributive groupoids. Matveev and Joyce showed that these structures determine the complement of a knot up to an orientation reversing homeomorphism. The date 1979 is common to both Joyce's thesis and Matveev's original talks. These works, it seems, were developed independently and at roughly the same time. Steve Winker's dissertation was completed in 1982 under the direction of Lou Kauffman. In the dissertation Winker shows that a knot is trivial if and only if its involutory quandle is trivial. Winker also creates a diagramming method for quandles analogous to the Cayley graph of a group and uses this method to determine that certain quandles are infinite while others are finite.

Brieskorn's vast article[9] appeared in the proceedings of a 1986 Santa Cruz conference on Braids.[7] It is notable for a couple of reasons. First, the volume itself marks the rebirth of braids and the braid group as important tools at the interface among algebra, topology, and analysis. This interface was due, in part, because of Jones's seminal discovery.[38] Second, Brieskorn's paper lays important foundations in which automorphic sets also play important roles among various branches of mathematics. Many of these connections remain unexplored. I encourage young researchers to examine this work closely.

The 1992 paper by Fenn and Rourke[29] also is very influential. At about that time, several of us were dabbling with the ideas of quandles (See for example[43]), but there were only a few results that were being achieved. A particularly, useful concept from the point of view of the current author was the notion of rack cohomology that was introduced.[30] See

`http://www.maths.sussex.ac.uk/Staff/RAF/Maths/`

where documents relating to the history of rack and quandle homology can be found. In works with other collaborators,[16] the definition of rack cohomology was modified to that of quandle cohomology, and we showed how to use quandle cocycles to produce invariants of classical knots and knotted surfaces in state-sum form.

Brian Sanderson spoke about trunks in the 1995 Warsaw conference, and I was in the audience. I am pretty sure that I mentioned it to Masahico in passing, but neither of us recalls specifically such correspondence. The electronic record might produce a "bit-trail," but for my part, I really did not think about the idea subsequently.

Shortly after Laurel Langford finished her dissertation, the three of us (Laurel, Masahico, and I) met in February to discuss the possibility of using Neuchl's cocycles to define knotted surface invariants. Upon return from a meeting of the AMS in Manhattan, Kansas, Masahico and I reformulated the idea in terms quandle cocycles. As we began writing down the material, I began to recall Brian Sanderson's talk. I had some notes from the conference and the cocycle notion is clearly stated therein.

In 1998, Seiichi Kamada and I spoke at length during the Knots in Hellas conference. Much of the pending discussion had to do with his upcoming travel arrangements to Mobile. He and Naoko arrived in Mobile in October of that year. Meanwhile, Dan Jelsovsky began pursuing his degree under Masahico's supervision. The original manuscript that contained the quandle cocycle invariant was posted[16] on the arXiv in March 1999.

A great deal of work by a number of authors in different contexts is based upon the quandle cocycle invariants and their generalizations. In the final section herein, I will summarize many of the results of which I know that use the cocycle invariants.

Next, I turn to the basic definitions and examples.

2. Fundamental Definitions and Results

Classical Quandle Theory

Definition 2.1. A *quandle*, X, is a set with a binary operation $(a, b) \mapsto a \triangleleft b$ such that

(I) For any $a \in X$, $a \triangleleft a = a$.
(II) For any $a, b \in X$, there is a unique $c \in X$ such that $a = c \triangleleft b$.
(III) For any $a, b, c \in X$, we have $(a \triangleleft b) \triangleleft c = (a \triangleleft c) \triangleleft (b \triangleleft c)$.

In Axiom (II), the element c that is uniquely determined from given $a, b \in X$ such that $a = c \triangleleft b$, is denoted by $c = a \triangleleft^{-1} b$. The reader may check that $\triangleleft^{-1}$ also defines a quandle structure. A function $\phi : X \to Y$ between quandles or racks is a *homomorphism* if $(a \triangleleft b)\phi = (a)\phi \triangleleft (b)\phi$ for any $a, b \in X$. In many contexts within the discussion herein, it will be convenient to write the function's name to the right of the argument, as above. Observe that in this notation, axioms (II) and (III) indicate that the quandle acts upon itself as a collection of *quandle automorphisms* — one-to-one onto maps that preserve the quandle operation. The *automorphism group* of a quandle X is the group $\text{Aut}(X)$ of all bijective quandle homomorphisms. The *inner automorphism group*, $\text{Inn}(X)$, is the subgroup of the automorphism group generated by $\{\phi_z : z \in X \;\; \& \;\; (x)\phi_z = x \triangleleft z\}$.

A quandle X is said to be *connected* if for any elements $x, y \in X$ there is an inner automorphism that act transitively on X; thus there is an inner-automorphism $\phi \in \text{Inn}(X)$ such that $(x)\phi = y$. A quandle is said to be *homogeneous* if $\text{Aut}(X)$ act transitively on X. If X is connected, it is homogeneous. A quandle is *involutory* if $\triangleleft = \triangleleft^{-1}$.

Example 2.1. The following are typical examples of quandles. A group $X = G$ with n-fold conjugation as the quandle operation: $a \triangleleft b = b^n a b^{-n}$ or $a \triangleleft b = b^{-n} a b^n$. The notation $\text{Conj}(G)$ indicates the quandle defined for a group G by $a \triangleleft b = bab^{-1}$. Any subset of G that is closed under such conjugation is also a quandle.

Any $\Lambda (= \mathbb{Z}[t, t^{-1}])$-module M is a quandle with $a \triangleleft b = ta + (1 - t)b$, $a, b \in M$, that is called an *Alexander quandle*. For a positive integer n, the ring $\mathbb{Z}_n[t, t^{-1}]/(h(t))$ is a quandle for a Laurent polynomial $h(t)$.

Let n be a positive integer, and for elements $i, j \in \{0, 1, \ldots, n - 1\}$, define $i \triangleleft j \equiv 2j - i \pmod{n}$. Then $\triangleleft$ defines a quandle structure called the *dihedral quandle*, R_n. This set can be identified with the set of reflections of a regular n-gon with conjugation as the quandle operation, and it is also isomorphic to the Alexander quandle $\mathbb{Z}_n[t, t^{-1}]/(t + 1)$. As a set of reflections of the regular n-gon, R_n can be considered as a subquandle of $\mathrm{Conj}(\Sigma_n)$ where Σ_n denotes the symmetric group on n letters.

Example 2.2. This example is due to Joyce[39] and Matveev.[45]

Let G be a group, H a subgroup, $s : G \to G$ an automorphism such that for each $h \in H$ $s(h) = h$. Define a binary operation $\lhd_s = \lhd$ on G by $a \lhd b = s(ab^{-1})b$. Then $\lhd$ defines a quandle structure on G. Axioms I and II are easily verified. For Axiom III, we have

$$(a \lhd b) \lhd c = s(s(ab^{-1})bc^{-1})c = s^2(a)s^2(b^{-1})s(b)s(c)^{-1}c$$

while

$$\begin{aligned}
&(a \lhd c) \lhd (b \lhd c) \\
&= s(s(ac^{-1})c(s(bc^{-1})c)^{-1})s(bc^{-1})c \\
&= s^2(a)s^2(c^{-1})s(c)[s^2(b)s^2(c^{-1})s(c)]^{-1}s(b)s(c^{-1})c \\
&= s^2(a)s^2(c^{-1})s(c)s(c)^{-1}[s^2(c^{-1})^{-1}]s^2(b)^{-1}s(b)s(c^{-1})c \\
&= s^2(a)s^2(b^{-1})s(b)s(c)^{-1}c.
\end{aligned}$$

This passes to a well-defined quandle structure on G/H that is given by $Ha \lhd Hb = Hs(ab^{-1})b$. In particular, if $z \in Z(H) \cap H$ where $Z(H) = \{z \in G : zh = hz \text{ for all } h \in H\}$, then $Ha \lhd Hb = Hab^{-1}zb$ defines a quandle structure. Let us denote the resulting quandle by (G, H, z).

Let X be any quandle. Let $F(X)$ denote the free group generated by X. Then $F(X)$ acts on X as follows. If $x \in X$ also denotes the generator in $F(X)$ and $\cdot$ denotes the action of $F(X)$ on X, then $y \cdot x = y \lhd x$; furthermore $y \cdot x^{-1} = y \lhd^{-1} x$. In general, a word $w = w_1 w_2 \in F(X)$ acts as $y \cdot (w_1 w_2) = (y \cdot w_1) \cdot w_2$. This, in turn, can be written in terms of $\lhd, \lhd^{-1}$ and the generators in X. Thus there is a map from $F(X)$ to a subgroup of the group of symmetries of X. Let K denote the kernel of this map. Then $F(X)/K$ is isomorphic to a subgroup of the automorphism group of the quandle. In fact, since the action of $F(X)$ is induced by the quandle product, $F(X)/K$ is isomorphic to $\mathrm{Inn}(X)$.

Define the *enveloping group* of a quandle to be the group, G_X that is generated by the elements of X subject to the set of relations $a * b = bab^{-1}$ for all $a, b \in X$. The action of $F(X)$ also passes to an action of G_X with kernel K', and so we have in general G_X/K' is isomorphic to a subgroup of $\mathrm{Inn}(X)$.

Note that for a quandle X we have have a map $\partial : X \to \mathrm{Aut}(X)$ given by $y \mapsto \phi_y$ where, $(x)\phi_y = x \lhd y$. Note also, that $\partial((x)\phi) = \phi^{-1}\partial(x)\phi$. Thus a quandle is also a crossed G-set where $G = \mathrm{Aut}(X)$, but it has additional structure.

Theorem 2.1. [39,45] *Let X denote a homogeneous quandle and suppose that z is a fixed element of X. Consider $H \subset \mathrm{Aut}(X)$ to be the stabilizer of z. That is $H = \{h \in \mathrm{Aut}(X) : (z)h = z\}$. Then there is a quandle isomorphism $X \cong (\mathrm{Aut}(X), H, z)$.*

Proof. Let $x \in X$, and chose ϕ_x so that $x = (z)\phi_x$. Given ϕ_x and ϕ'_x that map z to x, we see that $(z)\phi'\phi^{-1} = z$. So $\phi'\phi^{-1} \in H$. On the other hand, if $h \in H$, then $x = (z)\phi_x = zh\phi_x$. Therefore, any element of $H\phi_x$ maps z to x. We have a function, $E : \mathrm{Aut}(X) \to X$, $E(\phi) = (z)\phi$. To see that E is a quandle homomorphism, compute:

$$E(\phi \triangleleft \psi) = (z)((\phi\psi^{-1}) \triangleleft z)\psi = (z)(\phi\psi^{-1})\psi \triangleleft (z)\psi = E(\phi) \triangleleft E(\psi).$$

The quandle map E passes to a bijection form $(\mathrm{Aut}(X), H, \phi_z)$ to X. This completes the proof.

Example 2.3. Let QS_4 denote the quandle, $\mathbb{Z}_2[t, t^{-1}]/(t^2+t+1)$. Observe that this example is the 4-element field, but its quandle structure is given by $a * b = ta + (1+t)b$. Letting $[0] = 0$, $[1] = 1$, $[2] = t$, and $[3] = t+1$, we have the following quandle table:

$R * C$	[0]	[1]	[2]	[3]
[0]	[0]	[3]	[1]	[2]
[1]	[2]	[1]	[3]	[0]
[2]	[3]	[0]	[2]	[1]
[3]	[1]	[2]	[0]	[3]

Now $\mathrm{Aut}(X) \subset \Sigma_4 = \Sigma_4(0, 1, 2, 3)$, is generated by (123), (032), (013), and (021). So $\mathrm{Aut}(X)$ is the alternating group, $A_4 = A_4(0, 1, 2, 3)$. Let $H = \mathrm{Stab}(\{0\}) = \{1, (132), (123)\}$. The action of [0] as an automorphism, is $\partial[0] = (123)$. The stabilizer of 0 in A_4 is $H = \{1, (132), (123)\}$. And the coset space is $\{H, H(031), H(02)(13), H(013)\}$.

Definition 2.2. Suppose that $k : M^n \to \mathbb{R}^{n+2}$ is an embedding which is either smooth or PL locally-flat. Let $N(k)$ denote an open tubular neighborhood of $k(M)$. The *fundamental quandle* $\pi_Q(k)$, of a codimension 2 embedding is defined to be the set of homotopy classes of maps $\alpha : ([0, 1], \{0, 1\}, 1) \to (\mathbb{R}^{n+2} \setminus N(k), \partial N(k) \cup \{y_0\}, \{y_0\})$ where y_0 is a fixed base point pretty close to the boundary of the tubular neighborhood; later on, the base point will be chosen to be on the neighborhood. The homotopies between such maps are required to have their bottom boundaries on $\partial N(k)$ and their top boundaries fixed at the base point ($H(s,0) \in \partial N(k)$ while $H(s,1) = y_0$ for all $s \in [0,1]$, and $H(i,t) = \alpha_i(t)$ for $i = 0,1$). If α and β are such paths, then there is a unique oriented meridian μ_β that passes through the initial point of β. The quandle product is defined to be the path composition $\alpha * \beta = \alpha\beta^{-1}\mu_\beta\beta$. Here μ_β is the meridian that intersects the path β at β_0.

A summary of the definition of the fundamental quandle is depicted in Fig. 1.

Assume that $n = 1$, so that the following discussion occurs in the classical case. The fundamental group $\pi_1(k)$ acts upon the fundamental quandle $\pi_Q(k)$ via path multiplication: given representatives $\alpha : [0,1] \to M\backslash \text{int}N(k)$ and $\gamma : ([0,1],\{0,1\}) \to (\text{int}N(k),*)$ of the corresponding elements, then first travel along the quandle path α from the tubular neighborhood to the base point and once around the loop γ. The result is a new path $[\alpha\cdot\gamma] \in \pi_Q$.

Assume that k is knotted. Let $G = \pi_1(S^3 \setminus k(S^1))$, and choose a base point on the boundary of the tubular neighborhood of $k(S^1)$. Then $G \subset \text{Aut}(\pi_Q(k))$ except in the case of the unknot. (The unknot is singular since its fundamental quandle is trivial; hence the automorphism group is trivial while the fundamental group is $\mathbb{Z}$.) Let $z \in \pi_Q$ denote the constant path, and let H denote the peripheral subgroup that is induced by the inclusion of the torus into the knot complement. Then the elements of H fix z since the endpoint of z can follow as it likes around a path on the tubular neighborhood that defines an element of H. See Fig. 2. Moreover these are precisely the elements that fix z because any element that fixes z would result in the initial point of z moving around the tubular neighborhood. So a loop that fixes z is homotopic to a loop on the tubular neighborhood. Using the above quandle isomorphism, we have

Theorem 2.2.[39,45] *The fundamental quandle is given as* $\pi_Q(k) \cong (\pi_1(S^3 \setminus k(S^1)), P, \mu_z)$ *where* μ_z *is the meridional element in the fundamental group that links* $k(S^1)$ *at the base point.*

Quandle Colorings

Definition 2.3. Let X be a fixed quandle. Let K be a given oriented classical knot or link diagram, and let $\mathcal{R}$ be the set of (over-)arcs. A vector perpendicular to an arc in the diagram is called a *normal* or *normal vector.*

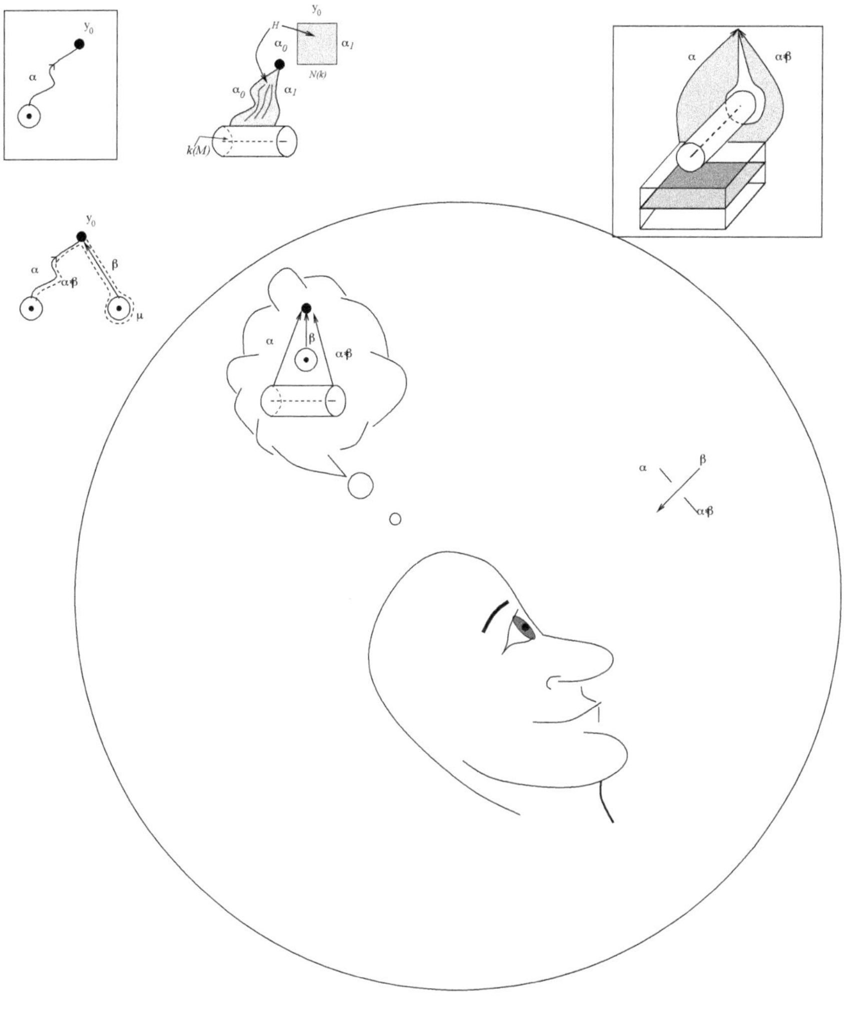

Fig. 1. The fundamental quandle

We choose the normal so that the ordered pair (tangent, normal) agrees with the orientation of the plane. This normal is called the *orientation normal.* A (quandle) *coloring* $\mathcal{C}$ is a map $\mathcal{C} : \mathcal{R} \to X$ such that at every crossing, the relation depicted in Fig. 3 holds. More specifically, let β be the over-arc at a crossing, and α, γ be under-arcs such that the normal of the over-arc points from α to γ. (In this case, α is called the *source arc* and γ is called the *target arc.*) Then it is required that $\mathcal{C}(\gamma) = \mathcal{C}(\alpha) \triangleleft \mathcal{C}(\beta)$. The color

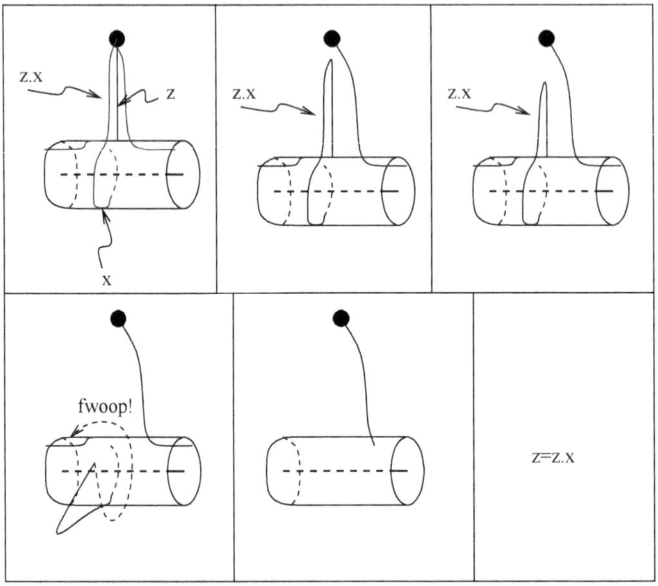

Fig. 2. The action of the peripheral subgroup

$\mathcal{C}(\gamma)$ depends only on the choice of orientation of the over-arc; therefore this rule defines the coloring at both positive and negative crossings. The colors $\mathcal{C}(\alpha)$, $\mathcal{C}(\beta)$ are called *source* colors.

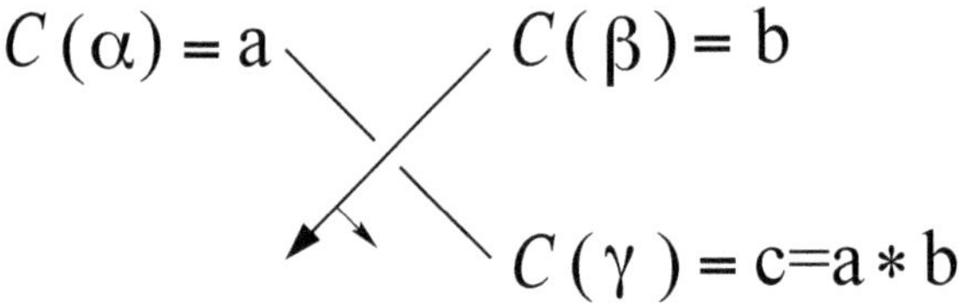

Fig. 3. Quandle coloring at a classical crossing

The definition has a strict analogue for diagrams of knotted surfaces (and indeed higher dimensional knottings), but in order to present this, I will need to spend a little time on the notion of a knotted surface diagram. Specifically, suppose that $k : F \to \mathbb{R}^4$ is a smooth embedding of a closed surface (compact without boundary) into 4-dimensional space. The image may be perturbed slightly so that the image $f = p(k(F))$ under a projection $p : \mathbb{R}^4 \to \mathbb{R}^3$ is a smooth surface in general position in 3-space. The general position assumption means that each point $f(x)$ (for $x \in F$) has a neighborhood in $\mathbb{R}^3$ in which the image equivalent to any one of the following four scenarios:

(1) $\{(x, 0, z) \in D^3 : |x| \leq 1,\ \&\ |z| \leq 1\}$; here D^3 represents a 3-dimensional ball of radius 1; the point $a \in F$ that maps to $(0, 0, 0)$ is called a *non-singular point*;
(2) $\{(x, y, 0) \in D^3 : |x| \leq 1,\ \&\ |y| \leq 1\} \cup \{(x, 0, z) \in D^3 : |x| \leq 1,\ \&\ |z| \leq 1\}$; the point $a \in F$ that maps to $(0, 0, 0)$ in this neighborhood is called a *double point*;
(3) $\{(x, y, 0) \in D^3 : |x| \leq 1,\ \&\ |y| \leq 1\} \cup \{(x, 0, z) \in D^3 : |x| \leq 1,\ \&\ |z| \leq 1\} \cup (0, y, z) \in D^3 : |y| \leq 1,\ \&\ |z| \leq 1\}$; the point $a \in F$ that maps to $(0, 0, 0)$ in this neighborhood is called a *triple point*;
(4) The image of f in a neighborhood of p resembles the cone on a figure-8; the point p in this case is called a *branch point*.

The notion of a *broken surface diagram* is analogous to a knot diagram and can be constructed from such a generic projection. Specifically, at the singular points some of the sheets of the surface are more distant from the 3-space into which the surface is being projected than the others are. Those that are farther away have a neighborhood of the double points removed within the diagrams. At a branch point, the differential of the function f is singular; at a branch point a broken sheet converges to an unbroken sheet. Figure 4 illustrates.

In the case that the surface F is oriented, an orientation normal in 3-space can be chosen consistently across the broken sheets in the diagram. A *quandle coloring of a broken surface diagram* is defined as an assignment of quandle elements to each sheet in a broken surface diagram such that the crossing rule depicted in Fig 5 is satisfied. Specifically, suppose that two lower sheets are broken colored a and c and are separated by a sheet labeled b with the normal to b pointing from a to c. Then $c = a \triangleleft b$.

In both the classical case and the knotted surface case, the quandle coloring is defined on a representative diagram of the given knot or knotted surface. If two classical diagrams differ by a Reidemeister move, and one of the diagrams has received a quandle coloring, then the quandle axioms allow us to color the second diagram in a unique fashion. In the knotted surface case, equivalent diagrams will differ by a finite sequence of Roseman moves.[56] If two broken surface diagrams differ from each other by a Roseman move and one has a quandle coloring, then there is a uniquely induced

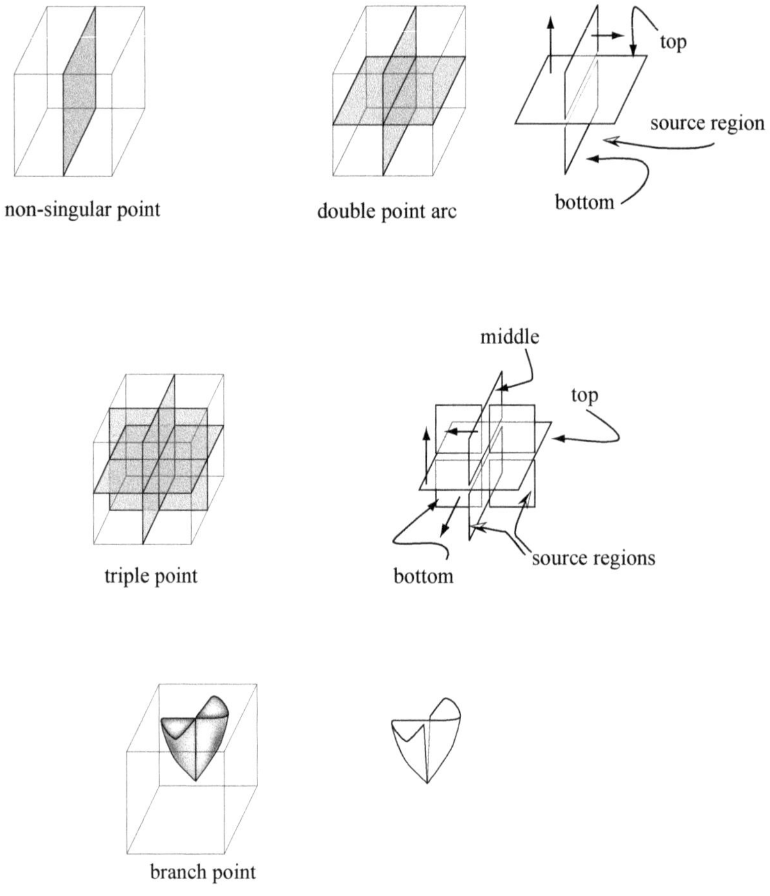

Fig. 4. Generic and broken surfaces

quandle coloring on the other. In either case, we can envision the quandle coloring as a quandle homomorphism from the fundamental quandle of the codimension 2 embedding to a test quandle.

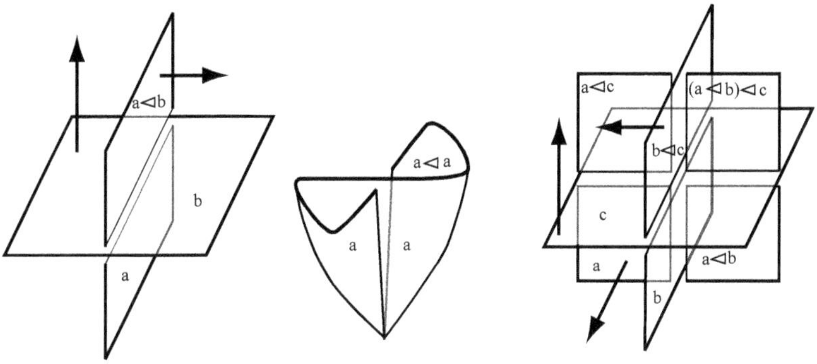

Fig. 5. Quandle coloring of broken surfaces

Quandle (Co)homology

Quandle cocycles and quandle homology in the general setting as it was originally introduced[3] was re-presented by my collaborators and me.[13] The discussion that follows is taken from our work.

Let X be a quandle. Let $\Omega(X)$ be the free $\mathbb{Z}$-algebra generated by $\eta_{x,y}$, $\tau_{x,y}$ for $x, y \in X$ such that $\eta_{x,y}$ is invertible for every $x, y \in X$. Define $\mathbb{Z}(X)$ to be the quotient $\mathbb{Z}(X) = \Omega(X)/R$ where R is the ideal generated by

$$\eta_{x \triangleleft y, z}\eta_{x,y} - \eta_{x \triangleleft z, y \triangleleft z}\eta_{x,z} \tag{1}$$
$$\eta_{x \triangleleft y, z}\tau_{x,y} - \tau_{x \triangleleft z, y \triangleleft z}\eta_{y,z} \tag{2}$$
$$\tau_{x \triangleleft y, z} - \eta_{x \triangleleft z, y \triangleleft z}\tau_{x,z} - \tau_{x \triangleleft z, y \triangleleft z}\tau_{y,z} \tag{3}$$
$$\tau_{x,x} + \eta_{x,x} - 1 \tag{4}$$

The algebra $\mathbb{Z}(X)$ thus defined is called the *quandle algebra* over X. In $\mathbb{Z}(X)$, we define elements $\overline{\eta_{z,y}} = \eta^{-1}_{z \triangleleft^{-1} y, y}$ and $\overline{\tau_{z,y}} = -\overline{\eta_{z,y}}\tau_{z \triangleleft^{-1} y, y}$. The convenience of such quantities will become apparent by examining type II moves.

A *representation* of $\mathbb{Z}(X)$ is an abelian group G together with (1) a collection of automorphisms $\eta_{x,y} \in \mathrm{Aut}(G)$, and (2) a collection of endomorphisms $\tau_{x,y} \in \mathrm{End}(G)$ such that the relations above hold. More precisely, there is an algebra homomorphism $\mathbb{Z}(X) \to \mathrm{End}(G)$, and we denote the image of the generators by the same symbols. Given a representation of $\mathbb{Z}(X)$ we say that G is a $\mathbb{Z}(X)$-module, or *a quandle module.* The action of $\mathbb{Z}(X)$ on G is written by the left action, and denoted by $(\rho, g) \mapsto \rho g (= \rho \cdot g = \rho(g))$, for $g \in G$ and $\rho \in End(G)$.

Example 2.4.[3] Let $\Lambda = \mathbb{Z}[t, t^{-1}]$ denote the ring of Laurent polynomials. Then any Λ-module M is a $\mathbb{Z}(X)$-module for any quandle X, by $\eta_{x,y}(a) = ta$ and $\tau_{x,y}(b) = (1-t)(b)$ for any $x, y \in X$.

Consider the enveloping group $G_X = \langle x \in X \mid x \triangleleft y = yxy^{-1} \rangle$ (a.k.a. the *associated group*[29]). For any quandle X, any G_X-module M is a $\mathbb{Z}(X)$-module by $\eta_{x,y}(a) = ya$ and $\tau_{x,y}(b) = (1 - x \triangleleft y)(b)$, where $x, y \in X$, $a, b \in M$.

When one ignores the terms involving κs in Figures 6 and 7, one can imagine a geometric motivation for the quandle module axioms. As the beads labeled a and c in Fig. 6 float up the strings past the crossings, they are acted upon by operators η and τ that are indexed by the colors on the arcs that cross. In Fig. 7, the results of the operators are indicated at the bead that is labeled d. The module axioms are the coefficients of a, b, and c — or rather the composition of operators upon these three markers. The remaining terms (those involving κs) yield the 2 cocycle condition for κ.

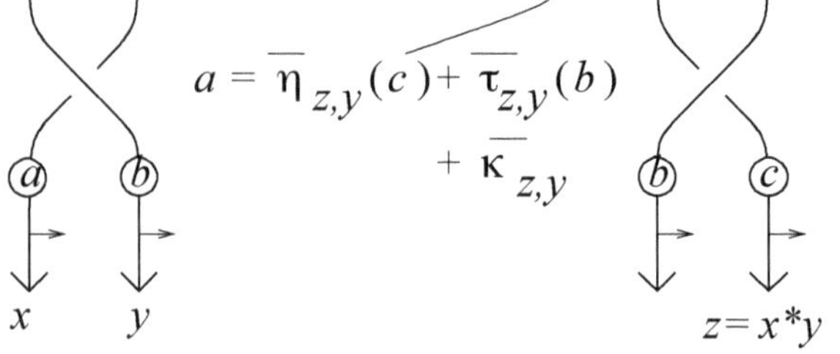

Fig. 6. The geometric notation at a crossing

Consider the free left $\mathbb{Z}(X)$-module $C_n(X) = \mathbb{Z}(X)X^n$ with basis X^n (for $n = 0$, X^0 is a singleton $\{x_0\}$, for a fixed element $x_0 \in X$). Boundary

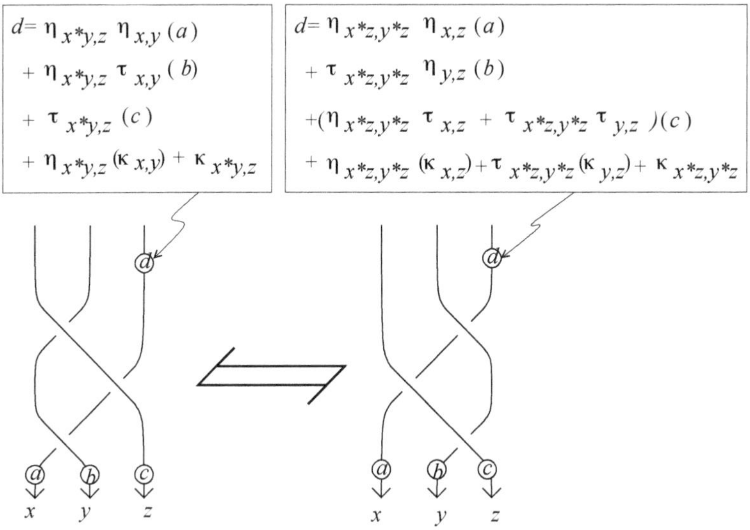

Fig. 7. Reidemeister moves and the quandle algebra definition

operators $\partial = \partial_n : C_{n+1}(X) \to C_n(X)$ are defined[3] by

$$\begin{aligned}
&\partial(x_1, \ldots, x_{n+1}) \\
&= (-1)^{n+1} \sum_{i=2}^{n+1} (-1)^i \eta_{[x_1,\ldots,\widehat{x_i},\ldots,x_{n+1}],[x_i,\ldots,x_{n+1}]}(x_1, \ldots, \widehat{x_i}, \ldots, x_{n+1}) \\
&\quad -(-1)^{n+1} \sum_{i=2}^{n+1} (-1)^i (x_1 \lhd x_i, \ldots, x_{i-1} \lhd x_i, x_{i+1}, \ldots, x_{n+1}) \\
&\quad +(-1)^{n+1} \tau_{[x_1,x_3,\ldots,x_{n+1}],[x_2,x_3,\ldots,x_{n+1}]}(x_2, \ldots, x_{n+1}),
\end{aligned}$$

where $\quad [x_1, x_2, \ldots, x_n] = ((\cdots(x_1 \lhd x_2) \lhd x_3) \lhd \cdots) \lhd x_n$

for $n > 0$, and $\partial_0(x) = -\tau_{x \bar{\lhd} x_0, x_0} x_0$ for $n = 0$. In particular, the condition to be a 2-cocycle for a 2-cochain $\kappa : C_2(X) \to \mathbb{Z}(X)$ (or $(\kappa : C_2(X) \to G) \in C^2(X; G) = \mathrm{Hom}_{\mathbb{Z}(X)}(C_2(X), G)$ when a representation $\mathbb{Z}(X) \to G$ is fixed to define the twisted coefficients) in this homology theory is written as $\kappa_{x,y}$

$$\eta_{x \lhd y, z}(\kappa_{x,y}) + \kappa_{x \lhd y, z} = \eta_{x \lhd z, y \lhd z}(\kappa_{x,z}) + \tau_{x \lhd z, y \lhd z}(\kappa_{y,z}) + \kappa_{x \lhd z, y \lhd z},$$

in $\mathbb{Z}(X)$ (or in G) for any $x, y, z \in X$, where $\kappa_{x,y}$ means $\kappa(x, y)$ for $(x, y) \in X^2$. This is called a *generalized (rack) 2-cocycle condition.* When κ further satisfies $\kappa_{x,x} = 0$ for any $x \in X$, it is called a *generalized quandle 2-cocycle.*

In case, $\eta_{x,y} = 1$ and $\tau_{x,y} = 0$ for all $x, y \in X$, then we recover the original quandle homology theory.[16]

The (rack) homology group is defined to be the quotient of the kernel of ∂_n modulo the image of ∂_{n+1}.

Dynamical cocycles

Let X be a quandle and S be a non-empty set. Let $\alpha : X \times X \to \mathrm{Fun}(S \times S, S) = S^{S\times S}$ be a function, so that for $x, y \in X$ and $a, b \in S$ we have $\alpha_{x,y}(a, b) \in S$.

Then it is checked by computations that $S \times X$ is a quandle by the operation $(a, x) \triangleleft (b, y) = (\alpha_{x,y}(a, b), x \triangleleft y)$, where $x \triangleleft y$ denotes the quandle product in X, if and only if α satisfies the following conditions:

(1) $\alpha_{x,x}(a, a) = a$ for all $x \in X$ and $a \in S$;
(2) $\alpha_{x,y}(-, b) : S \to S$ is a bijection for all $x, y \in X$ and for all $b \in S$;
(3) $\alpha_{x\triangleleft y,z}(\alpha_{x,y}(a, b), c) = \alpha_{x\triangleleft z,y\triangleleft z}(\alpha_{x,z}(a, c), \alpha_{y,z}(b, c))$ for all $x, y, z \in X$ and $a, b, c \in S$.

Such a function α is called a *dynamical quandle cocycle.*[3] The quandle constructed above is denoted by $S \times_\alpha X$, and is called the *extension* of X by a dynamical cocycle α. The construction is general, as Andruskiewitsch and Graña show:

Lemma 2.1.[3] *Let $p : Y \to X$ be a surjective quandle homomorphism between finite quandles such that the cardinality of $p^{-1}(x)$ is a constant for all $x \in X$. Then Y is isomorphic to an extension $S \times_\alpha X$ of X by some dynamical cocycle on the set S such that $|S| = |p^{-1}(x)|$.*

Example 2.5.[3] Let G be a $\mathbb{Z}(X)$-module for the quandle X, and κ be a generalized 2-cocycle. For $a, b \in G$, let

$$\alpha_{x,y}(a, b) = \eta_{x,y}(a) + \tau_{x,y}(b) + \kappa_{x,y}.$$

Then it can be verified directly that α is a dynamical cocycle. In particular, even with $\kappa = 0$, a $\mathbb{Z}(X)$-module structure on the abelian group G defines a quandle structure $G \times_\alpha X$.

We[13] used these definitions of quandle cocycles to define invariants of knots and knotted surfaces in the case that the quandle module M was a module over the enveloping group G_X of the quandle, and the actions of η and τ were given as: $\eta_{x,y}(a) = ya$ and $\tau_{x,y}(b) = (1 - x \triangleleft y)(b)$, where $x, y \in X$, $a, b \in M$.

I think that the best plan to explain the quandle cocycle invariants is to give the original definition (in the case $\eta_{x,y} = 1$ and $\tau_{x,y} = 0$ for all $x, y \in X$), and leave the curious to read the original sources.

Cocycle invariants of classical knots and knotted surfaces

Let a quandle X be given, and assume that $\phi : X \times X \to A$ is a quandle 2-cocycle that takes values in an Abelian group A. Thus

$$\phi(a,b) + \phi(a \triangleleft b, c) = \phi(a,c) + \phi(a \triangleleft c, b \triangleleft c),$$

and

$$\phi(a,a) = 0$$

for all $a, b, c \in X$.

Consider a knot diagram $\mathcal{D}(k)$ of a knot $k : S^1 \to \mathbb{R}^3$. Choose a quandle coloring of $\mathcal{D}(k)$ by X. In a neighborhood of a crossing, consider the under-arc away from which the normal to the over-arc points. The quandle color, a, on this arc is called the *source color of the under-arc*. The color, b, on the over-arc is its source color. Consider the value of the cocycle $\pm\phi(a,b)$; this is called the *Boltzmann weight* of the crossing. The sign coincides with the sign of the crossing. The weight of the coloring is the sum (in A) over all crossings of the diagram $\mathcal{D}(k)$: *i.e.* $B(\phi, C) = \sum_{\text{Crossings}} \pm\phi(a,b)$ Then the *quandle cocycle invariant* of k is the multiset of values $\{B(\phi, C)\}_C$ taken over all colorings C of the diagram.

In the case of knotted surfaces, the definition of the cocycle invariant is similar. The diagram of a triple point is considered. There are four sheets that are labeled as under sheets, two sheets that are labeled middle sheets, and one sheet that is labeled a top sheet. The normal to the top sheet points away from one of the middle sheets; that middle sheet is a *source region*. The normal to both the top and middle point away from one of the four bottom sheet; this is also a *source region*.

We assume that the knotted surface diagram has been colored by a test quandle X. Let the color of the source region to the bottom be a, let that of the source region in the middle be b, and let that of the color of the top sheet be c. The *sign of the triple point* is the orientation class of the triple (normal to top, normal to middle, normal to bottom) in comparision to the right handed orientation of 3-space.

A quandle 3-cocycle $\theta : X \times X \times X \to A$ is a function that takes values in an Abelian group A and that satisfies:

$$\theta(a,b,c)+\theta(a\triangleleft c, b\triangleleft c, d)+\theta(a,c,d) = \theta(a,b,d)+\theta(a\triangleleft b, c, d)+\theta(a\triangleleft d, b\triangleleft d, c\triangleleft d)$$

and

$$\theta(a,a,b) = \theta(a,b,b) = 0$$

At each triple point, the cocycle is evaluated on the ordered triple: (the color on the source region of the lower sheet, the color on the source region of the middle sheet, and the color on top sheet). The values of the cocycles, $\pm\theta(a,b,c)$ are summed over the set of triple points where the sign is chosen to coincide with the sign of the triple point. And as in the classical case, the multiset of values over all possible colorings is the value of the cocycle invariant on the knotted surface.

In both the classical case and the knotted surface case, the cocycle conditions, and the sign choices are exactly what are needed in order to show that the value of the invariant does not depend upon the choice of diagram representing the knotting.

Some non-trivial cocycles. Unfortunately, we do not have a lot of high-tech methods (such as projective resolutions or applications of homological algebra) to produce non-trivial cocycles. Some early non-trivial cocycles were determined by means of computer calculations. One of these was generalized by Mochizuki[46] via methods from algebraic geometry. Newer results will be summarized in the last section of the paper, but it is worth mentioning here that in the case of dihedral quandles, Niebrzydowski and Przytycki[52–54] have developed methods and conjectures about the higher order homology in the dihedral cases.

Example 2.6. For the Alexander quandle $QS_4 = \mathbb{Z}[t]/(2, t^2+t+1)$, the function,

$$\phi(x,y) = \begin{cases} 0 \text{ if } \ x=[3], y=[3], & \text{or } \ x=y \\ 1 & \text{otherwise} \end{cases}$$

that takes values in $\mathbb{Z}_2$ is a quandle 2-cocycle.

Example 2.7. In the case of the dihedral quandle $R_p = \{0,1,2,\ldots,p-1\}$ of order p when p is a prime ($a \triangleleft b = 2b - a \pmod p$), a generator of $H^3_{\mathrm{Q}}(R_p, \mathbb{Z}_p) \cong \mathbb{Z}_p$ is represented by the function:

$$\theta_p(x,y,z) = (x-y)\frac{(2z-y)^p + y^p - 2z^p}{p}.$$

The cocycle θ_p is Shin Satoh's simplification of Mochizuki's original construction.

Quandles with Good Involutions

An interesting class of quandles consist of quandle that have a good involution. Let me start with an example. Consider a subset Q of a group that is closed under conjugation and inversion. For elements $a, b \in Q$ define $a \triangleleft b = bab^{-1}$, and define $\rho(a) = a^{-1}$. The map ρ is an involution, but not necessarily a quandle map. We have that $\rho(a \triangleleft b) = \rho(a) \triangleleft b$, and $a \triangleleft \rho(b) = a \triangleleft^{-1} b$. In general, an involution ρ on a quandle X that satisfies these two properties will be called *a good involution,* and if X possesses such an involution, then (X, ρ) will be said to be *a symmetric quandle.* The material of the current section is a summarization of material that originally appeared in our paper.[21]

The associated group, $G_{(X,\rho)}$ of a symmetric quandle (X, ρ) was defined[40,42] by $G_{(X,\rho)} = \langle x \in X : x \triangleleft y = y^{-1}xy, \ \rho(x) = x^{-1} \rangle$. For a symmetric quandle (X, ρ), an (X, ρ)-set is a set Y equipped with a right action of the associated group $G_{(X,\rho)}$. The notation yg or $y \cdot g$ indicates the image of an element $y \in Y$ under the action of $g \in G_{(X,\rho)}$. The following three formulas hold: $y \cdot (x_1x_2) = (y \cdot x_1) \cdot x_2$, $y \cdot (x_1 \triangleleft x_2) = y \cdot (x_2^{-1}x_1x_2)$, and $y \cdot (\rho(x_1)) = y \cdot (x_1^{-1})$, for $x_1, x_2 \in X$ and $y \in Y$.

In the case of a symmetric quandle, the homology theory of a quandle may be modified as follows to construct a symmetric homology theory.

Let Y be an (X, ρ)-set which may be empty. Let $C_n(X)_Y$ be the free abelian group generated by $(y, x_1, \ldots, x_n)$, where $y \in Y$ and $x_1, \ldots, x_n \in X$. For a positive integer n, let $C_0(X)_Y = \mathbb{Z}(Y)$, the free abelian group generated by Y, and set $C_n(X)_Y = 0$ otherwise. (If Y is empty, then define $C_0(X) = 0$). Define the boundary homomorphism $\partial_n : C_n(X)_Y \longrightarrow C_{n-1}(X)_Y$ by

$$\partial_n(y, x_1, \ldots, x_n) = \sum_{i=1}^{n} (-1)^i [(y, x_1, x_2, \ldots, x_{i-1}, \hat{x}_i, x_{i+1}, \ldots, x_n) - (y \cdot x_i, x_1 \triangleleft x_i, x_2 \triangleleft x_i, \ldots, x_{i-1} \triangleleft x_i, \hat{x}_i, x_{i+1}, \ldots, x_n)]$$

for $n \geq 1$ and $\partial_n = 0$ for $n \leq 1$. Then $C_*(X)_Y = \{C_n(X)_Y, \partial_n\}$ is a chain complex.[31] Let $D_n^Q(X)_Y$ be the subgroup of $C_n(X)_Y$ generated by $\cup_{i=1}^{n-1}\{(y, x_1, \ldots, x_n) \mid x_i = x_{i+1}\}$, and let $D_n^\rho(X)_Y$ be the subgroup of $C_n(X)_Y$ generated by n-tuples of the form

$$(y, x_1, \ldots, x_n) + (y \cdot x_i, x_1 \triangleleft x_i, \ldots, x_{i-1} \triangleleft x_i, \rho(x_i), x_{i+1}, \ldots, x_n)$$

where $y \in Y$, $x_1, \ldots, x_n \in X$, and $i \in \{1, \ldots, n-1\}$. Then $\{D_n^Q(X)_Y, \partial_n\}$ and $\{D_n^\rho(X)_Y, \partial_n\}$ are subcomplexes of C_n,[42] and chain complexes $C_*^R(X)_Y$, $C_*^Q(X)_Y$, $C_*^{R,\rho}(X)_Y$, $C_*^{Q,\rho}(X)_Y$ are defined, respectively, from chain groups $C_n^R(X)_Y = C_n(X)_Y$, $C_n^Q(X)_Y = C_n(X)_Y/D_n^Q(X)_Y$, $C_n^{R,\rho}(X)_Y = C_n(X)_Y/D_n^\rho(X)_Y$, $C_n^{Q,\rho}(X)_Y = C_n(X)_Y/(D_n^Q(X)_Y + D_n^\rho(X)_Y)$. Their respective homology groups[42] are denoted by $H_*^R(X)_Y$, $H_*^Q(X)_Y$, $H_*^{R,\rho}(X)_Y$, and $H_*^{Q,\rho}(X)_Y$, respectively. When $Y = \emptyset$, this subscript is dropped. Corresponding cohomology groups are defined as usual, as well as (co)homology groups with other coefficient groups.

Let a surface diagram D of a surface-knot F be given. We cut the diagram further into *semi-sheets* by considering the upper sheets also to be broken along the double point arcs. Observe that in the local picture of a branch point there are two semi-sheets, at a double point there are 4 semi-sheets, and at a triple point, there are 12 semi-sheets.

Let (X, ρ) denote a symmetric quandle, and let Y denote an (X, ρ)-set. The surface diagram D has elements of X assigned to the sheets and elements of Y assigned to regions separated by the projection such that the following conditions are satisfied.

- (Quandle coloring rule on over-sheets) Suppose that two adjacent semi-sheets coming from an over-sheet of D about a double curve are labeled by x_1 and x_2. If the normal orientations are coherent, then $x_1 = x_2$, otherwise $x_1 = \rho(x_2)$.
- (Quandle coloring rule on under-sheets) Suppose that two adjacent under-sheets e_1 and e_2 are separated along a double curve and are labeled by x_1 and x_2. Suppose that one of the two semi-sheets coming from the over sheet of D, say e_3, is labeled by x_3. We assume that a local normal orientation of e_3 points from e_1 to e_2. If the normal orientations of e_1 and e_2 are coherent, then $x_1 \triangleleft x_3 = x_2$, otherwise $x_1 \triangleleft x_3 = \rho(x_2)$.
- (Region colors) Suppose that two adjacent regions r_1 and r_2 which are separated by a semi-sheet, say e, are labeled by y_1 and y_2, where $y_1, y_2 \in Y$. Suppose that the semi-sheet e is labeled by x. If the normal orientation of e points from r_1 to r_2, then $y_1 \cdot x = y_2$.
- An equivalence relation (of a local normal orientation assigned to each semi-sheet and a quandle element associated to this local orientation) is generated by the following rule (*basic inversion*): Reverse the normal orientation of a semi-sheet and change the element x assigned the sheet by $\rho(x)$.

A *symmetric quandle coloring*, or an $(X,\rho)_Y$*-coloring,* of a surface-knot diagram is such an equivalence class of symmetric quandle colorings. See Fig. 8.

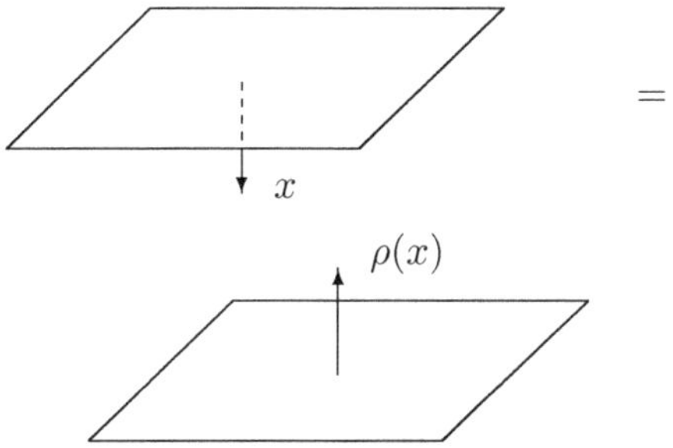

Fig. 8. A basic inversion

Let (D, C_D) be an $(X,\rho)_Y$-colored diagram of an $(X,\rho)_Y$-colored surface-link (F, C). For a triple point of D, define the *weight* as follows: Choose one of eight 3-dimensional regions around the triple point and call the region a *specified region.* There exist 12 semi-sheets around the triple points. Let S_T, S_M and S_B be the three of them that face the specified region, where S_T, S_M and S_B are in the top sheet, the middle sheet and the bottom sheet at the triple point, respectively. Let n_T, n_M and n_B be the normal orientations of S_T, S_M and S_B which point away from the specified region. Consider a representative of C_D such that the normal orientations of S_T, S_M and S_B are n_T, n_M and n_B and let x_1, x_2 and x_3 be the labels assigned the semi-sheets S_B, S_M and S_T, with the normal orientations n_B, n_M and n_T, respectively. Let y be the label assigned to the specified region. The weight is $\epsilon(y, x_1, x_2, x_3)$, where ϵ is $+1$ (or -1) if the triple of the normal orientations (n_T, n_M, n_B) does (or does not, respectively) match the orientation of $\mathbb{R}^3$. See Figure 9. The sum $\sum_\tau \epsilon(y, x_1, x_2, x_3)$ of the weight over all triple points of a colored diagram (D, C_D) represents a 3-cycle $[c_D] \in C_3^{Q,\rho}(X)_Y$.[42] Colored diagrams and cycles represented by colored diagrams are similarly defined when region colors are absent, or equivalently, when $Y = \emptyset$.

We[21] demonstrated (among other things) that for each positive integer N there is a non-orientable surface embedded in a 3-manifold $\times[0, 1]$ such

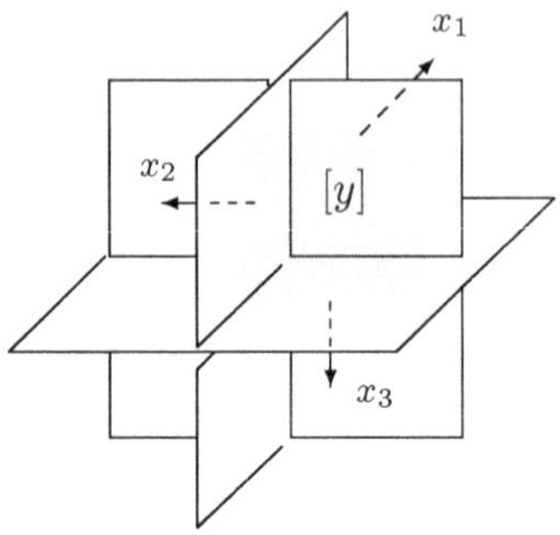

$+(y, x_1, x_2, x_3)$

Fig. 9. The weight of a triple point

that the projection to the 3-manifold is a generic surface with more than N triple points.

In addition, we produced the following family of quandles:

Theorem 2.3.[21] *For each positive integer n, there is an extension $\tilde{R}_{2n+1}$ of R_{2n+1} that is connected and that is not involutory but that has a non-trivial good involution ρ.*

Let e_j denote the column vector in $\mathbb{R}^m$ whose jth entry is 1 and the remaining entries are each 0. A *signed permutation matrix* is a square matrix of size m matrix whose columns are of the form $(\pm e_{\sigma(1)}, \pm e_{\sigma(2)}, \ldots, \pm e_{\sigma(m)})$ where $\sigma \in \Sigma_m$ is a permutation. The set of signed permutation matrices form a group H_m of order $2^m m!$ that is called the *hyper-octahedral group*. Define the group SH_m to be the signed permutation matrices of determinant 1.

To avoid extra subscripts, we write $(\pm e_{\sigma(1)}, \pm e_{\sigma(2)}, \ldots, \pm e_{\sigma(m)})$ as $(\pm\sigma(1), \pm\sigma(2), \ldots, \pm\sigma(m))$. Then the matrix multiplication, in this notation, is written as

$$\begin{aligned}&(\epsilon_1 \cdot \sigma(1),\ \ldots,\ \epsilon_m \cdot \sigma(m)) \cdot (\delta_1 \cdot \tau(1),\ \ldots,\ \delta_m \cdot \tau(m))\\ &= (\epsilon_{\tau(1)}\delta_1 \cdot \sigma(\tau(1)),\ \ldots,\ \epsilon_{\tau(m)}\delta_m \cdot \sigma(\tau(m))),\end{aligned}$$

where $\epsilon_i = \pm 1$ and $\delta_j = \pm 1$ for $i, j = 1, \ldots, m$. For example, $(1, 5, 4, -3, -2) \cdot (5, 1, 2, 3, 4) = (-2, 1, 5, 4, -3)$ while $(5, 1, 2, 3, 4) \cdot (1, 5, 4, -3, -2) = (5, 4, 3, -2, -1)$.

We identify a subgroup of SH_m that maps onto the dihedral group. Let $m = 2n + 1$. Consider the subgroup G_{2n+1} of SH_{2n+1} that is generated by the pair of elements $a = (1, 2n+1, 2n, \ldots, n+2, -(n+1), \ldots, -3, -2)$ and $b = (2n+1, 1, 2, \ldots, 2n)$.

The dihedral group, $D_{2(2n+1)}$, will be identified with the image of its faithful representation in permutation matrices. Specifically, we identify the reflection x with $(1, 2n+1, 2n, \ldots, 2)$, the rotation y with $(2n+1, 1, 2, \ldots, 2n)$, and $D_{2(2n+1)}$ with the subgroup of permutation matrices generated by these two elements. Then the map that takes each matrix (a_{ij}) to $(|a_{ij}|)$ defines a group homomorphism onto the dihedral group: $f : G_{2n+1} \to D_{2(2n+1)}$, such that $f(a) = x$ and $f(b) = y$.

Lemma 2.2.[21] *The order of* G_{2n+1} *is* $(2n+1) \cdot 2^{2n+1}$. *The centralizer of* a, $C(a) = \{c \in G_{2n+1} : ac = ca\}$, *is generated by the elements*

$$(1,\ \epsilon_2\ (2n+1),\ \epsilon_3\ (2n),\ \ldots,\ \epsilon_{n+1}\ (n+2),\ -\epsilon_{n+1}\ (n+1),\ \ldots,\ -\epsilon_2\ (2))$$

where $\epsilon_j = \pm 1$ *for* $j = 2, \ldots, n+1$. *In particular,* $|C(a)| = 2^{n+1}$.

Our paper[21] contains the proof.

Let $H = C(a)$ denote the centralizer of a. Consider the quandle structure $\tilde{R}_{2n+1} = (G_{2n+1}, H, a)$ given by $Hu \lhd Hv = Huv^{-1}av$. From the preceding lemma, we have $|\tilde{R}_{2n+1}| = (2n+1)2^n$.

Lemma 2.3. *There is a surjective quandle homomorphism* $f : \tilde{R}_{2n+1} \to R_{2n+1}$.

The good involution on $\tilde{R}_{2n+1} = (G_{2n+1}, H, a)$ is induced by the map $\rho : \tilde{R}_{2n+1} \to \tilde{R}_{2n+1}$ by

$$\rho(Hu) = \begin{cases} H(-1, -2, \ldots, -(n+1), n+2, \ldots, 2n+1)u & \text{if } n \text{ is an odd number,} \\ H(1, -2, \ldots, -(n+1), n+2, \ldots, 2n+1)u & \text{if } n \text{ is an even number.} \end{cases}$$

Let $\mathcal{I}$ be the kernel of the map $f : G_{2n+1} \to D_{2(2n+1)}$ in Lemma 2.3, i.e.,

$$\mathcal{I} = \{I_{\vec{\epsilon}} = (\epsilon_1(1), \ldots, \epsilon_{2n+1}(2n+1)) \mid \epsilon_i = \pm 1,\ \prod_{i=1}^{2n+1} \epsilon_i = 1\}.$$

Define the maps $f_a : \mathcal{I} \to \mathcal{I}$, $f_b : \mathcal{I} \to \mathcal{I}$ and $f_{b^{-1}} : \mathcal{I} \to \mathcal{I}$ as follows:

$$\begin{aligned} f_a[(\epsilon_1(1), \ldots, \epsilon_{2n+1}(2n+1))] &= (\epsilon_1(1), \epsilon_{2n+1}(2), \ldots, \epsilon_2(2n+1)), \\ f_b[(\epsilon_1(1), \ldots, \epsilon_{2n+1}(2n+1))] &= (\epsilon_{2n+1}(1), \epsilon_1(2), \ldots, \epsilon_{2n}(2n+1)), \text{ and} \\ f_{b^{-1}}[(\epsilon_1(1), \ldots, \epsilon_{2n+1}(2n+1))] &= (\epsilon_2(1), \ldots, \epsilon_{2n+1}(2n), \epsilon_1(2n+1)), \end{aligned}$$

where the order of $(\epsilon_2, \ldots, \epsilon_{2n+1})$ is reversed for f_a, and $(\epsilon_1, \ldots, \epsilon_{2n+1})$ is cyclically permuted for f_b and $f_{b^{-1}}$. We can easily see that f_a, f_b and $f_{b^{-1}}$

are automorphisms of $\mathcal{I}$, $f_a^{-1} = f_a$ and $f_b^{-1} = f_{b^{-1}}$. Moreover, their actions on the diagonal matrices correspond to the dihedral actions of the elements x, y, and y^{-1}.

We also consider distinguished elements

$$\begin{aligned} I_+ &= (1, \ldots, n, -(n+1), n+2, \ldots, 2n, -(2n+1)), \text{ and} \\ I_- &= (-1, 2, \ldots, n+1, -(n+2), n+3, \ldots, 2n+1), \end{aligned}$$

which have exactly two minus signs. Observe that $f_b(I_+) = I_-$. Then the following equalities hold:

$$I_{\vec{\epsilon}}\, a = a f_a(I_{\vec{\epsilon}}), \quad I_{\vec{\epsilon}} b = b f_b(I_{\vec{\epsilon}}), \quad I_{\vec{\epsilon}} b^{-1} = b^{-1} f_b^{-1}(I_{\vec{\epsilon}}),$$

$$ba = ab^{-1} I_+, \quad \text{and} \quad b^{-1}a = abI_-.$$

Consequently, we have the following relations:

$$b^{\pm j} a = ab^{\mp j} \left[\prod_{k=0}^{j-1} f_b^{\mp k}(I_\pm)\right], \quad \text{and} \quad I_{\vec{\epsilon}} b^{\pm j} = b^{\pm j} f_b^{\pm j}(I_{\vec{\epsilon}}).$$

If $1 \leq j \leq n$, then $\prod_{k=0}^{j-1} f_b^{-k}(I_+)$ is a diagonal matrix in $\mathcal{I}$ that has two blocks of j contiguous $(-)$-signs; the first block ends at $n+1$, and the second block ends at $2n+1$. In particular, $\prod_{k=0}^{n-1} f_b^{-k}(I_+)$ has exactly one $(+)$-sign at position (1). Similarly, $\prod_{k=0}^{n-1} f_b^{k}(I_-)$ has exactly one $(+)$-sign at position $(n+1)$.

For $i = 1, \ldots, 2n+1$, let $I(i)$ denote the diagonal matrix that has exactly one $(+)$-sign at position (i,i) and (-1)s at the other positions along the diagonal.

Lemma 2.4. *The following product formulas hold:*

(1) $$(b^i I_{\vec{\epsilon}})(b^j I_{\vec{\delta}}) = b^{i+j} f_b^j(I_{\vec{\epsilon}}) I_{\vec{\delta}},$$

(2) $$(b^i I_{\vec{\epsilon}})(ab^j I_{\vec{\delta}}) = ab^{j-i} \left[\prod_{k=0}^{i-1} f_b^{j-k}(I_+)\right] f_b^j(f_a(I_{\vec{\epsilon}})) I_{\vec{\delta}},$$

(3) $$(ab^i I_{\vec{\epsilon}})(ab^j I_{\vec{\delta}}) = a^2 b^{j-i} \left[\prod_{k=0}^{i-1} f_b^{j-k}(I_+)\right] f_b^j(f_a(I_{\vec{\epsilon}})) I_{\vec{\delta}},$$

(4) $$(ab^i I_{\vec{\epsilon}})(b^j I_{\vec{\delta}}) = ab^{i+j} f_b^j(I_{\vec{\epsilon}}) I_{\vec{\delta}}.$$

Proof. The calculations follow directly from the formulas above. □

Lemma 2.4 gives normal forms for the elements of G_{2n+1}. We[21] used these forms to calculate that $\tilde{R}_{2n+1}$ is connected.

A very interesting result therein is that the symmetric homology of $\tilde{R}_3$ contains an infinite cyclic summand.

3. Applications of Quandles and Quandle Cocycles

Here, I will give short descriptions of the myriad of results in computations of quandles, their homology and cohomology, and applications of their cocycles and cocycle invariants. The contents and organization of this section are based upon a talk that Masahico Saito gave at Knots in Washington. His original beamer presentation is available at

`http://shell.cas.usf.edu/~saito/talks/talk09DCquandle.pdf`

Quandles with a few elements and their (co)homologies. Computations of quandles, biracks, and their (co)-homology have largely been done via computer programs. Some algebraic methods also have been developed, but the general resolution/derived functor approach is lacking for quandles. While analogues between group and quandle cohomology theory are lacking for high degrees, we do know that extensions of quandles by abelian groups are quantified by quandle 2-cocycles. That is if $\phi : X \times X \to A$ represents a quandle 2-cocycle on X that takes values in an Abelian group A, then the product $(a, x) \triangleleft (b, y) = (a + \phi(x, y), x \triangleleft y)$ defines a quandle structure on $A \times X$. Similar constructions almost always hold in more general contexts;[3,14,21] see the discussion above.

A complete list of quandles with up to 6 elements is given in our book,[20] and computations of their quandle homology is provided. However, there are some errors in the homology computations since Uegaki's program did not distinguish between torsion groups of the form $\mathbb{Z}_p \oplus \mathbb{Z}_p$ and $\mathbb{Z}_{p^2}$. These errors have been corrected by Niebrzydowski and Przytycki. In the context of his dissertation, Niebrzydowski wrote a program that computed the higher homology groups of dihedral and Alexander quandles. At the limits of the computations, they were helped by Shumakovich who used his access to a supercomputer to compute the Smith normal form of some very large matrices. The computations led to the delayed Fibonacci conjecture. Przytycki recently sent me the following information:

- for the quandle $X = QS_4$ (Example 2.3) that is the last four element quandle listed on p. 172,[20] they computed:

  $$H_4^{\mathrm{Q}}(X) \cong \mathbb{Z}_4 \oplus \mathbb{Z}_2^4,$$

  and

  $$H_5^{\mathrm{Q}}(X) \cong \mathbb{Z}_4 \oplus \mathbb{Z}_2^5.$$

  In addition, they computed $H_6^{\mathrm{Q}}(X) \cong \mathbb{Z}_4 \oplus \mathbb{Z}_2^9$.

- For the quandle $X = QS_6$ (up to a labeling of elements is the quandle of 4-cycles in the symmetric group on 4 elements:

$$QS_6 = \{(1234), (1243), (1324), (1342), (1423), (1432)\}),$$

they compute

$$H_3^Q(X) \cong H_4^Q(X) = \mathbb{Z}_3 \oplus \mathbb{Z}_8.$$

This is the quandle of Theorem 2.3 for $n = 1$.

Sam Nelson and his students[24,49,50] have extended the tables of quandles in a number of directions, and they have provided python code to implement calculations of quandles via the matrix that represents the multiplication table. See for example,

`http://www.esotericka.org/cmc/quandles.html`

Bartholomew and Fenn[6] extended these results to biquandles and applied the results to reprove that the Burau representation is not faithful and to provide invariants of welded knots.

Graña and his student Vendramin are developing GAP programs to recognize quandles, racks, and to compute their homology. See

`http://mate.dm.uba.ar/~lvendram/rig/index.html`

On the algebraic side, Nelson has an classification of finite Alexander Quandles, and Graña has a classification of indecomposable racks of order p^2.[32] The idea of extensions[14] via quandle 2-cocycles is brought to its full generality;[3] however, there are still only few examples that are known. The interest in quandles among algebraists has to do with their relationships to Yetter-Drinfeld modules and to solutions to the Yang-Baxter equations.[3,26] A solution to the set-theoretic YBE is morally equivalent to a biquandle. See also Fenn, Jordan-Santana, and Kauffman[28] which is an excellent article on biquandles.

Niebrzydowski[51] began the problem of classifying Burnside keis. Joyce's and Matveev's classification of quandles (Theorem 2.1) as an operation on a coset space has been used[21] to construct a family of quandles that have interesting properties; see Theorem 2.3.

In terms of homology and cohomology there are still very limited results. The appendix to our book[20] gives calculations for low dimensional quandle homology for quandles with up to six elements (see corrections above). We[13] were able to find non-trivial 3-cocycles, via a computer search, for a quandle in the extended Andruskiewitsch and Graña theory. Niebrzydowski and Przytycki[54] computed up to H_{12} for the dihedral quandle of order 3. They proposed the "delayed Fibonacci conjecture" for R_p namely $H^n_{\mathrm{Q}}(R_p) = \mathbb{Z}_p^{f_n}$ where the sequence f_n satisfies the recurrence relation $f_n = f_{n-1} + f_{n-3}$, and $f(1) = f(2) = 0$, $f(3) = 1$. Nosaka has given a proof of this in certain cases. The Japanese summary of his results is found here:

`www.f.waseda.jp/taniyama/math-of-knots/knots2009proc.pdf`

See also the work of Clauwens.[23]

An early result on quandle homology is given by Litherland and Nelson[44] is that the rank of the integral quandle, rack, and degenerate homology is determined from the number of orbits of the inner automorphism group of the quandle. Etingof and Graña[27] prove a similar result and go on to show that the second rack cohomology can be expressed as $H^2(X, A) \cong H^1(G_X, \mathrm{Fun}(X, A))$ where G_X denotes the enveloping group, and $\mathrm{Fun}(X, A)$ denotes the set of functions from the quandle X to the abelian group A. Both pairs of authors improved upon our results.[19]

Mochizuki[46] computed the second and third homology groups of dihedral quandles of prime order. Ameur and Saito[1] found a large number of non-trivial cocycles for Alexander quandles by generalizing the polynomial formulas for Mochizuki cocycles. Some very promising techniques developed by Niebrzydowski and Przytycki[53,54] include homology operations to the chain groups that are algebra versions of twist spinning and partial derivatives defined at the chain level.

Computations of cocycle invariants beyond the original papers[16,17] began with one of our papers.[18] Another group of collaborators and I also computed invariants associated to the Andruskiewitsch and Graña cocycles.[13] Nelson and his students have developed more subtle invariants from quandles and biquandles, and they have applied these results to virtual knots and their generalizations (For full bibliographic information on Nelson and his collaborators, please search the ArXiv and MathSciNet). Ameur and Saito[1] developed used polynomial valued cocycles to study the tangle embedding problem. At about the same time, Niebrzydowski was studying similar problems using cocycle techniques. The link

`http://shell.cas.usf.edu/quandle/`

will send you to the database compiled by Masahico's master's degree student Chad Smudde. Included among the associated pages are several Maple programs that are used find and compute cocycles and their invariants.

Cocycles are used to show that certain tangles cannot be embedded into specific knots.[2]

Applications. Applications of the quandle 3-cocycle invariants were developed by various authors in various combinations. Infinite families of non-invertible 2-knots were found by Asami and Satoh;[5] Iwakiri[36] used cocycle invariants to determine lower bounds for the number of 1-handles that are needed to remove all triple points as well as bounds on the triple point number. Computations of the 3-cocycle invariant of 2-twist-spun 2-bridge knots are given.[37] Asami and Kuga[4] classify the non-trivial colorings of torus knots by Alexander quandles, and use Satoh's method[58] of twist-spinning to compute cocycle invariants for selected twist spins of these knots. Satoh and Shima[60] used cocycle invariants to show that the 2-twist-spun trefoil has at least 4 triple points. These techniques were applied by Hatakanaka[33] to show that the 2-twist-spun figure 8 knot (and the $(2,5)$-torus knot) each have at least 6-triple points.

Mohamad and Yashiro compute minimal triple point numbers for twist-spun knots that are colored by the dihedral quandle of order 5. An abstract of that talk is found here:

`http://faculty.ms.u-tokyo.ac.jp/~topology/abstracts.pdf`

The 3-cocycles defined over symmetric quandle homology (see above) were used by Oshiro and Kamada and Oshiro[42,48] to determine lower bounds for triple point numbers of linked surfaces in which at least one component is non-orientable. We[21] found families of non-orientable connected surfaces in thickened 3-manifolds whose projections contain as many triple points as you might like (see also the material above). It is quite likely that there are knotted (hence connected) non-orientable surfaces in 4-space whose projections will have an arbitrarily large number of triple points, and I would guess that symmetric quandle 3-cocycle can be used to detect them.

There is also a 3-cocycle invariant for classical knots. Its definition is similar to the 2-cocycle invariant, but 2-dimensional regions are also colored by quandle elements, normals to arcs point from a region colored a to a region colored $a \triangleleft b$ where b is the color on the arc that separates the two regions. The 3-cocycle invariant for classical knots can be shown to detect the chirality of the trefoil.[30]

In private communications, Shin Satoh has informed us that he can detect chirality of spacial graphs using quandle cocycle invariants. Ishii and

Iwakiri[35] have used quandle cocycles to distinguish flowed oriented spacial graphs. In particular, they can distinguish whether or not two knotted spacial graphs represent the same knotted handle body. By using cocycle-like techniques, my collaborators and I were able to show that certain knot isotopies required at least so-many type III Reidemeister moves.[15]

A classical knot is non-trivial if and only if the second quandle cohomology group of the fundamental quandle is non-trivial.[25] Eisermann goes on to show more. In general, $H_2(\pi_Q(k)) \cong \mathbb{Z}$ and a choice of generator is tantamount to a choice of orientation class. Thereby, the classification of the knot complement via the quandle is complete.

Masahico Saito[57] showed that Mochizuki's cocycle could be used to give necessary conditions for p colorable knots (with p a prime number larger than 7) to use at least 5 colors. Shin Satoh and Masahico Saito[59] used cocycle invariants to give bounds for the minimal number of unbroken sheets in knotted surface diagrams. For example, any diagram of the spun trefoil has at least 4 sheets. The three of us used quandle 2-cocycles to determine that certain surfaces are not ribbon condordant.[22]

In Japanese, there is an announcement that Kabaya and Inoue relate the quandle cocycle invariants to the Chern-Simon invariants.

`http://faculty.ms.u-tokyo.ac.jp/~topology/abstracts.pdf`

This result is a follow-up to[34] which announces that hyperbolic volume can also be understood as a quandle cocycle invariant. Additional topological applications are related to Lefschetz and elliptic fibrations via the Dehn quandle.[41,62]

As mentioned above, there is interest among algebraists in quandles and racks because of their relationships to point Hopf algebras, set theoretic solutions to the Yang-Baxter equations, and their roles in Yetter-Drinfeld modules. In addition, there are connections among categorical groups, Lie 2-algebras, and other notions of categorification or categorical internalizations. We have been involved in collaborations[11,12] that discuss these connections, and I anticipate forth-coming work on related topics.

4. Conclusion

Racks, quandles, their generalizations and their cohomology theories are finding wide applications in topology and the classifications of algebraic structures. The generalizations are far-reaching because the quandle structure is a natural structure to consider from several points of view. The current paper is a broad-brush painting of some of the current trends and applications that have been found. In as far as I am able, I hope that I pointed the reader to the best, and most up-to-date, bibliographic sources.

To the authors whose work I neglected to mention, please accept my apologies; the current paper is being prepared in the face of some deadlines, and in the face of other projects of my own and my collaborators that remain incomplete. I recognize, as should the reader, that the range of results and applications in the age of rapid information dissemination presents an ever-moving target. Since there is some lag between time of submission and publication, I will, if possible include summaries of other results if you call my attention to them.

I would like to complete this work with acknowledgements to my many collaborators on these projects, and in particular to my long time collaborator Masahico Saito. When faced with the task of writing this summary, he asked to be excluded only because he too is involved in many other projects. Much of the mathematics that is developed herein is due in large part to his creativity. And as I prepared the manuscript, he was very helpful with bibliographic references.

References

1. Ameur, K.; Saito, M., *Polynomial cocycles of Alexander quandles and applications.* J. Knot Theory Ramifications 18 (2009), no. 2, 151–165.
2. Ameur, K.; Elhamdadi, M.; Rose, T.; Saito, M.; Smudde, C., *Tangle embeddings and quandle cocycle invariants.* Experiment. Math. 17 (2008), no. 4, 487–497.
3. Andruskiewitsch, N.; Graña, M., *From racks to pointed Hopf algebras.* Adv. Math. 178 (2003), no. 2, 177–243.
4. Asami, S.; Kuga, K., *Colorings of torus knots and their twist-spuns by Alexander quandles over finite fields.* J. Knot Theory Ramifications 18 (2009), no. 9, 1259–1270.
5. Asami, S.; Satoh, S., *An infinite family of non-invertible surfaces in 4-space.* Bull. London Math. Soc. 37 (2005), no. 2, 285296.
6. Bartholomew, A.; Fenn, R.; *Biquandles of small size and some invariants of virtual and welded knots*, posted at arxiv.org/abs/1001.5127v1.
7. Birman, J. S.; Libgober, A., " Braids," Proceedings of the AMS-IMS-SIAM Joint Summer Research Conference in Mathematical Sciences on Artin's Braid Group held at the University of California, Santa Cruz, California, July 13–26, 1986." Contemporary Mathematics, 78. American Mathematical Society, Providence, RI, 1988.
8. Blaschke, W.; Bol, G., *Geometrie der Gewebe.* Springer, Berlin, 1938.
9. Brieskorn, E., *Automorphic sets and braids and singularities.* Braids (Santa Cruz, CA, 1986), 45–115, Contemp. Math., 78, Amer. Math. Soc., Providence, RI, 1988.

10. Burstin, C.; Mayer, W., *Distributive Gruppen*, J. Reine Angew. Math. 160 (1929), 111130.
11. Carter, J. S.; Crans, A. S.; Elhamdadi, M.; Saito, M., *Cohomology of categorical self-distributivity.* J. Homotopy Relat. Struct. 3 (2008), no. 1, 13–63.
12. Carter, J. S.; Crans, A. S.; Elhamdadi, M.; Saito, M., *Cohomology of the adjoint of Hopf algebras.* J. Gen. Lie Theory Appl. 2 (2008), no. 1, 19–34
13. Carter, J. S. ; Elhamdadi, M. ; Graña, M.; Saito, M. , *Cocycle knot invariants from quandle modules and generalized quandle cohomology.*, Osaka J. Math. 42 (2005), 499–541.
14. Carter, J. S. ; Elhamdadi, M. ; Nikiforou, M. A. ; Saito, M. , *Extensions of quandles and cocycle knot invariants.*, J. of Knot Theory and Ramifications 12 (2003) 725–738.
15. Carter, J. S.; Elhamdadi, M.; Saito, M.; Satoh, S., *A lower bound for the number of Reidemeister moves of type III.* Topology Appl. 153 (2006), no. 15, 2788–2794.
16. Carter, J. S. ; Jelsovsky, D.; Kamada, S. ; Langford, L. ; Saito, M. , *Quandle cohomology and state-sum invariants of knotted curves and surfaces*, Trans. Amer. Math. Soc. 355 (2003), 3947–3989 posted at arXiv:math/9903135 .
17. Carter, J. S.; Jelsovsky, D.; Kamada, S.; Langford, L.; Saito, M., *State-sum invariants of knotted curves and surfaces from quandle cohomology.* Electron. Res. Announc. Amer. Math. Soc. 5 (1999), 146–156 (electronic).
18. Carter, J. S.; Jelsovsky, D.; Kamada, S.; Saito, M., *Computations of quandle cocycle invariants of knotted curves and surfaces.* Adv. Math. 157 (2001), no. 1, 36–94.
19. Carter, J. S.; Jelsovsky, D.; Kamada, S.; Saito, M., *Quandle homology groups, their Betti numbers, and virtual knots.* J. Pure Appl. Algebra 157 (2001), no. 2-3, 135–155.
20. Carter, S.; Kamada, S.; Saito, M., "Surfaces in 4-space." Encyclopaedia of Mathematical Sciences, 142. Low-Dimensional Topology, III. Springer-Verlag, Berlin, 2004.
21. Carter, J. S.; Oshiro, K.; Saito, M., *Symmetric Extensions of Dihedral Quandles and Triple Points of Non-orientable Surfaces.* Top. and Its Appl. 157 (2010), 857-869.
22. Carter, J. S.; Saito, M.; Satoh, S., *Ribbon concordance of surface-knots via quandle cocycle invariants.* J. Aust. Math. Soc. 80 (2006), no. 1, 131–147.
23. , Clauwens, F.J. B.J., *The algebra of rack and quandle cohomology*, posted at http://arxiv.org/pdf/1004.4423
24. Creel, C.; Nelson, S., *Symbolic computation with finite biquandles.* J. Symbolic Comput. 42 (2007), no. 10, 992–1000.
25. Eisermann, M., *Homological characterization of the unknot.* J. Pure Appl. Algebra 177 (2003), no. 2, 131–157.
26. Eisermann, M., *Yang-Baxter deformations of quandles and racks.* Algebr. Geom. Topol. 5 (2005), 537–562.
27. Etingof, P.; Graña, M., *On rack cohomology.* J. Pure Appl. Algebra 177 (2003), no. 1, 49–59.
28. Fenn, R.; Jordan, M.; Kauffman, L., *Biquandles and virtual links,* Topology and its Applications, 145 (2004) 157–175.

29. Fenn, R.; Rourke, C., *Racks and links in codimension two.* J. Knot Theory Ramifications 1 (1992), no. 4, 343–406.
30. Fenn, R.; Rourke, C.; Sanderson, B., *The rack space.* Trans. Amer. Math. Soc. 359 (2007), no. 2, 701–740.
31. Fenn, R.; Rourke, C.; Sanderson, B., *An introduction to species and the rack space.* Topics in knot theory (Erzurum, 1992), 33–55, NATO Adv. Sci. Inst. Ser. C Math. Phys. Sci., 399, Kluwer Acad. Publ., Dordrecht, 1993.
32. Graña, M., *Indecomposable racks of order* p^2. Beiträge Algebra Geom. 45 (2004), no. 2, 665–676.
33. Hatakenaka, E., *An estimate of the triple point numbers of surface-knots by quandle cocycle invariants.* Topology Appl. 139 (2004), no. 1-3, 129–144.
34. Inoue, A., *Quandle and hyperbolic volume.* arxiv.org.0812.0425.
35. Ishii, A.; Iwakiri, M., *Quandle cocycle invariants for spatial graphs and knotted handlebodies*, available at

    `http://www.math.tsukuba.ac.jp/~aishii/files/paper017.pdf`

36. Iwakiri, M., *Triple point cancelling numbers of surface links and quandle cocycle invariants.* Topology Appl. 153 (2006), no. 15, 28152822.
37. Iwakiri, M. *Calculation of dihedral quandle cocycle invariants of twist spun 2-bridge knots.* J. Knot Theory Ramifications 14 (2005), no. 2, 217229.
38. Jones, V. F. R. *A polynomial invariant for knots via von Neumann algebras.* Bull. Amer. Math. Soc. (N.S.) 12 (1985), no. 1, 103–111.
39. Joyce, D., *A classifying invariant of knots, the knot quandle*, J. Pure Appl. Alg. 23 (1982), 37–65.
40. Kamada, S., *Quandles with good involutions, their homologies and knot invariants.* Intelligence of low dimensional topology 2006, 101–108, Ser. Knots Everything, 40, World Sci. Publ., Hackensack, NJ, 2007.
41. Kamada, S.; Matsumoto, Y.; *Enveloping monoidal quandles.* Topology Appl. 146/147 (2005), 133–148.
42. Kamada, S.; Oshiro, K., *Homology Groups of Symmetric Quandles and Cocycle Invariants of Links and Surface-links.* arXiv:0902.4277
43. Kauffman, L. H., "On knots." Annals of Mathematics Studies, 115. Princeton University Press, Princeton, NJ, 1987.
44. Litherland, R. A.; Nelson, S., *The Betti numbers of some finite racks.* J. Pure Appl. Algebra 178 (2003), no. 2, 187–202.
45. Matveev, S., *Distributive groupoids in knot theory (Russian)*, Math. USSR-Sbornik 47 (1982), 73–83.
46. Mochizuki, T. *Some calculations of cohomology groups of finite Alexander quandles.* J. Pure Appl. Algebra 179 (2003), no. 3, 287–330.
47. Moufang, R., *Zur Struktur von Alternativekorpern*, Math. Ann. 110 (1935), 416430.
48. Oshiro, K., *Triple point numbers of surface-links and symmetric quandle cocycle invariants.*, preprint.
49. Murillo, G.; Nelson, S., *Alexander quandles of order 16.* J. Knot Theory Ramifications 17 (2008), no. 3, 273–278. Erratum: Alexander quandles of order 16 J. Knot Theory Ramifications 18 (2009), no. 5, 727)

50. Murillo, G.; Nelson, S.; Thompson, A. *Matrices and finite Alexander quandles.* J. Knot Theory Ramifications 16 (2007), no. 6, 769–778.
51. Niebrzydowski, M., *On colored quandle longitudes and its applications to tangle embeddings and virtual knots.* J. Knot Theory Ramifications 15 (2006), no. 8, 1049–1059.
52. Niebrzydowski, M., Przytycki, J., *Burnside kei.* Fund. Math. 190 (2006), 211–229.
53. Niebrzydowski, M.; Przytycki, J. *Homology of dihedral quandles.* J. Pure Appl. Algebra 213 (2009), no. 5, 742–755.
54. Niebrzydowski, M.; Przytycki, J. *Homology operations on homology of quandles.* arXiv:0907.4732.
55. Niebrzydowski, M.; Przytycki, J. *The quandle of the trefoil knot as the Dehn quandle of the torus.* arXiv:0805.2743.
56. Roseman, D., *Reidemeister-type moves for surfaces in four-dimensional space.*, in "Knot theory (Warsaw, 1995)" 347–380 Banach Center Publ., 42, Polish Acad. Sci., Warsaw (1998).
57. Saito, M., *Minimal Numbers of Fox Colors and Quandle Cocycle Invariants of Knots*, http://arxiv.org/pdf/0905.4486 (to appear Journ. of Knot Theory and its Ramif.).
58. Satoh, S., *Surface diagrams of twist-spun 2-knots.* Knots 2000 Korea, Vol. 1 (Yongpyong). J. Knot Theory Ramifications 11 (2002), no. 3, 413–430.
59. Saito, M.; Satoh, S., *The spun trefoil needs four broken sheets.* J. Knot Theory Ramifications 14 (2005), no. 7, 853–858.
60. Satoh, S.; Shima, A., *Triple point numbers and quandle cocycle invariants of knotted surfaces in 4-space.* New Zealand J. Math. 34 (2005), no. 1, 71–79.
61. Takasaki, M., *Abstraction of symmetric transformations*, Tohoku Math. J. 49 (1943), 145–207.
62. Zablow, J., *On relations and homology of the Dehn quandle.* Algebr. Geom. Topol. 8 (2008), no. 1, 19–51.

COMBINATORICS OF VASSILIEV INVARIANTS

Sergei Chmutov

ABSTRACT. This paper is an introductory survey of the combinatorial aspects of the Vassiliev theory of knot invariants following the lectures delivered at the Advanced School on Knot Theory and its Applications to Physics and Biology in the ICTP, Trieste (Italy), May 2009. The exposition is based on the forthcoming book [CDM], where the reader may find further details, examples, and developments.

1. DEFINITION

Vassiliev knot invariants, also known as *finite type invariants*, were introduced independently by Victor Vassiliev (Moscow) [Vas] and by Mikhail Goussarov (St. Petersburg) [Gus] at the end of the 1980's. The principal idea of the combinatorial approach to the theory is to extend a knot invariant v to singular knots with double points according to the following rule, which we will refer to as *Vassiliev skein relation*:

(1)

The main definition is:

Definition 1.1. A knot invariant is said to be a *Vassiliev invariant* of order (or degree) $\leqslant n$ if its extension vanishes on all singular knots with more than n double points.

Denote by $\mathcal{V}_n$ the set of Vassiliev invariants of order $\leqslant n$ with values in the field of complex numbers $\mathbb{C}$. The definition implies that, for each n, the set $\mathcal{V}_n$ forms a complex vector space. Moreover, $\mathcal{V}_n \subseteq \mathcal{V}_{n+1}$, so we have an increasing filtration

Key words and phrases. Vassiliev invariants, chord diagrams, weight systems, the Kontsevich integral.

$$\mathcal{V}_0 \subseteq \mathcal{V}_1 \subseteq \mathcal{V}_2 \subseteq \cdots \subseteq \mathcal{V}_n \subseteq \cdots \subseteq \mathcal{V} := \bigcup_{n=0}^{\infty} \mathcal{V}_n \ .$$

It will be shown that the spaces $\mathcal{V}_n$ have finite dimension, and that the quotients $\mathcal{V}_n/\mathcal{V}_{n-1}$ admit a nice combinatorial description. The study of these spaces is the main purpose of the combinatorial Vassiliev invariant theory. The exact dimension of $\mathcal{V}_n$ is known only for $n \leqslant 12$:

n	0	1	2	3	4	5	6	7	8	9	10	11	12
$\dim \mathcal{V}_n$	1	1	2	3	6	10	19	33	60	104	184	316	548
$\dim \mathcal{V}_n/\mathcal{V}_{n-1}$	1	0	1	1	3	4	9	14	27	44	80	132	232

For higher values of n there are only asymptotic lower (O. Dasbach [Das]) and upper (D. Zagier [Zag]) bounds

$$e^{n/\log_a n} \lesssim \dim \mathcal{V}_n/\mathcal{V}_{n-1} \lesssim n!/a^n$$

for any constant $a < \pi^2/6$.

Along with well known invariants of finite order, there are classical invariants which are not of finite order. Some of them belong to the *closure* of the Vassilev invariants $\bar{\mathcal{V}}$ which can be defied as follows (S. Duzhin). The space $\bar{\mathcal{V}}$ consists of all knot invariants which distinguish only the knots distinguishable by a finite order invariant. In other words, $v \in \bar{\mathcal{V}}$ if and only if for any two knots K_1 and K_2 with $v(K_1) \neq v(K_2)$ there is a Vassiliev invariant v_n (depending on K_1 and K_2) of finite order $\leqslant n$ which also distinguishes these knots, $v_n(K_1) \neq v_n(K_2)$. Below are some very important open problems of the theory:

- prove or disprove that any knot invariant belongs to the closure $\bar{\mathcal{V}}$,
- decide whether Vassiliev invariants detect the unknot,
- decide whether Vassiliev invariants detect orientation of knots.

2. Examples of Vassiliev invariants

2.1. The Conway polynomial. The Conway polynomial may be defined by the skein relation and its normalization on the unknot:

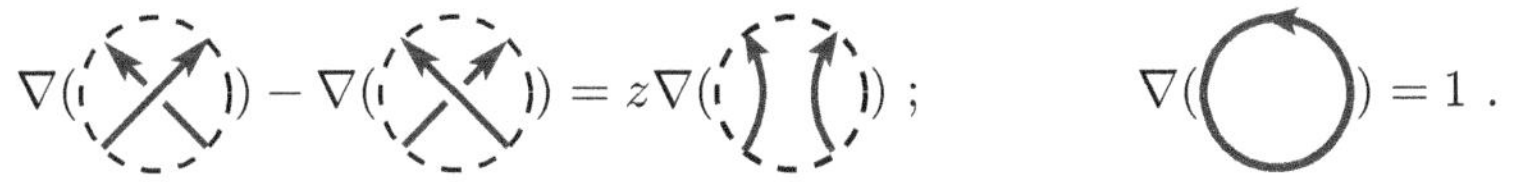

For a knot K the Conway polynomial is always even

$$\nabla(K) = 1 + c_2(K)z^2 + c_4(K)z^4 + \dots \ .$$

Its coefficients are Vassiliev invariants. Indeed, comparing its skein relation with the Vassiliev one (1) we conclude that the Conway polynomial of a knot with a singular point is divisible by z:

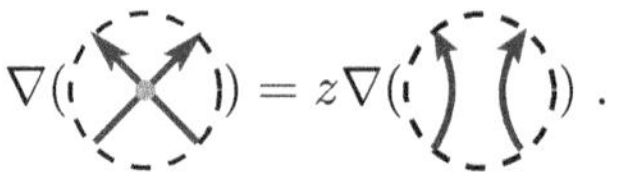

Consequently, the Conway polynomial of a knot with $k > n$ double points is divisible by z^k. Thus its coefficient $c_n(K)$ at z^n vanishes on singular knots with $> n$ double points. Therefore, it is a Vassiliev invariant of order $\leqslant n$.

Despite the fact that the coefficients are Vassiliev invariants, the degree of the Conway polynomial is not of finite type. Nevertheless, it belongs to the closure $\bar{\mathcal{V}}$.

2.2. The Jones polynomial. The skein relation for the Jones polynomial with the same initial condition for the unknot are

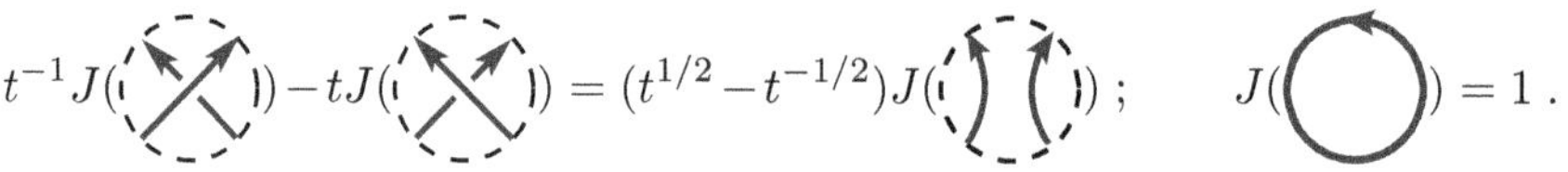

Making a substitution $t := e^h$ and then taking the Taylor expansion into a formal power series in h, we can represent the Jones polynomial of a knot K as a power series

$$J(K) = \sum_{n=0}^{\infty} j_n(K) h^n .$$

We claim that the coefficient $j_n(K)$ is a Vassiliev invariant of order $\leqslant n$. Indeed, substituting $t = e^h$ into the skein relation gives

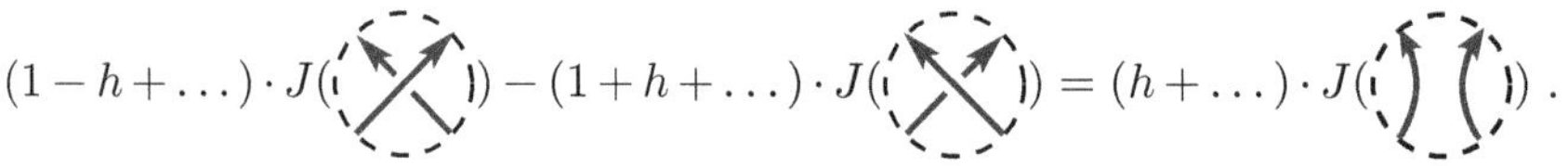

From which we get

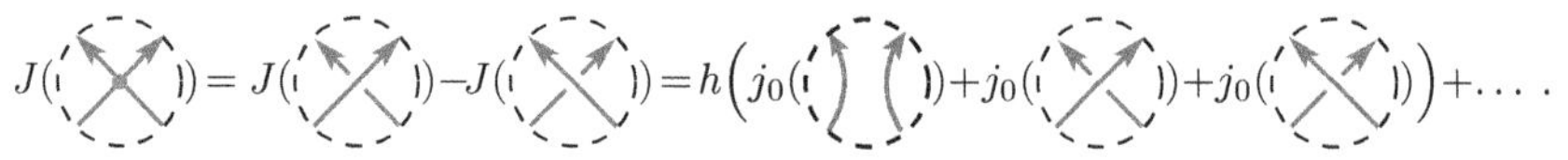

This means that the value of the Jones polynomial on a knot with a single double point is congruent to 0 modulo h. Therefore, the Jones polynomial of a singular knot with $k > n$ double points is divisible by h^k, and thus its nth coefficient vanishes on such a knot.

2.3. The HOMFLYPT polynomial. The skein relation for the HOMFLYPT polynomial is

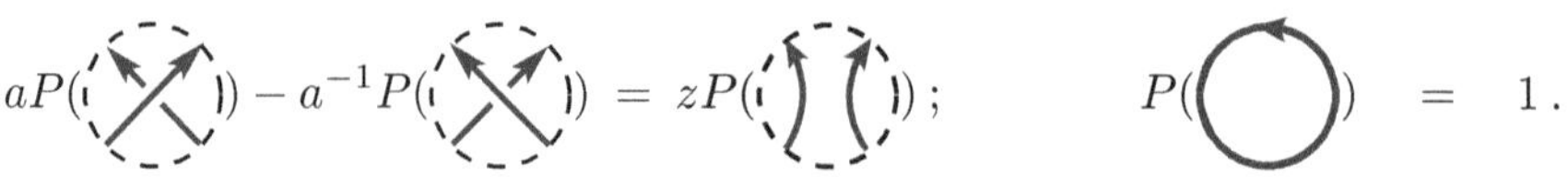

Making a substitution $a = e^h$ and taking the Taylor expansion in h, we represent $P(K)$ as a Laurent polynomial in z and a power series in h, $P(K) = \sum p_{k,l}(K)\, h^k z^l$. Similarly to the case of the Jones polynomial, one can show that $p_{k,l}(K)$ is a Vassiliev invariant of order $\leqslant k + l$.

2.4. Quantum knot invariants. J. Birman and X.-S. Lin proved [BL] that all quantum invariants produce Vassiliev invariants in the way similar to the previous examples. Namely, making a substitution $q = e^h$, one can show that the coefficient of h^n in the Taylor expansion of a quantum invariant is a Vassiliev invariant of order $\leqslant n$. Thus all quantum invariants belong to the closure $\bar{\mathcal{V}}$.

3. Chord diagrams, symbols, and weight systems

Let v be a Vassiliev invariant of order $\leqslant n$. It turns out that a value of v on a singular knot K with n double points does not depend on the specific knotedness of K. It depends only on the combinatorial arrangement of double points along the knot, which can be encoded by a *chord diagram* of K.

Definition 3.1. A *chord diagram* of order n (or *degree* n) is an oriented circle with a distinguished set of n disjoint pairs of distinct points, considered up to orientation preserving homeomorphisms of the circle.

This means that only the mutual combinatorial positions of the ends of chords are important. Their precise geometrical locations on the circle are irrelevant. By a chord diagram of a singular knot K we mean a circle parameterizing K with the two preimages of each double point connected by a chord. Here are two examples:

 ,  .

The value of $v(K)$ on a knot K with n double points depends only on the chord diagram of K. Indeed, let K_1 and K_2 be two singular knots with the same chord diagrams. If we place K_1, K_2 in $\mathbb{R}^3$ so that the corresponding double points (and both branches of the knot in their vicinity) coincide,

$K_1 =$  $K_2 =$  ,

then we can deform the arcs of K_1 into the corresponding arcs of K_2 so that the only singularities introduced or removed (one at a time) in the process are double points. By the Vassiliev skein relation, in each of these events the value of v does not change. So $v(K_1) = v(K_2)$.

Definition 3.2. The *symbol* of v is a restriction of v to the set of knots with precisely n double points, considered as a function on the set of chord diagrams.

It is obvious that if v_1 and v_2 are two Vassiliev invariants of order $\leqslant n$ with the same symbols, then their difference $v_1 - v_2$ is an invariant of order $\leqslant (n-1)$. Thus the space $\mathcal{V}_n/\mathcal{V}_{n-1}$ coincides with the space of all symbols of Vassiliev invariants of order $\leqslant n$. The set of chord diagrams with n chords is finite. So the space of all ($\mathbb{C}$-valued) functions on this set, and hence $\mathcal{V}_n$, is finite dimensional.

The symbol of an invariant is not an arbitrary function on chord diagrams; it satisfies certain relations. The so called *one-term relation* (1T) is very easy to see; the symbol of an invariant v always vanishes on a chord diagram with a short isolated chord (not intersecting other chords)

symb(v)() $= 0$.

This follows from the fact that we can choose a singular knot representing such a chord diagram as having a small kink parameterized by the arc between the end-points of the isolated chord. Then the Vassiliev skein relation (1) gives

symb(v)() $=$ v() $=$ v() $-$ v() $= 0$,

because the two knots on the right-hand side are isotopic. Another kind of relation is called the *four-term relation* (4T):

symb(v)() $-$ symb(v)() $+$ symb(v)() $-$ symb(v)() $= 0$.

Here it is assumed that the diagrams in the pictures may have other chords with end-points on the dotted arcs, while all the end-points of the chords on the solid portions of the circle are explicitly shown. For example, this means that in the first and second diagrams the two bottom end-points are adjacent. The chords omitted from the pictures should be the same in all four cases. To prove this relation we represent each chord diagram by a singular knot with a special fragment containing two double points. Then, resolving one of them for every knot, we will get eight terms which cancel out.

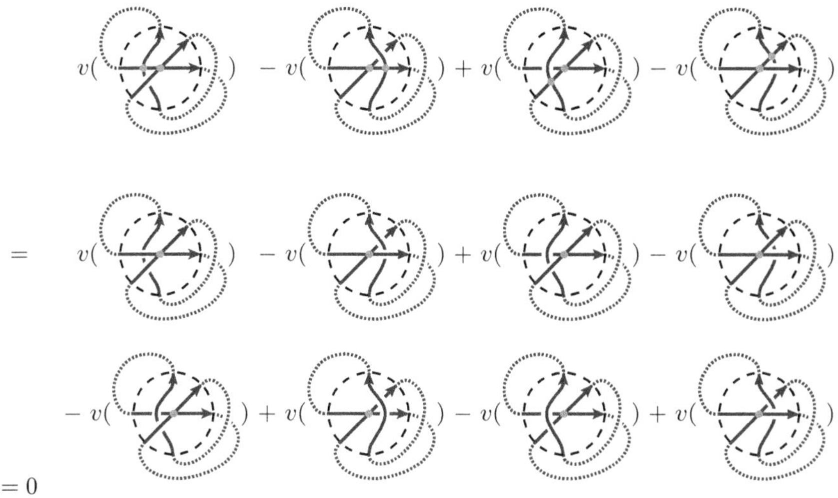

$= 0$

Definition 3.3. A *weight system* of order n is a function on the set of chord diagrams with n chords which satisfies one- and four-term relations.

It turns out that all other relations for symbols of Vassiliev invariants are consequences of (1T) and (4T) relations. This is the content of a fundamental theorem of M. Kontsevich.

Theorem 3.4 (M. Kontsevich [Kon]). *Any weight system is a symbol of an appropriate Vassiliev invariant of order* $\leqslant n$.

Together with the discussion above this theorem implies that the quotient space $\mathcal{V}_n/\mathcal{V}_{n-1}$ is isomorphic to the space of all ($\mathbb{C}$-valued) functions on the set of chord diagram with n chords satisfying the relations (1T) and (4T).

Let us introduce a $\mathbb{C}$-vector space $\mathcal{A}_n$ as the space spanned by chord diagrams modulo (1T) and (4T) relations:

(1T)  $= 0$

(4T) 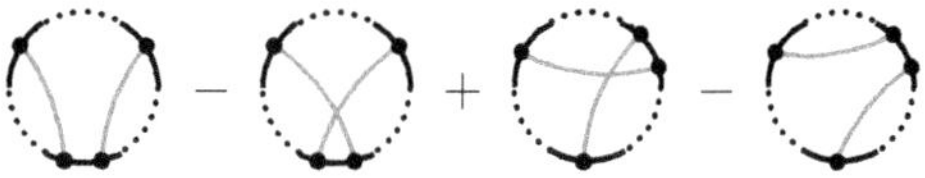 $= 0$.

Then, $\mathcal{V}_n/\mathcal{V}_{n-1} \cong (\mathcal{A}_n)^*$.

Examples 3.5. For $n = 2$, among the two possible chord diagrams,  and , the first one is equal to zero by (1T). Thus $\mathcal{A}_2 = \langle$  $\rangle$ and has dimension 1.

For $n = 3$ there are five chord diagrams: , , , , .

The first three diagrams are equal to zero by (1T). There is also a (4T) relation involving the last two

which gives  $= 2$ . Thus $\mathcal{A}_3 = \langle$  $\rangle$ and also has dimension 1.

4. Symbols of knot polynomials

The Kontsevich theorem reduces the study of Vassiliev knots invariants to the study of their symbols. Here we describe the symbols of coefficients of classical knot polynomials as functions on chord diagrams.

4.1. The Conway polynomial. For a chord diagram D with n chords, double every chord as shown

Let $|D|$ be equal to the number of components of the obtained (possibly self-intersecting) curve. Then

$$\operatorname{symb}(c_n)(D) = \begin{cases} 1, & \text{if } |D| = 1 \\ 0, & \text{otherwise} . \end{cases}$$

Note that doubling of a chord changes the number of component by 1 (either adding 1, or subtracting 1). Therefore $\operatorname{symb}(c_n)$ is automatically zero for odd n.

4.2. The Jones polynomial. For a chord diagram D with n chords, *a state* s is an arbitrary function on the set chords of D with values in the set $\{1,2\}$. With each state s we associate an immersed plane curve obtained from D by resolving (either doubling or deleting) all its chords according to s:

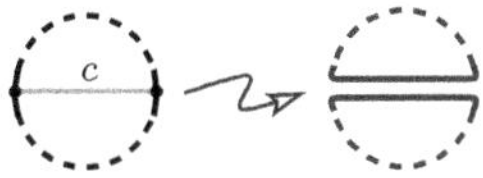

, if $s(c)=1$;
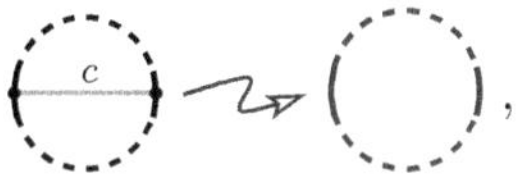

, if $s(c)=2$.

Let $|s|$ denote the number of components of the curve obtained in this way. Then

$$\operatorname{symb}(j_n)(D)=\sum_s\Big(\prod_c s(c)\Big)\ (-2)^{|s|-1}\ ,$$

where the product is taken over all n chords of D, and the sum is taken over all 2^n states for D.

For example, to compute the value of the symbol of j_3 on the chord diagram , we must consider 8 states:

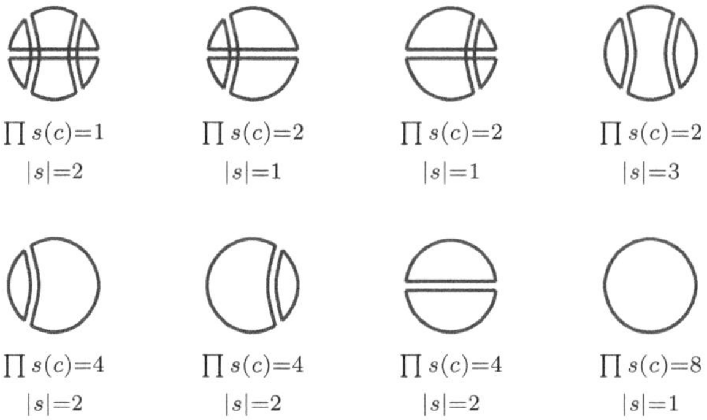

Therefore,

symb(j_3)(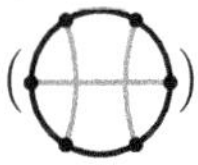) $=-2+2+2+2(-2)^2+4(-2)+4(-2)+4(-2)+8=-6\,.$

Similarly one can compute the values of symb(j_3) on all chord diagrams with three chords. Here is the result:

D					
symb(j_3)(D)	0	0	0	-6	-12

4.3. The HOMFLYPT polynomial. In this case, for a chord diagram D with $k+l$ chords, *a state* s will be a subset of chords of D. As above, we use the notation $|s|$ for the number of curves obtained by doubling all chords in s. Denote by $||s||$ the number of chords of s. A state s is called *admissible* for $p_{k,l}$ if $||s|| - |s| + 1 = l$. Then

$$\mathrm{symb}(p_{k,l})(D) = (-1)^{k+l} 2^k \sum_{\substack{s \\ \text{admissible}}} (-1)^{||s||} .$$

For example, if we would like to evaluate $\mathrm{symb}(p_{1,2})$ on the chord diagram , then out of the 8 states above only the first 3 are admissible:

$||s||=3,\ |s|=2$ $\quad$ $||s||=2,\ |s|=1$ $\quad$ $||s||=2,\ |s|=1$

Therefore,

$\mathrm{symb}(p_{1,2})($  $) = -2$.

In fact, $j_3 = 3p_{1,2}$.

5. Bialgebra of chord diagrams

Before explaining the idea of a proof of the Kontsevich theorem we need some algebraic structure on the vector spaces $\mathcal{A}_n$.

The vector space $\mathcal{A} := \bigoplus_{n \geqslant 0} \mathcal{A}_n$ has a natural algebra structure.

Definition 5.1. The *product* of two chord diagrams D_1 and D_2 is defined to be their connected sum:

Though the connected sum itself depends on the places where the two diagrams are joined, modulo (4T), the result does not depend on them. Extended by linearity, this product endows $\mathcal{A}$ with the structure of graded commutative algebra with a unit $\iota : \mathbb{C} \to \mathcal{A}$, $\iota(1) :=$ 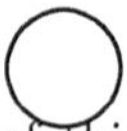 $\in \mathcal{A}_0$ represented by a chord diagram without chords. For a proof of this fact we refer to [BN, CDM].

Definition 5.2. The *coproduct* $\delta : \mathcal{A}_n \to \bigoplus_{k+l=n} \mathcal{A}_k \otimes \mathcal{A}_l$ is defined on chord diagrams by the sum of all ways to split the set of chords into two disjoint parts. Namely, for $D \in \mathcal{A}_n$ we set

$$\delta(D) := \sum_{J \subseteq [D]} D_J \otimes D_{\overline{J}},$$

the summation taken over all subsets J of the set of chords of D. Here D_J is the diagram consisting of the chords that belong to J, and $\overline{J} = [D] \setminus J$ is the complementary subset of chords. The operator δ is extended by linearity to the entire space $\mathcal{A}$.

Here is an example:

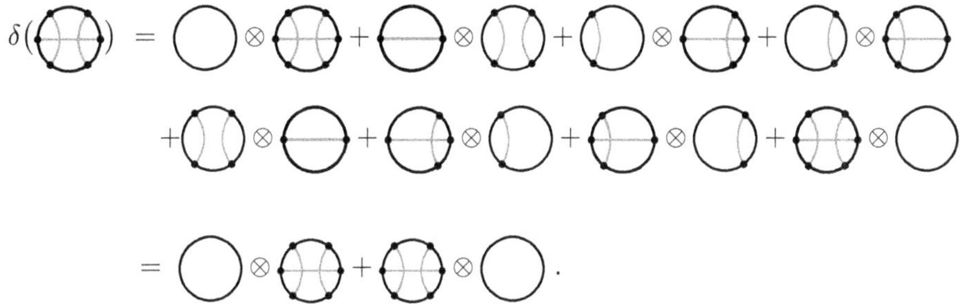

This operation endows $\mathcal{A}$ with a structure of a cocommutative coalgebra with counit $\varepsilon : \mathcal{A} \to \mathbb{C}$, $\varepsilon(x$  $+\ldots$"higher order terms"'$\ldots) := x$. The *cocommutativity* here means that the image of δ belongs to the symmetric tensor subalgebra of the tensor square $\mathcal{A}^{\otimes 2}$.

The operations of multiplication and comultiplication together turn $\mathcal{A}$ to be a *bialgebra.* For bialgebras, the *primitive space* $\mathcal{P}$ plays an important role. By definition, $\mathcal{P}(\mathcal{A})$ is the space of elements $D \in \mathcal{A}$ with the property $\delta(D) = 1 \otimes D + D \otimes 1$. It is also a graded vector space $\mathcal{P}(\mathcal{A}) = \bigoplus_{n \geqslant 1} \mathcal{P}_n$.

For example, both basic diagrams in degree 2 and 3,  and  are primitive, however in degree 4 there is a non primitive element , which is the square of the primitive element of degree 2. A basis for $\mathcal{P}_4$ consists of two elements 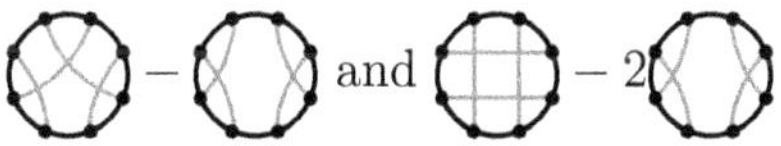 .

A graded commutative, cocommutative bialgebra $\mathcal{A}$ with a unit, counit, and one-dimensional zero-degree space $\mathcal{A}_0$ may be turned into a *Hopf algebra* by defining the *antipode* $S : \mathcal{A} \to \mathcal{A}$ to be an antiautomorphism acting by multiplication by (-1) on the primitive space. For such Hopf algebras $\mathcal{A}$, the classical Milnor–Moore [MiMo] theorem states that $\mathcal{A}$ is isomorphic to the symmetric tensor algebra of the primitive space:

$$\mathcal{A} \cong \mathcal{S}(\mathcal{P}(\mathcal{A})) .$$

This means that if we choose a basis $p_1, p_2, \ldots$ for the primitive space $\mathcal{P}(\mathcal{A})$ then any element of $\mathcal{A}$ can be uniquely represented as a polynomial in commuting variables $p_1, p_2, \ldots$. Theoretically this implies that for the purposes of combinatorics of Vassiliev invariants it is enough to study the primitive spaces $\mathcal{P}_n$ only. In particular, here is the table of dimensions of these spaces up to $n \leqslant 12$ calculated by J. Kneissler [Kn]:

n	1	2	3	4	5	6	7	8	9	10	11	12
$\dim \mathcal{P}_n$	0	1	1	2	3	5	8	12	18	27	39	55

The table of dimensions $\dim(\mathcal{V}_n/\mathcal{V}_{n-1}) = \dim \mathcal{A}_n$ on page 56 can be derived from this table.

6. The Kontsevich integral

Roughly speaking, an idea of the proof of the Kontsevich theorem is to construct an element $Z(K) \in \mathcal{A}$ of the algebra of chord diagram for every knot K. Having a weight system w we can apply it to $Z(K)$ and prove that $w(Z(K))$ is a Vassiliev invariant whose symbol is w. However, when we try to realize this idea, several complications occur.

The first one is that $Z(K)$ is going to be an element of the *graded completion* $\widehat{\mathcal{A}}$ of the algebra $\mathcal{A}$, or in other words, it is going to be an infinite sum of elements of $\mathcal{A}_n$ for all values of n, like a formal power series. The second one is that $Z(K)$ is not quite an invariant of knots. We will have to correct it before applying the weight system to it.

6.1. The construction. Let $z \in \mathbb{C}$ and $t \in \mathbb{R}$ be coordinates (z,t) in $\mathbb{R}^3$. The planes $t = \text{const}$ are thought of being horizontal. We define the Kontsevich integral for strict Morse knots, i.e. knots with the property that the coordinate t restricted to the knot has only non-degenerate critical points with distinct critical values.

Definition 6.1. The *Kontsevich integral* $Z(K)$ of a strict Morse knot K is given by the following formula

$$Z(K) := \sum_{m=0}^{\infty} \frac{1}{(2\pi i)^m} \int\limits_{\substack{t_{\min}<t_1<\cdots<t_m<t_{\max} \\ t_j \text{ are noncritical}}} \sum_{P=\{(z_j,z'_j)\}} (-1)^{\downarrow_P} D_P \bigwedge_{j=1}^{m} \frac{dz_j - dz'_j}{z_j - z'_j},$$

where

- the numbers $t_{\min}$ and $t_{\max}$ are the minimum and the maximum of the function t on K;
- the integration domain is the set of all points of the m-dimensional simplex $t_{\min} < t_1 < \cdots < t_m < t_{\max}$ none of whose coordinates

t_i is a critical value of t; the m-simplex is divided by the critical values into several *connected components*;

- the number of summands in the integrand is constant in each connected component of the integration domain, but can be different for different components; in each plane $\{t = t_j\} \subset \mathbb{R}^3$ choose an unordered pair of distinct points (z_j, t_j) and (z'_j, t_j) on K, so that $z_j(t_j)$ and $z'_j(t_j)$ are continuous functions; we denote by $P = \{(z_j, z'_j)\}$ the set of such pairs for $j = 1, \ldots, m$ and call it a *pairing*; the integrand is the sum over all choices of the pairing P;
- for a pairing P, the symbol '$\downarrow_P$' denotes the number of points (z_j, t_j) or (z'_j, t_j) in P where the coordinate t decreases as one goes along K;
- for a pairing P, consider the knot K as an oriented circle and connect the points (z_j, t_j) and (z'_j, t_j) by a chord; we obtain a chord diagram with m chords (thus, intuitively, one can think of a pairing as a way of inscribing a chord diagram into a knot in such a way that all chords are horizontal and are placed on different levels);
- over each connected component, z_j and z'_j are smooth functions in t_j; by $\bigwedge_{j=1}^{m} \frac{dz_j - dz'_j}{z_j - z'_j}$ we mean the pullback of this form to the integration domain of the variables $t_1, \ldots, t_m$; the integration domain is considered with the orientation of the space $\mathbb{R}^m$ defined by the natural order of the coordinates $t_1, \ldots, t_m$;
- by convention, the term in the Kontsevich integral corresponding to $m = 0$ is the (only) chord diagram of order 0 taken with coefficient one; it is the unit of the algebra $\widehat{\mathcal{A}}$.

Example 6.2. We exemplify the integration domain for a strict Morse knot with two local maxima and two local minima for the case of $m = 2$ chords.

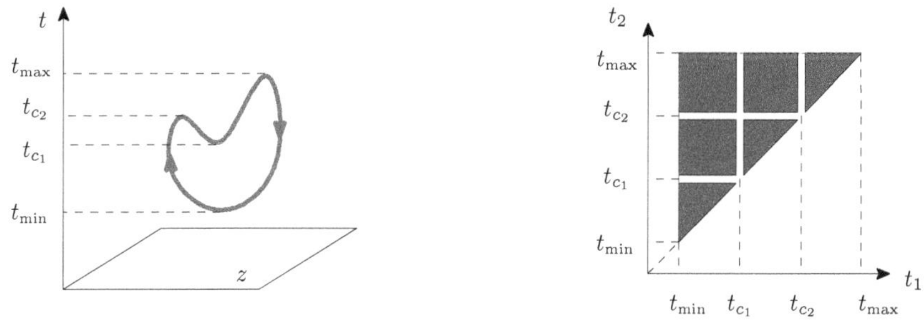

For each connected component of the integration domain, the number of summands corresponding to different choices of the pairing, a typical pairing P, and the corresponding chord diagram $(-1)^{\downarrow_P} D_P$ are shown in the picture.

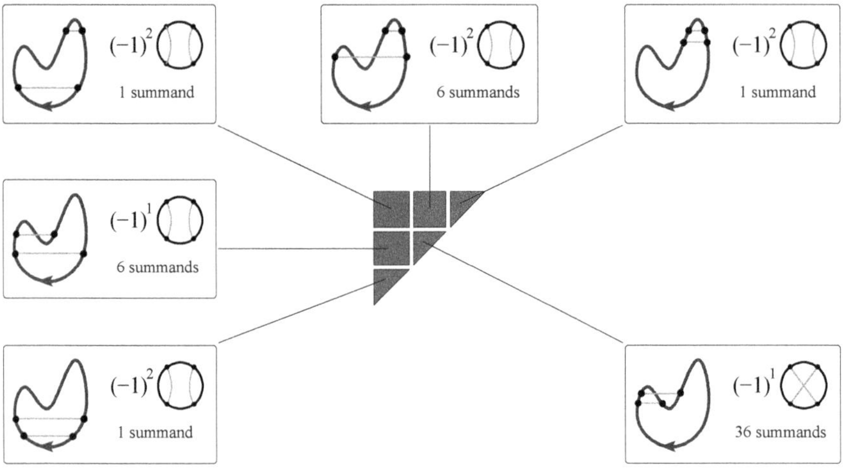

Now let us calculate the coefficient of the chord diagram  in $Z\Big($$\Big)$.

Out of the 51 pairings, the following 16 contribute to the coefficient:

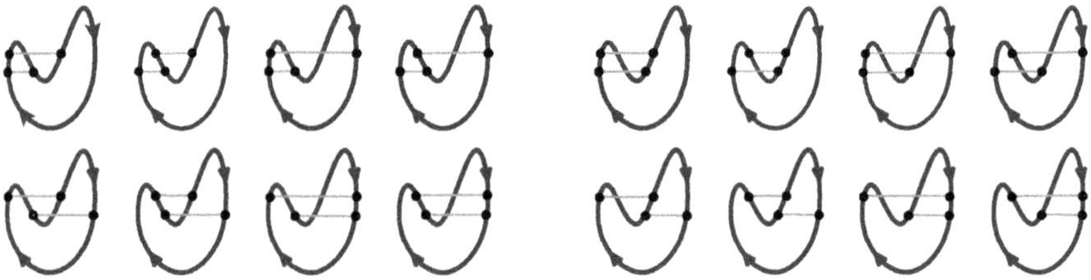

All of them appear on the middle triangular component, $t_{c_1} < t_1 < t_2 < t_{c_1}$ of the integration domain. To handle the integral which appears as the coefficient at , we denote the z-coordinates of the four points in the pairings on the level $\{t = t_1\}$ by a_1, a_2, a_3, a_4. Correspondingly, we denote the z-coordinates of the four points in the pairings on the level $\{t = t_2\}$ by

b_1, b_2, b_3, b_4:

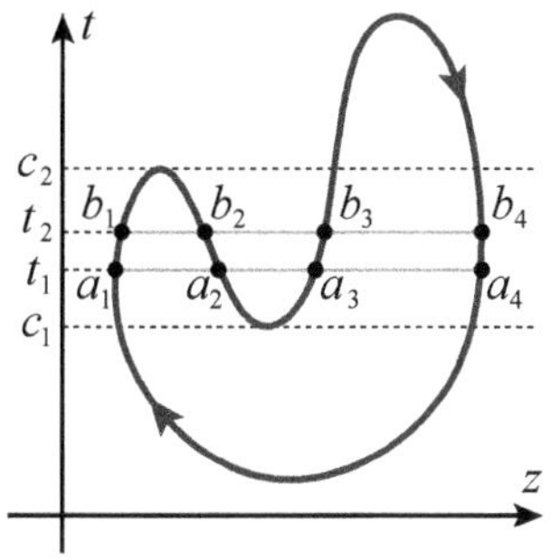

Then each of the four possible pairings $z_1 - z_1'$ on the level $\{t = t_1\}$ will look like $a_{jk} := a_k - a_j$ for $(jk) \in A := \{(12), (13), (24), (34)\}$. Similarly, each of the four possible pairings $z_2 - z_2'$ on the level $\{t = t_2\}$ will look like $b_{lm} := b_m - b_l$ for $(lm) \in B := \{(13), (23), (14), (24)\}$. The integral we are interested in now can be written as

$$\frac{1}{(2\pi i)^2} \int_\Delta \sum_{(jk)\in A} \sum_{(lm)\in B} (-1)^{j+k+l+m} d\ln a_{jk} \wedge d\ln b_{lm}$$

$$= -\frac{1}{4\pi^2} \int_\Delta \left. \sum_{(jk)\in A} (-1)^{j+k+1} d\ln a_{jk} \right)$$

$$\wedge \left. \sum_{(lm)\in B} (-1)^{l+m-1} d\ln b_{lm} \right)$$

$$= -\frac{1}{4\pi^2} \int_\Delta d\ln \frac{a_{12}a_{34}}{a_{13}a_{24}} \wedge d\ln \frac{b_{14}b_{23}}{b_{13}b_{24}} .$$

The change of variables

$$u := \frac{a_{12}a_{34}}{a_{13}a_{24}}, \qquad v := \frac{b_{14}b_{23}}{b_{13}b_{24}}$$

transforms the component Δ of the integration domain into the standard triangle Δ'

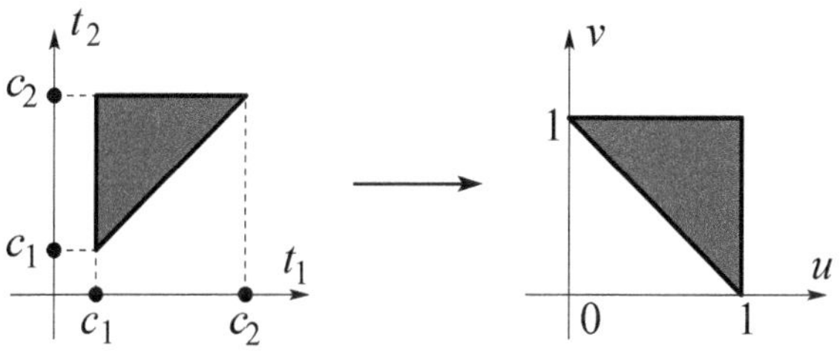

Since it changes the orientation of the triangle (has a negative Jacobian), our integral becomes

$$\begin{aligned}
\frac{1}{4\pi^2}\int_{\Delta'} d\ln u \wedge d\ln v &= \frac{1}{4\pi^2}\int_0^1 \left(\int_{1-u}^1 d\ln v \right)\frac{du}{u} \\
&= -\frac{1}{4\pi^2}\int_0^1 \ln(1-u)\frac{du}{u} = \frac{1}{4\pi^2}\sum_{k=1}^{\infty}\int_0^1 \frac{u^k}{k}\frac{du}{u} \\
&= \frac{1}{4\pi^2}\sum_{k=1}^{\infty}\frac{1}{k^2} = \frac{\zeta(2)}{4\pi^2} = \frac{1}{24}\,.
\end{aligned}$$

Therefore,

$Z\Big($  $\Big) = 1 + \frac{1}{24}$  $+ \dots ,$

where the free term 1 stands for the unit in the algebra $\mathcal{A}$ of chord diagrams,

$1 =$ 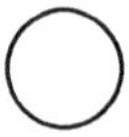 $\in \mathcal{A}.$

The following terms of this integral are of degree 4:

$Z\Big($  $\Big) = 1 + \frac{1}{24}$  $+ \frac{1}{5760}$ 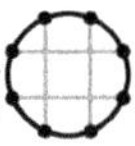 $- \frac{1}{1152}$ 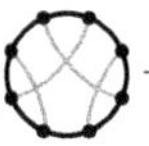 $+ \frac{1}{720}$  $+ \dots .$

The problem of the exact calculation of all the terms of this integral remained opened for a long time. Finally it was solved in [BLT]; see also the explanation of this work in [CDM]. However even the formulation of the results requires a different language for representing elements of $\mathcal{A}$. The coefficients in this integral are related to the Bernoulli numbers.

6.2. The universal Vassiliev invariant. Here we deal with the second complication in the proof of the Kontsevich theorem. The Kontsevich integral possesses several basic properties:

- $Z(K)$ converges for any strict Morse knot K.
- It is invariant under the deformations of the knot in the class of Morse knots with the same number of critical points.
- It behaves in a predictable way under the deformations that add a pair of new critical points to a Morse knot.

We will not prove these properties here referring the reader to [BN, CDM]. The last property may be specified as follows. Let H stand for the unknot from Example 6.2:

$$H := $$  .

If a knot isotopy creates one extra local minimum and one local maximum, the Kontsevich integral is multiplied by $Z(H)$:

$$Z\Big($$ 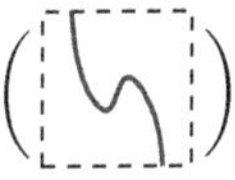 $$\Big) = Z(H)\cdot Z\Big($$  $$\Big),$$

where the multiplication is understood in sense of the formal infinite series from the graded completion of the algebra of chord diagrams $\widehat{\mathcal{A}}$.

Definition 6.3. With the help of this formula, we can define the *universal Vassiliev invariant* by either

$$I(K) := \frac{Z(K)}{Z(H)^{c/2}} \qquad \text{or} \qquad I'(K) := \frac{Z(K)}{Z(H)^{c/2-1}},$$

where c denotes the number of critical points of K in an arbitrary strict Morse representation, and the quotient means division in the algebra $\widehat{\mathcal{A}}$: $(1+a)^{-1} = 1 - a + a^2 - a^3 + \ldots$.

Both $I(K)$ and $I'(K)$ are invariants of topological knots with values in $\widehat{\mathcal{A}}$. The version $I'(K)$ has the advantage of being multiplicative with respect to the connected sum of knots; in particular, it vanishes (more precisely, takes the value 1) on the unknot. However, the version $I(K)$ is also used as it has a direct relationship with the quantum invariants. In particular,

$$I\Big($$  $$\Big) := \frac{1}{Z(H)} = 1 - \frac{1}{24}$$  $$- \frac{1}{5760}$$ 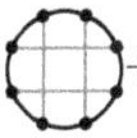 $$+ \frac{1}{1152}$$ 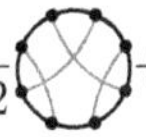 $$+ \frac{1}{2880}$$ 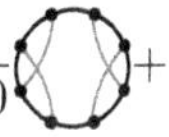 $$+ \ldots .$$

The central importance of $I(K)$ (as well as $I'(K)$) in the theory of finite-type invariants is that it is a universal Vassiliev invariant in the following sense. Consider a weight system w of order n (that is, a function on the set of chord diagrams with n chords satisfying one- and four-term relations). Applying w to the n-homogeneous part of the series $I(K)$, we get a numerical knot invariant $v := w(I(K))$. This invariant is a Vassiliev invariant of order $\leqslant n$ and $\mathrm{symb}(v) = w$. Moreover, any Vassiliev invariant is a sum of Vassiliev invariants obtained in this way. Indeed, any Vassiliev invariant v has a symbol $\mathrm{symb}(v)$ which is a weight system. So $v(K)$ and $\mathrm{symb}(v)(I(K))$ are two Vassiliev invariants with the same symbol. Their difference is of order $\leqslant (n-1)$ and we can repeat the same process by induction. This argument proves the Kontsevich theorem.

To prove that $v : K \longmapsto w(I(K))$ is a Vassiliev invariant with symbol w, it is enough to show that $I(K_D) = D +$ (terms of order $> n$) for a singular knot K_D representing a chord diagram D of order n. Let $p_1, \dots p_n$ be the double points of K_D.

Since the denominator of $I(K)$ starts with the unit of the algebra $\widehat{\mathcal{A}}$, it is sufficient to prove that

$$Z(K_D) = D + (\text{terms of order} > n) \ .$$

By definition, $Z(K_D)$ is the alternating sum of 2^n values of Z on the complete resolutions of the singular knot K_D at all its double points $p_1, \dots p_n$. To see what happens at a single double point p_j, let us look at the difference of Z on two knots $K_{j,+}$ and $K_{j,-}$ where an overpassing in $K_{j,+}$ is changed to an underpassing in $K_{j,-}$. By an isotopy we have:

$Z(K_{j,+}) - Z(K_{j,-}) =$ 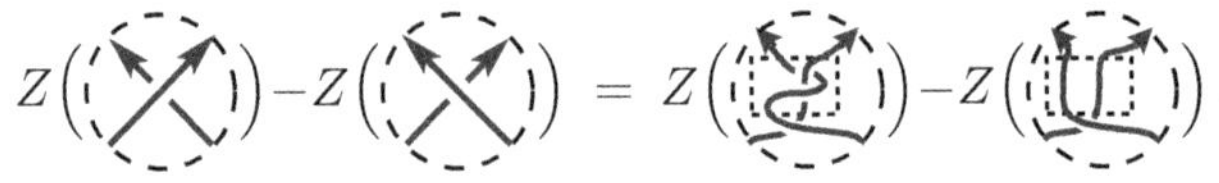

These two knots differ only in that the first knot has one string that makes a small curl around another string. We assume that this curl is very thin and enclose it in a box of height δ to consider it as a tangle $T_{j,+}$. Let $T_{j,-}$ be the similar tangle for $K_{j,-}$ consisting of two parallel vertical strings:

$T_{j,+} :=$  , $\qquad T_{j,-} :=$  .

The difference $Z(K_{j,+}) - Z(K_{j,-})$ comes only from the difference of the Kontsevich integrals of tangles $Z(T_{j,+}) - Z(T_{j,-})$. For the tangle $T_{j,-}$, both functions $z(t)$ and $z'(t)$ are constant, hence $Z(T_{j,-})$ consists of only one summand: the trivial chord diagram $D_-^{(0)}$ on $T_{j,-}$ (here 'trivial' means with an empty set of chords) with coefficient one. The zero degree term $D_+^{(0)}$ of $Z(T_{j,+})$ is also the trivial chord diagram on $T_{j,+}$ with coefficient one. So $D_-^{(0)} = D_+^{(0)}$ as chord diagrams. Thus the difference $Z(T_{j,+}) - Z(T_{j,-})$ starts with a first degree term which comes from the chord connecting the two strings of $T_{j,+}$.

Hence, $Z(K_D)$ starts from degree n. Moreover, the term of degree n is proportional to a chord diagram whose j-th chord connects the two strings of the tangle $T_{j,+}$ corresponding to the jth double point p_j in K_D. This chord diagram is precisely D. Now to calculate the order n part of $Z(K_D)$ we must compute the coefficient of D. It is equal to the product of coefficients of one chord terms in $Z(T_{j,+})$ over all j. We will show that all of them are equal to 1. Indeed, it is possible to choose the coordinates z and t is such a way that $z'(t) \equiv 0$ for one string, and for another one, the point $z(t)$ makes one complete turn around zero when t varies from 0 to δ:

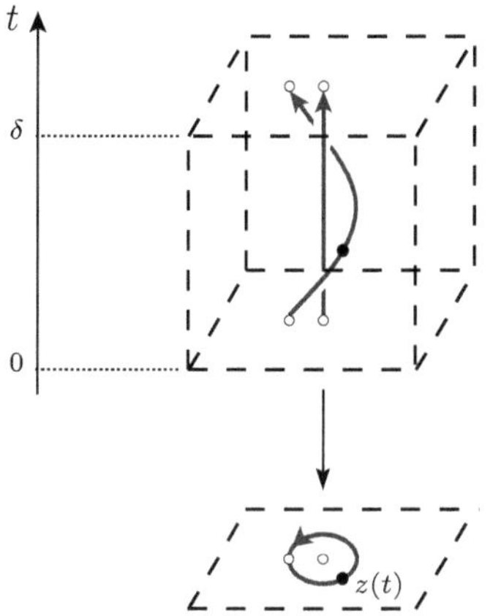

So we have

$$\frac{1}{2\pi i}\int_0^\delta \frac{dz-dz'}{z-z'} = \frac{1}{2\pi i}\oint_{|z|=1}\frac{dz}{z} = 1$$

by the Cauchy theorem.

7. Polyak–Viro formulas

Another universal way of representing Vassiliev invariants was suggested by M. Polyak and O. Viro [PV]. Its universality was proved by M. Goussarov in [GPV]. In general, Polyak–Viro formulas are far reaching generalization of the classical expression of the linking number of two curves as the sum of local writhes of the crossings of these curves.

7.1. Simplest example. The simplest example of the Polyak–Viro formula is a formula for the second coefficient of the Conway polynomial, also known as the *Casson invariant*. The invariant can be defined by the skein relation using the linking number lk:

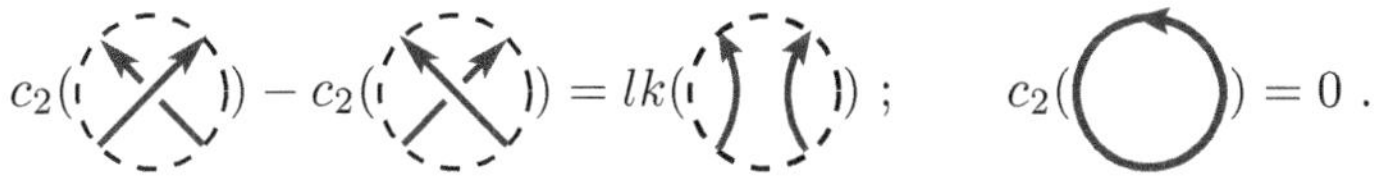

The Polyak–Viro formula for it is $c_2(K) = \sum_{\substack{i\,j\,i\,j \\ o\,u\,u\,o}} \varepsilon_i\varepsilon_j$, where the summation runs over all pairs of crossings (i,j) which appear in the order i, j, i, j when traveling along the knot and the first appearance of i is overpassing while the first appearance of j is underpassing; ε_i is the writhe of the ith crossing. For example, for the trefoil 3_1,

$3_1 =$ 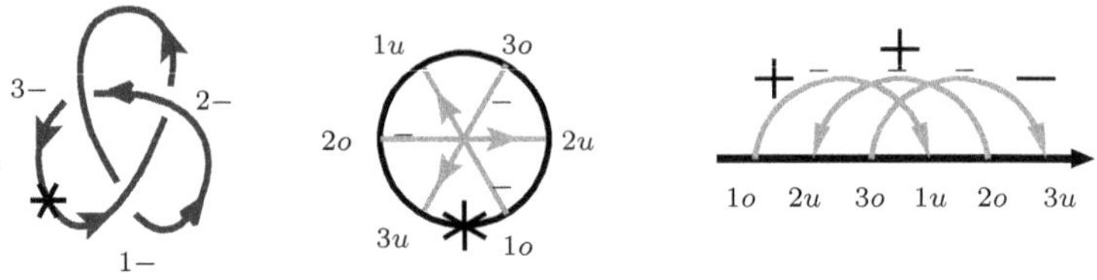

there is only one such pair $\begin{array}{|c|c|c|c}\hline 1\,2 & 3 & 1\,2 & 3\\ ou & o & uo & u\\\hline\end{array}$. So $c_2(3_1) = 1$.

An appropriate language for expressing Polyak–Viro formulas is the language of Gauss diagrams.

Definition 7.1. A *Gauss diagram* is an oriented circle with a distinguished set of distinct points divided into ordered pairs, each pair carrying a sign ± 1.

Graphically, an ordered pair of points on a circle can be represented by a chord with an arrow connecting them and pointing, say, to the second point. Gauss diagrams are considered up to orientation-preserving homeomorphisms of the circle. Sometimes, an additional basepoint is marked on the circle and the diagrams are considered up to homeomorphisms that keep the basepoint fixed. In this case, we speak of *based Gauss diagrams.*

To a plane knot diagram one can associate a Gauss diagram as follows. Pairs of points on the circle correspond to the values of the parameter where the diagram has a self-intersection, each arrow points from the overcrossing to the undercrossing and its sign is equal to the local writhe at the crossing.

The Gauss diagram of the trefoil above is shown next to it. Here is an example of the figure eight knot:

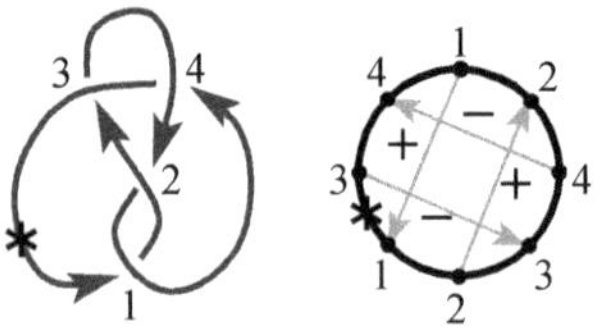

Sometimes we will draw based Gauss diagrams as line diagrams assuming that the base point is at infinity; such a diagram for the trefoil is above.

Definition 7.2. For two Gauss diagrams A and G we denote by $\langle A, G\rangle$ the number of occurrences of A as a subdiagram of G.

For example,

$\langle$  ,  $\rangle = 1$.

Define a pairing of a "non-signed" Gauss diagram A with a Gauss diagram G as the sum over all 2^n sign assignments $\varepsilon_1, \ldots, \varepsilon_n$ to the arrows of A, $A_{\varepsilon_1,\ldots,\varepsilon_n}$, of pairings $\langle A_{\varepsilon_1,\ldots,\varepsilon_n}, G\rangle$ with products of the signs $\varepsilon_1 \ldots \varepsilon_n$ as coefficients:

$$\langle A, G\rangle := \sum_{\varepsilon_1,\ldots,\varepsilon_n} (\varepsilon_1 \ldots \varepsilon_n)\langle A_{\varepsilon_1,\ldots,\varepsilon_n}, G\rangle \ .$$

In this notation, for example, the Polyak–Viro formula for c_2 above will be

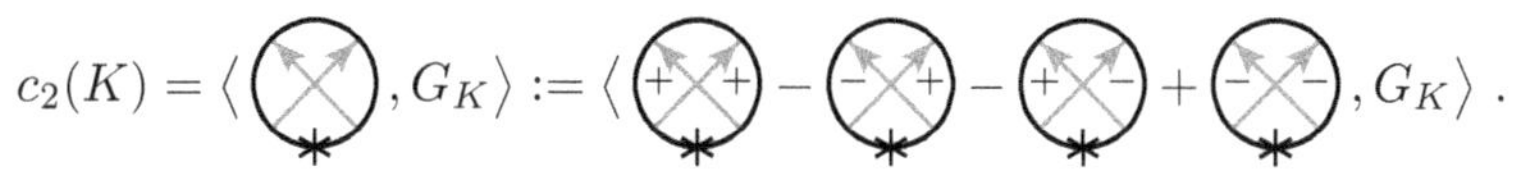

7.2. The Goussarov theorem. For a more formal definition, let us introduce a free $\mathbb{Z}$-module $\mathbb{Z}\mathbf{GD}$ spanned by the set of all Gauss diagrams $\mathbf{GD}$. We define the map $I : \mathbb{Z}\mathbf{GD} \to \mathbb{Z}\mathbf{GD}$ by simply sending a diagram to the sum of its subdiagrams:

$$I(D) := \sum_{D' \subseteq D} D'$$

and continuing this definition to the whole of $\mathbb{Z}\mathbf{GD}$ by linearity. For example, we have

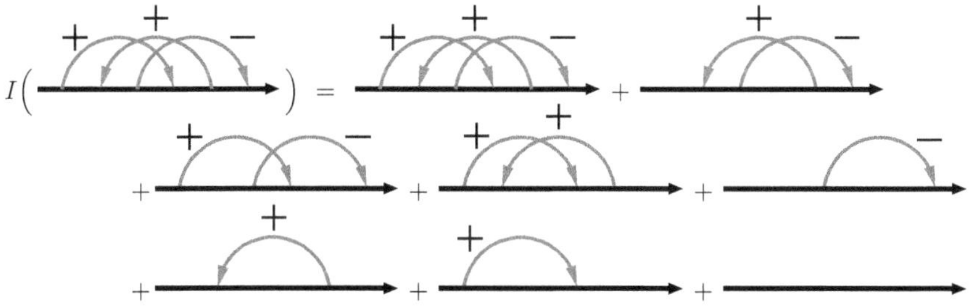

If we combine like terms in the expression $I(D)$ we get the sum over different Gauss diagrams,

$$I(D) = \sum_{A \in \mathbf{GD}} \langle A, D\rangle A.$$

Having a linear function $c : \mathbb{Z}\mathbf{GD} \to \mathbb{Z}$ we can construct a different function $I^*(c) : \mathbb{Z}\mathbf{GD} \to \mathbb{Z}$ as follows:

$$I^*(c)(D) := (c \circ I)(D) = \sum_{A \in \mathbf{GD}} \langle A, D\rangle c(A). \tag{2}$$

Gauss diagrams that encode long classical knots, or *realizable* diagrams, form a subset $\mathbf{GD}^{re} \subset \mathbf{GD}$. Any integer-valued knot invariant v gives rise to a function $\mathbf{GD}^{re} \to \mathbb{Z}$. By linearity, it extends to a function, also denoted by $v : \mathbb{Z}\mathbf{GD}^{re} \to \mathbb{Z}$, on the free $\mathbb{Z}$-module $\mathbb{Z}\mathbf{GD}^{re}$ spanned by the set $\mathbf{GD}^{re}$.

Theorem 7.3 (Goussarov). *For each integer-valued Vassiliev invariant v of classical knots of order $\leqslant n$ there exists a linear function $c_v : \mathbb{Z}\mathbf{GD} \to \mathbb{Z}$ such that $v = I^*(c_v)|_{\mathbb{Z}\mathbf{GD}^{re}}$ and such that c_v is zero on each Gauss diagram with more than n arrows.*

We may think about the expression $I^*(c_v)$ in the equation (2) as the most general form of the Polyak–Viro formula for a Vassiliev invariant v. For our example of $v = c_2$, we get the following function c_v:

$c_v($  $) = 1,$ $c_v($  $) = -1,$ $c_v($  $) = -1$ $c_v($  $) = 1,$

and zero on the other Gauss diagrams.

We refer to the paper [GPV] (and its explanation in [CDM]) for the proof of the Goussarov theorem.

7.3. Polynomials. The Polyak–Viro formulas for the coefficients c_n of the Conway polynomial were found in [CKR]. These results were generalized to the HOMFLYPT polynomial in [CP]. Below we give the formulas for $p_{k,l}$ for small degree. According to the Goussarov theorem there is a linear combination $A_{k,l}$ of (possibly unsigned) Gauss diagrams such that

$$p_{k,l}(K) = \langle A_{k,l}, G_K \rangle \ ,$$

where G_K is the Gauss diagram of a knot K.

For small k and l, these linear combinations look as follows

$A_{0,2} =$  ; $\qquad A_{2,0} = 0$;

$$A_{0,3} = 0 \ ; \qquad A_{3,0} = -4A_{1,2} \ ;$$

$A_{1,2} = -2\Big($  $+$  $+$  $+$  $+$  $-$  $+$ 

$+$ 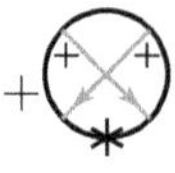 $-$  $\Big)$;

$A_{0,4} =$ 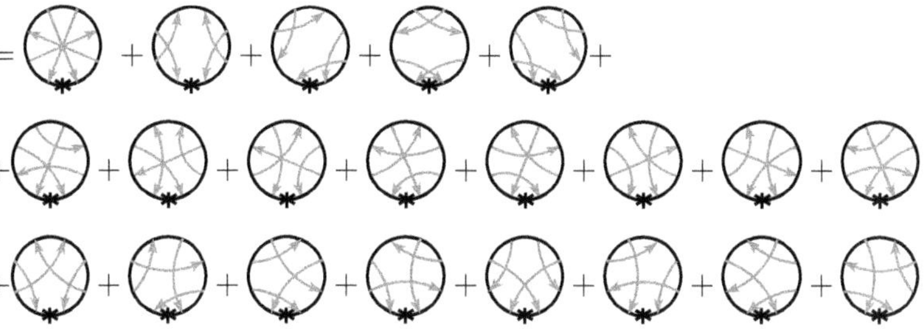

The formula for $A_{2,2}$ contain already 78 terms. Note that the formula for $A_{0,2}$ also gives the Polyak–Viro formula for c_2, however it differs from the one we gave before by reversion of all arrows. This is the common situation for all Conway coefficients of classical knots.

Acknowledgments.

This survey was written during my stay at the Mathematisches Forschungsinstitut Oberwolfach within the program "Reserach in Pairs" in March – April 2011. I am grateful to S. Lambropoulou for very useful discussions, and to the MFO institute for the excellent working conditions. I am thankful to the OSU MRI Stimulus Program for supporting my travel to Germany.

References

[BN] D. Bar-Natan, *On the Vassiliev knot invariants*, Topology, **34** (1995) 423–472.

[BLT] D. Bar-Natan, T. Q. T. Le and D. P. Thurston, *Two applications of elementary knot theory to Lie algebras and Vassiliev invariants*, Geom. Topol. **7** (2003) 1–31. `arXiv:math.QA/0204311`.

[BL] J. S. Birman and X.-S. Lin, *Knot polynomials and Vassiliev's invariants*, Invent. Math. **111** (1993) 225–270.

[CDM] S. Chmutov, S. Duzhin, J. Mostovoy, *Introduction to Vassiliev Knot invariants.* `arXiv:1103.5628v1 [math.GT]`.

[CKR] S. Chmutov, M. Khoury, A. Rossi, *Polyak-Viro formulas for coefficients of the Conway polynomial*, Journal of Knot Theory and its Ramifications **18** (2009) 773–783. `arXiv:math.GT/0810.3146}`.

[CP] S. Chmutov and M. Polyak, *Elementary combinatorics of the HOMFLYPT polynomial*, Int. Math. Res. Not. IMRN (2010) 480–495. `arXiv:math.GT/0810.4105`.

[Das] O. Dasbach. *On the combinatorial structure of primitive Vassiliev invariants. III. A lower bound*, Commun. Contemp. Math. **2** (2000) 579–590. `arXiv:math.GT/9806086`.

[Gus] M. Goussarov, *On n-equivalence of knots and invariants of finite degree*, Zap. Nauchn. Sem. S-Peterburg. Otdel. Mat. Inst. Steklov (POMI) **208** (1993) 152–173. English translation: Topology of manifolds and varieties (O. Viro, editor), Adv. in Soviet Math. **18** 173–192, Amer. Math. Soc., Providence, RI 1994.

[GPV] M. Goussarov, M. Polyak, O. Viro, *Finite type invariants of classical and virtual knots*, Topology **39** (2000) 1045–1068. `arXiv:math.GT/9810073v2`.

[Kn] J. Kneissler, *The number of primitive Vassiliev invariants up to degree twelve.* `arXiv:math.QA/9706022`.

[Kon] M. Kontsevich, *Vassiliev's knot invariants*, Adv. in Soviet Math. **16** (1993) 137–150.

[MiMo] J. Milnor and J. Moore, *On the structure of Hopf algebras*, Ann. of Math. (2) **81** (1965) 211–264.

[PV] M. Polyak and O. Viro, *Gauss diagram formulas for Vassiliev invariants*, Int. Math. Res. Notes **11** (1994) 445–454.

[Vas] V. A. Vassiliev, *Cohomology of knot spaces*, Theory of Singularities and Its Applications (ed. V. I. Arnold), Adv. in Soviet Math. **1** (1990) 23–69.

[Zag] D. Zagier, *Vassiliev invariants and a strange identity related to the Dedekind eta-function*, Topology **40** (2001) 945–960.

Sergei Chmutov
Department of Mathematics
The Ohio State University, Mansfield
1760 University Drive
Mansfield, OH 44906
chmutov@math.ohio-state.edu

BRAID ORDER, SETS, AND KNOTS

Patrick Dehornoy*
Laboratoire de Mathématiques Nicolas Oresme,
Université de Caen, 14032 Caen, France
**E-mail: dehornoy@math.unicaen.fr*
www.math.unicaen.fr/~dehornoy

We survey two of the many aspects of the standard braid order, namely its set theoretical roots, and the known connections with knot theory, including results by Netsvetaev, Malyutin, and Ito, and very recent work in progress by Fromentin and Gebhardt.

Keywords: Braid groups; Braid order; Self-distributivity; Elementary embedding; Large cardinal; Braid closure; Floor; Prime knot; Conjugacy problem.

Introduction

It has been known since 1992 (Refs. 1,2) that Artin's braid groups B_n are left-orderable, by an ordering that has several remarkable properties. In particular, it was proved by Laver[3] that its restriction to the braid monoid B_n^+ is a well-ordering—actually, a very long one, see Ref. 4 for recent paradoxical applications. Many subsequent results were established using different approaches, and the subject has developed so as to become the whole content of the monograph Ref. 5. The first main point is that many different approaches, some of them algebraic or combinatorial, others geometrical of topological, developed by a number of researchers, in particular Burckel, Dynnikov, Fenn, Fromentin, Funk, Greene, Larue, Rolfsen, Rourke, Short, Wiest, and the author, lead to one and the same distinguished ordering on braid groups, the so-called Dehornoy ordering, hereafter referred as the *D-ordering*. The second main point is that the family of all left-invariant braid orderings turns out to be an interesting space in which the D-ordering plays a significant role, as shown in works by Clay, Ito, Navas, Rolfsen, Short, Wiest.

The purpose of the current survey paper is not to repeat the material that is developed in Ref. 5, nor even to give a comprehensive introduction to that text, but rather to point out some aspects that are not, or not fully, described there, namely the set-theoretical roots of the D-ordering, as well as its known connections with knot theory. These aspects are alluded to in Sections III.2 and IV.5 of Ref. 5, but the latter is a quite short introduction, whereas the former is already obsolete due to several recent developments.

1. Connections with Set Theory

It turns out that the first proof of the result that Artin's braid groups are orderable groups and the construction of the specific D-ordering of braids stem from questions of Set Theory involving so-called large cardinal axioms. Here we shall briefly explain this connection. However, before starting, let us insist that the connection is historical rather than logical, as no set-theoretical axiom has ever been used in the construction of the braid order, the latter appearing precisely at the moment when set-theoretical axioms disappeared from the landscape.

1.1. *Braid groups*

We recall that, for $n \geqslant 2$, Artin's braid group B_n is the group that admits the finite presentation

$$\left\langle \sigma_1, ..., \sigma_{n-1} \;\middle|\; \begin{array}{ll} \sigma_i\sigma_j = \sigma_j\sigma_i & \text{for } |i-j| \geqslant 2 \\ \sigma_i\sigma_j\sigma_i = \sigma_j\sigma_i\sigma_j & \text{for } |i-j| = 1 \end{array} \right\rangle. \tag{1}$$

The elements of B_n are called *n-strand braids*. The braid group on infinitely many strands, denoted B_∞, is defined by a presentation with infinitely many generators $\sigma_1, \sigma_2, ...$ subject to the same relations. The identity mapping on $\{\sigma_1, ..., \sigma_{n-1}\}$ extends into an injective homomorphism of B_n to B_{n+1} and, therefore, we can identify B_n with the subgroup of B_∞ generated by $\sigma_1, ..., \sigma_{n-1}$.

According to the above definition, every braid admits decompositions in terms of the generators σ_i and their inverses. A word on the letters $\sigma_1, ..., \sigma_{n-1}$ and their inverses is called an *n-strand braid word*. If the braid β is the equivalence class of the braid word w, we say that w *represents* β, or is an *expression* of β. We say that two braid words are *equivalent* if they represent the same braid, *i.e.*, if they are equivalent with respect to the least equivalence relation that contains the relations of (1) and is compatible with multiplication.

It is well-known that B_n can be interpreted as the group of isotopy classes of n-strand braid diagrams, see for instance Ref. 6. Under this correspondence, the generator σ_i corresponds to the (isotopy class of the) elementary diagram in which the strand initially at position $i+1$ crosses over the strand initially at position i, see Figure 1.

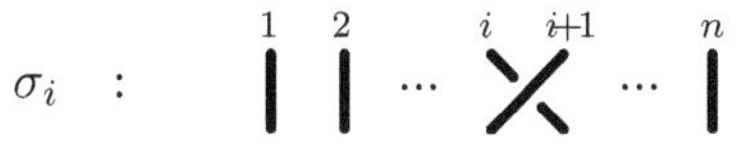

Fig. 1. Interpretation of σ_i as an elementary braid diagram—here the strands have an overall vertical direction

It is also well-known[6] that B_n can be interpreted as the mapping class group of a disk with n punctures. Under this interpretation, σ_i corresponds to the (isotopy class of the) half Dehn twist that exchanges the ith and the $i+1$st punctures clockwise, and keeps the other punctures as well as every point on the boundary circle fixed, see Figure 2.

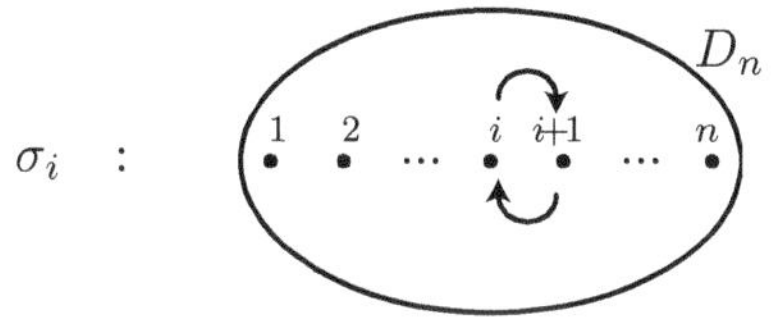

Fig. 2. Interpretation of σ_i as a Dehn half-twist in an n punctured disk

1.2. *The D-ordering of braids*

We start from the following notion.

Definition 1.1. We say that a nonempty braid word w is *σ-positive* if all letters σ_i with minimal i that occur in w have positive exponents.

For instance, the braid word $\sigma_1\sigma_2\sigma_1^{-1}$ is not σ-positive, since the letter σ_i with minimal i occurring in that word, namely σ_1, occurs both with exponent $+1$ and with exponent -1. On the other hand, the braid word $\sigma_2^{-1}\sigma_1\sigma_2$—which turns out to represent the same braid as the previous word—is σ-positive since, in this word, σ_1 occurs with exponent $+1$ only.

Saying that a braid word w is σ-positive means that, in the braid diagram encoded by w, all bottom crossings are positive—according to the convention that the strands are drawn horizontally and numbered from bottom up, see Figure 3.

The main result now is:

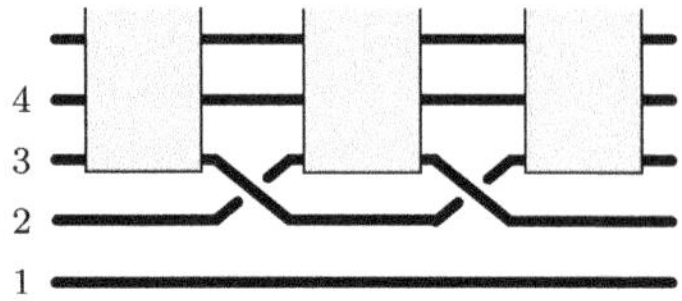

Fig. 3. A σ-positive braid diagram—here strands are drawn horizontally and numbered from bottom

Theorem 1.1 (D.[1,2]). *For β, β' in B_∞, declare that $\beta < \beta'$ holds if $\beta^{-1}\beta'$ can be represented by a σ-positive word. Then $<$ is a left-invariant linear ordering on B_∞.*

By definition, an order is called *linear* if any two elements are comparable, and, assuming that the domain is a group, it is called *left-invariant* if it is compatible with multiplication on the left, *i.e.*, $\beta < \beta'$ implies $\alpha\beta < \alpha\beta'$ for every α.

For instance, let us consider the braids $\beta = \sigma_1$ and $\beta' = \sigma_2\sigma_1$. Then we find $\beta^{-1}\beta' = \sigma_1^{-1}\sigma_2\sigma_1$, *i.e.*, the braid word $\sigma_1^{-1}\sigma_2\sigma_1$ is one of the possible expressions of the braid $\beta^{-1}\beta'$. We observed above that this braid word is not σ-positive, nor is its inverse either, so we cannot conclude anything. Now we also observed above that the braid word $\sigma_2\sigma_1\sigma_2^{-1}$ is another expression of the same braid, and that this word is σ-positive. So, by definition, we have $\beta < \beta'$.

It should be clear that, in order to prove Theorem 1.1, several things are to be proved. We shall not try to do it now, but rather ask

Question 1.1. Where does Theorem 1.1 come from?

And we claim that the answer is the following set-theoretical result of 1986:

Theorem 1.2 (D.[7]). *If j is an elementary embedding of a self-similar rank, then the LD-structure of* Iter(j) *implies $\mathbf{\Pi}_1^1$-determinacy.*

Our aim in the rest of this section will be to explain (a little) the meaning of Theorem 1.2 and, mainly, the connection between the latter (mysterious) statement and the existence of a braid ordering.

1.3. *Braid colorings*

The path from Theorem 1.2 to Theorem 1.1 goes through a major auxiliary idea, namely using braid colorings, which directly leads to the self-distributivity law.

Assume that S is a nonempty set and that we wish to use the elements of S as sorts of colours applied to the strands of a braid diagram. The principle we shall use is to attach colors to the left (initial) ends of the strands in a braid diagram, then propagate the colors through the diagram, and compare the right (final) colors with the initial colors to extract information about the braid represented by the diagram.

If the colors do not change when strands cross, then the final sequence of colors is simply the image of the initial sequence of colors under some permutation, which corresponds to the standard fact that each n-strand braid induces a well defined permutation of the integers $\{1, ..., n\}$.

A more interesting option is to assume that colors may change at crossings. In terms of complexity, the simplest case corresponds to the case when one of the strands, for instance the back one, keeps its color, whereas the front strand may change, but the new color only depends on the old colors of the two strands involved in the crossing. This case has been been considered by many authors, in particular Brieskorn[8], Joyce[9], and Matveev[10]. It amounts to saying that the set of colors S is equipped with a binary operation $*$, and that colors change according to the rule Now, as was said

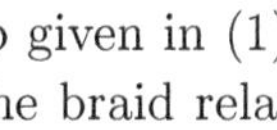

above, we wish to get information not about the braid diagram, but about the braid it represents. This means that we want that the final colors only depend on the isotopy class of the considered diagram. According to the presentation of the braid group given in (1), this is true if and only if the colors are not changed when the braid relations are applied. It is obvious to check that the relations $\sigma_i\sigma_j = \sigma_j\sigma_i$ with $|i-j| \geqslant 2$ lead to no problem. So the point is to have compatibility with the non-commutative braid relations.

Lemma 1.1. *Colors are preserved under the braid relations* $\sigma_i\sigma_j\sigma_i = \sigma_j\sigma_i\sigma_j$ *with* $|i-j| = 1$ *if and only if, for all colors* x, y, z, *we have*

$$x * (y * z) = (x * y) * (x * z). \tag{2}$$

Equality (2) is called the *left self-distributivity law—LD law* for short—as it asserts that the operation $*$ is distributive with respect to itself on the left. The easy proof of Lemma 1.1 is shown in Figure 4.

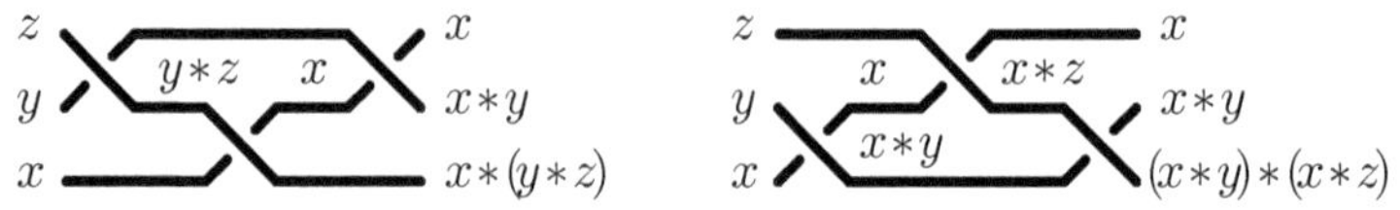

Fig. 4. Proof of Lemma 1.1: when the same initial colors are given to the two diagrams, the final colors are the same provided the self-distributivity condition is satisfied

1.4. *LD-systems*

So we are naturally led to look for algebraic systems $(S, *)$ where $*$ is a binary operation on S that satisfies the LD law. Such systems will be called *LD-systems.*

There exist classical examples of LD-systems, and using them to colour the strands of a braid leads to standard results about braid groups. The first famaily consists of the (trivial) operation $x * y = y$ on an arbitrary set S. This amounts to not changing colors in crossings. As was said above, it leads to associate with every n-strand braid a permutation of $\{1, ..., n\}$, thus yielding the well known surjective homomorphism of the braid group B_n onto the symmetric group $\mathfrak{S}_n$.

Another family consists in starting with a group G and using the conjugacy operation $x*y = xyx^{-1}$, which is easily seen to satisfy the LD law. This approach is closely connected with the theory of racks[11] and quandles[9]—yet self-distributive systems are used here in a different spirit, since the idea is not to associate a specific system with each braid or knot, but rather to fix a system and then use it for investigating all braids simultaneously. A particular case is specially interesting, namely when the involved group G is a free group and the initial colors attributed to the strands make a basis of this free group. Then the final colors also make a basis of the free group, thus associating with every n-strand braid β an automorphism $\phi(\beta)$ of the free group of rank n. It can be shown that ϕ is an injective homomorphism, known as the Artin representation of the braid group B_n.

A third family of examples arises when E is an R-module and the operation $x * y = (1-t)x + ty$, where t is a fixed invertible element of the base ring R. Then the final colors are a linear combination of the initial ones, which leads to associating with every n-strand braid an invertible $n \times n$ matrix with entries in R. In this way, we obtain a linear representation of the braid group B_n in the linear group $\mathrm{GL}_n(R)$. A typical case is when R is the ring of all Laurent polynomials $\mathbb{Z}[t, t^{-1}]$, in which case the representation is the Burau representation.

Definition 1.2. We say that an LD-system $(S, *)$ is *orderable* if there is a linear ordering $<$ on S satisfying $x < x * y$ for all x, y.

Note that all the LD-systems listed so far are certainly *not* orderable. Indeed, all satisfy the *idempotency law*, *i.e.*, $x*x = x$ always holds, whereas, in an orderable LD-system, we must have in particular $x < x * x \neq x$ for each x. However, the following result was established in 1991.

Theorem 1.3 (D.[1,2]). *There exist orderable LD-systems (namely free LD-systems).*

1.5. *From Theorem 1.3 to Theorem 1.1*

We claim that Theorem 1.1 (existence of the braid ordering) is a direct and natural consequence of Theorem 1.3. It is not hard to see that Theorem 1.1 relies on two main results.

Proposition 1.1. *A σ-positive braid word is never trivial.*

Proposition 1.1 implies that the relation $<$ involved in Theorem 1.1 is actually an ordering,

Proposition 1.2. *This ordering is linear, i.e., any two braids are comparable.*

Once the existence of an orderable LD-system is granted, it is easy to understand why Proposition 1.1 and 1.2 are true.

Sketch of proof of Proposition 1.1. Assume that $(S, *)$ is an orderable LD-system, and that w is a σ-positive braid diagram. We wish to prove that w is not trivial. To this end, we use colors from S. Then the diagram looks as follows—here we assume, without real loss of generality, that the bottom crossings are σ_1's:

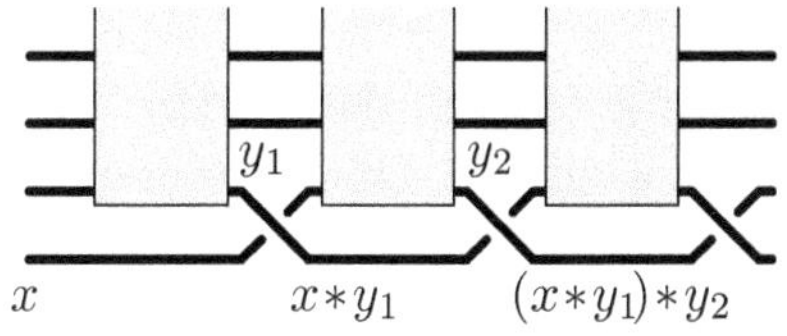

It follows from the definition of an orderable LD-system that we have

$$x < x * y_1 < (x * y_1) * y_2 < ...,$$

i.e., the colors on the bottom strand keep increasing. Hence the final bottom color cannot be the initial bottom color and, therefore, the braid diagram cannot be trivial. □

Sketch of proof of Proposition 1.2. Assume that $(S, *)$ is again an orderable LD-system and w, w' are two braid diagrams that we wish to compare. Then we put the same initial colors $x_1, x_2, ...$ from S in both diagrams. If $(y_1, y_2, ...)$ and $(y'_1, y'_2, ...)$ are the corresponding final colors, then we can use the order of S to lexicographically compare the two sequences $(y_1, y_2, ...)$ and $(y'_1, y'_2, ...)$:

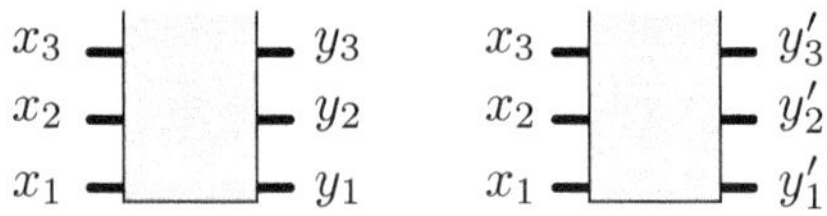

Then one of the two sequences is certainly less than the other, unless they are equal, which can happen only if the braid represented by w and w' are equal, as there exists an effective formula that computes the braid represented by $w^{-1}w'$ from the sequences of colours $(y_1, y_2, ...)$ and $(y'_1, y'_2, ...)$, and this formula returns the trivial braid when the sequences are equal. □

Remark 1.1. The above sketches are not formally correct. In the case of Proposition 1.1, we hid the question of coloring the negative crossings σ_i^{-1}. If one wishes that every such crossing may be colored, then one has to require that the left translations of the color set are bijective—thus the latter LD-system is a rack in the sense of Ref. 11—which actually discards orderable LD-systems. However, it can be shown with some work that, even if the latter case, one obtains well defined partial colorings, this meaning that every initial sequence of colors need not be eligible for coloring a given braid diagram but, for each such diagram, there always exist eligible sequences of colors. So the philosophy of the above sketch of proof is correct. Similarly, in the case of Proposition 1.2, we correctly argued that one obtains a linear ordering of braids, but we did not justify the fact that the resulting ordering coincides with the one of Theorem 1.1. This again requires some additional developments about orderable LD-systems, but the result is true and our sketch of proof correctly reflects the global situation.

So the reader should agree that the orderability of braid groups naturally comes from the existence of orderable LD-systems. But then the following question arises:

Question 1.2. Why care about the existence of orderable LD-systems?

1.6. *Large cardinal axioms*

The answer to Question 1.2 is: Because Set Theory told us.

Set theory is the mathematical study of infinity. By Gödel's incompleteness theorem, every axiomatic system for sets, in particular the standard Zermelo-Fraenkel system ZF, is incomplete, *i.e.*, there are statements that cannot be proved or disproved by that system. Then the question naturally arises of finding additional axioms that improve our description and understanding of sets (and infinity). Over the years, a consensus has emerged in the Set Theory community about the opportunity to consider a family of axioms that assert the existence of larger and larger infinities, according to a suggestion of Gödel known as Gödel's program. These axioms asserting the existence of "hyper-large" sets are usually called *large cardinal axioms.*

It is standard that a set X is infinite if and only if there exists an injection j of X into itself that is not surjective. A typical "hyper-infinite" set—the usual word is *self-similar*—is a set X such that there exists an injection j of X into itself that is not surjective and, in addition, j is a homomorphism with respect to every notion that can be defined from the membership relation $\in$. Such an injection is called a (non-trivial) *elementary embedding* of X.

For instance, the set $\mathbb{N}$ of all natural numbers is infinite, as the shift mapping $j : n \mapsto n+1$ is a non-surjective injection of $\mathbb{N}$ into itself. But this injection is not an elementary embedding, since it is not a homomorphism with respect to the addition of natural numbers and, as explained in textbooks of Set Theory, the latter is definable from the membership relation. Actually, it is easy to check that the only injection of $\mathbb{N}$ into itself that preserves $+$, and $<$, both definable from $\in$, is the identity mapping, which shows that $\mathbb{N}$ is not self-similar. In fact, a self-similar set has to be really huge.

1.7. *Ranks*

Here comes the most bizarre and counter-intuitive notion.

Definition 1.3. A *rank* is a set R such that every function of R into itself is an element of R.

In the way the above definition is stated, it is not clear that nonempty ranks exist. Actually, the technical definition is not exactly the one above, but, once again, we are not cheating the reader and the spirit of the statement is correct.

Now, let us mix the two notions of a self-similar set and of a rank. So assume that R is a self-similar rank, and that i, j are elementary embeddings of R. By hypothesis, i applies to each element of R. As j is a mapping of R to R, then, by the defining property of a rank, j belongs to R and, therefore, i applies to j, thus yielding a new object $i(j)$. Then, j is a function, being a function is definable in Set Theory, and i preserves every property that is definable in Set Theory, so $i(j)$ is again a function. Moreover, being an elementary embedding turns out to be also definable in Set Theory, hence $i(j)$ is not only a function, but even an elementary embedding. In other words, the mapping

$$i, j \mapsto i(j)$$

defines a binary operation on the family of all elementary embeddings of R.

Next, assume that i, j, k, ℓ are elementary embeddings of R, and we have $\ell = j(k)$. "Being the image under" is definable in Set Theory, hence, as i is an elementary embedding, ℓ being is the image of k under j implies that $i(\ell)$ is the image of $i(k)$ under $i(j)$, *i.e.*, $\ell = j(k)$ implies $i(\ell) = i(j)(i(k))$. In other words, we have

$$i(j(k)) = i(j)(i(k)).$$

This means that the application operation on elementary embedding satifies the LD law, and that the family of all elementary embeddings of R equipped with the application operation is an LD-system.

In particular, if we start with one elementary embedding j of a self-similar rank, we can consider the closure of $\{j\}$ under the application operation, *i.e.*, the smallest family that contains j and is closed under the application operation. This family consists of applying j to itself in all possible ways, iteratively. Its elements are called the *iterates* of j, and the family of all iterates of j is denoted $\mathrm{Iter}(j)$. Typical iterates of j are $j(j)$, $j(j(j))$, $j(j)(j)$, etc. The previous discussion implies

Proposition 1.3. *If j is an elementary embedding of a self-similar rank, then* $\mathrm{Iter}(j)$ *equipped with the application operation is an LD-system.*

1.8. *A bizarre situation*

Let us recall that our purpose is to explain why some attention had been given to the question of the existence of orderable LD-systems. The answer lies in the conjunction of two results, established independently and almost simultaneously in 1989.

Theorem 1.4 (D.[12]). *If there exists at least one orderable LD-system, then the word problem of the LD law is algorithmicatlly solvable, i.e., there exists an algorithm that, starting with two formal terms t, t', decides whether t and t' become equal when the LD law is assumed.*

Theorem 1.5 (Laver[13]). *If j is an elementary embedding of a self-similar rank, then* Iter(j) *is an orderable LD-system.*

A consequence of these results is that, *if there exists a self-similar rank*, then the word problem of the LD law is solvable. This created a very strange situation. Indeed, the existence of a self-similar rank is a large cardinal axiom, thus an unprovable strong logical assumption, whereas the word problem of the LD law is an algorithmic question that only involves finite objects, and there exists no visible connection between the set theoretical assumption and a problem of algorithmic combinatorics.

The reasons for looking for an orderable LD-system should now be clear: it was natural to look for such an algebraic object because it was needed to establish the decidability of the word problem of the LD law without appealing to an unprovable set theoretical axiom. That was done in Ref. 2 by introducing a certain group that describes the geometry of the LD law in some sense, and which is an analogue of Thomson's groups F and V—see Ref. 14 for more details. As this group turns out to be an extension of Artin's braid group B_∞, the braid applications, in particular the existence of the D-order, came as a bonus.

1.9. *Is the braid order an application of Set Theory?*

We have nearly completed our journey and explained the connection between sets and braids. The last question to raise in view of Theorems 1.4 and 1.5 is: Why care about the LD-systems Iter(j)? One answer to this last question is precisely Theorem 1.2. We shall not explain what Π^1_1-determinacy means, but it is enough to say here that this is a strong set-theoretical property. So Theorem 1.2 clearly showed that the LD-systems Iter(j), having strong consequences, must be non-trivial structures worth a closer investigation—other results in this direction were also established by Laver[15].

We thus now obtained a continuous path from Theorem 1.2, a result about sets, to Theorem 1.1, a result about braids.

As a last question, we can wonder whether the braid order described in Theorem 1.1 is or is not an application of Set Theory. Formally, it is not: braids appeared when sets disappeared, and they appeared precisely in order to avoid using sets. But one may also argue that, in essence, the answer is positive. Orderable LD-systems have been investigated because Set Theory showed they might exist and be involved in deep phenomena, and one may doubt that such algebraic systems would have been investigated without the motivation provided by Set Theory.

To conclude this part, let us propose an analogy with a situation that is rather common with physics, namely using physical assumptions to guess some statement that is subsequently passed to the mathematician for a formal rigorous proof. The situation here is quite similar: using logical assumptions (here the existence of a self-similar rank), one guesses some statement (here the existence of an orderable LD-system), and then passes it to the mathematician for a formal rigorous proof, *i.e.*, a proof that involves no extra logical axiom.

2. Connection with Knots

We turn to the second part, namely the recently discovered connections between the D-ordering of braids and knots and links. We shall mention two lines of research. The first one, initiated by Malyutin and Netsvetaev[16] and Malyutin[17], and recently developed by Matsuda and Ito, consists in showing that, if a braid β is very small or very large in the braid ordering, then the properties of its closure $\widehat{\beta}$ can be read easily. So, the family of knots and links that are closures of very small or very large braids with respect to the D-ordering makes an interesting family for further investigation. The second approach, which is still at a very preliminary step, consists in using some recently introduced normal forms on braid monoids to investigate the braid conjugacy relation and, conjecturally, Markov equivalence, *i.e.*, the equivalence relation on braids generated by conjugacy and Markov's stabilisation-destabilisation transformations.[6]

2.1. *The floor*

As well as any linear ordering, the D-ordering makes the braid group B_n into a line. Some natural milestones exist in this line. Indeed, let us recall that Garside's fundamental braid Δ_n is defined inductively for each n by $\Delta_1 = 1$, $\Delta_{n+1} = \Delta_n \sigma_n \sigma_{n-1} ... \sigma_1$. Then it is known[6] that Δ_n^2 generates the centre of the group B_n, and it is not hard to see that the powers of Δ_n^2 are unbounded in $(B_n, <)$ and, therefore, they partition B_n into intervals.

Definition 2.1. (See Figure 5.) For β in B_n, the *floor* $\lfloor\beta\rfloor$ is the unique integer m for which we have $\Delta_n^{2m} \leqslant \beta < \Delta_n^{2m+2}$.

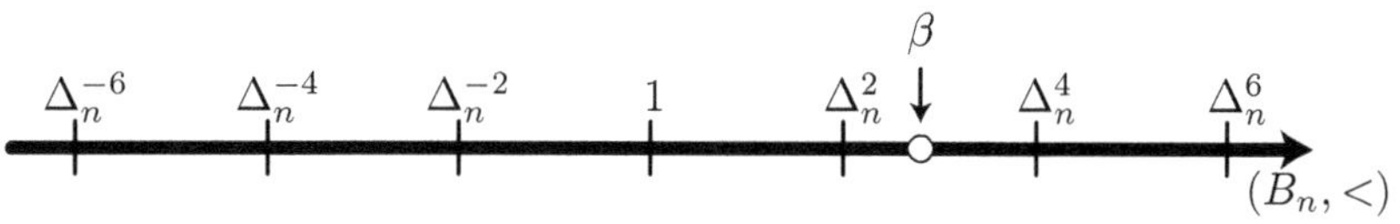

Fig. 5. The powers of Δ_n^2 partition B_n into a sequence of intervals. The floor of a braid β specifies to which interval β belongs: for instance, the floor of β above is 1

It follows from the properties of the D-ordering and from those of Δ_n^2 that the floor is nearly a homomorphism of the group B_n to the group of integers.

Proposition 2.1 (Malyutin–Netsvetaev[16]). *Define the* stable floor $\lfloor\beta\rfloor_s$ *of a braid* β *to be* $\lim_p \lfloor\beta^p\rfloor/p$. *Then the stable floor is a pseudo-character on* B_n *with defect* 1, *i.e., for all braids* β, β', *one has* $\lfloor\beta^p\rfloor_s = p\,\lfloor\beta\rfloor_s$, *and*

$$\left|\lfloor\beta\beta'\rfloor_s - \lfloor\beta\rfloor_s - \lfloor\beta'\rfloor_s\right| \leqslant 1.$$

Actually the stable floor is the only pseudo-character on B_n that is positive on braids larger than 1 in the D-ordering and is 1 on Δ_n^2. It is known that the space of pseudo-characters on B_n is infinite-dimensional, but very few concrete examples are known, except the exponent sum (which has zero defect, *i.e.*, that is a homomorphism) and the signature (which, according to Gambaudo–Ghys[18] has defect n).

2.2. *Very small and very large braids*

As was said above, the general philosophy is that, when a braid is very small or very large in the D-ordering, *i.e.*, if the absolute value of the floor is large enough, then some properties of the closure are easily readable from the braid.

The point of the argument is that, for every template move M in the sense of Birman–Menasco[19], *i.e.*, every transformation pattern that involves a finite number of strands such as an exchange move or a flype, there exists a number r_M such that, if the absolute value of the floor of a braid β is larger than r_M, then the closure of β is not eligible for M.

Theorem 2.1 (Malyutin–Netsvetaev[16]). *If an n-strand braid β satisfies $\beta < \Delta_n^{-4}$ or $\beta > \Delta_n^4$, then the link $\widehat{\beta}$ is prime and non-trivial.*

Idea of the proof. If χ a pseudo-character on B_n with defect d whose restriction to B_{n-1} is zero, then $|\chi(\beta)| > d$ implies that $\widehat{\beta}$ is not eligible for the exchange move, whereas the Birman–Menasco theory implies that a non-prime knot can be transformed into a composite knot by a sequence of exchange moves. Applying this to the stable floor gives the result. □

Theorem 2.2 (Malyutin–Netsvetaev[16]). *For each n, there exists $r(n)$ such that, if an n-strand braid β satisfies $\beta < \Delta_n^{-2r(n)}$ or $\beta > \Delta_n^{2r(n)}$, then the link $\widehat{\beta}$ is represented by a unique conjugacy class in B_n, i.e., $\widehat{\beta'}$ can be isotopic to $\widehat{\beta}$ only if β' is conjugated to β.*

Idea of the proof. Again by the Birman-Menasco Markov Theorem Without Stabilization[19], there exist finitely template moves for each fixed n, so the supremum of the corresponding integers r_M is finite. When the absolute value of the floor is larger than this supremum, the only way to transform the braid is to apply conjugacy. □

It is proved in Malyutin–Netsvetaev[16] that $r(3) \leqslant 3$ holds. Then Matsuda (personal communication) announced $r(4) \leqslant 4$, Ito[20] announced $r(3) = 2$—which is optimal—and he conjectured that $r(n) \leqslant n - 1$ might hold for each n. More recently, Ito announced (personal communication) a proof of the inequality $r(n) \leqslant n + 1$ for every n.

In the same vein, Ito announced (personal communication) a proof of the following statement: if β is an n-strand braid satisfying $\beta < \Delta_n^{-2k-2}$ or $\beta > \Delta_n^{2k+2}$, then the braid index of $\widehat{\beta}$ is at least the minimum of k and n. The reason is that $\widehat{\beta}$ cannot be eligible for any template move connecting the closure of an n-strand braid and that of an k-strand braid.

Finally, we mention two more results illustrating the principle that every braid β that is very small or very large in the D-ordering admits a closure whose properties can be read from those of β easily. The first one involves the genus of the closure.

Theorem 2.3 (Ito[21]). *If an n-strand braid β satisfies $\beta < \Delta_n^{-2k-2}$ or $\beta > \Delta_n^{2k+2}$, then the genus of $\widehat{\beta}$ is larger than $(k(n+2)-2)/4$.*

The second result involves the Nielsen–Thurston classification which takes a very simple form for very small and very large braids.

Theorem 2.4 (Ito[21]). *If an n-strand braid β satisfies $\beta \leqslant \Delta_n^{-4}$ or $\beta \geqslant \Delta_n^4$ and $\widehat{\beta}$ is a knot, then β is periodic (resp. reducible, resp. pseudo-Anosov) iff $\widehat{\beta}$ is a torus knot (resp. a satellite knot, resp. a hyperbolic knot).*

The above simple equivalence fails in general: for instance, the trefoil knot (a torus knot) is the closure of σ_1^3 (a periodic braid), $\sigma_1\sigma_2\sigma_3\sigma_1\sigma_2$ (a reducible braid), and $\sigma_1^3\sigma_2^{-1}$ (a pseudo-Anosov braid).

2.3. *Two functions*

We now come to another, much more speculative approach relying on the following deep result about the D-ordering.

Theorem 2.5 (Laver[3]). *For every braid β and every i, one has $\beta^{-1}\sigma_i\beta > 1$.*

This property, called the *Subword Property*, implies in particular that, for each n, the restriction of the D-order to the monoid B_n^+ (the submonoid of the braid group B_n generated by $\sigma_1, ..., \sigma_{n-1}$) is a well-order, *i.e.*, every nonempty subset admits a (unique) minimal element. It is then natural to introduce the following functions.

Definition 2.2. For β in B_n^+, put

$$\mu(\beta) = \min\{\beta' \in B_n^+ \mid \beta' \text{ is conjugate to } \beta\}, \tag{3}$$

$$\nu(\beta) = \min\{\beta' \in B_n^+ \mid \beta' \text{ is Markov equivalent to } \beta\}. \tag{4}$$

Any method for computing the function μ (*resp.* ν) would lead to a solution to the braid conjugacy problem (*resp.* the link isotopy problem). However, this approach is of little use as long as no such practical method is known.

2.4. *The alternating normal form*

The D-ordering of braids is a complicated object. For instance, this order is not Archimedian, *i.e.*, there exist β, β' larger than 1 satisfying $\beta^p < \beta'$ for each p, and it is not even Conradian, *i.e.*, there exist braids β, β' larger than 1 satisfying $\beta < \beta'\beta^p$ for each p. Therefore, whatever this exactly means, controlling the order effectively is not so easy. For instance, there is no simple connection between the order and the standard normal form of braids, namely the so-called greedy normal form(s) associated with the Garside structure(s) of braid groups, see for instance Refs. 14,22.

However, the good news are the recent introduction of two new normal forms, called the alternating and the rotating normal form, for braids. Contrary to the greedy normal form, these normal forms admit a simple connection with the D-order, bringing a reasonable hope to compute the function μ, and even possibly ν, in the future.

The principle underlying the alternating normal form is the observation that, for each braid β in the monoid B_n^+ of positive n-strand braids, there exists a unique finite sequence $(..., \beta_2, \beta_1)$ of braids in B_{n-1}^+, called the Φ_n-*splitting of* β, such that, using Φ_n for the n-flip automorphism of B_n that exchanges σ_i and σ_{n-i} for each i, one has

$$\beta = ... \cdot \Phi_n(\beta_4) \cdot \beta_3 \cdot \Phi_n(\beta_2) \cdot \beta_1$$

and, for each k, the braid β_k is the maximal right-divisor of $... \cdot \Phi_n(\beta_{k+1}) \cdot \beta_k$ that lies in B_{n-1}^+—see Figure 6.

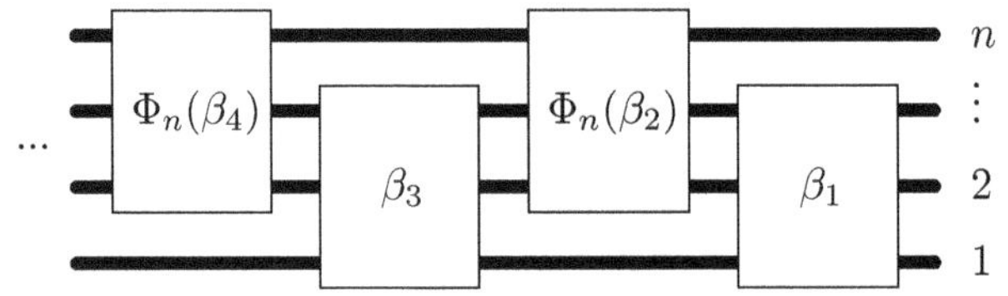

Fig. 6. The Φ_n-splitting of a braid of B_n^+: a distinguished decomposition of every positive n-strand braid into a finite sequence of $(n-1)$-strand braids. One extracts the maximal right fragment that involves the strands 1 to $n-1$, and 2 to n, alternatively

By iterating the splitting process, one easily obtains a unique expression, called the alternating normal form, for each braid of B_n^+. For our purpose, the important point is that the alternating normal form is connected with the D-order in a simple way.

Theorem 2.6 (D.[23], building on Burckel[24]). *The D-order of B_n^+ is the ShortLex-extension of the D-order of B_{n-1}^+ associated with the Φ_n-splitting: for β, β' in B_n^+, the relation $\beta < \beta'$ holds if and only if the Φ_n-splitting of β is shorter than that of β' or they have the same length and the splitting of β is lexicographically smaller.*

Remark 2.1. For Theorem 2.6 to be readily true, one has to replace the D-ordering with its Δ_n-conjugated version where one compares $\Phi_n(\beta)$ and $\Phi_n(\beta')$ instead of β and β'. In the latter ordering, σ_1 is smaller than σ_n, whereas, in the original D-order, σ_1 is larger than σ_n. Of course, the two orders are essentially equivalent.

2.5. *The rotating normal form*

The rotating normal form is analogous to the alternating normal form but, instead of involving the submonoid B_n^+ of B_n generated by the Artin generators σ_i, it appeals to the so-called dual braid monoid B_n^{+*} generated by the Birman–Ko–Lee generators of Ref. 25.

Definition 2.3 (Birman–Ko–Lee[25]). *The* dual *braid monoid B_n^{+*} is the submonoid of the braid group B_n generated by the elements $(a_{i,j})_{1\leqslant i<j\leqslant n}$ with*

$$a_{i,j} = \sigma_{j-1}^{-1}...\sigma_{i+1}^{-1}\sigma_i\sigma_{i+1}...\sigma_{j-1}.$$

The braid $a_{i,j}$ corresponds to a diagram where the jth strand crosses over the ith strand behind all intermediate strands. It is known that the monoid B_n^{+*} admits the same sort of algebraic structure as the monois B_n^+, namely what is now called a *Garside structure.* Then, there exists for B_n^{+*} an analogue of the Φ_n-splitting and the alternating normal form. In the current case, the flip automorphism Φ_n has to be replaced with the so-called rotating automorphism ϕ_n that maps $a_{i,j}$ to $a_{i+1,j+1}$, where indices are taken mod n. Provided braids are drawn on a cylinder rather than on a rectangle, ϕ_n corresponds to a $(2\pi/n)$-rotation, and the ϕ_n-splitting corresponds to the scheme of Figure 7.

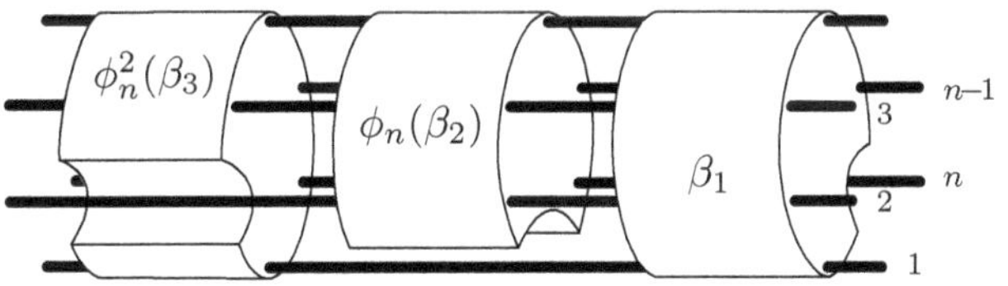

Fig. 7. The ϕ_n-splitting of a braid of B_n^{+*}: another distinguished decomposition into a finite sequence of $(n-1)$-strand braids

In this way, one obtains a normal form on B_n^{+*}. Once again, the nice point is the following connection with the D-order.

Theorem 2.7 (Fromentin[26,27]). *The D-order of B_n^{+*} is the ShortLex-extension of the D-order of B_{n-1}^{+*} associated with the ϕ_n-splitting.*

Theorem 2.7 is more than a counterpart of Theorem 2.6, because the rotating normal form has some nice combinatorial properties that the alternating normal does not share. Due to the highly redundant character of the Birman–Ko–Lee generators, selecting a distinguished expression with respect to these generators is more difficult than doing it with respect to the Artin generators, and, therefore, it is not surprising that the resulting normal form turns out to be a more powerful tool. Let us also mention Ref. 28 for further results along this line of research.

2.6. *A conjecture*

With the alternating and rotating normal forms of braids, we now have practical ways of controlling the braid order, making it reasonable to hope for explicit computational formulas expressing parameters connected with the D-ordering in terms of the normal forms. For instance, a typical result in this direction is that, for β in B_n^+, the floor of β defined in Section 2.1 is twice the length of the Φ_n-splitting of β diminished by 2 (except for very small values of the latter).

The computation of the μ-function of Section 2.3 has not yet been completed, but it seems now accessible. Let us mention the following simple formula, that has been checked experimentally for a large number of braids.

Conjecture 2.1 (D., Fromentin, Gebhardt). *For β in B_3^+, one has*

$$\mu(\beta\Delta_3^2) = \sigma_1\sigma_2^2\sigma_1 \cdot \mu(\beta) \cdot \sigma_1^2.$$

Establishing this formula and various similar computational rules should lead to the practical computation of the μ function on B_3^+, and subsequently on B_n^+. When this is done, addressing similar questions for the ν function (the one where conjugacy is replaced with Markov equivalence) might become a reasonable goal.

Acknowledgment

The author thanks Tetsuya Ito for introducing him to his recent results.

References

1. P. Dehornoy, Deux propriétés des groupes de tresses, *C. R. Acad. Sci. Paris* **315** (1992) 633–638.
2. P. Dehornoy, Braid groups and left distributive operations, *Trans. Amer. Math. Soc.* **345-1** (1994) 115–151.
3. R. Laver, Braid group actions on left distributive structures and well-orderings in the braid group, *J. Pure Appl. Algebra* **108-1** (1996) 81–98.

4. L. Carlucci, P. Dehornoy, A. Weiermann, Unprovability statements involving braids, *Proc. London Math. Soc.* (2010), doi:10.1112/plms/pdq016.
5. P. Dehornoy, with I. Dynnikov, D. Rolfsen, B. Wiest, Ordering Braids, Mathematical Surveys and Monographs vol. 148, Amer. Math. Soc. (2008).
6. J. Birman, Braids, Links, and Mapping Class Groups, Annals of Math. Studies **82** Princeton Univ. Press (1975).
7. P. Dehornoy, Pi$^1_1$-complete families of elementary sequences, *Ann. P. Appl. Logic* **38** (1988) 257–287.
8. E. Brieskorn, Automorphic sets and braids and singularities, *Braids, Contemporary Maths AMS* **78** (1988) 45–117.
9. D. Joyce, A classifying invariant of knots: the knot quandle, *J. of Pure and Appl. Algebra* **23** (1982) 37–65;
10. S. V. Matveev, Distributive groupoids in knot theory, *Math. Sbornik* **119, 1-2** (1982) 73–83.
11. R. Fenn & C. P. Rourke, Racks and links in codimension 2, *J. of Knot Theory and its Ramifications* (1992) 343–406;
12. P. Dehornoy, Sur la structure des gerbes libres, *C. R. Acad. Sci. Paris* **309** (1989) 143–148.
13. R. Laver, The left distributive law and the freeness of an algebra of elementary embeddings, *Advances in Math.* **91-2** (1992) 209–231.
14. P. Dehornoy, Braids and Self-Distributivity, Progress in Math. vol. 192, Birkhäuser (2000)
15. R. Laver, Elementary embeddings of a rank into itself, *Abstracts Amer. Math. Soc.* **7** (1986) 6.
16. A. Malyutin & N. Netsvetaev, Dehornoy's ordering on the braid group and braid moves, *St. Peterburg Math. J.* **15** (2004) 437–448.
17. A. Malyutin, Twist number of (closed) braids, *St. Peterburg Math. J.* **16** (2005) 791–813.
18. J.M. Gambaudo & E. Ghys, Braids and signatures, *Bull. Soc. Math. France* **133-4** (2005) 541–579.
19. J.S. Birman & W.W. Menasco, Stabilization in the Braid Groups (with applications to transverse knots), math.GT/0203227.
20. T. Ito, Braid ordering and the geometry of closed braids, Geom. & Topol., to appear, math.GT/0805.1447, Geom. & Topol., to appear
21. T. Ito, Braid ordering and knot genus, J. Knot Th. Ramif., to appear, math.GT/0805.2042, J. Knot Th. Ramif., to appear
22. D. Epstein, with J. Cannon, D. Holt, S. Levy, M. Paterson & W. Thurston, Word Processing in Groups, Jones & Bartlett Publ. (1992).
23. P. Dehornoy, Alternating normal forms for braids and locally Garside monoids, *J. Pure Appl. Algebra* **212-11** (2008) 2416–2439.
24. S. Burckel, The wellordering on positive braids, *J. Pure Appl. Algebra* **120-1** (1997) 1–17.
25. J. Birman, K.H. Ko & S.J. Lee, A new approach to the word problem in the braid groups, *Advances in Math.* **139-2** (1998) 322-353.
26. J. Fromentin, A well-ordering of dual braid monoids, *Comptes Rendus Mathématiques* **346** (2008) 729–734.

27. J. Fromentin, The well-order on dual braid monoids, *J. Knot Th. Ramifications* **19–5** (2010) 631–654.
28. J. Fromentin, Every braid admits a short sigma-definite expression, *J. Europ. Math. Soc.*, to appear; math.GR/0811.3902

FINDING KNOT INVARIANTS FROM DIAGRAM COLOURING

Roger Fenn
Sussex

A traditional method of finding topological invariants of a knot in 3-dimensional euclidean space is to colour the arcs of a planar diagram and use this to extract invariants of the knot such as the fundamental group. This method can be extended in many ways to other kinds of codimension 2 embeddings. Moreover the new algebraic objects which arise quite naturally from these generalisations such as quandles, racks, biquandles etc are of interest in their own right. In this paper we shall give a brief overview of the methods and the new algebraic structures.

Keywords: Knot colouring, quandle, biquandle, rack, birack, classical, virtual and welded knots.

1 Introduction

The notion of developing a knot invariant by colouring or labelling the arcs of a diagram is an old one. The very first usable knot invariant was "3-colouring" which is a representation of the fundamental group. The fundamental group itself can be thought of, via the Wirtinger presentation, as a knot invariant obtained by labelling the arcs of a diagram by group generators.

The universal application of these ideas leads to the notions of rack or quandle. A rack is the algebraic distillation of two of the Reidemeister moves. It is a set with a binary operation such that right multiplication is an automorphism. If in addition the first Reidemeister move is shadowed then the result is a quandle.

Any codimension two link has a fundamental rack or quandle which contains more information than the fundamental group. They provide an elegant and complete algebraic framework in which to study links and knots in 3–manifolds, and also for the 3–manifolds themselves. Racks and quandles have been studied by several authors and have been called a variety of names. The fundamental quandle is important because it is a complete invariant for classical links.

The first section is an updated distillation of [**FR**]. After the history of these ideas we develop the theory of racks and quandles. In later sections we extend the notion of a rack to a birack (and biquandle). This leads to important invariants of virtual links, that is, links in a surface cross an interval modulo surgery and to welded knots which can be interpreted as tori in $\mathbb{R}^4$.

This is only an overview. To include all proofs, results and examples would take a book length paper so only statements and references to other publications are given.

2 Previous work

An early, 1929, work on quandles (the earliest?) is by Burstin and Mayer, [**BM**]. The definition of a kei was given by Takasaki in 1943, see [**Tak**]. Other early birds are Conway and Wraith [**CW**], and we are indebted to these authors for a copy of their (unpublished) correspondence. They used the name **wrack** for the concept and this name was adopted in [**FR**], not merely because it is the oldest name, but also because it is a simple English word which (to their knowledge) has no other mathematical meaning. The more common spelling however has been adapted. Rack is used in the same sense as in the phrase "rack and ruin". The context of Conway and Wraith's work is the conjugacy operation in a group and they regarded a rack as the wreckage of a group left behind after the group operation is discarded and only the notion of conjugacy remains. They studied the basic algebra of racks in a special case (the quandle case) but also were aware of the general case and the main topological application (the fundamental quandle of a knot in a 3–manifold).

A comprehensive published study of quandles in a toplogical context is due to Joyce [**J**]. He invented the name **quandle**. Joyce establishes the basic algebra of quandles, giving several examples, and defines augmented quandles and the associated group (which he calls *Adconj*). He defines the fundamental quandle of a knot in S^3 giving both the topological definition in terms of "nooses" and the definition in terms of the presentation which can be read from a diagram of the knot. He proves the equivalence of the two definitions. His main result is that the fundamental quandle classifies the knot. Joyce's work was largely duplicated independently by Matveev [**Mat**] who used the name **distributive groupoid**. But note that a rack is not a groupoid in the usually accepted sense.

Kauffman [**K2**] defines racks in full generality using the name **crystal** for the concept. Kauffman defines the fundamental rack of a knot in S^3 and applies Joyce's theorem to prove that it is a classifying invariant. He also extends Joyce's work on the Alexander quandle to racks and defines an associated R-matrix.

An extensive algebraic survey of racks is given by Brieskorn [**Br**], who is particularly interested in the context of braids and singularities. In his introduction Brieskorn writes:

> *"Whilst preparing this survey, I found an extremely simple concept unifying many investigations on this subject as well as classical results of E.Artin, A.Hurwitz and W.Magnus. This is the notion of an automorphic set."*

The definition of an automorphic set coincides with that of a rack or crystal. Although "automorphic set" is the mathematically correct terminology for the concept, it has not generally been adopted. Brieskorn's paper contains a wealth of algebraic material about racks.

Winker [**Win**] extends Joyce's work and defines an analogue of the Cayley graph for a quandle and Krüger [**Kr**], independently and simultaneously with [**FR**], defines free products of racks and investigates the automorphism group of the free rack. There is an interesting connection of racks with computer theory. Roscoe [**Ros**] studies an algebraic object which satisfies just one of the two rack laws (the rack identity) in the context of computer information updates. Finally there is also a connection with some problems in logic, see Dehornoy, Jech and Laver [**D, Je, La**]

Apart from myself this paper is based on the work of several people. The main collaborations come from joint papers by myself and A. Bartholomew,

S. Budden, M. Jordan, L. Kauffman, R. Rimanyi, C. Rourke, B. Sanderson and V. Turaev.

3 Knot Diagrams and Associated Algebras

A familiar notion is that of a **classical knot diagram** consisting of an immersed oriented circle in the plane or 2-sphere with double points which are either over or under points. The generalisation to links of more than one component is easily seen. Remove the under points. The resulting components are called **arcs** of the diagram. A **colouring** of the arcs is an assignment of an element of an algebraic object (the colour) so that the three colours at any crossing are related in some way to make an invariant. The invariance will be governed by the colour changes under the Reidemeister moves, R1, R2 and R3, illustrated below.

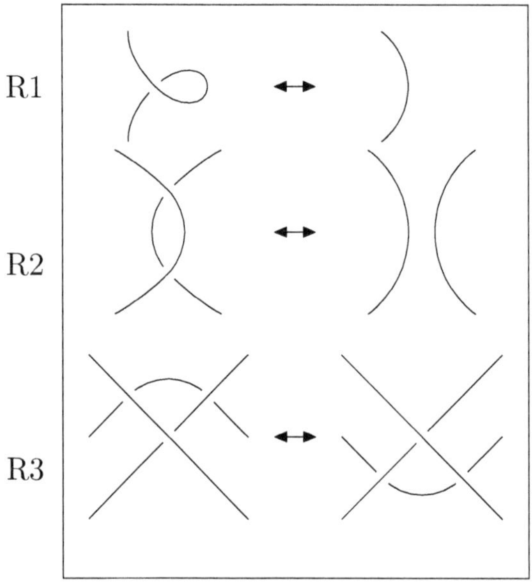

In the illustration the orientation of the immersed circle is suppressed as is the orientation of space. So for example the negative curl shown by R1 can be made into a positive curl by reflection in the plane.

Hence the relations of the three colours at a crossing must respect these moves.

As an illustration consider the illustrated crossing below with the three arcs coloured by a, b and c.

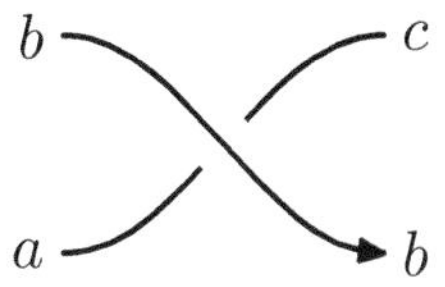

If we think of $a, b, c, \ldots$ as group generators and the relations at each vertex given by $c = b^{-1}ab$ then we get a presentation of the fundamental group of the complement of the knot. If there are n double points then there are n generators and n relations although one relation is redundant.

Let us now universally generalise this idea. We consider sets X with a binary operation which we shall write exponentially

$$(a, b) \longrightarrow a^b.$$

This will be the analogue of conjugation, $c = b^{-1}ab$, considered earlier. Note that the condition is independent of the orientation of the under arc but depends on the orientation of space. So the diagram could represent either a positive or negative crossing.

There are several reasons for writing the operation exponentially.

(1) The operation is unbalanced and should be thought of as an action, i.e. think of a^b as meaning the result of b, which is above, acting or operating on a, which is below.
(2) In group contexts exponentiation signifies conjugation. A group with conjugation is one of the principal examples of a rack — indeed this was the source for one strand of the earlier work on racks [**CW**]. A rack is an algebraic object which has just *some* of the properties of a group with conjugacy as the operation.
(3) Finally, and conveniently, exponential notation allows brackets to be dispensed with, because there are standard conventions for association with exponents. In particular

$$a^{bc} \text{ means } (a^b)^c \text{ and } a^{b^c} \text{ means } a^{(b^c)}.$$

Definition: Racks, Quandles and Keis

Consider the following axioms, which are related to the Reidemeister moves.

Axiom 1 Given $a \in X$ then $a^a = a$.
Axiom 2 Given $a, b \in X$ there is a unique $c \in X$ such that $a = c^b$.
Axiom 3 Given $a, b, c \in X$ the formula $a^{bc} = a^{cb^c}$ holds.

The formula in axiom 3 is called the **rack identity (first form).**

A **rack** is a non-empty set X with a binary operation satisfying axioms 2 and 3. A **quandle** is a rack satisfying axiom 1.

Several consequences flow from these axioms.

The second axiom implies that, for each $b \in X$ the function $f^b(x) := x^b$ is a bijection of X to itself, and this fits with the idea that the operation is a (right) action of X on itself.

We shall write $a^{b^{-1}} = (f^b)^{-1}(a)$ for the element c given by axiom 2. So, we have a second binary operation

$$(a, b) \longrightarrow a^{b^{-1}}.$$

which is, in effect, an inverse to the other binary operation in the sense that $a^{bb^{-1}} = a^{b^{-1}b} = a$ for all $a, b \in X$.

We can now give a meaning to any expression of the form x^w where $w = w(a, b, \ldots)$ is a word in $F(X)$, the free group on X, namely the result of repeatedly acting on x by $f^a, (f^a)^{-1}, f^b, (f^b)^{-1}$ etc. The word w is *not* to be regarded as an element of X, but as an **operator** on X. Shortly, we shall formalise this by introducing the **operator group**.

A **Kei** is a quandle which satisfies

Axiom K Given $a, b \in X$ then $a^{b^2} = a$. In other words $a^{b^{-1}} = a^b$.

The rack identity is a right self-distributive law as can be seen if we temporarily use the notation $a \cdot b$ for a^b:

$$(a \cdot b) \cdot c = (a \cdot c) \cdot (b \cdot c).$$

Axioms 2 and 3 are equivalent to the statement that *right multiplication is an automorphism.*

Substituting $d = a^c$ in the rack identity and then changing d back to a gives the alternative form:

Axiom 3′ Given $a, b, c \in X$ the formula

$$a^{b^c} = a^{c^{-1}bc}$$

holds. This is the **rack identity (second form).**

In other words b^c operates like $c^{-1}bc$, which makes clear the connection between the rack operation and conjugacy in a group.

At this point note that given a rack X there is a mirror rack $\overline{X}$ with binary operation $a^{b^{-1}}$.

There are many examples of racks and quandles to be found in [**FR**]. Here are some examples.

Examples of Racks, Quandles and Keis

Example 1 *The Conjugation Quandle*

Let G be a group, then conjugation in G i.e. $g^h := h^{-1}gh$ defines a rack operation on G. This makes G into the **conjugation rack** written $\mathrm{conj}(G)$ or alternatively G_{conj}.

Example 2 *The Dihedral Quandle*

Any union of conjugacy classes in a group forms a rack with conjugation as operation. In particular let R_n be the set of reflections in the dihedral group D_n of order $2n$ (which we regard as the symmetry group of the regular n-gon). Then R_n forms a rack of order n.

Example 3 *The Reflection Rack*

Let P, Q be points of the plane and define P^Q to be P reflected in Q (i.e. $2Q - P$ in vector notation).

It is elementary to show that this is a rack operation. This example can be generalised by replacing the plane by any geometry with point symmetries satisfying certain general conditions (see Joyce for details). Examples include the natural geometries of S^n and P^n. Interesting subracks of these latter racks are given by the action of Coxeter groups on root systems.

Example 4 *The Alexander Rack*

Let Λ be the ring of Laurent polynomials $\mathbb{Z}[t, t^{-1}]$ in the variable t. Any $\Lambda-$ module M has the structure of a rack with the rule $a^b := ta + (1-t)b$.

For example, letting M be the plane and the action of t multiplication by -1, yields the reflection rack of example 3.

The Quandle Condition

All the above examples have satisfied the identity $a^a = a$ for all $a \in X$, and are therefore quandles, indeed examples 2 and 3 are keis.

Here is an example of a non-quandle rack:

Example 5 *The Cyclic Rack*

The **cyclic rack of order** n, is given by $C_n = \{0, 1, 2, \ldots n-1\}$, the residues modulo n, with operation $i^j := i + 1 \mod n$ for all $i, j \in C_n$.

This example can be generalised: Let X be any G–set and choose a fixed element $g \in G$, then $a^b := a \cdot g$ for all $a, b \in X$ defines a rack structure on X.

Example 6 *The Free Rack*

The **free rack** $\mathrm{FR}(S)$ on a given set S is defined, as a set, to be $S \times F(S)$. We write the pair (a, w) as a^w, i.e.

$$\mathrm{FR}(S) = \{a^w \mid a \in S, w \in F(S)\}.$$

The rack operation is defined by

$$(a^w)^{(b^z)} = a^{wz^{-1}bz}.$$

The free rack has the universal property that any function $S \longrightarrow X$, where X is a given rack, extends uniquely to a rack homomorphism $\mathrm{FR}(S) \longrightarrow X$.

There is also the free quandle $\mathrm{FQ}(S)$ in which the additional conditions

$$a^{aw} = a^{a^{-1}w} = a^w$$

hold.

The Operator Group

In expressions such as $a^{bc^{-1}}$, a is referred to as being at **primary** level and b, c^{-1} as at **operator** level. The second form of the rack identity makes clear that we do not need any "higher" level operators. Expressions involving repeated operations can always be resolved into one of the form a^w where $a \in X$ is at the primary level and w, lying in the free group $F(X)$ on X, is at the operator level.

In this way we have an action by the group $F(X)$ on X. In general if G acts on X, written $(a, g) \mapsto a \cdot g$ and if $\mathrm{d} : X \longrightarrow G$ is a map satisfying $\mathrm{d}(a \cdot g) = g^{-1}\mathrm{d}(a)g$ then X has the structure of a rack given by $a^b := a \cdot \mathrm{d}(b)$. In many situations this is the most convenient method of describing the rack operation. The similarity with crossed modules should be clear.

To make operators precise we define **operator equivalence** by:

$$w \equiv z \iff a^w = a^z \text{ for all } a \in X$$

where $w, z \in F(X)$.

The equivalence classes form the **Operator Group** $Op(X)$ which could also be defined as $F(X)/N$ where N is the normal subgroup

$$N = \{w \in F(X) \mid w \equiv 1\}.$$

As with any group operation we can define stabilisers and orbits.

Examples of operator equivalence

Assume we have a rack. Since $b^{a^a} = b^{a^{-1}aa}$ (by the rack identity) $= b^a$ for all $a, b \in X$. We have

$$a^a \equiv a \quad \text{for all} \quad a \in X.$$

More generally if a^{b^n} means $a^{bb\ldots b}$ (n repeats) then $a \equiv a^{a^n}$.

In terms of operator equivalence, the rack identity can again be restated:

Axiom 3 $''$ Given $a, b \in X$ we have

$$a^b \equiv b^{-1}ab.$$

This is the **rack identity (third form).**

Examples of operator groups

The operator group in the conjugacy quandle is the group of inner automorphisms of G and the orbits are the conjugacy classes. Given $g, h \in G$ then $g \equiv h$ if and only if gh^{-1} is in the centre of G.

The dihedral quandle has operator group D_n, the dihedral group.

The operator group of the free rack is the free group $F(S)$ whilst the set of orbits is in bijective correspondence with the elements of S and all stabilizers are trivial.

The Associated Group

We can now see that a rack is a set X with an action of $F(X)$ (or its quotient $Op(X)$) on X satisfying the rack identity. We shall see that there is another group naturally associated to a rack, lying between $F(X)$ and $Op(X)$, called the **associated group**, which therefore also acts on X. The associated group is particularly important because it has a universal property.

If we interpret the operation of a rack as conjugation (i.e. read a^w as $w^{-1}aw$) then we obtain a group $As(X)$ called the **associated group**.

More precisely let $As(X) = F(X)/K$ where K is the normal subgroup of $F(X)$ generated by the words $a^b b^{-1} a^{-1} b$ where $a, b \in X$. So $As(X)$ is the biggest quotient of $F(X)$ with the property that, when considered as a rack via conjugation, the natural map from $F(X)$ to $As(X)$ is a rack homomorphism.

Given a rack homomorphism $f : X \longrightarrow Y$, then there is an induced group homomorphism $f_\sharp : As(X) \longrightarrow As(Y)$; thus we have an **associated group functor** As from the category of racks to the category of groups.

Theorem 3.1 *(Universal Property of the Associated Group) Let X be a rack and let G be a group. Given any rack homomorphism $f : X \longrightarrow G_{\text{conj}}$ there exists a unique group homomorphism $f_\sharp : As(X) \longrightarrow G$ which makes the following diagram commute:*

$$\begin{array}{ccc} X & \xrightarrow{\eta} & As(X) \\ \downarrow f & & \downarrow f_\sharp \\ G_{conj} & \xrightarrow{id} & G \end{array}$$

where η is the natural map. Moreover any group with the same universal property is isomorphic to $As(X)$. $\square$

Corollary 3.2

The functor As is a left adjoint to the conjugation functor. This means there is a natural identification

$$\text{Hom}(As(X), G) \cong \text{Hom}(X, \text{conj}(G))$$

of group homomorphisms with rack homomorphisms. $\square$

Example The cyclic rack C_n has operator group $\mathbb{Z}/n$ and associated group $\mathbb{Z}$.

This example makes it clear that the operator group is in general a non-trivial quotient of the associated group.

The associated quandle There is a natural inclusion of the category of quandles in the category of racks and there is a functor from racks to quandles $X \mapsto X_q$. Here X_q is called the **associated quandle** defined as follows: Let $\sim$ be the smallest congruence on X satisfying $a^a \sim a$ for all $a \in X$. Then $X_q := X/\sim$.

This functor is a retraction because it is clearly the identity for a rack which is already a quandle. For an explicit example, consider the Coxeter rack X with binary operation

$$\mathbf{u}^{\mathbf{v}} := \mathbf{u} - \frac{2(\mathbf{u}, \mathbf{v})}{(\mathbf{v}, \mathbf{v})}\mathbf{v},$$

defined on the unit sphere $S = (\mathbf{u}, \mathbf{u}) = 1$. Then the associated quandle is the **projectivisation** of X , defined by quotienting S by ± 1 .

Racks with an explicit group (augmented racks)

The concept of rack can be generalised to make the group action explicit.

An **augmented rack** comprises a set X with an action by a group G , which we write

$$(x, g) \mapsto x \cdot g \quad \text{where} \quad x,\ x \cdot g \in X \quad \text{and} \quad g \in G,$$

and a function $\mathrm{d} : X \longrightarrow G$ satisfying the **augmentation identity**:

$$\mathrm{d}(a \cdot g) = g^{-1}(\mathrm{d}a)g \quad \text{for all} \quad a \in X, g \in G,$$

which is precisely the same as saying that d is a G–map when the action of G on itself is taken to be conjugation.

We can now define an operation of X on itself by defining a^b to be $a \cdot \mathrm{d}b$. Then the augmentation identity implies

$$\mathrm{d}(a^b) = (\mathrm{d}b)^{-1}\mathrm{d}a\mathrm{d}b$$

So an augmented rack is an ordinary rack with the extra structure of an explicit operator group.

The fundamental rack (to be defined in the next section) has a natural structure as an augmented rack with the fundamental group of the link acting as a group of operators.

Homomorphisms and congruences

There are obvious notions of rack **homomorphism**, **isomorphism** and **subrack**.

An equivalence relation $\sim$ on X is called a **congruence** if it respects the rack operation, i.e.

$$a \sim b,\ c \sim d \implies a^c \sim b^d$$

$$a \sim b,\ c \sim d \implies a^{c^{-1}} \sim b^{d^{-1}}.$$

The equivalence classes form a rack $X/\sim$ with operations defined by $[a]^{[b]} := [a^b]$ and $[a]^{[b]^{-1}} := [a^{b^{-1}}]$, where $[a]$ denotes the equivalence class of a.

A homomorphism $f : X \longrightarrow Y$ of racks defines a congruence by $a \sim b \iff f(a) = f(b)$. Then the quotient $X/\sim$ is isomorphic to $f(X)$. This is an analogue of the first isomorphism theorem for groups.

Examples of Augmented Racks

(1) The Lie rack

Hyman Bass has pointed out the following important class of augmented racks. Let G be a Lie group and $\mathcal{G}$ the associated Lie algebra. Let $\mathrm{d} : \mathcal{G} \longrightarrow G$ be the exponential map and let G act on $\mathcal{G}$ via the adjoint action. Then the augmentation identity follows readily from definitions. Therefore the Lie algebra $\mathcal{G}$ is an augmented rack, with group the corresponding Lie group.

(2) *Gauge transformations*

Let E be a principal G–bundle and f a **gauge transformation** of E, that is an automorphism of E as a G–bundle. Then associated to f is an augmented rack structure on E with group G.

Let $p \in E$ then we can write $f(p) = p \cdot \mathrm{d}(p)$ where $\mathrm{d}(p) \in G$. This defines the function $\mathrm{d} : E \longrightarrow G$. It is easily seen that

$$\mathrm{d}(p \cdot g) = g^{-1}\mathrm{d}(p)g.$$

An augmented rack is a plain rack if we ignore or forget about the explicit group action. Conversely, there is a natural way to regard a plain rack as an augmented rack by taking as group $G = As(X)$ (the associated group) with d the natural map. Thus we can regard the category of racks as a subcategory of the category of augmented racks and then the forgetful functor is a retraction of the larger category onto the smaller.

Crossed modules

A **crossed module** is an augmented rack in which

(1) X is a group

(2) d is a homomorphism

(3) and we have the **crossed module identity**:

$$a \cdot \mathrm{d}b = b^{-1}ab \text{ for all } a, b \in X,$$

where the left-hand side is the G action and the right-hand side is multiplication in X.

Note that condition (3) implies that the rack operation in X is conjugation. Thus crossed modules correspond precisely to conjugation racks.

Crossed modules occur naturally in topology: the second homotopy group $\pi_2(X, A)$ of a pair of topological spaces is a crossed module with group $G = \pi_1(A)$ (see Whitehead [**Wh**]).

The associated crossed module

The associated group for a plain rack becomes the associated crossed module for an augmented rack. More precisely notice that if X is an augmented rack then its group G acts on $F(X)$ in the obvious way. The action of G on $F(X)$ induces an action on $As(X)$.

Thus $As(X)$ is a G–set and we also have the induced homomorphism $\mathrm{d}_\sharp : As(X) \longrightarrow G$. It can be readily checked that this gives $As(X)$ the structure of a crossed module, the **associated crossed module** to the augmented rack X.

4 The Fundamental Rack of a Link

This is the most important rack of all and is the *raison d'être* of the whole theory. In this section the fundamental rack of a framed link and the fundamental quandle of an unframed link are defined. The three most important results are stated, namely the topological definition, the interpretation as a presentation and the complete invariant of classical knots.

A (codimension two) **link** is defined to be a codimension two embedding $L : M \subset Q$ of one manifold in another. We shall assume that the embedding is proper at the boundary if necessary, that M is non-empty, that Q is connected and that M is **transversely oriented** in Q. In other words we assume that each normal disc to M in Q has an orientation which is locally and globally coherent.

The link is said to be **framed** if there is given a cross section (called a **framing**) $\lambda : M \longrightarrow \partial N(M)$ of the normal disk bundle. Denote by M^+ the image of M under λ. We call M^+ the **parallel** manifold to M.

We consider homotopy classes Γ of paths in $Q_0 = \text{closure}(Q - N(M))$ from a point in M^+ to a base point. During the homotopy the final point of the path at the base point is kept fixed and the initial point is allowed to wander at will on M^+. †

The set Γ has an action of the fundamental group of Q_0 defined as follows: let γ be a loop in Q_0 representing an element g of the fundamental group. If $a \in \Gamma$ is represented by the path α define $a \cdot g$ to be the class of the composite path $\alpha \circ \gamma$.

We can use this action to define a rack structure on Γ. Let $p \in M^+$ be a point on the framing image. Then p lies on a unique meridian circle of the normal circle bundle. Let m_p be the loop based at p which follows round the meridian in a positive direction. Let $a, b \in \Gamma$ be represented by the paths α, β respectively. Let $\mathrm{d}(b)$ be the element of the fundamental group determined by the loop $\overline{\beta} \circ m_\beta \circ \beta$. (Here $\overline{\beta}$ represents the reverse path to β and m_β is an abbreviation for $m_{\beta(0)}$ the meridian at the initial point

† This reverses the more usual dog wagging tail convention where the initial point of a path stays fixed. However the tail wagging dog convention fits in more comfortably with operations on the right.

of β.) The **fundamental rack** of the framed link L is defined to be the set $\Gamma = \Gamma(L)$ of homotopy classes of paths as above with operation

$$a^b := a \cdot \mathrm{d}(b) = [\alpha \circ \overline{\beta} \circ m_\beta \circ \beta].$$

If L is an unframed link then we can define its **fundamental quandle.** The definition is very similar. Let $\Gamma_q = \Gamma_q(L)$ be the set of homotopy classes of paths from the boundary of the regular neighbourhood to the base point where the initial point is allowed to wander during the course of the homotopy over the *whole* boundary. The rack structure on Γ_q is similar to that defined on Γ.

Theorem 4.3 *The fundamental rack of a framed link satisfies the axioms of a rack. The fundamental quandle of an unframed link satisfies the axioms of a quandle. In the unframed case the fundamental quandle of the corresponding link is the associated quandle of the fundamental rack.*

□

The main result of this section is the following classification theorem and uses a deep result of Waldhausen, [**Wal**].

Theorem 4.4 *(Classification theorem) The fundamental quandle is a complete invariant for non-split classical links up to orientation. The fundamental rack is a complete invariant for non-split framed classical links up to orientation.*

□

What this means is that if two non-split links have isomorphic quandles then there is a homeomorphism of space taking one link to the other. This result can be generalised in two ways: either to irreducible links in a 3-manifold or by making the homeomorphism orientation preserving. This latter result needs the notion of rack/quandle homology, see [**FRS1**] and [**FRS2**].

Theorem 4.5 *The fundamental quandle or rack of a compact codimension 2 embedding in euclidean space $\mathbb{R}^n$ is finitely presented.*

□

In the classical case this is the generalisation of the fundamental group presentation

5 Using the fundamental quandle or rack to find knot invariants

We have seen that the fundamental rack or quandle is a complete invariant of classical knots, at least up to orientation. Moreover it has a presentation given by any diagram. We can find invariants from this information by colouring the arcs with any rack or quandle which fits. In other words we find a non-trivial homomorphism from the fundamental quandle or rack to the given one.

Important examples of knot invariants are the fundamental group and the Alexander module which can be defined using the fundamental group. These can both be interpreted as invariants of the fundamental quandle. The Alexander quandle gives a presentation of a $\mathbb{Z}[t, t^{-1}]$-module. The determinant of this presentation is an invariant but unfortunately not a useful one as it is zero. However if we delete a row and a column and now take the determinant we get the Alexander polynomial. This is the start of a process defined as follows. Take the gcd of the codimension i minors of the presentation matrix. This is a Laurent polynomial Δ_i. We have $\Delta_0 = 0$, Δ_1 is the Alexander polynomial and

$$0 = \Delta_0 | \Delta_1 | \cdots | \Delta_{n-1} = 1$$

These polynomials are only defined up to multiplication by $t^{\pm 1}$.

6 Biquandles and Biracks

The idea of a biquandle or birack is founded on the notion of two binary operations; one, as before, for the action of the over crossing on the under crossing and another which is the action of the under crossing on the over crossing. The obvious reaction is "why bother?" The fundamental rack or quandle is already a complete invariant. "Why complicate matters?"

This is indeed true for the theory of classical knots but if we look at knots in different 3-manifolds and higher dimensional codimension 2 embeddings then this concept becomes very useful.

First of all let us generalise the notion of a diagram.

Virtual and Welded Knot diagrams

A **virtual (welded) knot diagram** is a knot diagram as before except that some of the under/over double points are replaced by virtual crossings (welds) as in the diagram.

Virtual Crossing A Weld

The new set of rules defining the generalised Reidemeister moves on virtual diagrams or welded diagrams is defined by the following diagram.

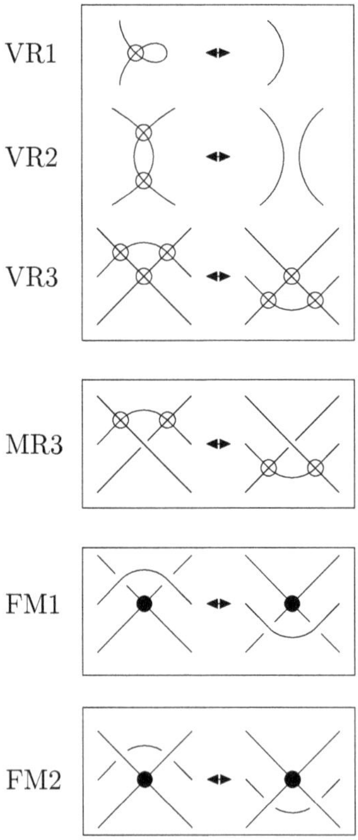

Generalised Reidemeister Moves

As before, the Reidemeister moves, R1, R2, R3 apply to any part of the diagram consisting only of classical crossings. The moves VR1, VR2, VR3 and MR3 (mixed Reidemeister 3) apply to virtual crossings but also when the virtual crossings are replaced by welds.

Consider the following equivalences, u-, v-, w- which are defined as follows. v-equivalence allows all the moves except FM1 and FM2. w-equivalence is v-equivalence plus FM1, the first "forbidden" move. u-equivalence is w-equivalence plus FM2, the second "forbidden" move.

A **virtual knot** is defined by a virtual knot diagram under v-equivalence. There are many interpretations of virtual knots. One is Gauss codes modulo Reidemeister moves. Another is knotted circles in a surface cross an interval modulo stable surgery, see [**K**], [**Ku**].

A **welded knot** is defined by a welded knot diagram under w-equivalence. Welded knots and links were defined to reflect quandles, see [**FRR**]. Another interpretation is as tori in four dimensions, see [**S**].

u-equivalence unknots any knot, see [**Kan, Nel**]

Colouring Virtual and Welded Diagrams

We use the notation and definitions given in [**FJK**], where more details can be found.

If a virtual (welded) knot diagram is broken at each kind of crossing the components are called **semi-arcs**. So if there are n crossings of every kind then there are $2n$ semi-arcs. We now colour each semi-arc and change the colouring at each crossing using two binary operations indicated in the following diagram.

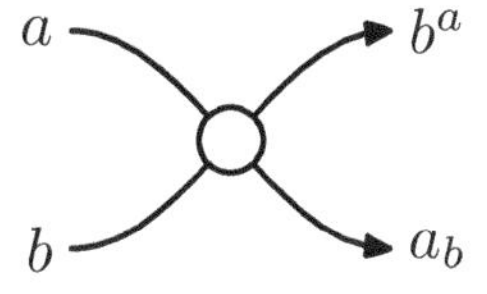

The crossing can be one of four types, a positive classical crossing, a negative classical crossing, a virtual crossing or a weld.

We have a set X of colours and two binary operations, up and down, written

$$(a,b) \longrightarrow a^b \text{ and } (a,b) \longrightarrow a_b$$

The convenience of the exponential and suffix notation is that brackets can be inserted in an obvious fashion and so are not needed, For example

$$a^{bc} = (a^b)^c,\ a^{b_c} = a^{(b_c)},\ a^b{}_c = (a^b)_c, \text{ etc.}$$

On the other hand expressions such as a^b_c are ambiguous and are not used.

Think of b as acting on a in both cases. We want these actions to be invertible. So there are binary operations

$$(a,b) \longrightarrow a^{b^{-1}} \text{ and } (a,b) \longrightarrow a_{b^{-1}}$$

satisfying

$$a^{bb^{-1}} = a^{b^{-1}b} = a \text{ and } a_{bb^{-1}} = a_{b^{-1}b} = a$$

It is convenient at this stage to introduce the function $S : X^2 \longrightarrow X^2$ defined by $S(a,b) = (b^a, a_b)$, known as a **switch**.

All these operations have to respect the Reidemeister moves and so we list the three axioms, B1-3 needed.

B1: For all $a \in X$ there is a unique $x \in X$ such that $a^x = x, a = x_a$ and there is a unique $y \in X$ such that $a_y = y, a = y^a$

B2: S is invertible. So there is a function $\overline{S} : X^2 \longrightarrow X^2$ such that $S\overline{S} = \overline{S}S = id$

B3: Let $S_1 = S \times id$ and $S_2 = id \times S$ then $S_1S_2S_1 = S_2S_1S_2$

Any pair satisfying B1, B2 and B3 defines a **biquandle**.. If only B2 and B3 are satisfied then we have a **birack**.

Axiom B1 implies that the infinite tower and the infinite well

$$x = a^{a^{a^{\cdot^{\cdot^{\cdot}}}}}, \quad y = a_{a_{a_{\ddots}}}$$

make sense.

We have $x = a_{a^{-1}}$, $y = a^{a^{-1}}$ and so B1 is equivalent to $a^{a_{a^{-1}}} = a_{a^{-1}}$ and $a_{a^{a^{-1}}} = a^{a^{-1}}$.

In [**Stan**] it is shown that only one half of B1 is necessary.

Using B2 write $\overline{S}(a,b) = (b_{\overline{a}}, a^{\overline{b}})$. This defines two binary operations

$$(a,b) \longrightarrow a^{\overline{b}} \text{ and } (a,b) \longrightarrow a_{\overline{b}}$$

In terms of the previous operations $\overline{a^b}$ acts down on b as a^{-1} and $\overline{a_b}$ acts up on b as a^{-1}. So all these new operations are defined by the two initial up and down operations.

B3 is sometimes called the set theoretic Yang-Baxter equation. If we follow the progress of the triple (a,b,c) through the two sides of the equation and swap variables we arrive at three relations true for all $a,b,c \in X$.

$$a^{c_b b^c} = a^{bc}, \quad a^b{}_{c^{b_a}} = a_c{}^{b_{c^a}}, \quad a_{c^b b_c} = a_{bc}$$

7 Examples

If the down operation is trivial, so $a_b = a$ for all $a,b \in X$, then a biquandle becomes a quandle. Symmetrically, this is also the case if the up operation is trivial. A birack with trivial down (up) operation is a rack.

The simplest biquandle, with both up and down operations trivial, is the **twist**, written $\mathbb{I}$. In this case $\mathbb{I}(a,b) = (b,a)$.

Linear Biquandles

A biquandle is said to be **linear** if it is determined by a 2×2 matrix $S = \begin{pmatrix} A & B \\ C & D \end{pmatrix}$, where A,B,C,D are elements of an associative ring related by certain equations, see [**BuF**].

The only example of a commutative linear biquandle is given by

$$a^b = \lambda a + (1 - \lambda\mu)b, \quad a_b = \mu a$$

where $a,b \in X$, a $\mathbb{Z}[\lambda^{\pm 1}, \mu^{\pm 1}]$-module [**Swa**]. This is called the **Alexander** biquandle and is denoted by $A_{\lambda\mu}(X)$. If $\mu = 1$ then resulting quandle is called the **Burau** quandle and is denoted by $B_\lambda(X)$. If $\mu = \lambda = 1$ then we have the twist.

Elements A, B of an associative K-algebra with unit R are said to satisfy the *fundamental* relation if they are invertible and

$$A^{-1}B^{-1}AB - B^{-1}AB = BA^{-1}B^{-1}A - A. \quad (1)$$

The following theorem is in [**F, BF, BuF**].

Theorem 7.1 If $A, B \in R$ satisfy the fundamental relation, $C = A^{-1}B^{-1}A(1-A)$ is invertible in R, and $D = 1 - A^{-1}B^{-1}AB$, then $S = \begin{pmatrix} A & B \\ C & D \end{pmatrix}$, is a biquandle.

□

There are many solutions to the fundamental relation. For example let $\mathbb{H}$ denote the quaternion algebra with standard generators $1, i, j, k$. If X is a left $\mathbb{H}$ module then

$$a^b = ja + (1+i)b, \quad a_b = -ja + (1+i)b$$

is a biquandle. This is called the **Budapest** biquandle and is a special case of the following theorem, see [**BuF**].

Theorem 7.2 *Suppose A, B are quaternions. Then A, B satisfy the fundamental relation, if and only if, A lies on the sphere $|A-1| = 1$ and B is orthogonal to A.*

□

Another result involving the quantum **Weyl algebra** appears in [**FT**]. The Weyl algebra is generated by two elements u, v and has one relation $uv - qvu = 1$, where q acts like an element of the ground ring.

Theorem 7.3 *Let $R = W_q$ and $A = v^{-1}u^{-1}$, $B = u$, $C = quvu^{-1}v^{-1}u^{-1}v^{-1}u^{-1}vu$, $D = 1 - q - u^{-1}v^{-1}$, where $uv - qvu = 1$. Then A, B satisfy the fundamental relation.*

Non-Linear Biquandles

One example of a non-linear biquandle is due to Wada, [**W**]. Let X be a group and put $S(a,b) = (a^2b, b^{-1}a^{-1}b)$.

Another is due to Silver and Williams, [**SW**]. Let G be a group with a $\mathbb{Z}^2$ action, $a \longrightarrow a \cdot x$ and $a \longrightarrow a \cdot y$ where x, y are a basis for $\mathbb{Z}^2$. Define $S(a,b) = (b \cdot y, (b \cdot xy)^{-1}(a \cdot x)b)$.

There are various lists of finite biquandles which are mostly non-linear. For a recent list of biquandles, biracks etc up to size 6 see [**BF2**].

8 Virtual and Weld Invariants

Each of the different kinds of crossings needs a biquandle switch attached to them. By convention we use the symbol S for a positive classical crossing. In view of B2 we attach $\bar{S} = S^{-1}$ to a negative classical crossing. The symbol T is reserved for a virtual crossing or a weld.

All the switches, S, S^{-1} and T define biquandles. We now need to consider relations between them which define their interaction.

A pair of biquandles (S,T) is a **virtual** invariant if

V . T^2 is the identity and $T_1S_2T_1 = T_2S_1T_2$

A virtual invariant pair is said to be a **weld** invariant if in addition the pair satisfy

$\mathbf{W_1}$. $T_1S_2S_1 = S_2S_1T_2$

This condition is called the first **forbidden** move and the pair is **essential** if the second forbidden move

$\mathbf{W_2}$. $S_1S_2T_1 = T_2S_1S_2$.

is NOT satisfied; think abelian and non-abelian. Since the use of both forbidden moves unknots any virtual knot, [**Kan, Nel**], it is important to find essential weld invariants. The following result can be found in [**BF2**].

Theorem 8.4 *Let S be a biquandle and let $T = \mathbb{I}$ be the twist acting on the set X. Then the pair (S,T) is a virtual invariant. It is a weld invariant if and only if the down operation is commutative and the down operation is trivial upstairs: in symbols $a_{bc} = a_{cb}$ and $a^{b_c} = a^b$. If S is linear then it is the Alexander biquandle and so the pair (S,T) cannot be an essential weld invariant.*

9 Finding Invariants

Any virtual or welded knot or link can be written as the closure of a virtual braid and so can be thought of as a word w in the variables σ_i, τ_i where $1 \leq i \leq n-1$ and n is some fixed positive integer. The variable σ_i corresponds to a positive classical crossing and its inverse σ_i^{-1} corresponds to a negative classical crossing. The variable τ_i corresponds to a virtual crossing or weld.

If a pair of biquandles (S,T) is a virtual or weld invariant pair let $S_i = (id)^{i-1} \times S \times (id)^{n-i-1}$ and similarly for T_i. Suppose w is a word representing some virtual braid and let K be its closure. Then the assignation $\sigma_i \longrightarrow S_i$, $\tau \longrightarrow T_i$ acting on the word w defines a permutation, $f_{(S,T)}$, of X^n which is an invariant of the braid. The classic example is the **Burau** map which uses the Burau quandle.

Any n-tuple, $\mathbf{x}$, of X^n fixed by $f_{(S,T)}$ defines a colouring of the arcs of K, the closure of the braid. Solutions of the equation $f_{(S,T)}\mathbf{x} = \mathbf{x}$ therefore define invariants of K.

For a finite X the simplest invariant of K is the number of fixed points of the equation. For example if X is finite with n elements then the unlink with c components has fixed point invariant n^c.

Invariants can also be obtained from finite biracks, see [**BF2**].

If the biquandle is linear then invariants can be defined by the non-commutative (Dieudonné) determinant, see [**BF, BuF, FT, F**] The theory follows in a similar fashion to the quandle situation considered earlier except that now the first determinant need not be zero and in the case of the Alexander biquandle defines a useful invariant polynomial of 2 variables, see [**Saw, BF**].

The Budapest biquandle can be used to show that the Kishino knot is non-trivial. The second polynomial obtained in this way is $1+(5/2)t^2+t^4$, see **[BF]**.

Final remarks: quandles and biquandles

There are interesting distinctions between quandles and biquandles. A quandle is topologically defined and has a well understood free object. Neither statement is true for biquandles. The topological definition of a quandle means that it can be defined for a general codimension 2 embedding. The fact that there is no topologically definition for a biquandle may make it a more subtle invariant for smooth surfaces embedded in $\mathbb{R}^4$. Free biquandles exist by general nonsense but understanding them is an interesting problem in logic.

References

[**BF**] **A. Bartholomew and Roger Fenn.** *Quaternionic Invariants of Virtual Knots and Links,* JKTR (2008), Vol 17; No 2, pages 231-251

[**BF2**] **A. Bartholomew and Roger Fenn**. *Biquandles of Small Size and some Invariants of Virtual and Welded Knots,* preprint

[**BM**] **C. Burstin, W. Mayer**, *Distibrutive Gruppen*, J. Reine Angew. Math. 160 (1929), 111130

[**Br**] **E.Brieskorn**, *Automorphic sets and singularities*, Contemporary maths., 78 (1988) 45-115

[**BuF**] **S. Budden and Roger Fenn.** *The equation*

$$[b, (a-1)(a, b)] = 0$$

and virtual knots and links, Fund. Math. 184, 1929 (2004).

[**CW**] **J.C.Conway and G.C.Wraith**, *correspondence,* (1959)

[**D**] **P.Dehornoy**, *Free distributive groupoids*, J. Pure and Applied Algebra, 16 (1989) 123-146

[**F**] **R. Fenn** *Quaternion Algebras and Invariants of Virtual Knots and Links I: The Elliptic Case* JKTR 17 (2008) 279-304

[**FJK**] **R. Fenn, M. Jordan, L. Kauffman,** *Biquandles and Virtual Links,* Topology and its Applications, 145 (2004) 157-175

[**FR**] **R.Fenn, C.Rourke**, *Racks and Links in Codimension Two*, Journal of Knot Theory and its Ramifications, Vol. 1 No.4 (1992) 343–406

[**FRS1**] **R.Fenn, C.Rourke and B.Sanderson**, *An introduction to species and the rack space*, Proceedings of the Topology Conference, Erzurum, 1992

[FRS2] **R.Fenn, C.Rourke and B.Sanderson**, *Trunks and classifying spaces*, Applied Categorical Structures, Volume 3 (1995) pages 321-356

[FRR] **R.Fenn, R.Rimanyi, C.Rourke** *The Braid–Permutation Group*, Topology, 18 (1979) 1-15

[FT] **R. Fenn and Vladimir Turaev** *Weyl Algebras and Knots,* Journal of Geometry and Physics. 57 (2007) 1313-1324.

[J] **D.Joyce**, *A classifying invariant of knots; the knot quandle*, J. Pure Appl. Alg., 23 (1982) 37-65

[Je] **T.Jech**, *A short proof of the irreflexivity conjecture*, preprint, Penn. State University

[Kan] **T. Kanenobu**, *Forbidden moves unknot a virtual knot*, J. Knot Theory Ramifications 10 (2001), 8996.

[K] **L.Kauffman.** *Virtual Knot Theory,* European J. Comb. Vol 20, 663-690, (1999)

[K2] **L.H.Kauffman**, *Knot-crystals – classical knot theory in modern guise*, Knots and Physics, World Sci.

[Ku] **G. Kuperberg,** *What is a Virtual Link?* Algebraic and Geometric Topology 587-591 (2003)

[Kr] **B.Krüger**, *Automorphe Mengen und die artinschen Zopfgruppen.* Dissertation, Rheinische Friedrich-Wilhelms-Universität Bonn, 1989 Bonn Mathematische Schriften, 207

[La] **R.Laver**, *The left distributive law and the freeness of an algebra of elementary embeddings*, Advances in Math., 91 (1992) 209-231

[Mat] **S. Matveev**, *Distributive groupoids in knot theory*, Math. USSR Sbornik, 47 (1984) 73-83

[Nel] **S. Nelson**, *Unknotting virtual knots with Gauss diagram forbidden moves*, J.Knot Theory Ramifications 10 (2001), 931935.

[Ros] **A.W.Roscoe**, *Maintaining consistency in distributed databases*, preprint, Oxford University Computing Laboratory

[S] **Shin Satoh** *Virtual knot presentation of ribbon torus-knots* JKTR Volume: 9, Issue: 4(2000) pp. 531-542

[Swa] **J. Sawollek,** *On Alexander-Conway polynomials for virtual knots and links,* arXiv:math.GT/9912173 21Dec (1999.) preprint

[SW] **D.S. Silver and S.G. Williams,** *Alexander Groups and Virtual Links,* JKTR. Vol 10 (2001). 151-160.

[Stan] **D. Stanovsky,** *On axioms of biquandles,* J. Knot Th. Ramifications 15/7 (2006). 931933.

[Tak] **M. Takasaki**, *Abstraction of symmetric transformations,* Tohoku Math. J. 40 (1943), 145-207

[W] **M. Wada.** *Twisted Alexander Polynomial for finitely presentable group.* Topology 33 No. 2. 241-256 (1994)
[Wal] **F. Waldhausen**, *Gruppen mit Zentrum und 3-dimensionale Mannigfaltigkeiten*, Topology 6 (1967) 505-517
[Wh] **J.H.C.Whitehead**, *On adding relations to homotopy groups*, Annals of Math. (2) 42 (1941) 409-428
[Win] **S. K. Winker**, *Quandles, Knot Invariants, and the n-fold Branched Cover*, Doctoral Thesis, Chicago Illinois, (1984)

EXCEPTIONAL DEHN FILLING

C. McA. Gordon

Department of Mathematics
The University of Texas at Austin
1 University Station – C1200
Austin, TX 78712-0257 U.S.A.
E-mail: gordon@math.utexas.edu

We consider triples $(M;\alpha,\beta)$, where M is a hyperbolic 3-manifold with boundary a disjoint union of tori, and α,β are distinct slopes on some boundary component such that the Dehn fillings $M(\alpha)$ and $M(\beta)$ are not hyperbolic. Although there are infinitely many such $(M;\alpha,\beta)$'s, we examine the question of whether they can all be obtained from a finite set by Dehn filling along some additional boundary components. If the distance $\Delta(\alpha,\beta)$ between the slopes is 1 or 2 then this is not the case, but we show that it might be true if $\Delta(\alpha,\beta)\geq 3$ and summarize what is known in this direction.

Keywords: Dehn filling; exceptional slope.

1. Introduction

Let M be a compact, orientable, irreducible 3-manifold, such that each component of ∂M is a torus, and let T_0 be a specified component of ∂M. For each slope α on T_0 we have the manifold $M(\alpha)$ obtained by α-Dehn filling on M, that is, $M(\alpha)=M\cup V_\alpha$, where V_α is a solid torus, attached to M by a homeomorphism $T_0\to\partial V_\alpha$ that sends α to the boundary of a meridian disk of V_α.

By Thurston's Geometrization Theorem for Haken manifolds [28], M is either hyperbolic (in the sense that int M has a complete hyperbolic structure of finite volume), toroidal (contains an essential torus), or a Seifert fiber space. In the last case, the manifolds $M(\alpha)$ are completely understood (see [21]), while the second case can more or less be reduced to the first. If M is hyperbolic, then a classical result of Thurston [28] is that the set of *exceptional slopes* $E(M,T_0)=\{\alpha : M(\alpha) \text{ is not hyperbolic}\}$ is finite. We will often abbreviate $E(M,T_0)$ to $E(M)$, the boundary component T_0 being understood, and call $(M;\alpha)$ *exceptional* if $\alpha\in E(M)$. In light of

Myers' result [27] that every compact, orientable 3-manifold with boundary a (possibly empty) disjoint union of tori is of the form $M(\alpha)$ for some hyperbolic manifold M, it is unreasonable to hope to classify all exceptional $(M;\alpha)$. On the other hand, it is relatively rare for a hyperbolic 3-manifold to have two distinct exceptional slopes α and β on T_0, so we define an *exceptional pair* $(M;\alpha,\beta)$ to be a hyperbolic 3-manifold M with $\alpha \neq \beta \in E(M,T_0)$. (It might be more mathematically correct to call $(M;\alpha,\beta)$ an exceptional *triple*, but we use *pair* to emphasize that we are dealing with a pair of slopes.) We may then ask

Question 1.1. *Can we classify all exceptional pairs* $(M;\alpha,\beta)$*?*

It turns out that if $\alpha,\beta \in E(M,T_0)$ then the *distance* $\Delta(\alpha,\beta)$ (that is, the minimal geometric intersection number of α and β on T_0) is usually small, and many of the results on exceptional Dehn filling take the form of universal upper bounds on $\Delta(\alpha,\beta)$ for various classes of non-hyperbolic manifolds $M(\alpha)$ and $M(\beta)$. Although it is almost certainly too optimistic to expect a positive answer to Question 1.1 in general, we shall see that such a classification may be possible provided $\Delta(\alpha,\beta) \geq 3$. In this brief survey we will summarize progress towards this goal and indicate the problems that remain.

2. Examples and some general results

Let us look at some examples. Define $\Delta(M) = \max\{\Delta(\alpha,\beta) : \alpha,\beta \in E(M,T_0)\}$.

Example 2.1. The simplest non-closed hyperbolic 3-manifold is M_8, the exterior of the figure-eight knot. Here $E(M_8) = \{1/0, 0, \pm1, \pm2, \pm3, \pm4\}$ [28], so that $\#E(M_8) = 10$ and $\Delta(M_8) = 8$.

The author conjectured in [15] that these values of $\#E(M)$ and $\Delta(M)$ are maximal, and this has recently been proved by Lackenby and Meyerhoff [23].

Theorem 2.1 (Lackenby-Meyerhoff). *For any hyperbolic 3-manifold* M, $\#E(M) \leq 10$, *and* $\Delta(M) \leq 8$.

Example 2.2. Let W^* be the exterior of the Whitehead sister link. Then (for a suitable parametrization of slopes) $E(W^*) = \{1/0, 0, 1, 2, 5/2, 3\}$; thus $\Delta(W^*) = 5$, realized by the slopes 0 and $5/2$.

For manifolds with more than one boundary component, this value of $\Delta(M)$ is maximal (see [14]).

Theorem 2.2 (Gordon). *If M is a hyperbolic 3-manifold with at least two boundary components then $\Delta(M) \leq 5$.*

Once one has established the optimal upper bound on Δ, one then wants to proceed to classify the pairs $(M;\alpha,\beta)$ that realize or almost realize that value. For manifolds with more than one boundary component this has been done for $\Delta = 5$ and 4 [20].

Theorem 2.3 (Gordon-Wu). (1) *$(W^*; 0, 5/2)$ is the only exceptional pair with more than one boundary component and $\Delta = 5$.*

(2) *There are exactly three exceptional pairs $(M;\alpha,\beta)$ where M has more than one boundary component and $\Delta = 4$.*

The pairs in (2) are explicitly described in [19]; two of the manifolds are exteriors of links in S^3, namely the Whitehead link and the 2-bridge link corresponding to the rational number 10/3.

Example 2.3. Let W be the exterior of the Whitehead link, and consider the following four manifolds obtained by Dehn filling on one boundary component of W: $W(1) = M_8$; $W(-5) = M_8^*$, the figure-eight sister manifold; $W(-5/2)$; and $W(2)$. These manifolds M are hyperbolic, and have $\Delta(M) = 8, 8, 7, 6$ respectively, realized by toroidal fillings $M(\alpha)$, $M(\beta)$; see [14]. In [15, Conjecture 3.4] the author made the following conjecture.

Conjecture 2.1. *The four manifolds M listed above are the only ones with $\Delta(M) \geq 6$.*

In this direction, Agol [1] has recently proved

Theorem 2.4 (Agol). *There are only finitely many hyperbolic 3-manifolds with $\Delta(M) \geq 6$.*

3. Ancestral finiteness

The strict finiteness statement of Theorem 2.4 ceases to hold when $\Delta(M) \leq 5$. To see this, let N be a hyperbolic 3-manifold with at least two boundary components, and let T be a component of $\partial N \setminus T_0$. As usual, let α and β be slopes on T_0. Using known results on Dehn filling it is not hard to show

Theorem 3.1. *If $(N;\alpha,\beta)$ is exceptional then $(N(\gamma);\alpha,\beta)$ is exceptional for infinitely many slopes γ on T.*

For instance, by Example 2.2, there are infinitely many hyperbolic manifolds $W^*(\gamma)$ obtained by Dehn surgery along one component of the Whitehead sister link with $\Delta(W^*(\gamma)) = 5$.

This leads to the following definition. A set $\mathcal{E}$ of exceptional pairs is *ancestrally finite* (AF) if there is a finite set $\mathcal{F}$ of exceptional pairs such that for all $(M;\alpha,\beta) \in \mathcal{E}$ there exists $(N;\delta,\varepsilon) \in \mathcal{F}$ such that $(M;\alpha,\beta) \cong (N(\gamma);\delta,\varepsilon)$, where $N(\gamma)$ is obtained by Dehn filling along some (possibly empty) subset of the components of $\partial N \backslash T_0$. In light of the above discussion, the appropriate finiteness question then seems to be

Question 3.1. *Is the set of all exceptional pairs ancestrally finite?*

Example 3.1. To illustrate Question 3.1, consider the Eudave-Muñoz knots. These are the infinite family of hyperbolic knots K in S^3 constructed in [11], each of which has a half-integral toroidal Dehn surgery $M_K(\alpha_K)$, where M_K denotes the exterior of K. Thus $\Delta(\alpha_K, \mu_K) = 2$ where μ_K is the meridian of K, and $(M_K; \mu_K, \alpha_K)$ is an exceptional pair. (By [18], these knots are the only hyperbolic knots in S^3 with non-integral toroidal surgeries.) Let Q be the exterior of the (appropriately twisted) 4-chain link. The quotient of Q under the involution induced by the strong inversion of the 4-chain link is a tangle in the 4-punctured 3-sphere; this is the tangle $\mathcal{Q}$ illustrated in [17, Figure 5.1]. It is easy to see that rational tangle fillings $1/0$ and $-1/2$ on the central puncture of $\mathcal{Q}$ give homeomorphic tangles $\mathcal{Q}(1/0) \cong \mathcal{Q}(-1/2)$ that contain a Conway disk. It follows by passing to the double branched cover that the corresponding Dehn fillings on Q, $Q(1/0)$ and $Q(1/2)$, are (homeomorphic) manifolds that are annular, i.e. contain an essential annulus. Also, it is easily seen from their tangle descriptions [11] that if K is a Eudave-Muñoz knot then there exists $\gamma_K = (\gamma_1, \gamma_2, \gamma_3)$ such that $(M_K; \mu_K, \alpha_K) \cong (Q(\gamma_K); 1/2, 1/0)$; see [17, Lecture 5]. Thus the infinite set of exceptional pairs $(M_K; \mu_K, \alpha_K)$ all come from the single exceptional pair $(Q; 1/2, 1/0)$; in particular this set is ancestrally finite.

Example 3.2. If $(M;\alpha,\beta)$ is an exceptional pair with $M(\alpha)$ reducible and $M(\beta)$ annular then $\Delta(\alpha,\beta) \leq 2$ [29]. Infinitely many examples $(M_n;\alpha_n,\beta_n)$, $n \geq 3$, with $M_n(\alpha_n)$ reducible, $M_n(\beta_n)$ annular, and $\Delta(\alpha_n,\beta_n) = 2$, were constructed by Eudave-Muñoz and Wu [13]. (Only the case n even is explicitly considered in [13], but, as pointed out in [24], the construction is valid for all $n \geq 3$.) Moreover, Lee showed [24] that these are the only reducible/annular pairs with $\Delta = 2$. Again these examples all come from the single exceptional pair $(Q; 1/0, 1/2)$:

$(M_n; \alpha_n, \beta_n) \cong (Q(-1/n, 1/n); 1/0, 1/2)$. Note that here of course the manifolds M_n have two boundary components.

It turns out that whether or not a set of exceptional pairs $(M; \alpha, \beta)$ is ancestrally finite depends only on the set of volumes of the manifolds M.

Proposition 3.1. *A set $\mathcal{E}$ of exceptional pairs is ancestrally finite if and only if the set of volumes $\{\operatorname{vol} M : (M; \alpha, \beta) \in \mathcal{E}\}$ is bounded above.*

Proof. If U is a disjoint union of tori, a *multislope* on U is a set γ of slopes obtained by picking a slope on each torus in some subcollection of the components of U. If the components of U are numbered, say $T_1, \ldots, T_m$, we write γ as an k-tuple of slopes, $0 \leq k \leq m$, in the obvious way. If N is a 3-manifold with ∂N a disjoint union of tori and γ is a multislope on ∂N, then $N(\gamma)$ denotes the manifold obtained from N by doing the Dehn filling determined by the slopes in γ along the corresponding boundary components.

The "only if" direction of the proposition follows from the fact that $\operatorname{vol} N(\gamma) \leq \operatorname{vol} N$ for all multislopes γ on ∂N.

To prove the "if" direction, suppose the set of volumes $\operatorname{vol} M$ is bounded above. Then by a theorem of Jørgensen (see [28]) there is a finite set $\mathcal{J}$ of hyperbolic manifolds such that if $(M; \alpha, \beta) \in \mathcal{E}$ then $M \cong N(\gamma)$ for some $N \in \mathcal{J}$ and some multislope γ on $\partial N \setminus T_0$. Also, if $\gamma' \subset \gamma$, then $N(\gamma')$ is also hyperbolic. It suffices, then, to consider the case of a single hyperbolic manifold N such that $(M; \alpha, \beta) \in \mathcal{E}$ implies $M \cong N(\gamma)$ for some γ on $\partial N \setminus T_0$. Furthermore, it clearly suffices to assume that each $\gamma = (\gamma_1, \gamma_2, \ldots, \gamma_k)$, where $T_1, \ldots, T_k$ are some fixed components of $\partial N \setminus T_0$, $k \geq 0$, and γ_i is a slope on T_i, $1 \leq i \leq k$. Note that $N(\gamma_i)$ is hyperbolic, $1 \leq i \leq k$.

We prove the proposition by induction on k.

<u>$k = 0$.</u> Here $M = N$, and there are only finitely many pairs (α, β) such that $(M; \alpha, \beta)$ is exceptional.

<u>$k > 0$.</u> By a result of Thurston [28] there are finite sets of slopes S_i on T_i, $0 \leq i \leq k$, such that if $N(\boldsymbol{\sigma})$ is not hyperbolic, where $\boldsymbol{\sigma} = (\sigma_0, \sigma_1, \sigma_2, \ldots, \sigma_k)$, σ_i a slope on T_i, $0 \leq i \leq k$, then $\sigma_i \in S_i$ for some i, $0 \leq i \leq k$. Note that if $(M; \alpha, \beta) \in \mathcal{E}$ then $N(\alpha, \gamma)$ and $N(\beta, \gamma)$ are not hyperbolic. Define

$$\mathcal{E}_0 = \{(M; \alpha, \beta) \in \mathcal{E} : \alpha, \beta \in S_0\}$$
$$\mathcal{E}_i = \{(M, \alpha, \beta) \in \mathcal{E} : \gamma_i \in S_i\}\ , \qquad 1 \leq i \leq k\ .$$

Then $\mathcal{E} = \bigcup_{i=0}^{k} \mathcal{E}_i$.

Now each M such that $(M; \alpha, \beta) \in \mathcal{E}_i$ for some $1 \leq i \leq k$ is a filling along $k-1$ boundary components on one of a finite set of (hyperbolic) manifolds $N(\gamma_i)$. Hence $\mathcal{E}_i$ $(1 \leq i \leq k)$ is AF by induction.

The following lemma completes the proof of the proposition. □

Lemma 3.1. *$\mathcal{E}_0$ is AF.*

Proof. Since there are only finitely many pairs (α, β) such that $\alpha, \beta \in S_0$, we can restrict attention to a fixed pair (α, β).

If $N(\alpha)$ and $N(\beta)$ are both non-hyperbolic, then $(N; \alpha, \beta)$ is exceptional and we are done.

So suppose without loss of generality that $N(\alpha)$ is hyperbolic. Then by [28] there are finite sets of slopes S_i' on T_i, $1 \leq i \leq k$, such that (since $N(\alpha, \gamma)$ is not hyperbolic) $\gamma_i \in S_i'$ for some $1 \leq i \leq k$. But $N(\gamma_i)$ is hyperbolic. Hence $\mathcal{E}_0$ is AF by induction on k. □

4. Some positive results

By the Geometrization Conjecture, recently proved by Perelman, if a 3-manifold is not hyperbolic then it is either reducible, toroidal, or a small Seifert fiber space. A small Seifert fiber space either has non-empty boundary, in which case it contains either an essential disk (if it is a solid torus) or an essential annulus, or is closed, in which case it is either S^3, a lens space, or has a Seifert fibered structure with orbit surface S^2 and exactly three exceptional fibers. Following Martelli and Petronio [26], let us call the last class of manifolds $\mathcal{Z}$. Thus a non-hyperbolic 3-manifold that is not in $\mathcal{Z}$ contains a surface of non-negative Euler characteristic that is either essential or a Heegaard surface. In the classification of exceptional pairs $(M; \alpha, \beta)$, the hardest cases have proved to be those where $M(\alpha)$ or $M(\beta)$ (or both) are in $\mathcal{Z}$. In fact, in all other cases we get good answers to Question 3.1 when $\Delta \geq 5$.

For example, when $\Delta \geq 6$, Conjecture 2.1 is known to hold provided we stay away from the class $\mathcal{Z}$. This follows from results due to several people; see [16, Table 2.1] for references.

Theorem 4.1. *The four exceptional pairs $(M; \alpha, \beta)$ in Example 2.3 are the only ones with $\Delta(\alpha, \beta) \geq 6$ and $M(\alpha), M(\beta) \notin \mathcal{Z}$.*

As we have said, for $\Delta = 5$ there are infinitely many exceptional pairs, and the appropriate question is now Question 3.1. If we assume that neither of the fillings is in $\mathcal{Z}$, we get a positive answer.

Theorem 4.2. *The answer to Question 3.1 is 'yes' for the set of exceptional pairs* $(M;\alpha,\beta)$ *such that* $\Delta(\alpha,\beta)=5$ *and* $M(\alpha),M(\beta)\notin\mathcal{Z}$.

The only cases with $\Delta = 5$ left open after the results discussed in [16] are: $M(\alpha)$ toroidal, and $M(\beta)$ either (1) a lens space, or (2) toroidal. Case (1) is dealt with by Lee [25], who shows that here $\Delta(\alpha,\beta)\leq 4$, and case (2) is done in [20]. There, the exceptional set $\mathcal{F}$ is explicitly described: it consists of seven pairs $(N;\alpha,\beta)$; one manifold (the Whitehead sister link exterior W^*) has two boundary components, the others, one.

When $\Delta = 4$, the toroidal/toroidal case is done in [20], so the only problem remaining here (apart from the case where one of the fillings belongs to $\mathcal{Z}$) is the possibility of a lens space and toroidal filling at distance 4.

Theorem 4.3. *The answer to Question 3.1 is 'yes' for the set set of exceptional pairs* $(M;\alpha,\beta)$ *such that* $\Delta(\alpha,\beta)=4$ *and* $M(\alpha),M(\beta)\notin\mathcal{Z}$, *and* $\{M(\alpha),M(\beta)\}$ *is not a lens space/toroidal pair.*

The corresponding exceptional set $\mathcal{F}$ is explicitly determined in [20]; again it consists of seven pairs $(N;\alpha,\beta)$. Three of the manifolds have two boundary components (they are the manifolds in part (2) of Theorem 2.3), the others, one.

5. Some negative results

Having adduced evidence for a positive answer to Question 3.1, we now point out that actually it has a negative answer in general. The first examples to show this were certain hyperbolic knots in S^3 with toroidal surgeries constructed by Eudave-Muñoz and Luecke [12].

Theorem 5.1 (Eudave-Muñoz-Luecke). *For every* C *there is a hyperbolic knot* K *in* S^3 *with* $\mathrm{vol}(S^3\backslash K)>C$ *such that* K *has an integral toroidal Dehn surgery.*

By Proposition 3.1, this shows that Question 3.1 has a negative answer for exceptional pairs with $\Delta(\alpha,\beta)=1$.

Other examples which show this are certain Berge knots. Recall that Berge [5] constructed some infinite families of hyperbolic knots in S^3 with lens space surgeries. (Such surgeries are necessarily integral by the Cyclic Surgery Theorem [10].) Baker has shown [2]

Theorem 5.2 (Baker). *There are hyperbolic Berge knots with arbitrarily large volume.*

Again this gives a negative answer to Question 3.1 when $\Delta(\alpha, \beta) = 1$. The knots in question belong to Berge's families XI and XII, i.e. the knots that lie on a fiber of either the trefoil or figure-eight knot. By contrast, Baker shows that all the other Berge knots come from surgery on the minimally twisted 5-chain link [3].

In fact Question 3.1 also has a negative answer when $\Delta(\alpha, \beta) = 2$. This follows immediately from the following result, proved in [16].

Theorem 5.3 (Gordon). *There are hyperbolic 3-manifolds M with arbitrarily many boundary components having toroidal Dehn fillings $M(\alpha)$ and $M(\beta)$ where $\Delta(\alpha, \beta) = 2$.*

Although these examples do give a negative answer to Question 3.1 when $\Delta(\alpha, \beta) = 2$, their construction is local in nature, so in fact they all arise from a single example, albeit in a different sense from that in Question 3.1: one removes solid tori rather than adding them. So maybe Question 3.1 should be modified to account for this phenomenon.

6. Distance at least 3

There are no known counterexamples to a positive answer to Question 3.1 when $\Delta(\alpha, \beta) \geq 3$.

Question 6.1. *Is the set of exceptional pairs with $\Delta(\alpha, \beta) \geq 3$ ancestrally finite?*

In the remainder of the paper we indicate the problems that remain in establishing this.

It follows from known results that if $(M; \alpha, \beta)$ is exceptional with $\Delta(\alpha, \beta) = 3$ and $M(\alpha), M(\beta) \notin \mathcal{Z}$, then $M(\beta)$ (say) is toroidal and $M(\alpha)$ is either

(a) toroidal; or
(b) reducible; or
(c) a lens space.

Example 6.1. Let N be the exterior of the 3-chain link. This is the *magic manifold* of [19], whose exceptional Dehn fillings are completely described in [26]. Then $N(0)$ and $N(3)$ are toroidal; see [26] (we choose the mirror image of the link illustrated there).

Regarding (b), Eudave-Muñoz and Wu [13] constructed an infinite family of hyperbolic 3-manifolds $M_n, n \geq 2$, with a single boundary component, having slopes α_n, β_n such that $M_n(\alpha_n)$ is reducible, $M_n(\beta_n)$ is toroidal, and $\Delta(\alpha_n, \beta_n) = 3$. Kang has shown that these are the only such examples [22].

Theorem 6.1 (Kang). *$(M_n; \alpha_n, \beta_n)$, $n \geq 2$ are the only reducible/toroidal exceptional pairs with $\Delta = 3$.*

On the other hand one can show that $(M_n; \alpha_n, \beta_n) \cong (N(-n, n+4); 0, 3)$; see [26, Table A.11]. So Question 6.1 has an affirmative answer in this case.

To answer Question 6.1 there are therefore three problems remaining:

(A) *Classify the toroidal/toroidal exceptional pairs at distance 3.*

This should be feasible. Note that Example 6.1 above shows that we now have to consider manifolds with three boundary components.

(B) *Show that a lens space filling and a toroidal filling have distance at most 3, and classify those at distance 3.*

Two infinite families of examples at distance 3 of the form $(N(\gamma_1, \gamma_2); 0, 3)$, where N is the 3-chain link exterior, are described in [26, Table A.11]. So these are all descendants of the exceptional pair $(N; 0, 3)$.

Another infinite family comes from the Whitehead link, whose exterior W is homeomorphic to $N(1)$. With the slope parametrization used in [26], $M_n = W(-2 + 1/n)$ $(n \neq 0, 1)$ is hyperbolic, while $M_n(-2)$ is a lens space and $M_n(1)$ contains an incompressible non-separating torus. See [26, Table A.5] and [4]. The members of this family are therefore descendants of the exceptional pair $(W; -2, 1)$ ($W(-2)$ is Seifert fibered and $W(1)$ is toroidal; see [26, Table A.1].)

(C) *Deal with the case where $M(\alpha)$ (say) is in $\mathcal{Z}$.*

One can further subdivide the class $\mathcal{Z}$ into three subclasses $\mathcal{Z}_S$, $\mathcal{Z}_E$ and $\mathcal{Z}_H$, according as the corresponding base orbifold $S^2(p, q, r)$ is spherical, Euclidean, or hyperbolic, i.e. according as $1/p + 1/q + 1/r$ is $> 1, = 1$, or < 1. In particular, $\mathcal{Z}_S$ is the class of manifolds in $\mathcal{Z}$ with finite fundamental group. Again, the first thing one wants to do is establish optimal upper bounds for Δ in the various cases. So far this has only been achieved in the following few cases, most involving $\mathcal{Z}_S$:

if $M(\alpha) \in \mathcal{Z}_S$ then
$\Delta(\alpha, \beta) = 1$ *if $M(\beta)$ is reducible* [7];
$\Delta(\alpha, \beta) \leq 2$ *if $M(\beta)$ is a lens space* [8];
$\Delta(\alpha, \beta) \leq 3$ *if $M(\beta) \in \mathcal{Z}_S$* [9];
if $M(\alpha) \in \mathcal{Z}_E$ and $M(\beta)$ is reducible then $\Delta(\alpha, \beta) \leq 2$ [6].

Of course, as far as Question 6.1 is concerned one is happy if one can show, for some class of exceptional pairs $(M;\alpha,\beta)$, that $\Delta(\alpha,\beta) \leq 2$.

Finally we remark that even though the answer to Question 3.1 is 'no', it may still be possible to classify certain sets of exceptional $(M;\alpha,\beta)$'s that are not ancestrally finite. For example, as noted above, the Berge knots have unbounded volume, but they can still be explicitly described.

Acknowledgements

I would like to thank Yoav Rieck for helpful discussions.

References

[1] I. Agol, *Bounds on exceptional Dehn filling II*, Geom. Top. **14** (2010), 1921–1940.

[2] K.L. Baker, *Surgery descriptions and volumes of Berge knots I: large volume Berge knots*, J. Knot Theory and its Ramifications **17** (2008), 1077–1097.

[3] K.L. Baker, *Surgery descriptions and volumes of Berge knots II: descriptions on the minimally twisted five chain link*, J. Knot Theory and its Ramifications, **17** (2008), 1099–1120.

[4] K.L. Baker, *Once-punctured tori and knots in lens spaces*, arXiv:0612525v2 [math.GT].

[5] J. Berge, *Obtaining lens spaces by surgery on knots*, unpublished manuscript.

[6] S. Boyer, M. Culler, P.B. Shalen and X. Zhang, *Characteristic subsurfaces, character varieties and Dehn fillings*, Geom. Top. **12** (2008), 233–297.

[7] S. Boyer, C.McA. Gordon and X. Zhang, *Reducible and finite Dehn fillings*, J. London Math. Soc. **79** (2009), 72–84.

[8] S. Boyer and X. Zhang, *Finite Dehn surgery on knots*, J. Amer. Math. Soc. **9** (1996), 1005–1050.

[9] S. Boyer and X. Zhang, *A proof of the finite filling conjecture*, J. Diff. Geom. **59** (2001), 87–176.

[10] M. Culler, C.McA. Gordon, J. Luecke and P.B. Shalen, *Dehn surgery on knots*, Ann. of Math. **125** (1987), 237–300.

[11] M. Eudave-Muñoz, *Non-hyperbolic manifolds obtained by Dehn surgery on hyperbolic manifolds*, in Geometric Topology, Studies in Advanced Mathematics, Vol. 2.1, W.H. Kazez, Editor, AMS and International Press, 1997, 35–61.

[12] M. Eudave-Muñoz and J. Luecke, *Knots with bounded cusp volume yet large tunnel number*, J. Knot Theory and its Ramifications **4** (1999), 437–446.

[13] M. Eudave-Muñoz and Y.-Q. Wu, *Nonhyperbolic Dehn fillings on hyperbolic 3-manifolds*, Pac. J. Math. **190** (2000), 261–275.

[14] C. McA. Gordon, *Boundary slopes of punctured tori in 3-manifolds*, Trans. Amer. Math. Soc. **350** (1998), 1713–1790.

[15] C. McA. Gordon, *Dehn filling: a survey*, in Knot Theory, Banach Center Publications 42, Institute of Mathematics, Polish Academy of Sciences, Warsaw, 1998, 129–144.

[16] C. McA. Gordon, *Small surfaces and Dehn filling*, Geometry and Topology Monographs Vol. 2, Proceedings of the Kirbyfest, 1999, 177–199.

[17] C. McA. Gordon, *Dehn Surgery and 3-Manifolds*, in Low Dimensional Topology, T.S. Mrowka and P.S. Ozsváth, Editors, IAS/Park City Mathematics Series Vol.15, American Mathematical Society, Providence, RI, 2009, 23–71.

[18] C. McA. Gordon and J. Luecke, *Non-integral toroidal Dehn surgeries*, Comm. Anal. Geom. **12** (2004), 417–485.

[19] C. McA. Gordon and Y.-Q. Wu, *Toroidal and annular Dehn fillings*, Proc. London Math. Soc. **78** (1999), 662–700.

[20] C. McA. Gordon and Y.-Q. Wu, *Toroidal Dehn fillings on hyperbolic 3-manifolds*, Memoirs Amer. Math. Soc. vol. 194, no. 909, American Mathematical Society, Providence, RI, 2008.

[21] W. Heil, *Elementary surgery on Seifert fiber spaces*, Yokohama Math J. **22** (1974), 135–139.

[22] S. Kang, *Reducible and toroidal Dehn fillings with distance 3*, Topology **47** (2008), 277–315.

[23] M. Lackenby and R. Meyerhoff, *The maximal number of exceptional Dehn surgeries*, arXiv:0808.1176v1 [math.GT].

[24] S. Lee, *Reducing and annular Dehn fillings*, Trans. Amer. Math. Soc. **359** (2007), 227–247.

[25] S. Lee, *Lens spaces and toroidal Dehn fillings*, Math. Z. **267** (2011), 781–802.

[26] B. Martelli and C. Petronio, *Dehn filling of the "magic" 3-manifold*, Comm. Anal. Geom. **14** (2006), 969–1026.

[27] R. Myers, *Simple knots in compact, orientable 3-manifolds*, Trans. Amer. Math. Soc. **273** (1982), 75–91.

[28] W.P. Thurston, *The Geometry and Topology of 3-manifolds*, Princeton University, 1978.

[29] Y.-Q. Wu, *Sutured manifold hierarchies, essential laminations, and Dehn surgery*, J. Diff. Geom. **48** (1988), 407–437.

GRAPH-LINKS

D. P. Ilyutko

Department of Mechanics and Mathematics, Moscow State University, Moscow, Russia
E-mail: ilyutko@yandex.ru
www.msu.ru

V. O. Manturov

Moscow, Russia
E-mail: vomanturov@yandex.ru

The present paper is a review of the current state of *Graph-Link Theory* (graph-links are closely related to *homotopy classes of looped interlacement graphs*): a theory suggested in,[1,2] see also,[3] dealing with a generalisation of knots obtained by translating the Reidemeister moves for links into the language of intersection graphs of chord diagrams. In this paper we show how some methods of classical and virtual knot theory can be translated into the language of abstract graphs, and some theorems can be reproved and generalised to this graphical setting. We construct various invariants, prove certain minimality theorems and construct functorial mappings for graph-knots and graph-links. In this paper, we first show non-equivalence of some graph-links to virtual links.

Keywords: Knot, virtual knot, graph-link, looped interlacement graph, chord diagram, intersection graph.

1. Introduction

It is well known that classical and virtual knots[4] can be represented by Gauss diagrams, and the whole information about the knot and its invariants can be read out of any Gauss diagram encoding it, see Fig. 1. Whenever a Gauss diagram does not describe any *embedded 4-valent graph* in $\mathbb{R}^2$ (just because the corresponding Gauss code is not planar) as in Fig. 2, one gets a virtual knot, where generic immersion points of intersections of edges are encircled.

It turns out that some information about the knot can be obtained from a more combinatorial data: intersection graphs of Gauss diagrams. The intersection (adjacency) graph is a *simple graph*, i.e. a graph without

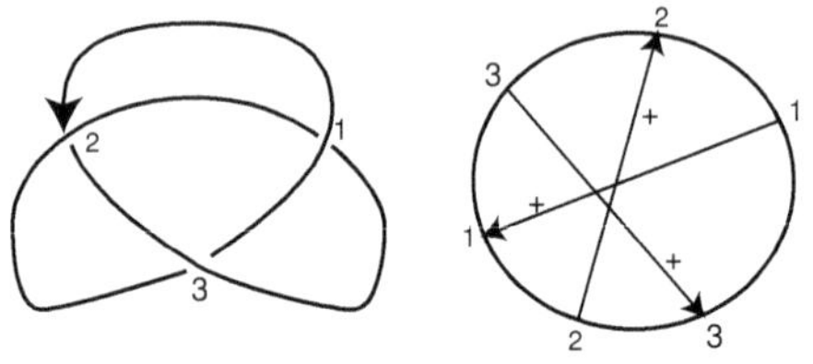

Fig. 1. The Right Trefoil and Its Gauss Diagram

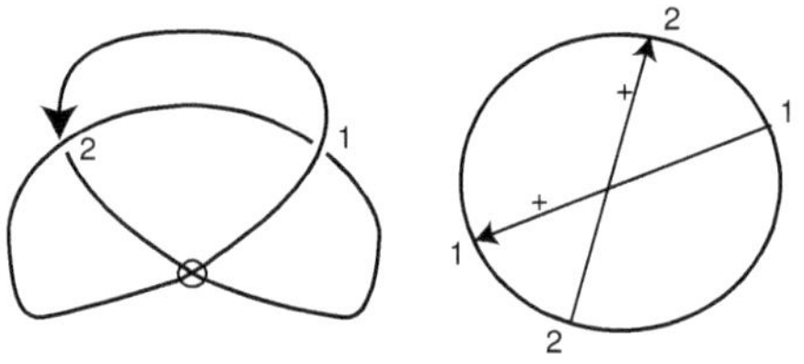

Fig. 2. The Virtual Trefoil and Its Gauss Diagram

loops and multiple edges, whose vertices are in one-to-one correspondence with chords of the Gauss diagram of the knot (the latter are, in turn, in one-to-one correspondence with crossings of the knot). Two vertices of the intersection graph are *adjacent* whenever the corresponding arrows of the Gauss diagram are *linked*, i.e. the corresponding chords "intersect" each other in the picture, see Fig. 3. Each vertex of the intersection graph is endowed with the local writhe number of the crossing. However, sometimes the Gauss diagram can be obtained from the intersection graph in a non-unique way, see Fig. 4, and some graphs (shown in Fig. 5) can not be represented by chord diagrams at all.[5]

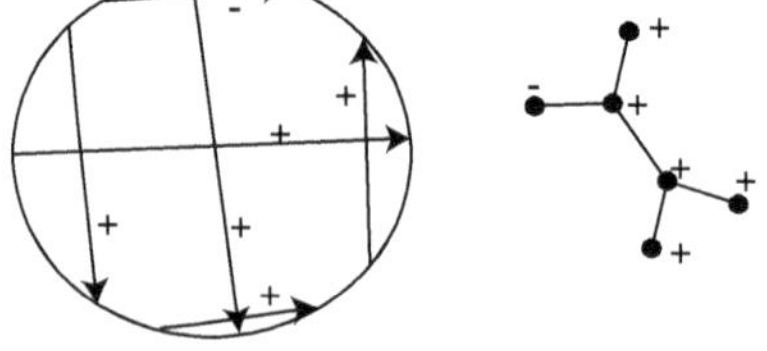

Fig. 3. A Gauss Diagram and its Labeled Intersection Graph

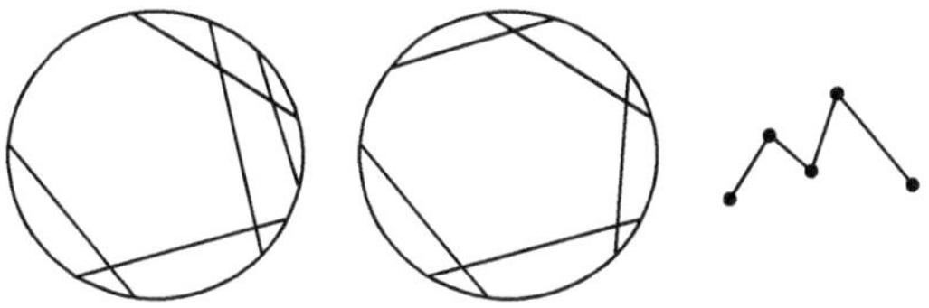

Fig. 4. A Graph not Uniquely Represented by Chord Diagrams

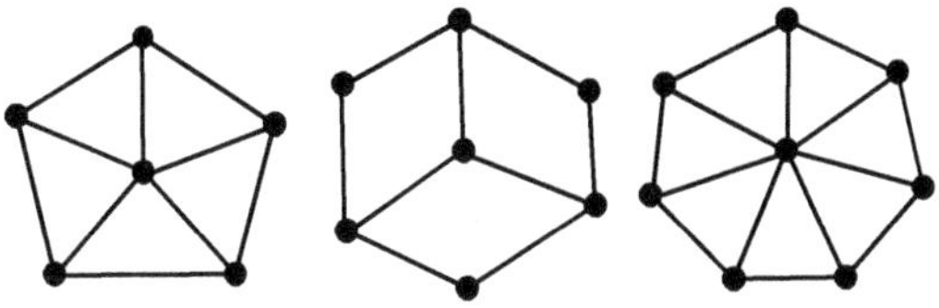

Fig. 5. Non-Realisable Bouchet Graphs

When passing to the intersection graph, we remember the writhe number information, but forget about the *right-left* (indicating a local orientation of the neighbourhood) information encoded by the arrows. In principle, it is possible to describe analogous objects when all information is saved in the intersection graph; however, already the writhe number information is sufficient to recover a lot of data, as we shall see. Even more, if we forget about the writhe number information and only have the structure of opposite edges, we shall get non-trivial objects (modulo Reidemeister's moves).

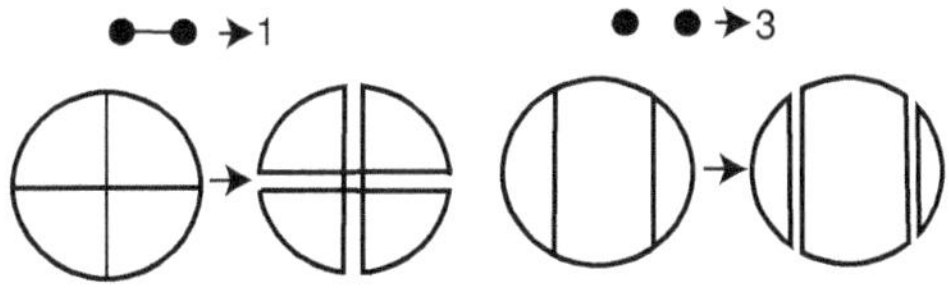

Fig. 6. Resmoothing along Two Chords Yields One or Three Circles

Probably, the simplest evidence that one can get some information out of the intersection graph is the number of circles one gets in a certain state after a smoothing, see Fig. 6. The simplest example shows that the number of circles obtained after resmoothings in two crossings gives three circles if the corresponding chords are unlinked or one circle if they are not. The general theorem (the circuit-nullity formula, see ahead) allows one to count the number of circles in Kauffman's states out of the intersection graph.

In particular, this means that graphs not necessarily corresponding to any knot admit a way of generalising the Kauffman bracket, which coincides with the usual Kauffman bracket when the graph is realisable by a knot. This was the initial point of investigation for L. Traldi and L. Zulli[3] (looped interlacement graphs): they constructed a self-contained theory of "non-realisable graphs" possessing lots of interesting knot theoretic properties. These objects are equivalence classes of (decorated) graphs modulo "Reidemeister moves" (translated into the language of intersection graphs). A significant disadvantage of this approach was that it had applications only to knots, *not links*: in order to encode a link, one has to use a more complicated object rather than just a Gauss diagram, a Gauss diagram *on many circles*. This approach was further developed in Traldi's works,[6,7] and it allowed one to encode not only knots but also links with any number of components by decorated graphs. The question arises here: whether or not every graph is Reidemeister equivalent to the looped interlacement graph of a virtual knot diagram. This question is important and up to the present paper there was no answer to this question.

We suggested another way of looking at knots and links and generalising them: whence a Gauss diagram corresponds to a transverse passage along a knot, one may consider a *rotating circuit* (or a *non-crossing Euler tour*[8]) which never goes straight and always turns right or left at a classical crossing. One can also encode the type of smoothing (Kauffman's A-smoothing or Kauffman's B-smoothing) corresponding to the crossing where the circuit turns right or left and never goes straight, see Fig. 7. We note that chords of the diagrams are naturally split into two sets: those corresponding to crossings where two opposite directions correspond to *emanating* edges with respect to the circuit and the other two correspond to *incoming* edges, and those where we have two consecutive (opposite) edges one of which is incoming and the other one is emanating.

Certainly, there is an intuitive way of transforming Gauss diagrams into rotating circuits and vice versa (in the case of knots), see Fig. 7, 8. The second important question which arises here is whether there is an equivalence between the set of homotopy classes of looped interlacement graphs introduced by L. Traldi and L. Zulli[3] and the set of graph-knots introduced by ourselves.[1,2] The equivalence of these two sets was proved in[9,10] and, moreover, the homotopy class of the looped interlacement graph and the graph-knot both constructed from a given virtual knot diagram are related by this equivalence. This construction will be described in detail in Section 2. Moreover, one can consider an arbitrary Euler tour (which

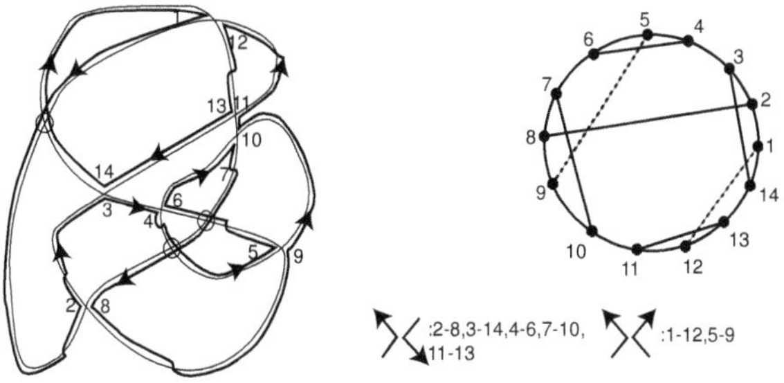

Fig. 7. Rotating Circuit Shown by a Thick Line; Chord Diagram

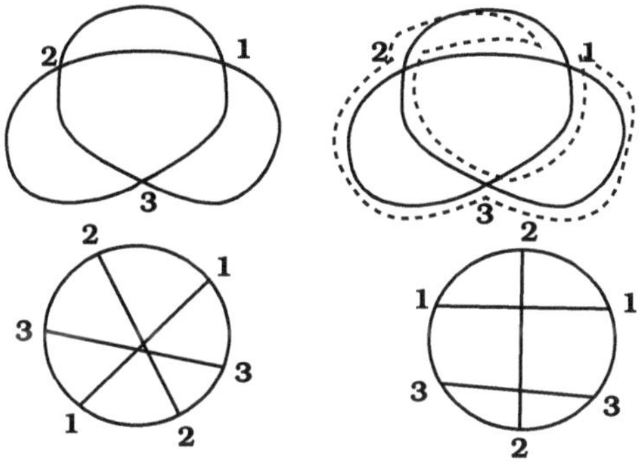

Fig. 8. A Gauss Circuit; A Rotating Circuit

was initiated by Traldi[6]). In this paper we do touch only on rotating circuits and transverse circuits. The way of translating "rotating circuits" into "transverse circuits" can be rewritten combinatorially and translated into the language of intersection graphs. In some sense it is obvious that *whenever an intersection graph is realisable in the sense of Gauss diagrams, the corresponding "rotating" intersection graph is realisable in the sense of rotating diagrams*, just because if one of these two graphs is realisable, the corresponding 4-valent graph can be just drawn on the plane (with virtual crossings) and the algorithm becomes a "real redrawing algorithm". The redrawing algorithm shows that the statement is true in the realisable case. In the general case the statement follows immediately from the theorems in Refs. 9 and 10. These theorems allow us to switch from Gauss diagrams to rotating diagrams and vice versa, whenever we are proving some non-realisability theorems. We shall answer the question of whether or not every graph is Reidemeister equivalent to the looped interlacement graph of a virtual knot diagram, and by using the equivalence we immediately get the answer to this question concerning graph-knots. We shall show that there are graphs which are not equivalent to intersection graphs of Gauss diagrams of knots by Reidemeister moves and it follows from the equivalence that the situation is the same for graph-links. The examples of "non-realisable" looped interlacement graphs first appeared in Refs. 11 and 12. So, the theory of graph-links is interesting for various reasons:

a) in some cases it exhibits purely combinatorial ways of extracting invariants for knots;

b) in some cases it produces heuristic approaches to new "knot theories";

c) it highlights some "graphical" effects which are hardly visible in usual or virtual knot theory.

We call a graph-link (resp., a homotopy class of looped interlacement graphs) *non-realisable* if it has no realisable representative. We conclude the introduction part by a couple of examples of non-realisable free graph-links (free homotopy classes of looped interlacement graphs). Here we do not indicate any crossing decoration because *any graph-link with this underlying graph is non-realisable.* The homotopy class of looped interlacement graphs generated by the looped graph depicted in Fig. 5 (left) gives us a non-realisable *free homotopy class of looped interlacement graphs.* Having a virtual link diagram, we may forget about over/under and right/left information and pay attention only to the 4-valent underlying graph with the structure of opposite edges. This leads to the notion of *free knots* and *free links* considered in Refs. 11 and 12. It turns out that some information about virtual knots can be caught just from the 4-valent graph, which proves non-triviality of many free knots and free links. The same trick works for graph-links and homotopy classes of looped interlacement graphs: however, here instead of the underlying 4-valent graph we consider the abstract graph which plays the role of the intersection graph of the *non-existing chord diagram.*

Certainly, the graph shown in Fig. 9 (left) is itself non-realisable (in the sense of looped interlacement graphs and Gauss diagrams), but what if we decorate its crossings in some way and then try to apply Reidemeister moves hoping to make it realisable. For some graph-links (homotopy classes of looped interlacement graphs) it is possible, see, e.g. Fig. 9. The homotopy class of looped interlacement graphs, generated by the looped graph shown in Fig. 9 (left), is realisable. Indeed, the second Reidemeister move translated into the language of Gauss diagrams is an addition/removal of two "parallel" chords. In the language of intersection graphs, chords correspond to vertices, and "parallel" chords correspond to vertices having the same set of adjacent vertices. So, the vertices A and A' in Fig. 9 have the same adjacencies, and the removal of these two vertices makes our graph realisable.

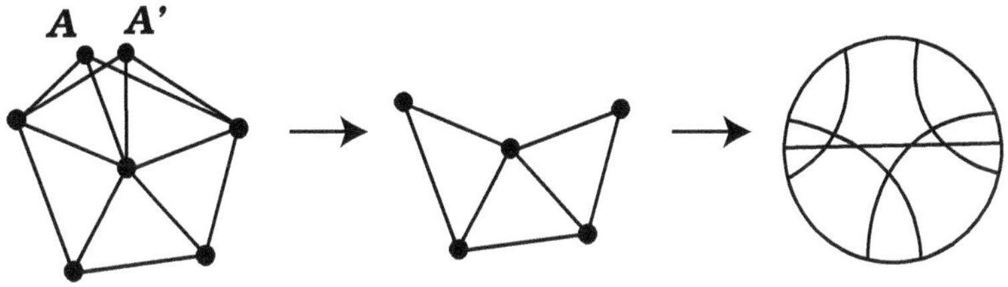

Fig. 9. A Non-Realisable Graph Representing a Trivial Homotopy Class

The problem of finding free graph-links and free homotopy classes of looped interlacement graphs having no representative realisable by a chord diagram is a problem similar to the problem of constructing a virtual knot which is not equivalent to any classical knot. Surprisingly, the solution in the case of free graph-links (free homotopy classes of looped interlacement graphs) can be achieved by using *parity considerations* (Theorem 3.4): all the vertices shown in Fig. 5 (left) are *odd* (each of them is adjacent to an odd number of other vertices) and there is no immediate way to contract any two of them by using a second Reidemeister move. This is indeed sufficient for a homotopy class of looped interlacement graphs to be non-realisable in a very strong sense: any graph of this homotopy class has *a spur* of the initial graph (we disregard the writhe number information), which is non-realisable, and, the homotopy class is, in turn, itself non-realisable. Also we have an example of the non-realisable homotopy class of looped interlacement graphs whose vertices are all even, see Fig. 10.

In fact, parity arguments allow one to prove very strong theorems in the realm of virtual knots, and we strongly recommend the reader to read the paper[13] in the present volume.

The present paper is organized as follows. We first give definitions of graph-links from two points of view: rotating circuits and Gauss diagrams, and describe their interactions.

In the third section we introduce parity and prove non-triviality results. In particular, parity arguments allow one to construct graph-valued invariants of graph-links.

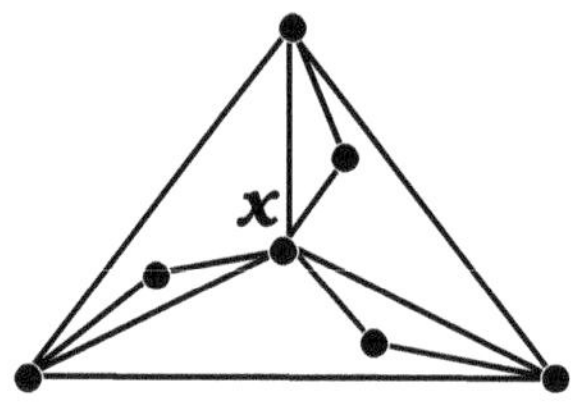

Fig. 10. An Even Non-Realisable Graph Representing a Non-Realisable Homotopy Class

In Section 4, we briefly describe the way of extending the Kauffman bracket and some other invariants. We also formulate some minimality results for graph-links.

2. Graph-Links and Looped Interlacement Graphs

2.1. *Chord diagrams and Framed 4-Graphs*

Throughout the paper all graphs are finite. Let G be a graph with the set of vertices $V(G)$ and the set of edges $E(G)$. We think of an edge as an equivalence class of the two half-edges. We say that a vertex $v \in V(G)$ has *degree* k if v is incident to k half-edges. A graph whose vertices have the same degree k is called k*-valent* or a k*-graph*. The free loop, i.e. the graph without vertices, is also considered as k-graph for any k.

Definition 2.1. A 4-graph is *framed* if for every vertex the four emanating half-edges are split into two pairs of (formally) opposite edges. The edges from one pair are called *opposite to each other.*

A *virtual diagram* is a framed 4-graph embedded into $\mathbb{R}^2$ where each crossing is either endowed with a classical crossing structure (with a choice for underpass and overpass specified) or just said to be virtual and marked by a circle. A *virtual link* is an equivalence class of virtual diagrams modulo generalised Reidemeister moves. The latter consist of usual Reidemeister moves referring to classical crossings and the *detour move* that replaces one arc containing only virtual intersections and self-intersections by another arc of such sort in any other place of the plane. A *projection* of a virtual diagram is a framed 4-graph obtained from the virtual diagram by considering classical crossings as vertices of the 4-graph and virtual crossings are just intersection points of images of different edges.

A virtual diagram is *connected* if its projection is connected. It is evident that any virtual link has a connected virtual diagram. The following statement is true: any two equivalent (in the class of all virtual diagrams) connected virtual diagrams are equivalent in the class of connected virtual diagrams.[1] Therefore, without loss of generality, *all virtual diagrams are assumed to be connected and contain at least one classical crossing.*[1,2]

A *chord diagram* is a cubic graph consisting of a selected cycle (the *core circle*) and several non-oriented edges (*chords*) connecting points on the circle in such a way that every point on the circle is incident to at most one chord. A chord diagram is *labeled* if every chord is endowed with a label (a, α), where $a \in \{0, 1\}$ is the *framing* of the chord, and $\alpha \in \{\pm\}$ is the *sign* of the chord. If no labels are indicated, we assume the chord diagram has all chords with label $(0, +)$. Two chords of a chord diagram are called *linked* if the ends of one chord lie in different connected components of the set obtained from the core circle by removing the end-points of the second chord.

Let D be a labeled chord diagram. We can construct a virtual link diagram $K(D)$ (up to *virtualisation*[13]) as follows. Let us immerse D in $\mathbb{R}^2$ by taking an embedding of the core circle and placing some chords inside the circle and the other ones outside the circle. After that we remove neighbourhoods of each of the chord ends and replace them by a pair of lines (connecting four points on the circle which are obtained after removing neighbourhoods) with a classical crossing if the chord is framed by 0 and a couple of lines with a classical crossing and a virtual crossing if the chord is framed by 1 (see Fig. 11) in the following way. The choice for underpass and overpass is specified as follows. A crossing can be smoothed in two ways: A : 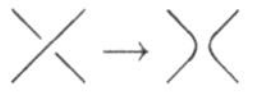 and B : 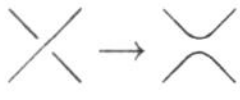 as in the Kauffman bracket polynomial; we require that the initial piece of the circle corresponds to the A-smoothing if the chord is positive and to the B-smoothing if it is negative.

Conversely, having a connected virtual diagram K, one can get a labeled chord diagram $D_C(K)$, see Fig. 7. Indeed, one takes a *circuit* C of K which is a map from S^1 to the projection of K. This map is bijective outside classical and virtual crossings, has exactly two preimages at each classical and virtual crossing, goes transversally at each virtual crossing and turns from an half-edge to an adjacent (non-opposite) half-edge at each classical crossing. Connecting the two preimages of a classical crossing by a chord we get a chord diagram, where the sign of the chord is $+$ (resp., $-$) if the circuit locally agrees with the A-smoothing (resp., the B-smoothing), and the framing of a chord is 0 (resp., 1) if two opposite half-edges have the opposite (resp., the same) orientations, see Fig. 12. It can be easily checked that this operation is indeed inverse to the operation of constructing a virtual

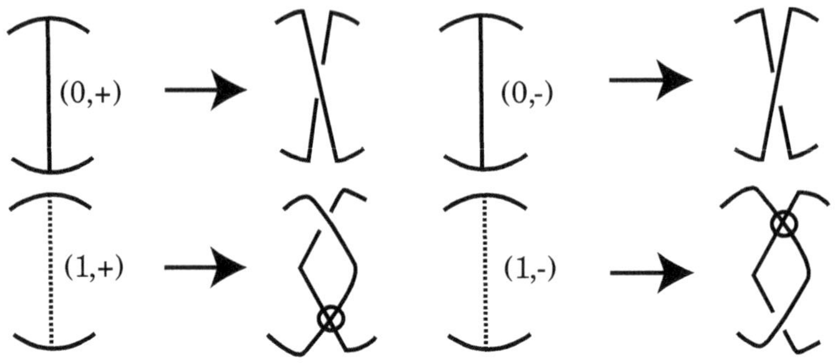

Fig. 11. Replacing a chord by a pair of lines

link out of a chord diagram: if we take a chord diagram D, and construct a virtual diagram $K(D)$ out of it, then for some circuit C the chord diagram $D_C(K(D))$ will coincide with D. The rule for setting classical crossings here agrees with the rule described above. This proves the following

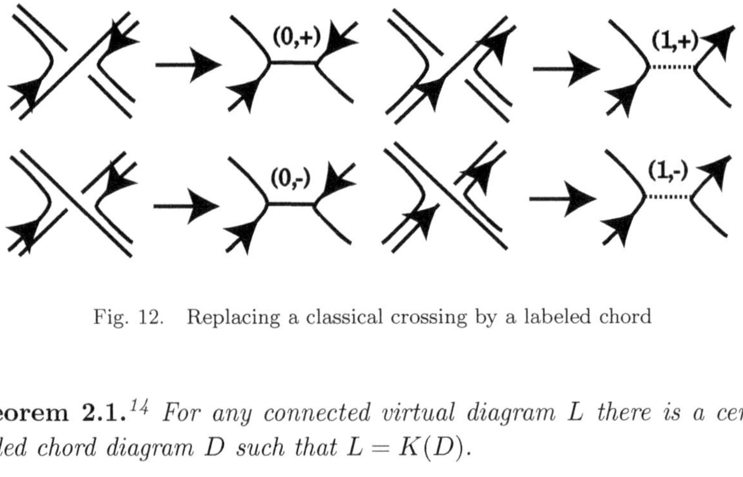

Fig. 12. Replacing a classical crossing by a labeled chord

Theorem 2.1.[14] *For any connected virtual diagram L there is a certain labeled chord diagram D such that $L = K(D)$.*

2.2. *Reidemeister Moves for Looped Interlacement Graphs and Graph-links*

Now we are describing moves on graphs obtained from virtual diagrams by using a rotating circuit[1,2] and the Gauss circuit.[3] These moves in both cases will correspond to the "real" Reidemeister moves when they are applied to graphs realisable by chord diagrams. Then we shall extend these moves to all graphs (not only to realisable ones). As a result we get two new objects, a *graph-link* and a *homotopy class of looped interlacement graphs*, in a way similar to the generalisation of classical knots to virtual knots: the passage from realisable Gauss diagrams (classical knots) to arbitrary chord diagrams leads to the concept of a virtual knot, and the passage from realisable (by means of chord diagrams) graphs to arbitrary graphs leads to the concept of two new objects, *graph-links* and *homotopy classes of looped interlacement graphs* (here 'looped' corresponds to the writhe number, if the writhe number is -1, then the corresponding vertex has a loop). To construct the first object we shall use simple labeled graphs, and for the second one we shall use (unlabeled) graphs without multiple edges, but loops are allowed.

Definition 2.2. A graph is *labeled* if every vertex v of it is endowed with a pair (a, α), where $a \in \{0, 1\}$ is the *framing* of v, and $\alpha \in \{\pm\}$ is the *sign* of v. Let D be a labeled chord diagram. The *labeled intersection graph*, cf.,[15,16] $G(D)$ of D is the labeled graph: 1) whose vertices are in one-to-one correspondence with chords of D, 2) the label of each vertex corresponding to a chord coincides with that of the chord, and 3) two vertices are connected by an edge if and only if the corresponding chords are linked. A simple graph H is called *realisable* if there is a chord diagram D such that $H = G(D)$.

The following lemma is evident.

Lemma 2.1. *A simple graph is realisable if and only if each of its connected components is realisable.*

Definition 2.3. Let G be a graph and let $v \in V(G)$. The set of all vertices adjacent to v is called the *neighbourhood of a vertex* v, and denoted by $N(v)$.

Let us define two operations on simple unlabeled graphs which are well known in graph theory.

Definition 2.4. (Local Complementation) Let G be a graph. The *local complementation* of G at $v \in V(G)$ is the operation which toggles adjacencies between $a, b \in N(v)$, $a \neq b$, and does not change the rest of G. Denote the graph obtained from G by the local complementation at a vertex v by $\mathrm{LC}(G; v)$.

(Pivot) Let G be a graph with distinct vertices u and v. The *pivoting operation* of a graph G at u and v is the operation which toggles adjacencies between x, y such that $x, y \notin \{u, v\}$, $x \in N(u)$, $y \in N(v)$ and either $x \notin N(v)$ or $y \notin N(u)$, and does not change the rest of G. Denote the graph obtained from G by the pivoting operation at vertices u and v by $\mathrm{piv}(G; u, v)$.

Example 2.1. In Fig. 13 the graphs G, $\mathrm{LC}(G; u)$ and $\mathrm{piv}(G; u, v)$ are depicted.

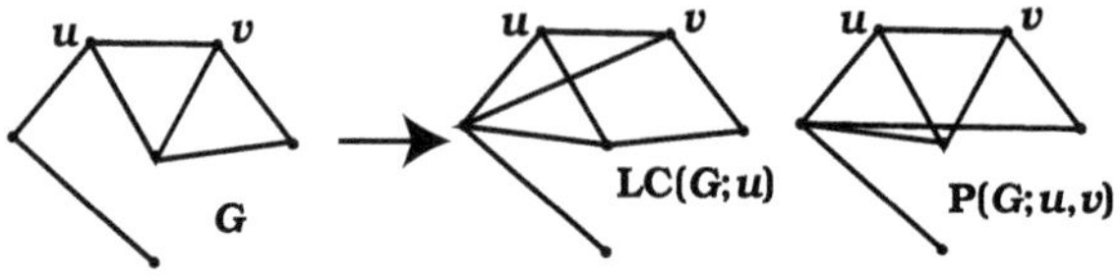

Fig. 13. Local Complementation and Pivot

The following lemma can be easily checked.

Lemma 2.2. *If u and v are adjacent, then there is an isomorphism*

$$\mathrm{piv}(G; u, v) \cong \mathrm{LC}(\mathrm{LC}(\mathrm{LC}(G; u); v); u).$$

Let us define graph-moves, i.e. moves on labeled graphs, by considering intersection graphs of labeled chord diagrams constructed by using *a rotating circuit.* These moves correspond to the Reidemeister moves on virtual diagrams. As a result we obtain a new object — an equivalence class of labeled graphs under formal moves. These moves were defined in Refs. 1 and 2.

Definition 2.5. $\Omega_g 1$. The first Reidemeister graph-move is an addition/removal of an isolated vertex labeled $(0, \alpha)$, $\alpha \in \{\pm\}$.

$\Omega_g 2$. The second Reidemeister graph-move is an addition/removal of two non-adjacent (resp., adjacent) vertices labeled $(0, \pm\alpha)$ (resp., $(1, \pm\alpha)$) and the same adjacencies with other vertices.

$\Omega_g 3$. The third Reidemeister graph-move is defined as follows. Let u, v, w be three vertices of G all having label $(0, -)$ so that u is adjacent only to v and w in G, and v and w are non-adjacent. Then we only change the adjacency of u with the vertices v, w and $t \in N(v)\backslash N(w) \bigcup N(w)\backslash N(v)$ (for other pairs of vertices we do not change their adjacency). In addition, we switch the signs of v and w to $+$. The inverse operation is also called the third Reidemeister graph-move.

$\Omega_g 4$. The fourth graph-move for G is defined as follows. We take two adjacent vertices u and v labeled $(0, \alpha)$ and $(0, \beta)$ respectively. Replace G with $\mathrm{piv}(G; u, v)$ and change signs of u and v so that the sign of u becomes $-\beta$ and the sign of v becomes $-\alpha$.

$\Omega_g 4'$. In this fourth graph-move we take a vertex v with the label $(1, \alpha)$. Replace G with $\mathrm{LC}(G; v)$ and change the sign of v and the framing for each $u \in N(v)$.

The comparison of the graph-moves with the Reidemeister moves yields the following theorem.

Theorem 2.2. *Let G_1 and G_2 be two labeled intersection graphs corresponding to virtual diagrams K_1 and K_2, respectively. If K_1 and K_2 are equivalent in the class of connected diagrams, then G_1 and G_2 are obtained from one another by a sequence of $\Omega_g 1 - \Omega_g 4'$.*

Definition 2.6. A *graph-link* is an equivalence class of simple labeled graphs modulo $\Omega_g 1 - \Omega_g 4'$ graph-moves.

Remark 2.1. Having a chord diagram, we can construct an *atom*[13] corresponding to the chord diagram, see Refs. 1 and 2. In fact, chord diagrams in the sense of rotating circuits with all chords having framing 0 encode all orientable atoms, see Subsection 5.3 of Ref. 17. Chord diagrams with all positive chords encode all atoms with one white cell: this white cell corresponds to the A-state of the virtual diagram, and chords indicate how this cell approaches itself in neighbourhoods of crossings (atom vertices).

For a graph-link having representatives with orientable *atoms*,[13] there are two formally different equivalence relations. The first relation is described in Ref. 1 (which includes only diagrams with orientable atoms) and the last one defines graph-links. We do not know whether these equivalence relations coincide for graph-links (they do for graph-knots). Nevertheless, we use the same term 'graph-link' for the object introduced in this paper.

Remark 2.2. Let us consider a simple realisable labeled graph and represent this graph as an intersection graph of a chord diagram D. To construct a virtual diagram $K(D)$, we have to restore the structure of opposite edges at each vertex. The first components of labels of vertices (which are called *framings* of vertices) are responsible for the structure of opposite edges. Therefore, we shall consider framed graphs of two types. The first type is framed 4-graphs defined in Definition 2.1. The second one is *free framed graphs* which are equivalence classes of simple labeled graphs with labels having only framings modulo $\Omega_g 4$ and $\Omega_g 4'$ graph-moves up to signs of labels (we disregard the sign of each vertex).

Definition 2.7. A *free graph-link* is an equivalence class of free framed graphs modulo $\Omega_g 1 - \Omega_g 3$ graph-moves up to signs of labels.

Definition 2.8. We call a free framed graph *realisable* if any of its representative is realisable by a chord diagram.

Remark 2.3. It is not difficult to show that free framed graph is realisable if and only if there exists a realisable representative of it, we just redraw the picture.

Let $D_G(K)$ be the Gauss diagram of a virtual diagram K. Let us construct the graph obtained from the intersection graph of $D_G(K)$ by adding loops to vertices corresponding to chords with the negative writhe number.[3] We refer to this graph as the *looped interlacement graph* or the *looped graph.* Let us describe the moves on looped graphs. These moves are similar to the moves for graph-links and also correspond to the Reidemeister moves on virtual diagrams.

Definition 2.9. The first Reidemeister move for looped interlacement graphs is an addition/removal of an isolated looped or unlooped vertex.

The second Reidemeister move for looped interlacement graphs is an addition/removal of two vertices having the same adjacencies with other vertices and, moreover, one of which is looped and the other one is unlooped.

The third Reidemeister move for looped interlacement graphs is defined as follows. Let u, v, w be three vertices such that v is looped, w is unlooped, v and w are adjacent, u is adjacent to neither v nor w, and every vertex $x \notin \{u, v, w\}$ is adjacent to either 0 or precisely two of u, v, w. Then we only toggle adjacencies between three vertices u, v, w. The inverse operation is also called the third Reidemeister move.

The two third Reidemeister moves do not exhaust all the possibilities for representing the third Reidemeister move on Gauss diagrams.[3] It can be shown that all the other versions of the third Reidemeister move are combinations of the second and third Reidemeister moves, see Ref. 18 for details.

Definition 2.10. We call an equivalence class of graphs (without multiple edges, but loops are allowed) modulo the three moves listed in Definition 2.9 a *homotopy class* of looped interlacement graphs. A *free homotopy class* is an equivalence class of simple graphs modulo the Reidemeister moves for looped interlacement graphs up to loops, i.e. we forget about loops.

Remark 2.4. The equivalence relation from Definition 2.10 is called *Reidemeister equivalence* in Ref. 3 and different from classical relation of link homotopy.

Looped interlacement graphs encode only knot diagrams but graph-links can encode virtual diagrams with any number of components. The approach using a rotating circuit has an advantage in this sense. In Ref. 6 L. Traldi introduced the notion of a marked graph by considering any directed Euler tour.

Definition 2.11. We call a (free) graph-link (resp., a (free) homotopy class of looped interlacement graphs) *realisable* if it has a realisable representative.

2.3. *Looped Interlacement Graphs and Graph-Links*

Let G be a labeled graph on $V(G) = \{v_1, \ldots, v_n\}$, and let $A(G)$ be the *adjacency matrix* of G over $\mathbb{Z}_2$ defined as follows: a_{ii} is equal to the framing of v_i, $a_{ij} = 1$, $i \neq j$, if and only if v_i is adjacent to v_j and $a_{ij} = 0$ otherwise. If G and G' represent the same graph-link, then $\operatorname{corank}_{\mathbb{Z}_2}(A(G) + E) = \operatorname{corank}_{\mathbb{Z}_2}(A(G') + E)$, where E is the identity matrix.

Definition 2.12. The *number of components* in a graph-link $\mathfrak{F}$ is $\operatorname{corank}_{\mathbb{Z}_2}(A(G) + E) + 1$, where G is a representative of $\mathfrak{F}$. A graph-link $\mathfrak{F}$ with $\operatorname{corank}_{\mathbb{Z}_2}(A(G) + E) = 0$ for any representative G of $\mathfrak{F}$ is called a *graph-knot*.

Assume $\operatorname{corank}_{\mathbb{Z}_2}(A(G) + E) = 0$, and set $B_i(G) = A(G) + E + E_{ii}$ for each vertex $v_i \in V(G)$, where E_{ii} is the matrix with the only one non-zero element equal to 1 in the i-th column and i-th. The *writhe number* w_i of G (with $\operatorname{corank}_{\mathbb{Z}_2}(A(G) + E) = 0$) at v_i is $w_i = (-1)^{\operatorname{corank}_{\mathbb{Z}_2} B_i(G)} \operatorname{sign} v_i$ and the *writhe number* of G is $w(G) = \sum\limits_{i=1}^{n} w_i$.

If G is realisable by a chord diagram and, therefore, by a virtual diagram, then w_i is the "real" writhe number of the crossing corresponding to v_i.

Definition 2.13. We say that an $n \times n$ matrix $A = (a_{ij})$ *coincides with an $n \times n$ matrix $B = (b_{kl})$ up to diagonal elements* if $a_{ij} = b_{ij}$, $i \neq j$.

Lemma 2.3. [9] *Let A be a symmetric matrix over $\mathbb{Z}_2$. Then there exists a symmetric matrix $\widetilde{A}$ with* $\det \widetilde{A} = 1$ *coinciding with A up to diagonal elements.*

Let G be a representative of a graph-knot $\mathfrak{F}$. Consider the simple graph H having the adjacency matrix which coincides with $(A(G) + E)^{-1}$ up to diagonal elements. Construct the graph $L(G)$ from H by just adding a loop to any vertex of H corresponding to a vertex of G with negative writhe number. Let us define the map χ from the set of graph-knots to the set of homotopy classes of looped interlacement graphs by $\chi(\mathfrak{F}) = \mathfrak{L}$, where $\mathfrak{L}$ is the homotopy class of $L(G)$. It turns out that the map χ is well defined.[10] The inverse map is defined as follows.[10] Let $\mathfrak{L}$ be the homotopy class of L. By using Lemma 2.3 we can construct a symmetric matrix $A = (a_{ij})$ over $\mathbb{Z}_2$ coinciding with the adjacency matrix of L up to diagonal elements and having $\det A = 1$. Let $G(L)$ be the labeled simple graph having the matrix $(A^{-1} + E)$ as its adjacency matrix. Therefore, the first component of the vertex label is equal to the corresponding diagonal element of $(A^{-1} + E)$, and the second component of the label of the vertex with the number i is $w_i(1 - 2a_{ii})$, where $w_i = 1$ if the vertex of L with the number i does not have a loop, and $w_i = -1$ otherwise.[9,10] We have $\chi^{-1}(\mathfrak{L}) = \mathfrak{F}$, where $G(L)$ is a representative of $\mathfrak{F}$. Therefore, we get

Theorem 2.3.[9,10] *There is a one-to-one correspondence between the set of all looped graphs and the set of all equivalence classes of labeled graphs G with $\operatorname{corank}_{\mathbb{Z}_2}(A(G) + E) = 0$ modulo fourth graph-moves. This correspondence generates an equivalence between the set of all homotopy classes of looped interlacement graphs and the set of all graph-knots. Moreover, the graph-knot and the homotopy class of looped interlacement graphs both constructed from a given virtual knot diagram are related by this correspondence.*

2.4. *The Circuit-Nullity Formula and its Corollaries*

Assume we are given a framed chord diagram (each chord of it is endowed only with framing). Define the *surgery over the set of chords as follows.* For every chord having framing 0 (resp., 1), we draw a parallel (resp., intersecting) chord near it and remove the arc of the circle between adjacent ends of the chords as in Fig. 14. By a small perturbation, the picture in $\mathbb{R}^2$ is transformed into a one-manifold in $\mathbb{R}^3$. This manifold $m(D)$ is the result of surgery, see Fig. 14. Surprisingly, the number of the connected components of $m(D)$ can be determined from the intersection graph.

Theorem 2.4.[19–21] *Let D be a labeled chord diagram, and let G be its labeled intersection graph. Then the number of connected components of $m(D)$ equals $\operatorname{corank}_{\mathbb{Z}_2} A(G) + 1$, where $A(G)$ is the adjacency matrix of G.*

Remark 2.5. Theorem 2.4 allows us to define 'the number of circles' after a "surgery" for a graph even when the given graph is not an intersection graph of any chord diagram. We refer to this theorem as the circuit-nullity

Fig. 14. A Surgery of the Circuit Along Chords and the Manifold $m(D)$

formula (in our previous works we referred to this theorem as Soboleva's theorem).

Throughout the rest of the paper, we shall use the following *main principle*: *Assume there is an equality concerning numbers of circles in some states of chord diagrams, which can be formulated in terms of the intersection graph. Then the corresponding equality usually holds even for non-realisable graphs.*

The reason for this principle to hold is the following: every time we have a picture where some two numbers of circles are equal to each other (or differ by a constant), this can be expressed in terms of the corresponding adjacency matrices, and the proof does not generally depend on the behaviour of the matrix outside of the crossings in question. This means that the equality holds true for generic matrices, thus, it works for general intersection graphs. This principle has lots of consequences. We shall demonstrate it for two examples.

The first example comes from Definition 2.12. It shows that the number of components defined for non-realisable graphs by the same formula as for realisable ones does not change under the Reidemeister moves and the writhe number of a graph does not change under the second and third Reidemeister graph-moves and changes by ± 1 under the first Reidemeister graph-moves.

The second example is as follows. Assume we have a framed 4-graph K with a vertex v and we would like to know whether this vertex belongs to one component or it belongs to different components of the corresponding graph-link. Then we may take the two *smoothings*, see ahead, K_a, K_b of the K at v and see how many components we get. If v belongs to two branches of the same component of K, then the number of components of one of K_a, K_b is equal to that of K, and the number of components of the other one is equal to that of K plus one. If v belongs to two different components of K, then the number of components of each of K_a, K_b is that of K minus one. Now, turning to graph-links, by taking appropriate matrix ranks, we may see whether each vertex belongs to one component or to two different components of the graph-link. This method is used for proving the invariance of the Kauffman bracket polynomial.

2.5. *Smoothing Operations and Turaev's* Δ

Let us define smoothing operations. Note that the idea of a "smoothing" of a 4-graph has been studied by Nash-Williams, Bouchet and others, who call it a *detachment*, see for instance.[22,23]

By a *smoothing* of a framed 4-graph at a vertex v we mean any of the two framed 4-graphs obtained by removing v and repasting the edges, see Ref. 13. Generally, a *smoothing* of a framed 4-graph in a collection of vertices is the framed 4-graph obtained by a sequence of smoothings.

Now, we can mimic this definition for the case of free graph-links. Let $\mathcal{G}$ be a free framed graph, i.e. an equivalence class of labeled simple graphs, and let v be a vertex of $\mathcal{G}$. Let us consider two cases. In the first case there exists a representative H of $\mathcal{G}$ for which v has either framing 1 or the degree greater than 0. It is not difficult to see that v has the same property for each representative of $\mathcal{G}$, and there are two representatives H_1 and H_2 of $\mathcal{G}$ which differ from each other by $\Omega_g 4$ or $\Omega_g 4'$ at v. By a *smoothing* of a free framed graph $\mathcal{G}$ at a vertex v we mean any of the two free framed graphs having the representatives $H_1 \backslash \{v\}$ and $H_2 \backslash \{v\}$, respectively. In the realisable case this means that a framed 4-graph has a rotating circuit having any of the two possible connection types at the vertex. Then the smoothing at the vertex of the framed 4-graph corresponds to a removal of the chord of the hord diagram (vertex of the intersection graph) corresponding to the vertex. If a smoothing of a framed 4-graph leads to a disconnected graph, this may be repaired by taking another representative of the same graph-link. In the language of $\mathcal{G}$ it looks as follows: v has framing 0 and is isolated for each representative of $\mathcal{G}$. Let H be a representative of $\mathcal{G}$. Let us construct the new graph H' obtained from H by adding a new vertex u with framing 0 to H which is adjacent only to v, see Fig. 15 for the case of realisable graphs (the dashed line is a rotating circuit). By a *smoothing* of a free framed graph $\mathcal{G}$ at a vertex v we mean any of the two free framed graphs having the representatives $H \setminus \{v\}$ and H', respectively. Generally, a *smoothing* of a free framed graph in a collection of vertices is the free framed graph obtained by a sequence of smoothings.

Proposition 2.1. *Assume $\mathcal{G}$, $\mathcal{G}'$ are free framed graphs and $\mathcal{G}$ can be obtained from $\mathcal{G}'$ as a result of smoothings at some vertices. Then if $\mathcal{G}'$ is realisable by a chord diagram, then so is $\mathcal{G}$.*

Fig. 15. One of the two smoothings at an isolated vertex

The proof is obvious: if $\mathcal{G}'$ is realisable and $\mathcal{G}$ is obtained from $\mathcal{G}'$ by applying a fourth graph-move and/or deleting a vertex, then one can draw the corresponding framed 4-graph, and take the corresponding resmoothing which will lead us to the chord diagram for $\mathcal{G}$. In the other case, if $\mathcal{G}'$ is realisable, then the realisability of $\mathcal{G}$ follows from Lemma 2.1.

Let $i > 1$ be a natural number. Define the set $\mathbb{Z}_2\mathcal{G}_i$ to be the $\mathbb{Z}_2$-linear space generated by the set $\mathcal{G}_i$ of free framed graphs having i components (i.e. for each $\mathcal{G} \in \mathcal{G}_i$ we have $\operatorname{corank}_{\mathbb{Z}_2}(A(\mathcal{G}) + E) = i - 1$) modulo the following relations:

1) the second graph-moves;

2) $\mathcal{G} = 0$, if $\mathcal{G}$ has two vertices with framing 0 which are adjacent only to each other.

The meaning of 2) is an addition of a free component to a graph-link.

For $i = 1$, we define $\mathbb{Z}_2\mathcal{G}_1$ analogously with respect to equivalence 1) and not 2).

Let us define the map $\Delta\colon \mathbb{Z}_2\mathcal{G}_1 \to \mathbb{Z}_2\mathcal{G}_2$, cf.[13] Given a free framed graph $\mathcal{G}$ with $\operatorname{corank}_{\mathbb{Z}_2}(A(\mathcal{G}) + E) = 0$, we shall construct an element $\Delta(\mathcal{G})$ from $\mathbb{Z}_2\mathcal{G}_2$ as follows. For each vertex v of $\mathcal{G}$, there are two ways of smoothing it. One way gives a graph from $\mathbb{Z}_2\mathcal{G}_1$, and the other smoothing gives a free framed graph $\mathcal{G}_v$ from $\mathbb{Z}_2\mathcal{G}_2$. We take $\mathcal{G}_v$ and set

$$\Delta(\mathcal{G}) = \sum_v \mathcal{G}_v \in \mathbb{Z}_2\mathcal{G}_2.$$

Theorem 2.5. *Δ is a well-defined mapping from $\mathbb{Z}_2\mathcal{G}_1$ to $\mathbb{Z}_2\mathcal{G}_2$.*

By using the main principle from Subsection 2.4 we can define whether a vertex belongs to "one component" or "different components" of a free framed graph. Namely, we call a vertex v_i of a free framed graph $\mathcal{G}$ *oriented* if $\operatorname{corank}_{\mathbb{Z}_2}(A(\mathcal{G}) + E) \leqslant \operatorname{corank}_{\mathbb{Z}_2}(B_i(\mathcal{G}))$. It is not difficult to show that $\operatorname{corank}_{\mathbb{Z}_2} B_i(\mathcal{G}) \neq \operatorname{corank}_{\mathbb{Z}_2} B(\mathcal{G} \setminus \{v_i\})$ if v_i is oriented.

By using the notion of an oriented vertex we can define the map Δ^i (iteration) by considering smoothings at oriented vertices and taking that smoothing which has more components than other in each step.

Corollary 2.1. *Δ^i is a well-defined mapping from $\mathbb{Z}_2\mathcal{G}_1$ to $\mathbb{Z}_2\mathcal{G}_{i+1}$.*

3. Parity, Minimality and Non-Triviality Examples

In the present section we consider parity for free graph-knots and free graph-links in the spirit of.[11,12] Since we have constructed the equivalence between the set of graph-knots and the set of homotopy classes of looped interlacement graphs it is sufficient to construct a parity for free homotopy classes of looped interlacement graphs and for graph-links with more than one components.

Consider the category of free homotopy classes of looped interlacement graphs.

Definition 3.1. For every looped interlacement graph L we call a vertex v *even* if v is adjacent to an even number of vertices distinct from v; otherwise we call v *odd*.

The parity axioms for framed 4-graphs described in Section 3 of the paper[13] can be straightforwardly generalised for the case of looped graphs (and, consequently, for graph-knots).

Theorem 3.1. *The parity defined on homotopy classes of looped interlacement graphs (resp., free homotopy classes) by using even and odd vertices satisfies the parity axioms.*

The proof[13] evidently follows from the consideration of the Reidemeister moves for looped interlacement graphs.

By using this parity we can define the map Δ^i_{odd} where the sum is taken over all odd oriented vertices or Δ^i_{even} where the sum is taken over all even oriented vertices. We have to define the notion of *even* and *odd* vertex for free framed graphs with many components.

We call a vertex v of $\mathcal{G}$ with one component *even* (resp., *odd*) if the vertex corresponding to v of the looped interlacement graph $\chi(\mathcal{G})$ is even (resp., odd). Let us consider the free framed graph $\mathcal{G}_{v_1,\dots,v_{k-1}}$ with k components which is obtained from $\mathcal{G}$ by smoothing $\mathcal{G}$ consequently at $v_1,\dots,v_{k-1}$; v_1 is an oriented vertex in $\mathcal{G}$ and v_i, $i=2,\dots,k-1$, is an oriented vertex in $\mathcal{G}_{v_1,\dots,v_{i-1}}$. An oriented vertex u of $\mathcal{G}_{v_1,\dots,v_{k-1}}$ is *even with respect to the smoothing at* $v_1,\dots,v_{k-1}$ (resp., *odd with respect to the smoothing at* $v_1,\dots,v_{k-1}$) if the number of oriented vertices in $\mathcal{G}_{v_1,\dots,v_{k-1}}$ which are incident to u in $\chi(\mathcal{G})$ is even (resp., odd).

Remark 3.1. We have defined even vertices only for those free framed graphs with many components which originate from free framed graphs with 1 component. This is sufficient to define the iteration $\Delta^i_{odd(even)}$.

Proposition 3.1. *Δ^i_{odd} is a well-defined mapping from $\mathbb{Z}_2\mathcal{G}_1$ to $\mathbb{Z}_2\mathcal{G}_{i+1}$.*

Let us consider another parity for graph-links with two components.

Definition 3.2. For a graph representing a two-component graph-link we call a vertex of it *even* if it is *oriented* and *odd* otherwise.

Theorem 3.2. *The parity defined above satisfies the parity axioms.*

The proof again follows from the definition of parity.

Now, having these parities in hand, we can define the brackets $[\cdot]$ for graph-knots and $\{\cdot\}$ for graph-links analogously to the case of Ref. 13. For a graph G representing a free graph-knot (resp., a graph H representing a free two-component graph-link), consider the following sums

$$[G] = \sum_{s\ even.,1\ comp} G_s \in \mathbb{Z}_2\mathcal{G}_1, \quad \text{and} \quad \{H\} = \sum_{s\ even.\ non-trivial} H_s \in \mathbb{Z}_2\mathcal{G}_2,$$

where the sums are taken over all smoothings at all *even* vertices (with respect to the definitions above), and only those summands are taken into account where $\operatorname{corank}_{\mathbb{Z}_2}(A(G_s)+E)=0$. In the second formula, H_s is not equivalent to any simple graph having two vertices with framing 0 which are adjacent only to each other. Thus, if G has k even vertices, then $[G]$ will contain at most 2^k summands, and if all vertices of G are odd, then we shall have exactly one summand, the graph G itself. The same is true for a graph H and $\{H\}$.

Now, we are ready to formulate the main theorems of this section:

Theorem 3.3. *If G and G' represent the same free graph-knot, then the following equality holds: $[G]=[G']$. If H and H' represent the same free graph-link with two components, then $\{H\}=\{H'\}$.*

The proof of Theorem 3.3 verbally reproduces the proof of the main theorem from Ref. 13 according to the *main principle* or, maybe, a slight modification of it.

Definition 3.3. We call a labeled graph G (resp., a looped graph L) *minimal* if there is no representative of the graph-link corresponding to G (resp., the homotopy class of L) having strictly smaller number of vertices than G (resp., L) has.

Theorem 3.4 (cf. Theorems 5.1, 5.2 from Ref. 13). *Let G (resp., H) be a simple labeled graph representing a free graph-knot (resp., a two-component graph-link) with all odd vertices in the sense of Definition 3.1 (resp., Definition 3.2), such that no decreasing second Reidemeister move is applicable to G (resp., H). Then there is no simple graph equivalent to G (resp., H) with strictly smaller number of vertices.*

As a consequence of this theorem we may deduce the following

Corollary 3.1. *The graph G (shown in Fig. 5 left) representing a free graph-knot is minimal; in particular, the corresponding free graph-knot is non-trivial and not realisable.*

The first claim of the corollary is trivial: we just check the conditions of the theorem and see that G is minimal; to see that the second claim indeed holds, we shall need to go through the proof of Theorem 3.3, where we see that any representative G' of the graph-knot $\mathfrak{F}$ has G as a *smoothing*, that is, G *lies inside* each representative G' of the same graph-knot, and if G is not realisable, then so is G' according to Proposition 2.1. Analogously, one sees that the free two-component graph-link $\mathfrak{H}$ with the representative H shown in Fig. 16 (left part) has no realisable representative because $\{H\} = H$. Note that H in Fig. 16 is equivalent to the Bouchet graph shown in the right part of the picture by $\Omega_g 4$ graph-move; so they represent the same two-component free graph-link. Let us consider one more example.

Proposition 3.2. *The looped graph K represented by the "Gauss diagram" shown in Fig. 10 is minimal and non-realisable.*

The proof consists of the following steps very similar to the example from.[13]

First, note that $\Delta(K)$ consists of 7 summands $L + \sum_i L_i$, where only one summand (corresponding to the vertex x) represents 2-component free graph-link and all its vertices are *odd*; for each of the remaining summands L_i, there is at least one even vertex. This is, indeed, very easy: the graph K has only one vertex x which is adjacent to all the remaining vertices, then the argument just repeats the argument from Ref. 13.

Now, the 2-component free graph-link generated by L has a rotating circuit diagram shown in Fig. 16; all framings of the vertices are 0. To see it, one should consecutively perform the following operations for K: first, we "smooth" its crossing x; it can be done only at the expense of changing our circuit to the rotating one at some vertex different from x (a Gauss

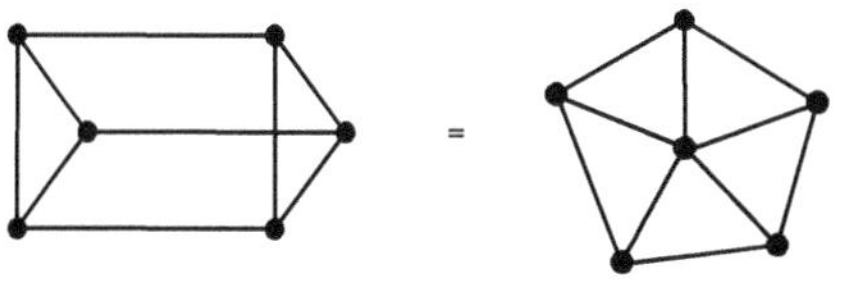

Fig. 16. The Two-Component Free Graph-Link

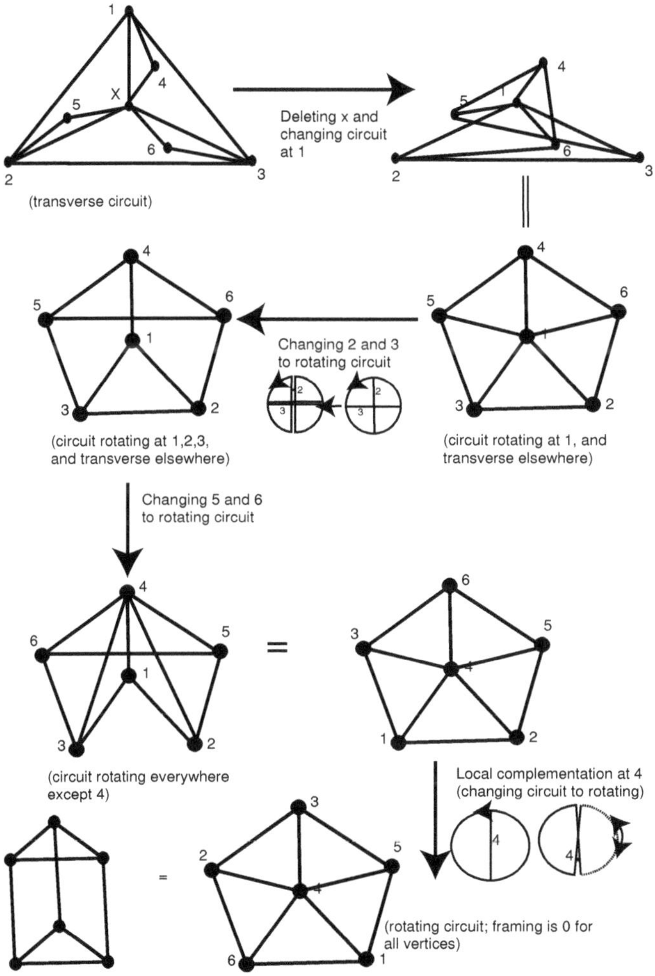

Fig. 17. The Two-Component Free Graph-Link

circuit can not represent a link); after that, we have to make the circuit rotating at all other vertices. All these steps are shown in Fig. 17.

Now, consider the bracket $\{\Delta(K)\} = L + \sum_i \{L_i\}$. Note that all graph-links generated by summands $\{L_i\}$ have representatives with strictly less than 6 vertices since each of L_i has at least one even vertex; on the other hand, the graph-link generated by L has no representative with less than 6 crossings; so, this element L is not canceled in the sum. Since it is not realisable, the free framed knot K is not realisable either.

4. Kauffman's Bracket Generalisation and Other Invariants. Minimality Theorems

We have already considered a minimality theorem, Theorem 3.4. In this section we present minimality theorems which use a generalisation of Kauffman's bracket.[24] Establishing minimal crossing number of a certain link is one of the important problems in classical knot theory.

Definition 4.1. The difference between the leading degree and the lowest degree of non-zero terms of a polynomial $P(x)$ is called the *span* of $P(x)$ and is denoted by $\operatorname{span} P(x)$. A classical link diagram is called *alternating* if while passing along every component of it we alternate undercrossings and overcrossings.

From the 'atomic' point of view, alternating link diagrams are those having atom genus (Turaev genus[25]) zero (more precisely, diagram has genus zero if it is a connected sum of several alternating diagrams).

Definition 4.2. A virtual link diagram D is called *split* if there is a vertex X of the corresponding atom (M, Γ) such that $\Gamma \backslash X$ is disconnected.

The main reason of these minimality theorems and further crossing estimates come from the following theorem which generalises the well-known Kauffman-Murasugi-Thistlethwaite Theorem:

Theorem 4.1. [14,26,27] *For a non-split virtual link diagram K on n crossings we have* $\operatorname{span}\langle K\rangle \leqslant 4n - 4g$, *where g is the genus of the corresponding atom. Here, $\langle K\rangle$ is the Kauffman bracket of K.*

All the minimality theorems for the case of graph-links in this section rely on the generalisation of the Kauffman bracket polynomial for graph-links. Now, let us generalise the notions defined above for the case of graph-links and define the Kauffman bracket polynomial for a labeled graph G.

Let $s \subset V(G)$ be a subset of the set $V(G)$ of vertices of G. Set $G(s)$ to be the induced subgraph of the graph G with the set of vertices $V(G(s)) = s$ and the set of edges $E(G(s))$ such that $\{u, v\} \in E(G(s))$, where $u, v \in s$, if and only if $\{u, v\} \in E(G)$.

Definition 4.3. We call a subset of $V(G)$ a *state* of G. The *A-state* is the state consisting of all the vertices of G labeled $(a, -)$, $a \in \{0, 1\}$, and no vertex labeled $(b, +)$, $b \in \{0, 1\}$. Analogously, the *B-state* is the state consisting of all vertices of G labeled $(b, +)$ and no vertex labeled $(a, -)$.

The *Kauffman bracket polynomial* of G is

$$\langle G \rangle (a) = \sum_s a^{\alpha(s) - \beta(s)} (-a^2 - a^{-2})^{\operatorname{corank} A(G(s))},$$

where the sum is taken over all states s of G, $\alpha(s)$ is equal to the sum of the vertices labeled $(a, -)$ from s and the vertices labeled $(b, +)$ from $V(G) \setminus s$, $\beta(s) = |V(G)| - \alpha(s)$.

Theorem 4.2. *The Kauffman bracket polynomial of a labeled graph is invariant under $\Omega_g 2 - \Omega_g 4'$ and gets multiplied by $(-a^{\pm 3})$ under $\Omega_g 1$.*

Definition 4.4. For a labeled graph G the *atom genus* (*Turaev genus*) is $1 - (k + l - n)/2$, where k and l are the numbers of circles in the A-state s_1 and the B-state s_2 of G, respectively, i.e. $k = \operatorname{corank}_{\mathbb{Z}_2} A(G(s_1)) + 1$ and $l = \operatorname{corank}_{\mathbb{Z}_2} A(G(s_2)) + 1$.

Note that this number agrees with the atom genus in the usual case: we just use $\chi = 2 - 2g$, where χ is the Euler characteristic, and count χ by using the number of crossings n, number of edges $2n$ and the number of 2-cells (A-state circles and B-state circles).

Definition 4.5. A labeled graph G on n vertices is *alternating* if its atom genus is equal to 0. G is *non-split* if it has no isolated vertices.

Theorem 4.3. *An alternating non-split labeled graph is minimal.*

Example. Consider the graph BW_3, see Fig. 5 (center), consisting of the 7 vertices with the following incidences: i is connected to j iff $i - j \equiv \pm 1 \pmod 6$, $i, j = 1, \ldots, 6$, and 7 is connected to $2, 4, 6$. Label all even vertices by $(0, +)$, and label all odd vertices by $(0, -)$. BW_3 is alternating. By Theorem 4.3, BW_3 is minimal. We conjecture that the graph-link represented by BW_3 is non-realisable. At least we know that it has no representative with seven or fewer crossings.

We conclude the present paper with one significant question which seems to be very interesting for us: is there a graph-link having two representatives realisable by chord diagrams, which are equivalent in the graph-link only by means of non-realisable graphs?

Acknowledgments

The authors are grateful to L. H. Kauffman, V. A. Vassiliev, and A. T. Fomenko for their interest in this work. The first author acknowledges the support of the program "Leading Scientific Schools" (project no. 3224.2010.1), the Russian Foundation for Basic Research (project no. 10-01-00748-a), the program "Development of the Scientific Potential of High School" (project no. 2.1.1.3704), the Federal Agency for Education NK-421P/108, Ministry of Education and Science of the Russian Federation 02.740.11.5213 and 14.740.11.0794. The second author acknowledges the support of the program "Leading Scientific Schools" (project no. 3224.2010.1), Ministry of Education and Science of the Russian Federation 14.740.11.0794, Analytic departmental task program "Development of the high school scientific potential".

References

1. D. P. Ilyutko, V. O. Manturov, Introduction to graph-link theory, *J. Knot Theory Ramifications* **18**:6 (2009), pp. 791-823.
2. D. P. Ilyutko, V. O. Manturov, Graph-links, *Doklady Mathematics* **80**:2 (2009), pp. 739–742 (Original Russian Text in *Doklady Akademii Nauk* **428**:5 (2009), pp. 591-594).
3. L. Traldi, L. Zulli, A bracket polynomial for graphs, arXiv:math.GT/0808.3392.
4. L. H. Kauffman, Virtual knot theory, *Eur. J. Combinatorics* **20**:7 (1999), pp. 662–690.
5. A. Bouchet, Circle graph obstructions, *J. Combinatorial Theory B* **60** (1994), pp. 107–144.
6. L. Traldi, A bracket polynomial for graphs. II. Links, Euler circuits and marked graphs, arXiv:math.GT, math.CO/0901.1451.
7. L. Traldi, A bracket polynomial for graphs. III. Vertex weights, arXiv:math.GT, math.CO/0905.4879.
8. M. Las Vergnas, Eulerian circuits of 4-valent graphs imbedded in surfaces, in *Algebraic methods in graph theory*, Szeged (1978), *Colloq. Math. Soc. Janos Bolyai* **25** (North-Holland, Amsterdam-New York, 1981), pp. 451–477.
9. D. P. Ilyutko, Framed 4-Valent Graphs: Euler Tours, Gauss Circuits and Rotating Circuits, arXiv:Math.CO/0911.5504.
10. D. P. Ilyutko, An Equivalence between the Set of Graph-knots and the Set of Homotopy Classes of Looped Graphs, arXiv:Math.GT/1001.0360.

11. V. O. Manturov, On Free Knots, arXiv:Math.GT/0901.2214.
12. V. O. Manturov, On Free Knots and Links, arXiv:Math.GT/0902.0127.
13. V. O. Manturov, Free Knots and Parity, ibid.
14. V. O. Manturov, *Teoriya Uzlov* (Knot Theory), (Moscow-Izhevsk, RCD), 2005 (512 pp.).
15. R. C. Read, P. Rosenstiehl, On the Gauss crossing problem, *Colloq. Math. Soc. Janos Bolyai*, North-Holland, Amsterdam and New-York, (1976), pp. 843–876.
16. S. V. Chmutov, S. V. Duzhin, S. K. Lando, Vassiliev Knot Invariants. I, II, III *Adv. Sov. Math.* **21** (1994), pp. 117–147.
17. V. O. Manturov, Embeddings of 4-Valent Framed Graphs into 2-Surfaces, *Doklady Mathematics* **79**:1 (2009), pp. 56–58 (Original Russian Text in *Doklady Akademii Nauk* **424**:3 (2009), pp. 308-310).
18. Olof-Petter Östlund, Invariants of knot diagrams and relations among Reidemeister moves, arXiv:math.GT/0005108.
19. M. Cohn and A. Lempel, Cycle decomposition by disjoint transpositions, *J. Combin. Theory Ser. A* **13** (1972), pp. 83–89.
20. E. Soboleva, Vassiliev Knot Invariants Coming from Lie Algebras and 4-Invariants, *J. Knot Theory Ramifications* **10**:1 (2001), pp. 161–169.
21. L. Traldi, Binary nullity, Euler circuits and interlace polynomials, arXiv:math.CO/0903.4405.
22. C. St. J. A. Nash-Williams, Connected detachments of graphs and generalized Euler trails, *J. London Math. Soc.* (2) **31**:1 (1985), pp. 17–29
23. A. Bouchet, Multimatroids. I. Coverings by independent sets, *SIAM J. Discrete Math.* **10**:4 (1997), pp. 626–646
24. L. H. Kauffman, State Models and the Jones Polynomial, *Topology* **26** (1987), pp. 395–407.
25. V. Turaev, A simple proof of the Murasugi and Kauffman theorems on alternating links, *L'Enseignement Mathématique* **33** (1987), pp. 203–225.
26. K. Murasugi, The Jones polynomial and classical conjectures in knot theory, *Topology* **26** (1987), pp. 187–194.
27. M. B. Thistlethwaite, Kauffman polynomial and alternating links, *Topology* **27** (1988), pp. 311-318.

DIAGRAMMATIC KNOT PROPERTIES AND INVARIANTS

S. V. Jablan and R. Sazdanović

The Mathematical Institute,
Belgrade 11001, Knez Mihailova 36, P.O.Box 367, Serbia
E-mail: sjablan@gmail.com

University of Pennsylvania
209 South 33rd Street
Philadelphia, PA 19104-6395, USA
E-mail: radmilas@math.upenn.edu

We analyze relations between several knot invariants such as the unknotting (unlinking) number, adequacy, and quasi-alternating property, and the corresponding values obtained from (minimal crossing number) knot diagrams. In particular, we define the BJ-unlinking number, and compute it for various families of knots and links for which the unlinking number is unknown.

Keywords: Link, knot diagram, writhe, unknotting number, adequacy, Khovanov homology, quasi-alternating.

1. Introduction

A knot K is an embedding of a circle in three-dimensional space R^3 (or S^3) while a link is an embedding of a collection of circles. Knots and links are often presented by two dimensional projections called *knot (link) diagrams.* More precisely, link diagram is a one-to-one projection of a link onto a plane except at the crossings where the lines in the projection intersect transversely. Isotopy between two links can be realized on link diagrams by Reidemeister moves. In particular, two different diagrams of a given knot are related by Reidemeister moves, but there is no way to know which moves are needed. In this paper we consider only minimal diagrams of a given knot or link – diagrams with the minimal number of crossings.

The following move on link diagrams, called *flype* and illustrated on Fig. 1, plays an important role in obtaining all minimal diagrams of alternating links according to the Tait's Flyping Conjecture, Theorem 1.1(iii). In particular, minimal diagrams of alternating knots are related by a sequence of flypes.

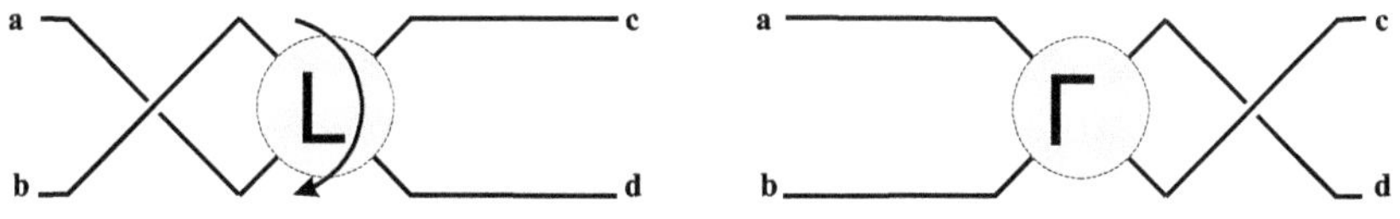

Fig. 1. Flype consists of fixing the endpoints of a tangle and then twisting it by a 2-fold rotation.

Theorem 1.1 (Tait's Conjectures).

- *(i) (Tait's First Conjecture) A reduced alternating diagram is the minimum diagram of its alternating link. Moreover, the minimum diagram of a prime alternating link can only be an alternating diagram. In other words, a non-alternating diagram can never be the minimum diagram of a prime alternating link.*
- *(iii) (Tait's Third Conjecture) Two reduced alternating diagrams of the same link are related by a finite series of flypes.*

The first Tait's conjecture[1] was proven independently in 1986 by L. Kauffman, K. Murasugi and M. Thistlethwaite.[2–6] The third Tait's conjecture (Tait's flyping conjecture) was proven by W. Menasco and M. Thistlethwaite.[1,7]

In this paper we utilize Conway notation[8] and the generalization of t_k-moves[9] referred to as families of knots and links.

Definition 1.1. Let S denote the set of numbers in the unreduced[a] Conway symbol $C(L)$ of a link L. Given $C(L)$ and an arbitrary (non-empty) subset $\tilde{S} = \{a_1, a_2, \ldots, a_m\}$ of S, the family $F_{\tilde{S}}(L)$ of knots or links derived from L is constructed by substituting each $a_i \in \tilde{S}$, $a_i \neq 1$ in $C(L)$ by $sgn(a_i)(|a_i| + n)$, for $n \in \mathbb{N}^+$.

For even integers $n \geq 0$ this construction preserves the number of components, i.e., we obtain (sub)families of links with the same number of components.

Note that classification of minimal diagrams is an open problem for non-alternating knots and links. Minimal diagrams of non-alternating knots and links in Conway notation were enumerated up to 12 crossings.

[a]The Conway notation is called *unreduced* if 1's denoting elementary tangles in vertices are not omitted in symbols of polyhedral links.

In this paper we consider prime knots and links and their families. We use Conway notation implemented in knot theory software *LinKnot* to say more about how writhe, unknotting number, adequacy, and quasi-alternating property relate to the values obtained from certain knot and link diagrams. We work under the assumption that the following conjecture is true: all diagrams determined by Conway symbols of a knot family starting with the Conway symbol denoting a minimal diagram of the source knot, are minimal.

2. Writhe

Writhe is one of the simplest interesting properties of knot and link diagrams. A crossing in an oriented knot diagram can be either positive or negative according to the convention shown on Fig. 2.

Definition 2.1. Given a diagram K' of an oriented knot or a link K we define *writhe*, denoted by $w(K')$, as the total number of positive crossings minus the total number of the negative crossings in a diagram K'.

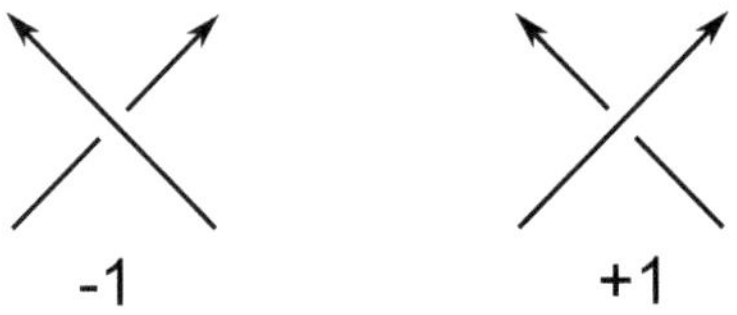

Fig. 2. Negative and positive crossings.

Writhe is clearly not a knot invariant since the first Reidemeister move changes it by ± 1. Next theorem states that all minimal diagrams of alternating knots have the same writhe.

Theorem 2.1 (Second Tait's Conjecture). *If K' and K'' are two reduced alternating knot diagrams of the same knot K, then $w(K') = w(K'')$.*[10]

However, non-alternating knots and links can have two distinct minimal diagrams with different writhes.

The smallest[b] non-alternating knot whose minimal diagrams $3 : -2\,0 : -2\,0$ and $2\,1 : -2\,0 : -2\,0$ have different writhes is the Perko knot (see Fig. 4). Knots with this property are scarce in knot theory literature and hard to find. According to our computations, there are only 29 knots with this property among all knots with at most 12 crossings, and interestingly, they are all non-algebraic:

10 crossings: the Perko knot 10_{161},
11 crossings: $11n_{116}$, $11n_{135}$, and $11n_{143}$,
12 crossings: $12n_{349}$, $12n_{382}$, $12n_{394}$, $12n_{398}$, $12n_{417}$, $12n_{430}$, $12n_{436}$, $12n_{519}$, $12n_{535}$, $12n_{552}$, $12n_{579}$, $12n_{594}$, $12n_{617}$, $12n_{624}$, $12n_{629}$, $12n_{638}$, $12n_{640}$, $12n_{644}$, $12n_{647}$,$12n_{650}$, $12n_{655}$, $12n_{739}$, $12n_{764}$, $12n_{850}$, and $12n_{851}$, total of 25 knots.

Computational results suggest that these knot diagrams can be extended to infinite subfamilies of knot diagrams whose writhes differ by two.

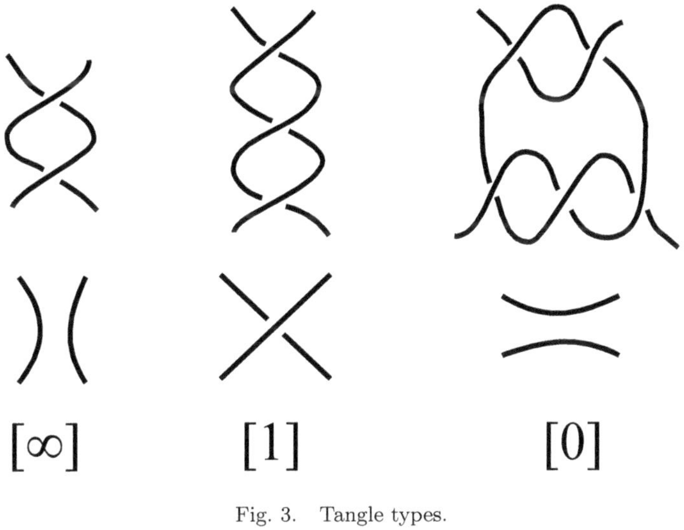

Fig. 3. Tangle types.

Example 2.1. Families of knot diagrams generated by two Perko knot diagrams $3 : -2\,0 : -2\,0$ and $2\,1 : -2\,0 : -2\,0$, determined by Conway

[b]The term "smallest" refers to the minimal crossing number diagram of a given knot or a link.

symbols $(2k+1):-2\,0:-2\,0$ and $(2k)\,1:-2\,0:-2\,0$, $k \in \mathbb{N}^+$, have writhes that differ by 2. The same is true for families of diagrams $2\,(2k):-2\,0:-2\,0$ and $2\,(2k-1)\,1:-2\,0:-2\,0$, starting with knots $11n_{135}$, $13n_{3546}$, and $15n_{114094}$, obtained for $k=1,2,3$ respectively.

Example 2.2. First, we introduce the notion of tangle types. Every 2-tangle consists of two strands connecting SW-SE and NW-NE ends, SW-NE and SE-NW ends, or SW-NW and SE-NE ends. According to this, every tangle will be of the [0]-type, [1]-type, or [∞]-type, respectively (Fig. 3)[c].

Consider two families of equivalent knot diagrams $t\,(k+1):-2\,0:-2\,0$ and $t\,k\,1:-2\,0:-2\,0$, where t is a positive rational tangle[d] whose Conway symbol does not contain the basic tangle 1 in the first position, and $k \in \mathbb{N}$. Notice that writhes of these diagrams differ by two and the first diagram corresponds to a knot if the tangle $t\,(k+1)$ is of type[8] [1] or [∞], and 2-component link if it is of type [0]. Hence, for a suitable choice of a tangle t, we obtain two families of minimal knot diagrams $t\,(k+1):-2\,0:-2\,0$ and $t\,k\,1:-2\,0:-2\,0$ whose writhes differ by 2. Hence, for every $n \geq 10$ there exist at least one prime non-alternating knot with this property.

Fig. 4. Perko pair: diagrams $3:-2\,0:-2\,0$ and $2\,1:-2\,0:-2\,0$ have writhe equal to 8, and 10, respectively.

[c]Switching between over–crossings and under–crossings does not change the tangle type and, consequently, the number of components of a link.
[d]A rational tangle is called positive if all numbers in its Conway symbol are positive.

The curious reader may try to approach, perhaps computationally, the following open questions:

- Do algebraic knots with two distinct minimal diagrams and different writhes exist?
- How many different values can writhe of an non-alternating prime knot take on its minimal diagrams?
- Is it possible to find two minimal diagrams of the same knot with n crossings whose writhes differ by an arbitrary number k ($k \leq n$ or $k \leq 2n$)? Find an upper bound on k.

The natural question is: can we modify the definition of writhe in such a way that it becomes an invariant of knots, not only of their diagrams? The naive attempt to take the minimum of the absolute values of writhe over all knot diagrams will not work. The first Reidemeister move changes writhe by ± 1. Hence, every knot has a diagram whose writhe is zero, and our invariant is trivial. Restriction to proper (reduced, loopless) diagrams will not work either, since the second Reidemeister move can be used to eliminate (hide) loops created by the first Reidemeister move. Instead, we can define the *minimum writhe* taken over all minimal diagrams.

A. Stoimenow[11] found an example of a positive knot (a knot admitting a diagram with all positive crossings) $11n_{183}$ with no positive minimal diagram. Computations show that among 29 knots with diagrams whose writhes differ by 2, six have one positive and one non-positive minimal diagram: Perko knot 10_{161} (see Fig. 4, with positive diagram on the right), $12n_{417}$, $12n_{655}$, $12n_{764}$, $12n_{850}$, and $12n_{851}$.

3. Unlinking number

Computing unknotting and unlinking numbers, also known as Gordian numbers, is one of the most difficult problems in knot theory.[12–16] General formulas for unknotting number are known only for some special classes of knots. For example, according to the famous Milnor conjecture[17] proved by P.B. Kronheimer and T.S. Mrowka,[18,19] the unknotting number of a torus knot $[p, q]$ is $\frac{(p-1)(q-1)}{2}$. Unknotting (unlinking) process consists of crossing changes, i.e., switching an overcrossing to undercrossing or *vice versa*.

There are two different yet equivalent algorithms[20] for computing the unlinking number of a link L:

(1) according to the *classical definition*, one is allowed to make a planar isotopy after each crossing change and then continue the unlinking process

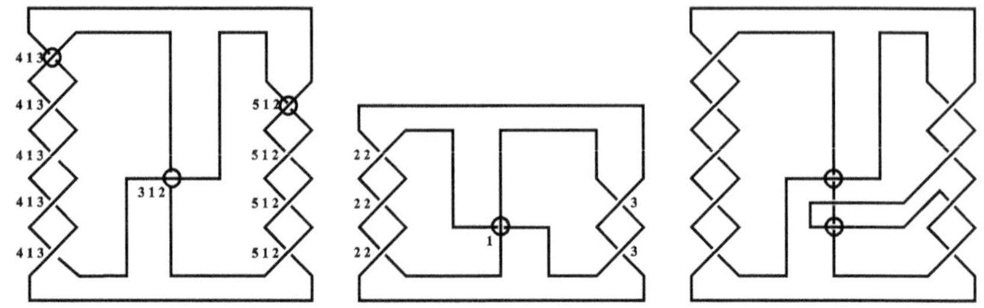

514 (a)   312 (b)   514 (c)

Fig. 5. The Nakanishi-Bleiler example: (a) the minimal diagram of the knot 5 1 4 that requires at least three simultaneous crossing changes to be unknotted; (b) Unknotting number one rational knot $5-14=3\,1\,2$: changing circled crossing yield an unknotting number 1 knot and switching any other crossing yields rational unknotting number 1 rational knots, whose symbol is written next to the crossing. (c) Non-minimal diagram of the knot 5 1 4 from which we obtain the correct unknotting number $u(5\,1\,4)=2$.

with the newly obtained diagram, until an unlink is obtained;

(2) the *standard definition* requires all crossing changes to be done simultaneously on a fixed diagram.

Y. Nakanishi[21] and S. Bleiler[22] have discovered the famous rational knot 10_8 (or 5 1 4 in Conway notation), showing that two methods for computing the unknotting number described above do not give the same result, and potentially not the correct unknotting number if we consider fixed minimal diagrams only. Standard definition applied only to minimal diagrams may not give the correct unknotting number: unknotting 10_8 takes at least three crossing changes on its unique minimal diagram (the crossings are denoted by circles on Fig. 5a). On the other hand, making a crossing change in the middle crossing of the diagram from Fig. 5a, followed by the reduction $5-1\,4=3\,1\,2$, we obtain the minimal diagram of the knot 3 1 2 (Fig. 5b) that can be unknotted by one crossing change. Hence, unknotting according the classical definition, using only minimal diagrams, but additionally allowing reducing diagrams after every crossing change, yields a correct unknotting number 2. The unknotting number can also be obtained from the non-minimal diagram of the knot 5 1 4 (Fig. 5c) using the standard definition. As it was shown by J. Bernhard,[23] the whole knot family $(2k+1)\,1\,(2k)$, $k\geq 2$ has this property.

Motivated by the Nakanishi-Bleiler example we introduce the *BJ*-

unlinking number which is computable due to the algorithmic nature of its definition.

Definition 3.1. Let D denote a diagram of a knot or a link L, v a crossing in D, and D_v the link diagram obtained from D by switching a sign of crossing v.

a) The *unlinking number* $u(D)$ of a link diagram D is the minimal number of crossing changes on the diagram required to obtain an unlink.
b) The *unlinking number* of a link L, denoted by $u(L)$, is the minimum of unlinking numbers of all diagrams D corresponding to the link L: $u(L) = \min_D u(D)$.
c) The *BJ-unlinking number* $u_{BJ}(D)$ of a diagram D is defined recursively in the following manner:
   (1) $u_{BJ}(D) = 0$ *iff* D represents an unlink.
   (2) Assume that the sets of diagrams $\mathcal{D}_k$ with $u_{BJ}(D) \leq k$ are already defined. A diagram D has $u_{BJ}(D) = k+1$ if $D \notin \mathcal{D}_k$ and there exists a crossing v on the diagram D such that $u_{BJ}(D'_v) = k$ where D'_v is a minimal link diagram representing the same link as diagram D_v obtained from D by a crossing change at v. Notice that $u_{BJ}(D)$ is well defined for every diagram D as $D \in \mathcal{D}_{n!}$ where n is the number of crossings in D.
d) $u_M(L) = \min_D u(D)$ where the minimum is taken over all minimal diagrams D representing L.
e) The *BJ-unlinking number* $u_{BJ}(L)$ of a link L is defined by $u_{BJ}(L) = \min_D u_{BJ}(D)$ where the minimum is taken over all minimal diagrams D representing L.[24]

J.A. Bernhard[23] and independently S. Jablan in 1995,[25] conjectured:

Conjecture 3.1. *(Bernhard-Jablan Conjecture) Unlinking number and BJ-unlinking number coincide for every link L:* $u(L) = u_{BJ}(L)$.

The proof of the next theorem by Y. Nakanishi[27] and A. Stoimenow[28] which says that the BJ-conjecture holds for the following class of rational knots and links relies on the work of T. Kanenobu and H. Murakami,[29] and P. Kohn.[15]

Theorem 3.1. *All rational knots and links with unlinking number one have an unlinking number one minimal diagram.*

Notice that if the BJ-conjecture does not hold for all knots and links, it may be true for some classes of links, such as rational ones. Moreover, BJ-unlinking number is a computable upper bound for the unlinking number.

We illustrate the importance of the BJ-conjecture by computing the BJ-unlinking numbers of knots and links in several families, and prove that in some cases it coincides with the unlinking number.

Proposition 3.1.

a) *For rational (2-bridge) knots with Conway symbol* $(2m+1)\ 1\ (2n+1)$, $(m \geq n)$ *BJ-unknotting number is* $u_{BJ} = n+1$;
b) *for pretzel knots* a, b, c $(1 < a \leq b \leq c)$[e], *where* a, b, c *are odd numbers:*

$$u_{BJ}(a,b,c) = \frac{a+b}{2}$$

The unknotting number of these knots is still unknown, except for the smallest knots, e.g., 3, 3, 3 with unknotting number 3, computed by B. Owens.[30]

Lemma 3.1. *Let $R_{[a,b]}$ denote a 2-bridge link with the Conway symbol $a\ b$. Then the following holds:*

a) *If a, b are both odd then for a link* $R_{[a,b]} = R_{[2m+1,2n+1]}$ *we have* $u_{BJ}(R_{[2m+1,2n+1]}) = u_M(R_{[2m+1,2n+1]}) = u(R_{[2m+1,2n+1]}) = \frac{a+b}{2} = m+n+1$.
b) *If a is odd and b is even then for a knot* $R_{[a,b]} = R_{[2m+1,2n]}$ *we have* $u_{BJ}(R_{[2m+1,2n]}) = u_M(R_{[2m+1,2n]}) = u(R_{[2m+1,2n+1]}) = n$.
c) *If a, b are both even then for a knot* $R_{[a,b]} = R_{[2m,2n]}$ *we have* $u_{BJ}(R_{[2m,2n]}) = u_M(R_{[2m,2]}) = min(m,n)$. [f]

Definition 3.2.

a) The *BJ-unlinking gap* of a diagram D denoted by $\delta_{BJ}(D)$ is the difference $\delta_{BJ}(D) = u(D) - u_{BJ}(D)$.
b) The *BJ-unlinking gap* of a link L denoted by $\delta_{BJ}(L)$ is defined by $\delta_{BJ}(L) = u_M(L) - u_{BJ}(L)$.[24]

Since flype can be viewed as a special case of mutation, according to the Theorem 1.1(iii) and Lemma 3.2, we can compute BJ-unlinking number from any minimal diagram of an alternating link.

---

[e]Proposition 3.1 also holds for rational knots obtained by setting the value $a = 1$.
[f]Determining the unknotting number $u(R_{[2m,2n]})$ is an open question for most of knots with $m, n > 1$.

Lemma 3.2. *If $m(D)$ is a mutation of a diagram D, then $u(D) = u(m(D))$.*[24]

Corollary 3.1. *For every prime alternating link L and its minimal diagrams D and D' the following holds: $u(D) = u(D') = u_M(L)$, $u_{BJ}(D) = u_{BJ}(D') = u_{BJ}(L)$, $\delta_{BJ}(D) = \delta_{BJ}(D') = \delta_{BJ}(L)$.*

However, we need to consider all minimal diagrams of non-alternating knots and links, since their minimal diagrams can have different BJ-unknotting numbers. A. Stoimenow's[32] example, obtained in the search for the counterexample to the BJ-Conjecture, supports the previous statement: two minimal diagrams (Fig. 6) of the 14-crossing knot $14n_{36750}$ (in the *Knotscape* notation) given by the Dowker codes

$$\{\{14\},\{4,14,18,12,28,26,-22,2,-24,10,8,-16,6,20\}\}$$
$$\text{and } \{\{14\},\{6,-10,26,20,-2,-18,-22,28,-24,8,-14,-12,4,16\}\}$$

have two different BJ-unknotting numbers, 1 and 2, respectively.

Fig. 6. Two minimal diagrams of the knot 14_{36750}.

Two different minimal diagrams of the knot 14_{36750} (Fig. 6) can be extended to the two-parameter families of minimal diagrams:

$$124^* - 1. - 1. - 1. - 1. - 1 : (-1, (-2l))\, 0 : (2k-1) : . - 1$$
$$\text{and } 124^* - 1. - 1. - 1. - 1. - 1 : (-1, (-2l))\, 0 : (2k-1) : . - 1$$

which satisfy the condition $l < 2k$ (Fig. 7)[g]. Based on the computational

[g] 124^* denotes the basic polyhedron $12D$ from the paper by A. Caudron.[33]

results, we conjecture that BJ-unknotting number of the first family of diagrams is k, and that the second family of diagrams has BJ-unknotting number equal to $k+1$.

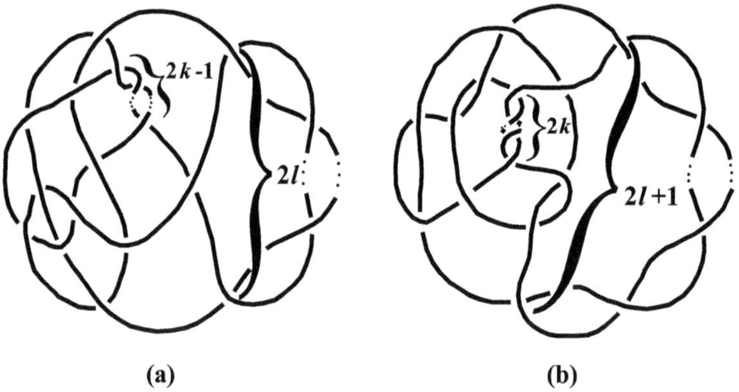

Fig. 7. Two diagrams (a) $124^* - 1. - 1. - 1. - 1. - 1 : (-1, (-2l))\,0 : (2k-1) : . - 1$; (b) $8^* - 2\,0 : (2l+1)\,0 : -2\,0.(-2k)\,0. - 1.2\,0$ of the same knot.

K. Murasugi[34] proved the lower bound for the unknotting number for knots $u(K) \geq \frac{|\sigma(K)|}{2}$ and links $u(K) \geq \frac{|\sigma(K)|+1}{2}$. Therefore, if $u_{BJ}(K) = \frac{|\sigma(K)|}{2}$ or $\frac{|\sigma(K)|+1}{2}$, then the BJ-unknotting number of a knot or link, respectively, is equal to the unknotting (unlinking) number. Additional bounds on unlinking numbers are given by P. Kohn.[16]

Unknotting numbers computed according to the BJ-conjecture coincide with unknotting numbers for all knots with up to 10 crossings given in the book *A Survey of Knot Theory* by A. Kawauchi (Appendix F)[14] and in Kawauchi's table updated with reference to recent unknotting number results,[26] if in all ambiguous cases (A=1 or 2; B=2 or 3) we take the bigger number. Hence, if any of these unknotting numbers is smaller than its maximal estimated value, this will be a counterexample for the BJ-conjecture. BJ-conjecture holds for all two-component links up to ten crossings, whose unlinking numbers were computed by P. Kohn.[16] The complete list of BJ-unknotting numbers for knots with 11 crossings computed by the authors using the program *LinKnot* is included in the *Table of Knot Invariants* by C. Livingston and J.C. Cha.[26] The results by B. Owens[30] and Y. Nakanishi[13] confirmed the unknotting number 3 for the knot 9_{35}, and the unknotting number 2 for the knots 10_{83}, 10_{97}, 10_{105}, 10_{108}, 10_{109}, and 10_{121}

computed according to the BJ-conjecture. For many knots with at most 10 crossings results of P. Ozsváth and Z. Szabó[36] confirmed that the unlinking number is equal to 2.

Based on the computational results we can say more about how BJ-unlinking number and BJ-unlinking behave inside families. For example, for all knots in families $p\,2$, $p\,1\,1\,p$, $(p+2)\,1\,1\,p$, $p\,2\,1\,2$, $p\,1\,1\,1\,2$, unknotting and BJ-unknotting numbers are equal to 1, while for knots of the family p $(p = 2k+1)$, they are equal to k.

The book *LinKnot – Knot Theory by Computer*[8] contains the complete list of alternating as well as several examples of non-alternating knots and links up to 12 crossings with non-zero BJ-unlinking gap.

4. Adequacy

We begin this section by reviewing properties of adequate, semi-adequate, and inadequate link diagrams, and their corresponding links.

Let D be a diagram of an unoriented, prime, framed link L in $\mathbb{R}^3$. A Kauffman state of a diagram D is a function from the set of crossings of D to the set of signs $\{+1, -1\}$ or $\{0, \infty\}$. On the level of link diagrams, Kauffman state can be viewed as a choice of smoothing or resolving (positive and negative markers) all crossings in a link diagram according to the convention illustrated in Fig. 8 which shows three diagrams D_+, D_0, and D_∞ which differ at a single crossing.

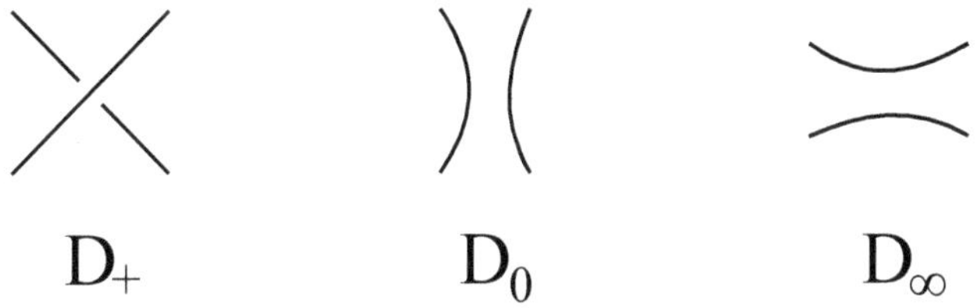

Fig. 8. Smoothings of a positive crossing D_+: vertically according to the positive markers D_0, and horizontally along the negative marker D_∞.

A *state diagram* D_s is a system of circles, called *state circles*, obtained by smoothing all crossings of D. Kauffman states with all positive or all negative resolutions are denoted by s_+ and s_-, and the corresponding state diagrams by D_{s_+} and D_{s_-}.

Definition 4.1.[37] A diagram D is s-adequate if no circle in D_s contains both arcs obtained by smoothing one crossing. In particular, a diagram D is + or −adequate if it is adequate for states s_+ or s_-.

- D it is called *semi-adequate* if it is only +adequate or −adequate.
- D is called *inadequate* if it is neither +adequate nor −adequate.
- D is called *adequate* if it is both +adequate and −adequate.[38,39]

Next, to every state diagram D_s we associate the graph G_s in the following way: vertices correspond to state circles of D_s, and edges connect those vertices whose corresponding circles contain points obtained by smoothing the same crossing in D. Now, Definition 4.1 can be reformulated: knot diagram D is s-adequate if the corresponding graph G_s is loopless.

Definition 4.2. A link is *adequate* if it has an adequate (+adequate and −adequate) diagram. A link is *semi-adequate* if it has a + or −adequate diagram. A link is *inadequate* if it is neither + or −adequate.[38,39]

Notice that if a diagram is s-adequate, its mirror image is $(-s)$-adequate, where $(-s)$ denotes the state obtained from s by switching positive and negative smoothings.

Definition 4.3. A link with two different diagrams, one +adequate one −adequate is called *weakly adequate.*

For example, Perko's knot[40] 10_{161} $3:-2\,0:-2\,0$ (Fig. 11), and knot $11n_{146}$ $9^*.-2:.-2$ that has −adequate 11-crossing diagram and +adequate 12-crossing diagram $6^*-2.2.-2.2.2\,0.-2\,0$, are both weakly adequate.

Theorem 4.1. *A reduced alternating knot or link diagram is adequate.*[35,38,39]

Hence, all alternating links are adequate.

Theorem 4.2. *An adequate diagram of a knot or a link is a minimal crossing number diagram.*[35,38,39]

Theorem 4.3. *Any reduced semi-adequate diagram represents a non-trivial link. Any connected semi-adequate diagram represents a non-split link.*[41]

Theorem 4.4. *Two adequate diagrams of a link have the same crossing number and the same writhe.*[35]

Given a 2-tangle t, there are two ways of connecting the endpoints without introducing additional crossings:

- numerator closure $N(t)$ obtained by connecting NW to NE and SW to SE corners,
- denominator closure $D(t)$ obtained by connecting NW to SW and NE to SE corners.

Definition 4.4. An alternating diagram of a marked 2-tangle t is called *strongly alternating* if both the numerator $N(t)$ and denominator closure $D(t)$ are irreducible.[35,38,39]

Theorem 4.5. *The non-alternating sum of two strongly alternating tangles is adequate.*[35,38,39]

As a corollary we get that all semi-alternating diagrams are adequate.[38,39]

According to Theorem 4.2, minimal diagrams can be used to determine if a link is adequate, but not to distinguish semi-adequate links from inadequate ones.

Theorem 4.6. *A link is inadequate if coefficients of the highest and lowest degree terms in its Jones polynomial are different from* ± 1.

The proof of this theorem for knots follows from results of W.B.R. Lickorish and M. Thistlethwaite.[38]

Using *Knotscape* tables of knots, A. Stoimenow detected all non-alternating adequate knots up to 16 crossings. We analyze adequacy of non-alternating links and their families.

Adequate non-alternating links with less than 10 crossings are given in Table 1. All of them, except polyhedral ones, either satisfy the conditions of Theorem 4.5 or can be obtained from pretzel links, which satisfy this theorem, by permuting their rational tangles.

Theorem 4.6 gives sufficient but not necessary conditions for recognizing inadequate links. For example, the first and last coefficient of Jones polynomial of the knot $11n_{95}$, with Conway symbol $2\,0.-2\,1.-2\,0.2$, are different from ± 1, hence it must be inadequate.[35]

For knots with at most 12 crossings every minimal diagram of a semi-adequate knot is semi-adequate. However, this is not true in general. The first example of a semi-adequate knot with a minimal inadequate diagram (Fig. 9) is the knot $13n_{4084}$ $10^{**}.-1.-1.-1:.-2.2.-2$ with the minimal Dowker-Thistlethwaite code

$$\{\{13\},\{6,-10,12,24,20,-18,-26,-22,-4,2,-16,8,-14\}\}.$$

Except this inadequate diagram whose writhe is 9, it has another semi-adequate minimal diagram[42,43] $11^{**}.-2::-2\,0:-1.-1.-1$ whose writhe

Table 1. Adequate non-alternating links with less than 10 crossings.

$n = 8$	2 links	
$2, 2, -2, -2$	$(2, 2) - (2, 2)$	
$n = 9$	4 links	
$3, 2, -2, -2$	$(3, 2) - (2, 2)$	$(2\,1, 2) - (2, 2)$
$. - (2, 2)$		
$n = 10$	3 knots	
$(3, 2) - (3, 2)$	$(3, 2) - (2\,1, 2)$	$(2\,1, 2) - (2\,1, 2)$
$n = 10$	22 links	
$3, 2\,1, -2, -2$	$3, 3, -2, -2$	$3, -2, 2\,1, -2$
$3, -2, 3, -2$	$4, 2, -2, -2$	$2, 2, 2, -2, -2$
$2\,2, 2, -2, -2$	$(4, 2) - (2, 2)$	$(3, 2\,1) - (2, 2)$
$(3\,1, 2) - (2, 2)$	$(2\,1, 2\,1) - (2, 2)$	$(3, 3) - (2, 2)$
$(2\,1\,1, 2) - (2, 2)$	$(2, -2, -2)\,(2, 2)$	$(2\,2, 2) - (2, 2)$
$(2, 2, 2) - (2, 2)$	$(2, 2), 2, -(2, 2)$	$. - (2, 2).2$
$. - (2, 2).2\,0$	$. - (2, 2) : 2\,0$	$. - (2, 2) : 2$
$103^* - 1. - 1. - 1. - 1 :: . - 1$		

is 7, with the Dowker-Thistlethwaite code

$$\{\{13\}, \{6, 12, -16, 23, 2, 17, 21, 26, 11, -4, -25, 7, 13\}\}.$$

First example of a semi-adequate knot with no minimal semi-adequate diagram has 15 crossings: knot $15n_{164563}$ has only one minimal diagram $10^{**} - 1. - 2\,0.2\,0 :: .2\,0.2\,0. - 2\,0$, which is inadequate (Fig. 10). However, it has a 16-crossing diagram $11^*2\,0. - 1. - 2. - 1.3\,0. - 1.2\,0 :: -1$ which is semi-adequate.[43] This example extends to the family of knot diagrams $10^{**} - 1. - 2\,0.(2k)\,0 :: .2\,0.2\,0. - 2\,0$ and $11^*(2k)\,0. - 1. - 2. - 1.3\,0. - 1.2\,0 :: -1$ $(k \geq 1)$ with the same properties, respectively.

Adequacy of all knots with at most 12 crossings was confirmed using their minimal diagrams. For all knots and links with 13 or more crossings a single minimal diagram given by their Conway symbol was used.

Recall that all knots having two distinct minimal diagrams with different writhes, given in Section 2, are weakly adequate. Since semi-adequate knots can have both +adequate and −adequate minimal diagrams, we make the following conjecture:

Conjecture 4.1. *(Jablan) Every knot which has minimal diagrams with different writhes is weakly adequate.*

At least for small number of crossings, most of non-alternating links are semi-adequate, so adequate and inadequate links represent a small portion of all non-alternating links. Hence, it is of interest to tabulate inadequate

Fig. 9. Semi-adequate knot $13n_{4084}$ with a minimal inadequate diagram $10^{**}. - 1. - 1. - 1 : . - 2.2. - 2$ and minimal semi-adequate diagram $11^{**}. - 2 :: -2\,0 : -1. - 1. - 1$.[42,43]

Fig. 10. Semi-adequate knot $15n_{164563}$ with only minimal diagram $10^{**} - 1. - 2\,0.2\,0 :: .2\,0.2\,0. - 2\,0$ which is inadequate, and non-minimal 16-crossing diagram $11^{*}2\,0. - 1. - 2. - 1.3\,0. - 1.2\,0 :: -1$ which is semi-adequate.[42,43]

and candidates for inadequate non-alternating links, and find criteria for adequacy of certain classes of links.

Links, families, or classes of links whose minimal diagrams are inadequate will be referred to as *candidates for inadequate links.* In some cases Theorem 4.6 can be used to confirm that they indeed are inadequate.

Among 202 non-alternating links with at most 10 crossings there are only 28 adequate links and 3 adequate knots. Links with inadequate minimal diagrams are even more rare – there are only 3 such links with less than 10 crossings: $(2, 2, -2)\,(2, -2)$, $2. - 2\,0. - 2.2\,0$, and $103^{*} - 1. - 1 :: -1. - 1$. Links $2. - 2\,0. - 2.2\,0$ and $103^{*} - 1. - 1 :: -1. - 1$ are inadequate according to

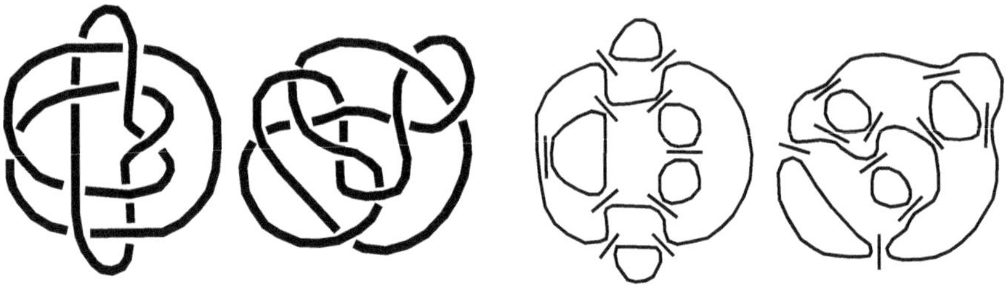

Fig. 11. Perko pair: weakly adequate knot with two minimal diagrams $6^*3 : -2\,0 : -2\,0$ and $6^* -2 -1. -1.2\,0. -1.2\,0. -1$ with the adequacy of different signs, where the first is +adequate, and the other −adequate.

Theorem 4.6, and the link $(2,2,-2)\,(2,-2)$, all of whose minimal diagrams are inadequate, is a candidate for inadequate link.

Knot $2\,0. -2\,1. -2\,0.2$ is inadequate according to Theorem 4.6, and since all minimal diagrams of the knot $2\,0. -3. -2\,0.2$ are inadequate, it is a candidate for inadequate knot.

Table 2 contains all inadequate knots among nineteen 12-crossing knots with inadequate minimal diagrams,

Table 2. Inadequate 12-crossing knots.

$2. -2\,0. -2.2\,1\,1\,0$	$2 : (-2, 2\,1)\,0 : -2\,0$	$2 : (2, -2\,-1)\,0 : -2\,0$
$2.2. -2.2\,0. -2\,-1$	$3. -2\,0. -2.2\,1\,0$	$3. -2\,-1\,0. -2.2\,0$
$8^*2\,0. -2\,0. -2\,0.2\,0$	$8^* -2\,-1.2\,0. -2$	$9^*. -2 : -2\,0. -2$
$101^* -2\,0 :: . -2\,0$	$102^* -2\,0 :: -2$	

while inadequacy of the remaining 8 knots from Table 3 remains unknown.

Table 3. Candidates for inadequate 11-crossings knots.

$2. -3\,0. -2\,-1.2\,0$	$8^* -2\,-1\,-1 :: -2\,0$	$8^*2 : . -2\,0 : . -2\,-1\,0$
$8^* -2\,-1 :: -3\,0$	$8^*2 : . -2\,-1\,0 : . -2\,0$	$8^* -2\,0.2 : -2\,-1\,0$
$8^* -2.2. -2\,0 : 2\,0$	$8^* -2\,0 : -2\,0 : -2\,0 : 2\,0$	

The following four 11-crossing links

- $(2\,1,2) -1\,-1\,(2,2)$
- $(2\,1,2) -1\,-1\,(2,2)$
- $6^*3. -2\,0. -2.2\,0$
- $6^*(2,-2).2. -2$

are inadequate according to Theorem 4.6, and the 8 links from Table 4 are candidates for inadequate links, since all of their minimal diagrams are inadequate.

Table 4. Table of candidates for inadequate links with 11 crossings.

$(-2\ -1,2)\,1\,1\,(2,-2)$	$(-2\ -1,2,2)\,(2,-2)$	$(2\,1,2,-2)\,(2,-2)$
$(2,2)\ -1\ -1\ -1\,(2,-2)$	$(2,2,-2)\,(-2\ -1,2)$	$(2,2,-2)\,(2\,1,-2)$
$6^{*}2\,1.-2\,0.-2.2\,0$	$6^{*}(2,-2),-2$	

Among 12-crossing links 63 are inadequate according to Theorem 4.6, and the remaining 232 links are candidates for inadequate links.[h]

Definition 4.5. A link diagram given by Conway symbol containing only tangles 1, −1, 2, or −2 is called a *source link diagram.*

Theorem 4.7. *Link diagrams which belong to a family obtained from a source link diagram which is +adequate (− adequate) are also +adequate (− adequate).*

Proposition 4.1. *The adequacy of a link diagram remains unchanged if we replace every positive rational tangle by 2, and every negative rational tangle by −2.*

Theorem 4.8. *A non-alternating Montesinos link $t_1, t_2, \ldots, t_n$ is semi-adequate if it contains exactly one rational tangle of one sign, and all the other rational tangles of the opposite sign. Otherwise, it is adequate.*

5. Quasi-alternating property

We present computational results addressing the classification of quasi-alternating links based on the thickness of Khovanov homology[44–46] $\overline{Kh}$, and odd Khovanov homology[47] $\overline{Kh}'$, and analyzing their minimal diagrams.

Definition 5.1. The set Q of quasi-alternating links (QA-links) is the smallest set of links such that

- the unknot is in Q,
- if the link L has a diagram D with a crossing c such that
  (1) both smoothings of c, L_0 and L_∞, are in Q;

[h]Tables of adequate non-alternating links with at most 12 crossings, in Conway notation, can be downloaded in the form of *Mathematica* notebook from the address: `http://www.mi.sanu.ac.rs/vismath/adequate.pdf`

(2) $\det(L)=\det(L_0)+\det(L_\infty)$,

then L is in Q. We say that a crossing c satisfying the properties above is a quasi-alternating crossing of the diagram D, or that D is quasi-alternating at the crossing c.[48,49]

C. Manolescu and P. Ozsváth show that both Khovanov homology $\overline{Kh}$, and Heegaard-Floer homology $\widehat{HFK}$ can be used to detect links which are not QA. QA-links are Khovanov homologically σ-thin (over $\mathbb{Z}$) and Floer homologically σ-thin (over $\mathbb{Z}/2\mathbb{Z}$).[50] The same property extends to odd Khovanov homology[47] $\overline{Kh}'$, and we use it as the first criterion to select knots that can not be QA from a database obtained by *KhoHo* and *LinKnot*.

If a knot or link is thin with respect to the following homology theories $\overline{Kh}$, $\widehat{HFK}$, and $\overline{Kh}'$, we say it is *homologically thin* (without qualification).

Recursive nature of this definition makes it difficult to determine if a knot is quasi-alternating. In particular, we were interested in finding homologically thin knots which are not QA. For a long time, knots $9_{46} = 3, 3, -3$ and $10_{140} = 4, 3, -3$ have been the main candidates. However, according to A. Shumakovitch's computations[51] they are not QA, being $\overline{Kh}'$-thick, although they are both $\widehat{HFK}$- and $\overline{Kh}$-thin.

Based on Theorem 5.1,[49] quasi-alternating links with a higher number of crossings can be obtained as extension of links which are already recognized as quasi-alternating.[49,52]

Consider the crossing denoted by c in Definition 5.1, as a 2-tangle with marked endpoints. Using Conway's notation for rational tangles, let $\varepsilon(c) = \pm 1$, depending on the sign of the crossing, i.e., whether the overstrand has positive or negative slope. We say that a rational 2-tangle $\tau = C(a_1, \ldots, a_m)$ extends c if τ contains c and $\varepsilon(c) \cdot a_i \geq 1$ for $i = 1, \ldots, m$. In particular, τ is an alternating rational tangle.

Theorem 5.1. *If L is a quasi-alternating link, let L' be obtained by replacing any quasi-alternating crossing c with an alternating rational tangle that extends c. Then L' is a quasi-alternating link.*[49]

We use Theorem 5.1 as a criterion for listing quasi-alternating links up to 12 crossings as well as examples of quasi-alternating links with at least two different minimal diagrams, where one is quasi-alternating and the other is not. We also provide examples of knots and links with up to 12 crossings which are homologically thin and have no minimal quasi-alternating diagrams.

Mathematica package *LinKnot* is used for deriving families of knots and links given in Conway notation and their distinct diagrams. Odd Khovanov homology $\overline{Kh}'$ is computed using *KhoHo* by A. Shumakovitch.[53] All flype-equivalent minimal diagrams of non-alternating knots and links up to 12 crossings are derived from alternating link diagrams, using software *LinKnot*. *KnotFind*, the part of the program *Knotscape*,[54] is used for recognition of knots, and the Jones and Kauffman polynomials are used for distinguishing links. In addition, we used the criterion that homologically thick knots are not quasi-alternating and that knots obtained by positive and negative smoothings of a quasi-alternating crossing must be homologically thin.

As a first criterion we use the fact that knots with thick odd Khovanov homology cannot be quasi-alternating. Table 5 gives an overview of the numbers of non-alternating knots with $8 \leq n \leq 12$ crossings and how many among them are $\overline{Kh}'$-thin:

Table 5. Overview of non-alternating and odd-homology thin knots with $8 \leq n \leq 12$ crossings.

No. of crossings	8	9	10	11	12
No. of non-alternating knots	3	8	42	185	888
No. of $\overline{Kh}'$-thin non-alternating knots	2	6	31	142	663

Knots in Table 6 are $\overline{Kh}$ and $\widehat{HFK}$-homologically thin, and have a minimal diagram which is not QA. Even columns in Table 6 contain Conway symbols of these knots. Four of them: 9_{46}, 10_{140}, $11n_{139}$, and $11n_{107}$, are $\overline{Kh}'$-thick,[55] so they are not quasi-alternating.

Table 6. $\overline{Kh}$- and $\widehat{HFK}$-homologically thin knots with $9 \leq n \leq 11$ crossings.

9_{46}	$3, 3, -3$	10_{140}	$4, 3, -3$	$11n_{139}$	$5, 3, -3$
$11n_{107}$	$-2\,1\,2, 3, 3$	$11n_{50}$	$-2\,2, 2\,2, 3$	$11n_{65}$	$(3, -2\,1)\,(2\,1, 2)$

Knot $11n_{65}$ has two minimal diagrams: $(3, -2\,1)\,(2\,1, 2)$ (Fig. 12a) and $6^*2.2\,1. - 2\,0. - 1. - 2$ (Fig. 12b). The first diagram is not QA, and the second is QA. By smoothing at the crossing denoted by c, second diagram resolves into unknot and QA link $(2, 2+) - (2\,1, 2)$, which further resolves into QA knots $3, 2\,1, -2$ and $2\,1\,1, 2\,1, -2$ by smoothing the crossing c_1 (Fig. 12c). Moreover, all minimal diagrams belonging to the families $6^*2.p\,1. - 2\,0. - 1. - 2$ and $(3, -2\,1)\,(p\,1, 2)$, $p \geq 2$ have the same property as their generating minimal diagrams of $11n_{65}$: the first is QA, and the other is not.

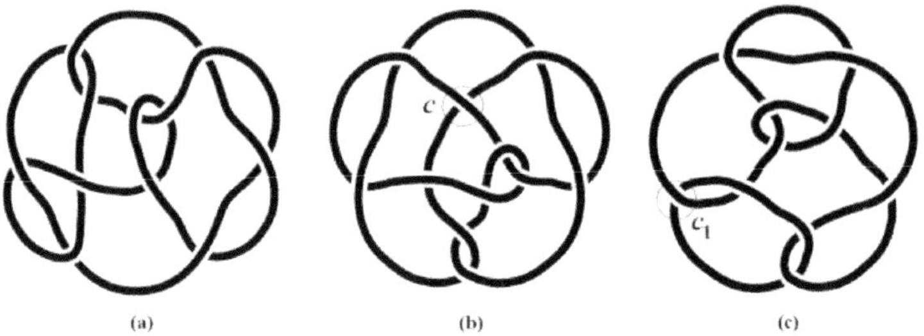

Fig. 12. (a) Non-QA minimal diagram $(3,-2\,1)\,(2\,1,2)$ of the knot $11n_{65}$; (b) the other quasi-alternating minimal diagram $6^{*}2.2\,1.-2\,0.-1.-2$ of the knot $11n_{65}$; (c) quasi-alternating link $(2,2+)-(2\,1,2)$.

J. Greene proved that the remaining Montesinos knot (Fig. 13a)

$$11n_{50} = -2\,2, 2\,2, 3 = M(0;(5,-2),(5,2),(3,1)) = M(1;(5,3),(5,2),(3,1))$$

is homologically thin non-QA knot.[56]

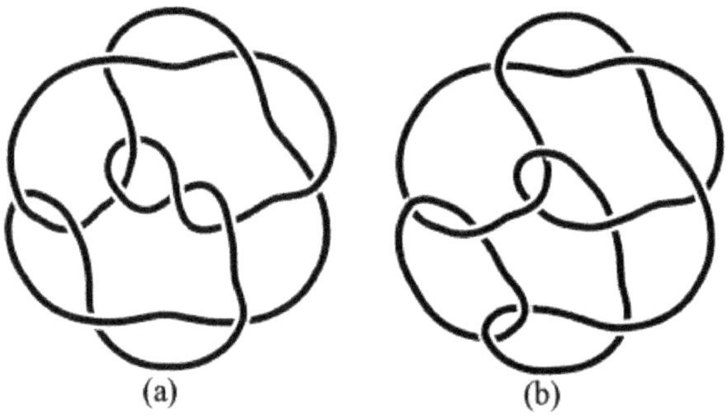

Fig. 13. Homologically thin non-QA knot $11n_{50}$ (a) and link $11n_{90}$ (b).

12-crossing knots $12n_{196} = (-3\,1,3)\,(2\,1,2)$, $12n_{393} = 8^{*}2.2\,0 : -2\,1\,0$ and $12n_{397} = 2\,1\,1 : -2\,1\,0 : 2\,0$ have a non-standard minimal diagram which is quasi-alternating. Knots given in Table 7 are candidates for homologically thin non-QA knots.

Notice that the only Montesinos knot in Table 7 is

$$12n_{145} = M(0;(5,-2),(5,2),(4,1)) = M(1;(5,3),(5,2),(4,1))$$

with Conway symbol $-2\,2, 2\,2, 4$.

Table 7. Candidates for homologically thin 12 crossing knots.

$12n_{139}$	$.2.(-2\,1,2).2$	$12n_{145}$	$-2\,2,2\,2,4$
$12n_{296}$	$-2\,2,2\,1\,2,3$	$12n_{331}$	$(3,2+)\,-\,(2\,1,3)$
$12n_{768}$	$2:-3\,1\,0:3\,0$	$12n_{838}$	$-2.-2.-2\,0.2.2.2\,0$

Moreover, J. Greene[56] proved that all Kanenobu knots $K(n, 3-n)$ except the knot $11n_{132} = K(1,2) = K(2,1)$ are homologically thin non quasi-alternating knots. Notice that Kanenobu knots $K(n, 3-n)$ are given by Conway symbols $2.2.-p\,0.-2.-2.-q\,0$, where $|p-q| = 3$, $n \geq 0$.

The remaining odd-homology thin knots, 141 eleven-crossing knots and 656 twelwe-crossing knots, are quasi-alternating.

As additional candidates for homologically thin knots which are not QA, we propose the following 13-crossing Montesinos knots with non-QA minimal diagrams[i]:

$$13n_{1408} = -3\,2,2\,2,2\,2 = M(0;(7,-3),(5,2),(5,2)) = M(1;(7,4),(5,2),(5,2)),$$

$$13n_{2006} = -3\,2,3\,2,3 = M(0;(7,-3),(7,3),(3,1)) = M(1;(7,4),(7,3),(3,1)),$$

and

$$13n_{3142} = -2\,2,3\,1\,2,3 = M(0;(5,-2),(11,4),(3,1)) = M(1;(5,3),(11,4),(3,1)).$$

We provide examples of quasi-alternating links with two distinct minimal diagrams, where the standard diagram is not quasi-alternating, and the non-standard is: $(3,-2\,1)\,(2,2) = .2.-3\,0.2$, $(-2\,1,4)\,(2,2) = 6^*-3.-2.2\,0:2.-1$.

The method developed by J. Greene[56] can be used to prove that

$$L11n_{90} = (2,2+)\,-\,(2\,1,3),$$

shown on Fig. 13b, is a non-QA homologically thin link, which makes it the smallest link with this property. Among 11-crossing and 12-crossing links with no minimal quasi-alternating diagrams there are four 11-crossing candidates for homologically thin non-QA links:

- $L11n_{77}$ with Conway symbol $6^*2.(2,-2):2\,0$,
- $L11n_{226}$ with Conway symbol $6^*-2.2.-2:2\,1$,
- $L11n_{432}$ with Conway symbol $8^*-2\,0:-2\,0:.2\,0$,
- $L11n_{437}$ with Conway symbol $6^*-2.2.-2:2\,1\,0$,

and eight 12-crossing candidates:

[i]For 13-crossing knots we considered only their standard minimal diagrams, i.e., one minimal diagram for each knot.

- $2\,1\,1, 2\,2, -2\,1\,1$,
- $(-2\,1, 2\,1\,1)\,(2, 2+)$,
- $(-2\,1, -4)\,(2, 2+)$,
- $(2\,1, 3), 2, (2, -2)$,
- $6^*2 : .(-2\,1, 3)\,0$,
- $6^*3.(2, -2) : 2\,0$,
- $8^*2. - 2\,1\,0 :: 2$,
- $8^*(2, -2) :: -2\,0$.

Moreover, knots and links in the following families are candidates for infinite families of non-QA homologically thin knots and links[j]:
$(p, 2+) - (2\,1, 3)$ and $6^*p.(2, -2) : 2\,0$ $(p \geq 2)$.

Members of the first family (beginning with the non-QA homologically thin link $L11n_{90} = (2, 2+) - (2\,1, 3)$ and knot $12n_{331} = (3, 2+) - (2\,1, 3)$) are links for even p and knots for odd p, and the members of the other family are 2-component links.

References

1. Menasco, W. W., Thistlethwaite, M. B. (1991) *The Tait flyping conjecture*, Bull. Amer. Math. Soc., **25**, 2, 403–412.
2. Kauffman, L. H. (1987) *State Models and the Jones polynomial*, Topology, **26**, 395–407.
3. Murasugi, K. (1987) Jones polynomials and classical conjectures in knot theory, *Topology*, **26**, 187–194.
4. Murasugi, K. (1987) Jones polynomials and classical conjectures in knot theory II, *Math. Proc. Cambridge Philos. Soc.*, **102**, 317–318.
5. Thislethwaite, M. B. (1987) *A spanning tree expansion for the Jones polynomial*, Topology, **26**, 297–309.
6. Thistlethwaite, M.B. (1988) Kauffman's Polynomial and Alternating Links, *Topology*, **27**, 311–318.
7. Menasco, W. W., Thistlethwaite, M. B. (1993) *The classification of alternating links*, Annals of Math., **138** 113–171.
8. Jablan, S.V., Sazdanović, R. (2007) LinKnot – Knot Theory by Computer, World Scientific, New Jersey, London, Singapore, (`http://math.ict.edu.rs/`).
9. Przytycki, J.H. (1988) t_k-moves on links, *Contemporary Math.*, Vol. 78, Braids – Proceedings of the Santa Cruz conference on Artin's braid groups (July 1986), 615–655, `arXiv:math/0606633v1 [math.GT]`.
10. Murasugi, K. (1996) Knot Theory and its Applications, Birkhäuser, Boston, Basel, Berlin.
11. Stoimenow, A. (2001) On the crossing number of positive knots and braids and braid index criteria of Jones and Morton-Williams-Franks, `arXiv:math/0110016v1 [math.GT]`.
12. Wendt, H. (1937) Die Gordische Auflösung von Knoten, *Math. Zeit.*, **42**, 680–696.

[j] Odd homology was computed for these families up to 16 crossings.

13. Nakanishi, Y. (1981) A note on unknotting number, *Math. Sem. Notes Kobe Univ.*, **9**, 99–108.
14. Kawauchi, A. (1996) A Survey of Knot Theory, Birkhäuser, Basel, Boston, Berlin.
15. Kohn, P. (1991) Two-bridge links with unlinking number one, *Proc. Amer. Math. Soc.*, **113**, 1135–1147.
16. Kohn, P. (1993) *Unlinking two–component links*, Osaka J. Math., **30** 741–752.
17. Milnor, J.W. (1968) Singular points of complex hypersurfaces, Annals of Mathematics Studies 61, Princeton University Press, Princeton.
18. Kronheimer, P.B., Mrowka, T.S. (1993) Gauge Theory for Embedded Surfaces. I., *Topology*, **32**, 773–826.
19. Kronheimer, P.B., Mrowka, T.S. (1995) Gauge Theory for Embedded Surfaces. II., *Topology*, **34**, 37–97.
20. Adams, C.C. (1994) The Knot Book, Freeman, New York.
21. Nakanishi, Y. (1983) Unknotting numbers and knot diagrams with the minimum crossings, *Math. Sem. Notes Kobe Univ.*, **11**, 257–258.
22. Bleiler, S.A. (1984) A note on unknotting number, *Math. Proc. Camb. Phil. Soc.*, **96**, 469–471.
23. Bernhard, J.A. (1994) Unknotting numbers and their minimal knot diagrams, *J. Knot Theory Ramifications*, **3**, 1, 1–5.
24. Jablan, S.V., Sazdanovic, R. (2007) Unlinking number and unlinking gap, *J. Knot Theory Ramifications*, **16**, 10, 1331–1355.
25. Jablan, S.V. (1998) Unknotting number and ∞-unknotting number of a knot, *Filomat* (Niš) **12**, 1, 113–120.
26. Livingston, C., Cha J.C. (2005) Table of knot invariants, `http://www.indiana.edu/~knotinfo/`
27. Nakanishi, Y. (1996) Unknotting number and knot diagram, *Rev. Mat. Univ. Complut. Madrid*, **9**, 2, 359–366.
28. Stoimenow, A. (2004) On unknotting numbers and knot triviadjacency, *Mathematica Scandinavica* **94**, 2, 227–248.
29. Kanenobu, T., Murakami, H. (1986) Two-bridge knots with unknotting number one, *Proc. Amer. Math. Soc.*, **98**, 499–502.
30. Owens, B. (2008) Unknotting information from Heegaard Floer homology, *Advances in Mathematics* **217** 5, 2353–2376, `arXiv:math/0506485v1 [math.GT]`.
31. Nakanishi, Y. (2005) A note on unknotting number, II *J. Knot Theory Ramifications*, **14**, 1, 3–8.
32. Stoimenow, A. (2001) Some examples related to 4-genera, unknotting numbers, and knot polynomials, *Jour. London Math. Soc.*, **63**, 2, 487–500.
33. Caudron, A. (1982) Classification des nœuds et des enlancements, Public. Math. d'Orsay 82. Univ. Paris Sud, Dept. Math., Orsay.
34. Murasugi, K. (1965) On a certain numerical invariant of link types, *Trans. Amer. Math. Soc.*, **117**, 387–422.
35. Cromwell, P. (2004) Knots and links, Cambridge University Press, Cambridge.
36. Ozsváth, P., Szabó, Z. (2005) Knots with unknotting number one and Hee-

gaard Floer homology, *Topology* **44**, 4, 705–745, `arXiv:math/0401426v2 [math.GT]`.
37. Pabiniak, M., Przytycki, J.H., Sazdanović, R. (2009) On the first group of the chromatic cohomology of graphs, *Geometriae Dedicata*, Vol 140, No. 1 19–48, `arxiv: math.GT/0607326`.
38. Lickorish, W.B.R., Thistlethwaite, M. (1988) Some links with non-trivial polynomials and their crossing-numbers, *Comment. Math. Helvetici* **63**, 527–539.
39. Lickorish, W.B.R. (1991) An Introduction to Knot Theory, Springer-Verlag, New York, Berlin, Heidelberg.
40. Stoimenow, A. (2007) Non-triviality of the Jones polynomial and the crossing numbers of amphicheiral knots, `arXiv:math/0606255v2 [math.GT]`.
41. Thistlethwaite, M. (1988) Kauffman polynomial and adequate links, *Invent. Math.* **93**, 2, 258–296.
42. Kidwell, M.E., Stoimenow, A. (2003) Examples Relating to the Crossing Number, Writhe, and Maximal Bridge Length of Knot Diagrams, *Michigan Math. J.* **51**, 3–12.
43. Stoimenow, A. (2008) On the crossing number of semiadequate links, preprint.
44. Khovanov, M. (2000) A categorification of the Jones polynomial, Duke Math. J. **101**, 359-426 `arXiv:math/9908171v2 [math.QA]`.
45. Asaeda, M., Przytycki, J.H. (2004) Khovanov homology: torsion and thickness, *Advances in topological quantum field theory*, 135-166, NATO Sci. Ser. II Math. Phys. Chem., 179 `arXiv:math/0402402v2 [math.GT]`.
46. Lowrance, A., The Khovanov width of twisted links and closed 3-braids, `arXiv:0901.2196v2 [math.GT]`.
47. Ozsváth, P., Rasmussen, J., Szabó, Z. (2007) Odd Khovanov homology, `arXiv:math/0710.4300v1 [math.QA]`.
48. Manolescu, C., Ozsváth, P. (2007) On the Khovanov and knot Floer homologies of quasi-alternating links, *Proceedings of the 14th Gökova Geometry-Topology Conference* 60–81, `arXiv:math/0708.3249v2 [math.GT]`.
49. Champanerkar, A., Kofman, I. (2008) Twisting quasi-alternating links, *Proceedings of the American Mathematical Society 137* 2451–2458, `arXiv:math/0712.2990v3 [math.GT]`.
50. Ozsváth, P., Szabó, Z. (2005) On the Heegaard Floer homology of branched double-covers, *Adv. Math.* 194 **1**, 1–33, `arXiv:math/0309170v1 [math.GT]`.
51. Shumakovitch, A. (2008) Private communication.
52. Widmer, T. (2009) Quasi-alternating Montesinos links, *J. Knot Theory Ramifications*, **18**, 10, 1459–1469, `arXiv:math/0811.0270v1 [math.GT]`.
53. Shumakovitch, A. (2008) KhoHo, `http://www.geometrie.ch/KhoHo/`
54. Hoste, J., Thistlethwaite, M. *Knotscape*, `http://www.math.utk.edu/~morwen/knotscape.html`
55. Jablan, S.V., Sazdanović, R. (2008) Quasi-alternating links and odd homology: computations and conjectures, `arXiv:math/0901.0075v1 [math.GT]`.
56. Greene, J. (2009) Homologically thin, non-quasialternating links, `arXiv:math/0906.2222v1 [math.GT]`.

HARD UNKNOTS AND COLLAPSING TANGLES

Louis H. Kauffman and Sofia Lambropoulou

1. Introduction

Classical knot theory is about the classification, up to isotopy, of embedded closed curves in three-dimensional space. Two closed curves embedded in three-dimensional space are said to be *isotopic* if there is a continuous family of embeddings starting with one curve and ending with the other curve. This definition of isotopy captures our intuitive notion of moving a length of elastic rope about in space without tearing it. Given a closed loop of rope in space, it can sometimes be deformed in this way to a simple flat circle. In such a case we say that the loop is *unknotted*.

So it is not absurd if we sometimes ask: Is this knot diagram unknotted? The question is not only non-contradictory, it is often hard. Look ahead to Figures 5 and 6. The shapes illustrated are unknot diagrams, but most people's intuition does not allow recognizing the unknotting, although it will present itself if one makes a rope model and shakes it. Recognizing unknot diagrams is a starting point for all of knot theory.

Corresponding to the physical models, one can make mathematical models of knots by formalizing the diagrams that one naturally draws for them, as in Figure 2. The study based on the diagrams is called *combinatorial knot theory*. The diagrams are well defined as mathematical structures, namely, planar graphs with four edges incident to each vertex and with extra structure: indication at each vertex of which strand crosses over and which under.

This paper studies unknot diagrams and the collapsing of diagrams to knotted diagrams. We present an infinite collection of examples of unknot diagrams that are *hard*, in the sense that they cannot be simplified by Reidemeister moves without being made more complicated. We pinpoint the smallest hard unknot diagrams (Theorem 8), and we give a beautiful characterization of a certain class of unkots in terms of continued fractions and their convergents (Theorems 6 and 7). Our methods of proof are elementary and so we make the paper as self-contained as possible. Along the way, we develop and use John H. Conway's theory of *rational tangles*. Rational tangles allow algebraic operations, leading to connections with matrices and continued fractions. We extend our previous results [28] about collapsing of closures of tangle sums to the unknot to collapses to arbitrary knots and links. Thus we produce "hard" versions of knots and links as well as hard unknots and characterize these also in terms of continued fractions (Theorems 9 and 10).

Given a knot or link one can apply our methods of collapse in reverse, producing complex configurations that collapse back to that knot or link. This configuration can be allowed to undergo a crossing switch that locks it into a new knot or link that does not collapse. Thus we can create towers of knots and links such that each stage of the tower will collapse to the previous stage under a single crossing switch.

We envision this paper as useful for research knot theorists, for students beginning in knot theory and for scientists such as biologists who would like an entrance into knot theory that leads to connections with other fields. In particular, we have ended the paper with two sections about such connections. Section 12 is about the connection of our results with molecular biology, and it discusses how the form of the sequence of knots and links in a process of processive recombination stabilizes so that eventually extra twists are being added to the same place in the diagrammatic form of the produced knots and links. Along with this stabilization process, there is the general form of the recombination process itself that is related to our study of collapse. Our results apply to the process of successive recombination in DNA where a molecular configuration may undergo a crossing switch of DNA strands that locks it into a new topological configuration. Thus, in successive recombination towers of knots and links are created, each one a single crossing switch away from the next. The patterns of long sequences of such recombinations are beyond the reach of experiment, but can be understood mathematically. The results of this paper could apply to future experiments with DNA recombination. The reader may be interested in comparing our results with those of Isabel Darcy in her papers [7, 8].

Perhaps the most intriguing of our results is formulated in Theorem 7 which tells us that the numerator closure of the sum of two rational tangles is an unknot if and only if the continued fractions for the two tangles are convegents of one another. This means that one continued fraction is a one-step truncate of the other, and it means that if P/Q and R/S are the reduced fractions, then $|PS - QR| = 1$. Fractions with this relationship come up in many mathematical contexts. Section 13 makes connection with other mathematics via this very relationship. Here we contact Farey series, Pick's Theorem and the Ford circles in the hyperbolic plane. This section demonstrates how far-reaching are the mathematical processes involved in the study of fundamental questions in topology.

Acknowledgements. The first author thanks the National Science Foundation for support of this research under NSF Grant DMS-0245588. It gives both authors pleasure to acknowledge the hospitality of the Mathematisches Forschungsinstitut Oberwolfach, the University of Illinois at Chicago and the National Technical University of Athens, Greece, where much of this research was conducted. We particularly thank Slavik Jablan for conversations and for helping us by using his computer program LinKnot.

2. The Culprit

Combinatorial knot theory got its start in the hands of Kurt Reidemeister [38], who discovered a set of moves on planar diagrams that capture the topology of knots and links embedded in three dimensional space. Reidemeister proved that if we have two knots or links in three-dimensional space, then they are isotopic if and only if corresponding diagrams for them can be obtained, one from the other, by a sequence of moves of the types shown in Figure 1.

The moves are local and they are performed on the diagrams in such a way that the rest of the diagram is left fixed. This means that when one searches for available Reidemeister moves on a diagram, one searches for one-sided, two-sided, and three-sided regions with the approriate crossing patterns as shown in Figure 1. It is important to note that knot and link diagrams are taken to be on the surface of a two dimensional sphere that is standardly embedded in three-dimensional space. Thus the outer region of a planar diagram is handled just like any other region in that diagram, and can be the locus of any of the moves. In this section, we discuss the unknotting of knot diagrams using the Reidemeister moves.

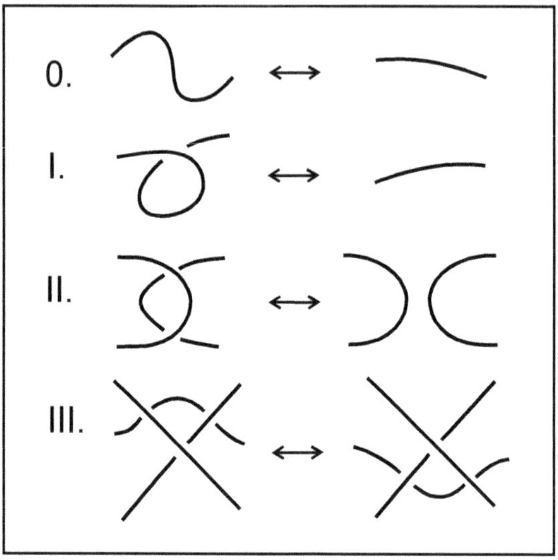

Fig. 1. The Reidemeister Moves

Definition 1. We define the *complexity* of a knot or link diagram K to be the number of crossings, $C(K)$, in the diagram. A Reidemeister move is said to be *simplifying* if it reduces the crossing number of the diagram. We say that a diagram *can be simplified* (to the unknot) if there is a sequence of simplifying Reidemeister moves, combined with type III Reidemeister moves that transform the diagram to the unknot diagram. We shall call a diagram of the unknot *hard* if it has the following two properties.

(1) There are no simplifying type I or type II moves on the diagram.

(2) There are no type III moves on the diagram.

Hard unknot diagrams must be made more complex before they will simplify to the unknot, if we use Reidemeister moves.

Here is an example of a hard unknot diagram (originally due to Ken Millett [34]), in Figure 2. We like to call this diagram "The Culprit". The Culprit is a hard unknot diagram. It has 10 crossings, and in order to be undone, we definitely have to increase the number of crossings before decreasing them to zero. The reader can verify this for himself by checking each region in the diagram of the Culprit. A simplifying Reidemeister II-move can occur only on a two-sided region, but no two-sided region in the diagram admits such a move. Similarly, on the Culprit diagram there are no simplifying Redeimeister I-moves and there are no Reidemeister III-moves (note that a III-move does not change the complexity of the diagram). Culprit is just our term for a hard unknot, and we usually refer to Millett's example as "The Culprit" in this paper because it was our detective work in tracking the properties of this culprit that led to our results and questions.

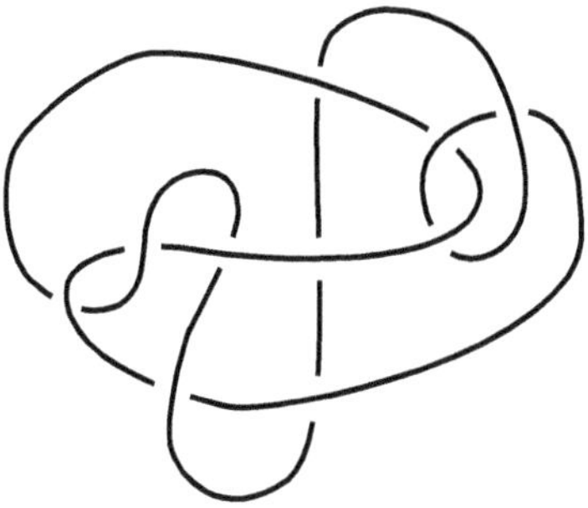

Fig. 2. The Culprit

View Figure 3 for an unknotting sequence for the Culprit. Notice that we undo it by swinging the arc that passes underneath most of the diagram outward, and that in this process the number of crossings in the intermediate diagrams increases. In the diagrams of Figure 3 the largest increase is to a diagram of 12 crossings. This is the best possible result for the initial diagram in the figure.

In contrast to the Culprit, view Figure 4 for an example of an easy unknot. This one simplifies after Reidemeister III-moves unlock further moves of type I and type II. The reader can probably just look at the diagram in Figure 4 and see at once that it is unknotted.

It is an unsolved problem just how much complexity can be forced by a hard unknot diagram.

Definition 2. Let K be a hard unknot diagram. Let K' be a diagram isotopic to K such that K' can be simplified to the unknot. For any unknotting

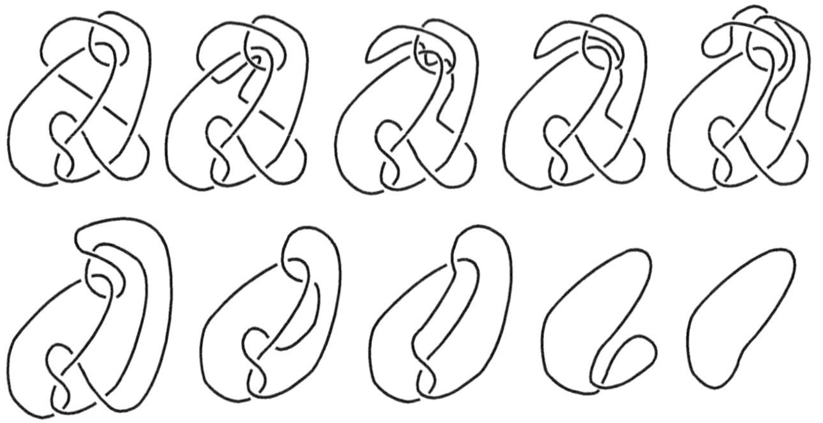

Fig. 3. The Culprit Undone

Fig. 4. An Easy Unknot

sequence of Reidemeister moves for K there will be a diagram K'_{max} with a maximal number of crossings. Let $Top(K)$ denote the minimum of $C(K'_{max})$ over all unknotting sequences for K. Let

$$R(K) = Top(K)/C(K)$$

be called the *recalcitrance* of the hard unknot diagram K. Very little is known about $R(K)$.

In the case of our Culprit, $Top(K) = 12$, while $C(K) = 10$. Thus $R(K) = 1.2$. A knot that can be simplified with no extra complexity has recalcitrance equal to 1. We shall have more to say about the recalcitrance later.

In [20] Hass and Lagarias show that there exists a positive constant c_1 such that, for each $n > 1$, any unknotted diagram K with n crossings can be transformed to the trivial knot diagram using less than or equal to $2^{c_1 n}$ Reidemeister moves. As a corollary to this result, they conclude that any unknotted diagram K can be transformed by Reidemeister moves to the trivial knot diagram through a sequence of knot diagrams each of which has at most $2^{c_1 n}$ crossings. In our language, this result says that

$$R(K) \leq \frac{2^{c_1 C(K)}}{C(K)}.$$

The authors of [20] prove their result for $c_1 = 10^{11}$, but remark that this is surely too large a constant. Much more work needs to be done in this domain. We should also remark that the question of the knottedness of a knot is algorithmically decideable, due to the work of Haken and Hemion [21]. This algorithm is quite complex, but its methods are used in the work of Hass and Lagarias.

In fact, we can now do considerably better in estimating $R(K)$. In [22] it is shown that if the diagram K represents the unknot and if K is presented with a height function in the plane (a Morse diagram) with $m(K)$ maxima and $c(K)$ crossings then one can unknot K by Reidemeister moves using a diagram K' with no more than $(c(K)+2m(K)-2)^2$ crossings. This means that

$$R(K) \leq \frac{(c(K)+2m(K)-2)^2}{c(K)}.$$

The problem of bounding the number of Reidemeister moves needed to unknot K is still difficult. A technique of working with large-scale moves is formalized in the work of Dynnikov [9] where he shows, using special grid diagrams and a complete set of moves distinct from Reidemeister moves, how to unknot diagrams without making the special diagrams more complex. Dynnikov's methods are used in proving the results in [22].

We mention the phenomenon shown in Figure 26. Here the first diagram illustrated is a hard unknot published by Goeritz [18] in 1934. We show how to unknot the Goeritz diagram in this figure by illustrating large-scale swing moves that are combinations of Reidemeister moves. Each swing move passes through diagrams that are larger than the original diagram before the swing is started. Note that in the figure there are a number of swing moves so that, using Reidemeister moves, the number of crossings in the diagram goes up and down a number of times. Furthermore, right in the middle of the process we arrive at another, smaller, hard unknot diagram (here labeled H). In this way we discovered H, the smallest possible hard unknot diagram.

The next example is illustrated in Figure 5, a three-dimensional rendering of an unknot. This example is interesting psychologically, because it looks knotted to most observers. The three-dimensional picture is a frame from a deformation of this example in the energy minimization program KnotPlot [40]. It is illuminating to run the program on this example and watch the knot self-repel and undo itself. We leave the verification that this is an unknot as an exercise for the reader.

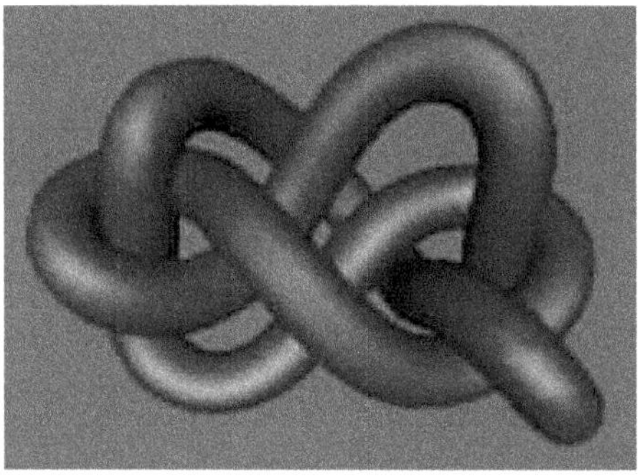

Fig. 5. Three-Dimensional Unknot

In Figure 6 we give an example of a hard unknot that is of a different type than the sort that we are considering in this paper. (It is not the closure of the sum of two rational tangles.) There are many knots and many hard unknots. We only scratch the surface of this subject.

For work on unknots related to braids see [35]. Another approach to detecting unknots is the use of invariants of knots and links. For example, it is conjectured that the Jones polynomial [24, 25] of a knot diagram is equal to one, only when that knot is unknotted.

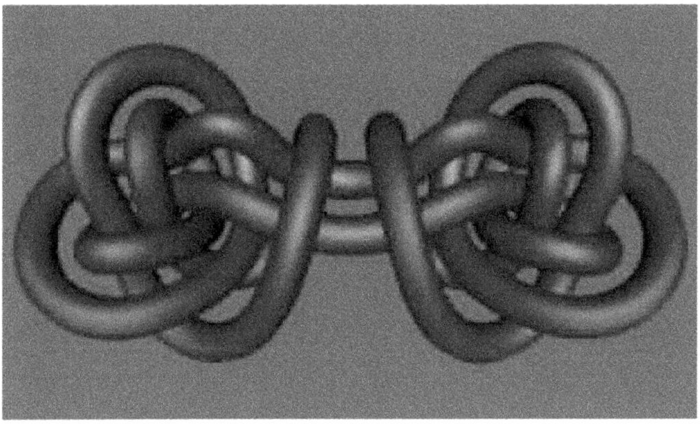

Fig. 6. Another Hard Unknot

3. Rational Tangles, Rational Knots and Continued Fractions

In this section we recall the subject of rational tangles and rational knots and their relationship with the theory of continued fractions. By the term "knots" we will refer to both knots and links, and whenever we really mean "knot" we shall emphasize it. Rational knots comprise the simplest class of knots. They are also known in the literature as Viergeflechte, four-plats or 2-bridge knots depending on their geometric representation. The theory of tangles was introduced in 1967 by Conway [5] in his work on enumerating and classifying knots and links.

A 2-*tangle* is a proper embedding of two unoriented arcs and a finite number of circles in a 3-ball B^3, so that the four endpoints lie in the boundary of B^3. A *tangle diagram* is a regular projection of the tangle on a cross-sectional disc of B^3. By "tangle" we will mean "tangle diagram". A *rational tangle* is a special case of a 2-tangle obtained by applying consecutive twists on neighbouring endpoints of two trivial arcs. Such a pair of arcs comprise the $[0]$ or $[\infty]$ tangles (see Figure 8), depending on their position in the plane. We shall say that such a tangle is in *twist form*. For an example see Figure 7. Conway defined the rational knots as "numerator" or "denominator" closures of the rational tangles. See Figure 7. Conway [5] also defined *the fraction* of a rational tangle to be a rational number or ∞, obtained via a continued fraction that is associated with the tangle. We discuss this construction below.

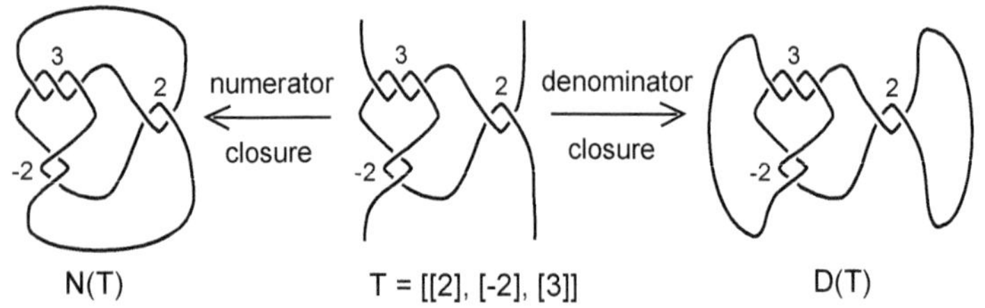

Fig. 7. A Rational Tangle and its Closures to Rational Knots

We are interested in tangles up to isotopy. Two tangles, T, S are *isotopic*, denoted by $T \sim S$, if and only if they have identical configurations of their four endpoints, and they differ by a finite sequence of the Reidemeister moves which take place in the interior of the projection disc. Of course, each twisting operation used in the definition of a rational tangle changes the isotopy class of the tangle to which it is applied. Rational tangles are classified by their fractions by means of the following theorem, different proofs of which are given in [36], [3, 4], [19] and [26].

Theorem 1 (Conway, 1975). *Two rational tangles are isotopic if and only if they have the same fraction.*

3.1. The Tangle Fraction

We shall now recall from [26] the main properties of rational tangles and of continued fractions, which illuminate the classification of rational tangles. The elementary rational tangles are displayed as either horizontal or vertical twists, and they are enumerated by integers or their inverses, see Figure 8.

The crossing types of 2-tangles (and of unoriented knots) follow the *checkerboard rule*: shade the regions of the tangle in two colors, starting from the left outside region with grey, and so that adjacent regions have different colors. Crossings in the tangle are said to be of "positive type" if they are arranged with respect to the shading as exemplified in Figure 8 by the tangle [+1], i.e. they have the region on the right shaded as one walks towards the crossing along the over-arc. Crossings of the reverse type are said to be of "negative type" and they are exemplified in Figure 8 by the tangle $[-1]$. The reader should note that our crossing type conventions are the opposite of those of Conway in [5]. Our conventions agree with those of Ernst and Sumners [12, 13], [43] which, in turn, follow the standard conventions of biologists.

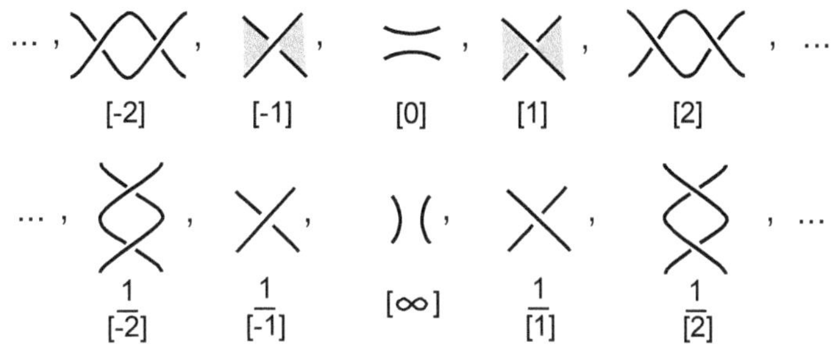

Fig. 8. The Elementary Rational Tangles and the Types of Crossings

In the class of 2-tangles we have the operations *addition* and *multiplication*, as illustrated in Figure 9, which are denoted by "+" and "$*$" respectively. These operations are well-defined up to isotopy. In order to read out a rational tangle we transcribe it as an algebraic expression using horizontal and vertical twists. For example, Figure 10 illustrates the rational tangle in twist form: $[1]+([1]*[3]*\frac{1}{[-3]})+[1]$, and also an isotopic form of the same tangle with expression $(([3]*\frac{1}{[-2]})+[2])$. The decomposition of this last tangle is shown in Figure 11.

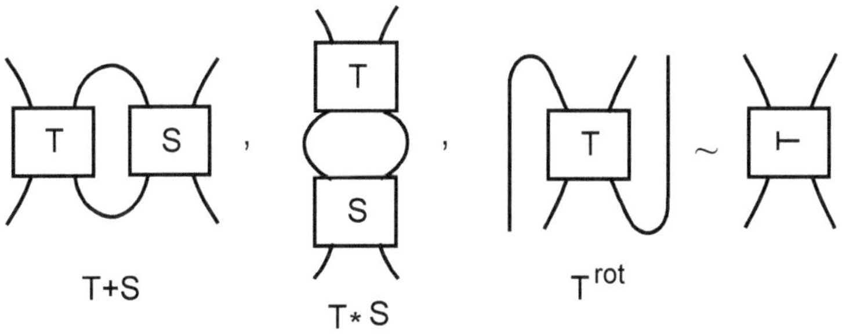

Fig. 9. Addition, Multiplication and Rotation of 2-Tangles

Note that addition and multiplication do not, in general, preserve the class of rational tangles (but every rational tangle can be constructed from the $[0]$ tangle or the $[\infty]$ tangle by using these two operations). For example, the 2-tangle $\frac{1}{[3]} + \frac{1}{[3]}$ is not rational. The sum (product) of two rational tangles is rational if and only if one of the two consists in a number of horizontal (vertical) twists.

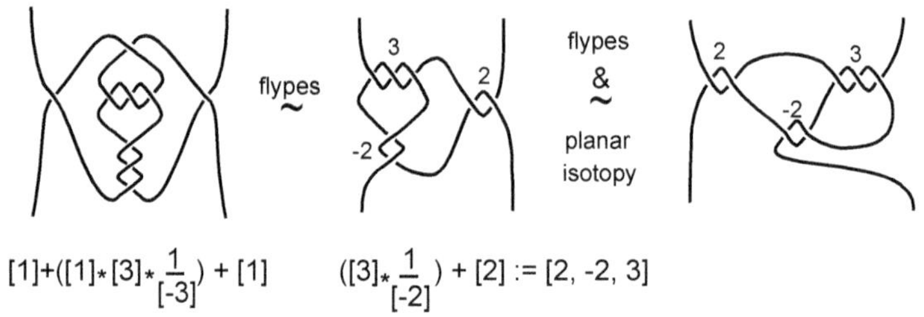

Fig. 10. A Rational Tangle in Twist Form Converted to Its Standard Form and to Its 3-Strand-Braid Representation

The *mirror image* of a tangle T, denoted $-T$, is T with all crossings switched. For example, $-[n] = [-n]$ and $-\frac{1}{[n]} = \frac{1}{[-n]}$. Then, the *subtraction* is defined as $T - S := T + (-S)$. The *rotation* of T, denoted T^{rot}, is obtained by rotating T on its plane counterclockwise by 90°. The *inverse* of T is defined to be $-T^{rot}$. Thus, inversion is accomplished by rotation and mirror image. Note that T^{rot} and the inverse of T are in general not isotopic to T and they are order 4 operations. But for rational tangles the inversion is an operation of order 2. (This follows from *the remarkable property of rational tangles that they are isotopic to their horizontal or vertical flip in space by* 180°. Note that a flip switches the endpoints of the tangle and, in general, a flipped tangle is not isotopic to the original one.) For this reason we shall denote the inverse of a rational tangle T by $1/T$, and hence the rotation of the tangle T will be denoted by $-1/T$. This explains the notation for the tangles $\frac{1}{[n]}$.

As we said earlier, there is a fraction associated to a rational tangle which characterizes its isotopy class (Theorem 1). In Figure 11, we illustrate how one obtains the fraction by using the decomposition of a rational tangle into sums and products. Notice, in that figure, how the tangle labeled x is rotated by 90° (becoming $-1/x$) and is then recognizable as $-1/x = -1/[3] + [2]$. The arithmetical operations for computing the tangle fraction are in direct parallel to this "topological arithmetic". Thus the fraction associated with x is $F(x) = -1/(-1/3 + 2)$ and, putting this together, the fraction for the tangle in the figure is $F(T) = 2 - 1/(-1/3+2)$.

In fact, the fraction is defined for any 2-tangle and always obeys the following three rules:

(1) $F([\pm 1]) = \pm 1$.
(2) $F(T+S) = F(T) + F(S)$.
(3) $F(T^{rot}) = -1/F(T)$.

If $T = [\infty]$, we assign $F([\infty]) := \infty = \frac{1}{0}$, as a formal expression.

These rules suffice for computing the fraction inductively for rational tangles. In this way, they serve as the definition of the tangle fraction for rational tangles. Read Figure 11 again and note that the computation follows exactly the rules above. Generalizing, the reader can easily deduce the formula below:

$$T * \frac{1}{[n]} = \frac{1}{[n] + \frac{1}{T}}$$

Indeed, rotate $T * \frac{1}{[n]}$ by $90°$ and note that it becomes $-[n] - \frac{1}{T}$. The original tangle is the negative reciprocal of this tangle. This formula implies that the two operations: addition of $[+1]$ or $[-1]$ and inversion between rational tangles suffice for generating the whole class of rational tangles. As for the fraction, we have the corresponding formula:

$$F(T * \frac{1}{[n]}) = \frac{1}{n + \frac{1}{F(T)}}.$$

For example, in Figure 11, the fraction associated with the tangle T is

$$F(T) = 2 + \frac{1}{-2 + \frac{1}{3}}.$$

Starting from a rational tangle in twist form, one can always bring the twists alternately to the right and to the bottom of the tangle using only isotopy moves on subtangles, called *flypes*. See Figure 12 for an abstract illustration of flypes. Note that in a flype the endpoints of the subtangle remain fixed; this is not true for a flip of the tangle in space. One can then associate to the rational tangle a vector of integers $(a_1, a_2, \ldots, a_n)$, where

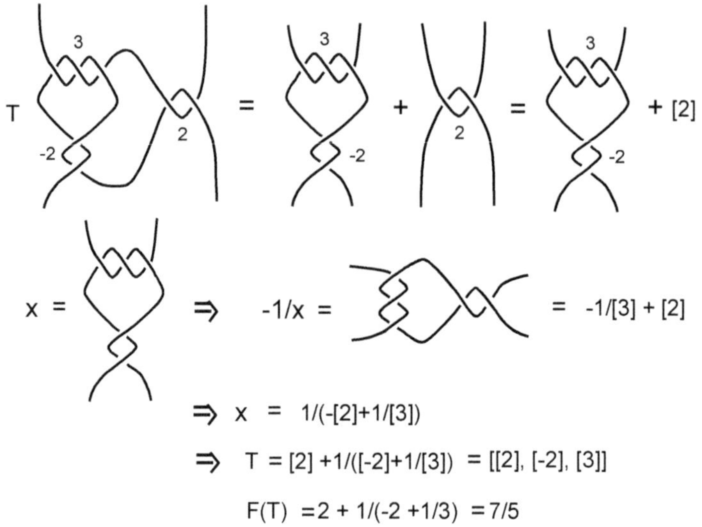

Fig. 11. Finding the Fraction

the first entry denotes the place where the tangle starts untwisting and the last entry where it begins to twist. For example the tangle of Figure 7 corresponds to the vector $(2, -2, 3)$. We then say that the rational tangle is in *standard form* and we denote it by $T = [[a_1], [a_2], \ldots, [a_n]]$. We also call this standard form the *continued fraction form* of the tangle, since it can literally be written in the form of a continued fraction involving the elementary integer tangles $[a_i]$.

From the above discussion it follows that for a rational tangle in standard form,

$$T = [[a_1], [a_2], \ldots, [a_n]] = [a_1] + \cfrac{1}{[a_2] + \cdots + \frac{1}{[a_{n-1}] + \frac{1}{[a_n]}}}$$

for $a_1 \in \mathbb{Z}$, $a_2, \ldots, a_n \in \mathbb{Z} - \{0\}$, its fraction is the numerical continued fraction

$$F(T) = [a_1, a_2, \ldots, a_n] := a_1 + \cfrac{1}{a_2 + \cdots + \frac{1}{a_{n-1} + \frac{1}{a_n}}}.$$

Since fractions and rational tangles correspond, we will, from now on, use the notation $[a_1, \ldots, a_n]$ to denote both the continued fraction and a rational tangle in continued fraction form.

Fig. 12. The Flype Moves

The subject of continued fractions is of perennial interest to mathematicians. See for example [30], [37], [33], [45]. Here we only consider continued fractions of the above type, with all numerators equal to 1. As in the case of rational tangles, we allow the term a_1 to be zero. Clearly, the two simple algebraic operations *addition of* $+1$ *or* -1 and *inversion* generate inductively the whole class of continued fractions, starting from zero.

If a rational tangle T changes by an isotopy, the associated continued fraction form may also change, while the fraction remains invariant. For example, $[2,-2,3]=[1,2,2]=\frac{7}{5}$, see Figure 13. Indeed, the rational tangles $[2,-2,3]$ and $[1,2,2]$ are isotopic as shown by the swing moves of Figure 13. A *swing move* is an isotopy move in which we swing an arc joining two consecutive crossings of opposite type across a subtangle box. See the first move in Figure 13. For the isotopy type of a rational tangle T with fraction $\frac{p}{q}$ we shall use the notation $[\frac{p}{q}]$.

The key to the exact correspondence of fractions and rational tangles lies in the construction of a canonical alternating form for the rational tangle. A tangle is said to be *alternating* if it is isotopic to a tangle where the crossings alternate from under to over as we go along any component or arc of the weave. Similarly, a knot is alternating if it possesses an alternating diagram. Using the swing moves and an induction argument, one shows that rational tangles (and rational knots) are alternating. We shall say that the rational tangle $S=[\beta_1,\beta_2,\ldots,\beta_m]$ is in *canonical form* if S is alternating and m is odd. From the above, S alternating implies that the β_i's are all of the same sign. It turns out that the canonical form for S is unique. In Figure 13 we bring our working rational tangle $T=[2,-2,3]$ to its canonical form $S=[1,2,2]$.

On the other hand, by Euclid's algorithm and keeping all remainders of the same sign, one can show that every continued fraction $[a_1,a_2,\ldots,a_n]$

Fig. 13. Reducing to Alternating Form Using the Swing Moves

can be transformed to a unique canonical form $[\beta_1, \beta_2, \ldots, \beta_m]$, where all β_i's are positive or all negative integers and m is odd. For example,

$$[2, -2] = [1, 1, 1] = \frac{3}{2}.$$

There is also an algorithm that can be applied directly to the initial continued fraction to obtain its canonical form, which works in parallel with the algorithm for the canonical form of rational tangles. Indeed, we have:

Proposition 1. *The following identity is true for continued fractions and it is also a topological equivalence of the corresponding tangles:*

$$[..., a, -b, c, d, e, ...] = [..., (a-1), 1, (b-1), -c, -d, -e, ...].$$

This identity gives a specific inductive procedure for reducing a continued fraction to all positive or all negative terms. In the case of transforming to all negative terms, we can first flip all signs and work with the mirror image. Note also that

$$[..., a, b, 0, c, d, e, ...] = [..., a, b+c, d, e, ...]$$

will be used in these reductions.

Proof. The identity in the Proposition follows immediately from the formula:

$$a + 1/(-b) = (a-1) + 1/(1 + 1/(b-1)).$$

If a and b are positive, this formula allows the reduction of negative terms in a continued fraction. □

For the study of rational knots it is easier to use another way of representing an abstract rational tangle in standard form, illustrated in Figure

10. This is the 3-*strand-braid representation.* As illustrated in Figure 10, the 3-strand-braid representation is obtained from the standard representation by planar rotations of the vertical sets of crossings, thus creating a lower row of horizontal crossings. By the checkerboard shading convention, the signs of the crossings in the lower row do not change under these rotations.

Note that the set of twists of a rational tangle may be always assumed *odd.* Indeed, let n be even and let the left-most twist $[a_1]$ be on the upper part of the 3-strand braid representation. Up to the ambiguity (upper or lower row) of the right-most crossing, the vector associated to a rational tangle is *unique*, i.e. $(a_1, a_2, \ldots, a_n) = (a_1, a_2, \ldots, a_n - 1, 1)$, if $a_n > 0$, and $(a_1, a_2, \ldots, a_n) = (a_1, a_2, \ldots, a_n + 1, -1)$, if $a_n < 0$.

Remark 1. We have omitted here the proof of the invariance of the fraction. The interested reader can consult [5], [19], [26] for various proofs of this fact. One definition of the fraction of a general 2-tangle is:

$$F(T) = i < N(T) > / < D(T) >,$$

where $i^2 = -1$ and $< L > = < L > (A)$ denotes the bracket polynomial [25] of a link L with $A = i^{1/2}$. Here $N(T)$ and $D(T)$ are the numerator and denominator closures of the tangle T, as illustrated in Figure 7.

3.2. Rational Knots and Continued Fractions

By joining with simple arcs the two upper and the two lower endpoints of a 2-tangle T, we obtain a knot called the *Numerator* of T, denoted $N(T)$. Joining with simple arcs each pair of the corresponding top and bottom endpoints of T we obtain the *Denominator* of T, denoted by $D(T)$, see Figure 7. Note that $N(T) = D(T^{rot})$ and $D(T) = N(T^{rot})$. As we shall see in the next section, the numerator closure of the sum of two rational tangles is still a rational knot. But the denominator closure of the sum of two rational tangles is not necessarily a rational knot. For example, consider the denominator of the sum $\frac{1}{[3]} + \frac{1}{[3]}$.

A rational knot is defined to be the numerator of a rational tangle. Given two different rational tangle types $[\frac{p}{q}]$ and $[\frac{p'}{q'}]$, when do they close to isotopic rational knots? More than one rational tangle can yield isotopic rational knots. The corresponding equivalence relation between the rational tangles is reflected by an arithmetic equivalence of their corresponding fractions. Rational knots and links are one of the few cases in knot theory where we have a complete classification story. Indeed, we have the following theorem:

Theorem 2 (Schubert, 1956). *Suppose that rational tangles with fractions $\frac{p}{q}$ and $\frac{p'}{q'}$ are given. Here p and q are relatively prime, similarly for p' and q'. If $K(\frac{p}{q})$ and $K(\frac{p'}{q'})$ denote the corresponding rational knots obtained by taking numerator closures of these tangles, then $K(\frac{p}{q})$ and $K(\frac{p'}{q'})$ are isotopic if and only if*

(1) $p = p'$ *and*
(2) either $q \equiv q' \bmod p$ *or* $qq' \equiv 1 \bmod p$.

Different proofs of Theorem 2 are given in [41], [3], [27]. The proof in [27] is the only published combinatorial proof of Schubert's Theorem.

Schubert classified rational knots by finding canonical forms via representing them as 2-bridge knots. In [27] we gave a new combinatorial proof of Theorem 2, by posing the question: given a rational knot diagram, at which places may one cut it so that it opens to a rational tangle? We then pinpoint two distinct categories of cuts that represent the two cases of the arithmetic equivalence of Schubert's theorem. The first case corresponds to the *special cut*, as illustrated in Figure 14. The two tangles $T = [-3]$ and $S = [1] + \frac{1}{[2]}$ are non-isotopic by the Conway Theorem (Theorem 1), since $F(T) = -3 = 3/(-1)$, while $F(S) = 1 + 1/2 = 3/2$. But they have isotopic numerators: $N(T) \sim N(S)$, the left-handed trefoil. Now $-1 \equiv 2 \bmod 3$, confirming Theorem 2. See [27] for a complete analysis of the special cut.

The second case of Schubert's equivalence corresponds to the *palindrome cut*, an example of which is illustrated in Figure 15. Here we see that the tangles

$$T = [2, 3, 4] = [2] + \frac{1}{[3] + \frac{1}{[4]}}$$

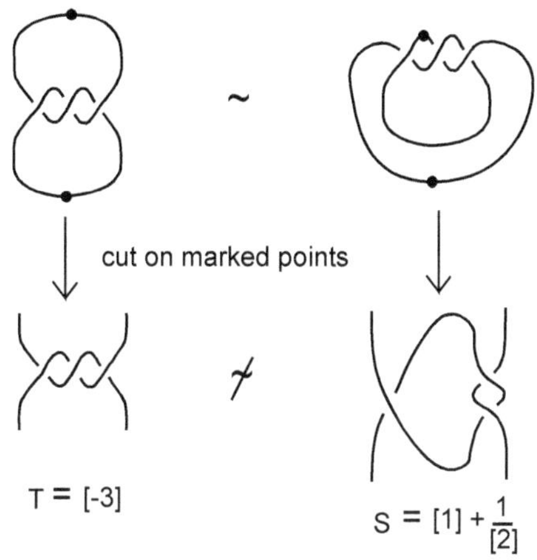

Fig. 14. An Example of the Special Cut

and

$$S = [4, 3, 2] = [4] + \frac{1}{[3] + \frac{1}{[2]}}$$

both have the same numerator closure. Their corresponding fractions are

$$F(T) = 2 + \frac{1}{3 + \frac{1}{4}} = \frac{30}{13} \quad \text{and} \quad F(S) = 4 + \frac{1}{3 + \frac{1}{2}} = \frac{30}{7}.$$

Note that $7 \cdot 13 \equiv 1 \bmod 30$, again confirming Theorem 2.

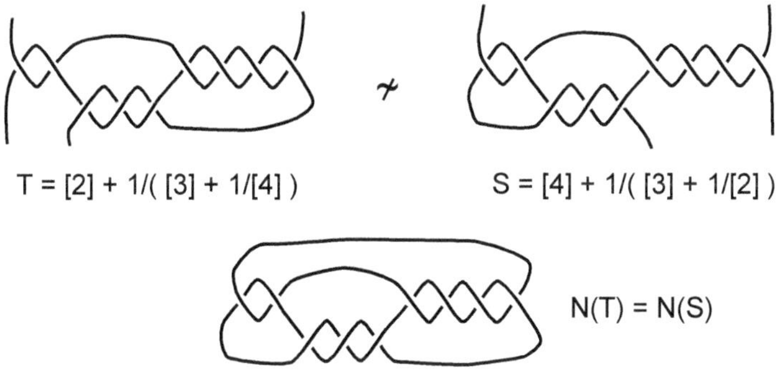

Fig. 15. An Instance of the Palindrome Equivalence

In the general case if $T = [a_1, a_2, \ldots, a_n]$, we shall call the tangle $S = [a_n, a_{n-1}, \ldots, a_1]$ *the palindrome of* T. Clearly these tangles have the same numerator. In order to check the arithmetic in the general case of the palindrome cut we need to generalize this pattern to arbitrary continued fractions and their palindromes (obtained by reversing the order of the terms).

Theorem 3 below is a known result about continued fractions. See [27], [42] or [29]. We shall give here our proof of this statement. For this we will first present a way of evaluating continued fractions via 2×2 matrices (compare with [17], [33]). This method of evaluation is crucially important for the rest of the paper. We define matrices $M(a)$ by the formula

$$M(a) = \begin{pmatrix} a & 1 \\ 1 & 0 \end{pmatrix}.$$

These matrices $M(a)$ are said to be the *generating matrices* for continued fractions, as we have:

Theorem 3 (The Palindorome Theorem). *Let* $\{a_1, a_2, \ldots, a_n\}$ *be a collection of* n *integers, and let*

$$\frac{P}{Q} = [a_1, a_2, \ldots, a_n]$$

and

$$\frac{P'}{Q'} = [a_n, a_{n-1}, \ldots, a_1].$$

Then $P = P'$ *and* $QQ' \equiv (-1)^{n+1} \bmod P$.

Moreover, for any sequence of integers $\{a_1, a_2, \ldots, a_n\}$ *the value of the corresponding continued fraction*

$$\frac{P}{Q} = [a_1, a_2, \ldots, a_n]$$

is given through the following matrix product

$$M = M(a_1)M(a_2)\ldots M(a_n)$$

via the identity

$$M = \begin{pmatrix} P & Q' \\ Q & U \end{pmatrix}$$

where this matrix also gives the evaluation of the palindrome continued fraction

$$[a_n, a_{n-1}, \ldots, a_1] = \frac{P}{Q'}.$$

Proof. Let

$$\frac{R}{S} = [a_2, a_3, \ldots, a_n].$$

Then

$$\frac{P}{Q} = [a_1, a_2, \ldots, a_n] = a_1 + \frac{1}{\frac{R}{S}} = a_1 + \frac{S}{R} = \frac{Ra_1 + S}{R}.$$

By induction we may assume that

$$M(a_2)M(a_3)\ldots M(a_n) = \begin{pmatrix} R & S' \\ S & V \end{pmatrix}.$$

Hence

$$M(a_1)M(a_2)\ldots M(a_n) = \begin{pmatrix} a_1 & 1 \\ 1 & 0 \end{pmatrix}\begin{pmatrix} R & S' \\ S & V \end{pmatrix} = \begin{pmatrix} a_1R + S & a_1S' + V \\ R & S' \end{pmatrix}.$$

This proves by induction that

$$M = M(a_1)M(a_2)\ldots M(a_n) = \begin{pmatrix} P & Q' \\ Q & U \end{pmatrix}.$$

To see the result about the palindrome continued fraction (and the terms P and Q' in the matrix), let M^T denote the transpose of M. Then

$$M^T = M(a_n)M(a_{n-1})\cdots M(a_1),$$

from which the statement about the palindrome follows by a repeat of the above calculation. Note also that $Det(M) = (-1)^n$ since M is a product of n matrices of determinant equal to -1, and $Det(M) = PU - QQ'$. Hence

$$PU - QQ' = (-1)^n.$$

This last equation implies the congruence stated at the beginning of the Theorem, and completes the proof of the Theorem. □

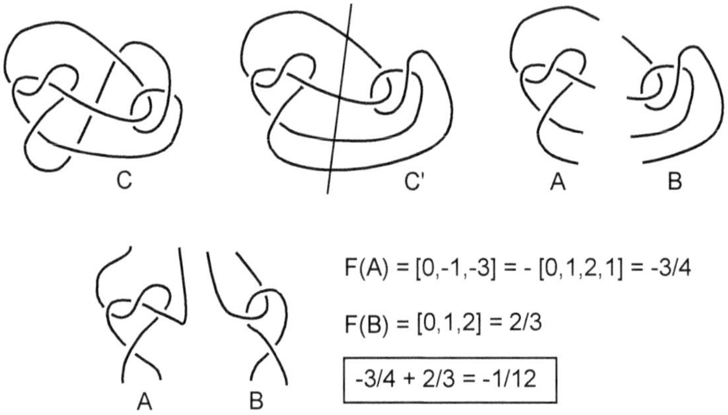

Fig. 16. The Culprit Analysis

4. The Return of the Culprit

In order to see the Culprit in a way that allows us to understand why it is unknotted, we shall use the language and technique of the theory of tangles as explained in the previous section.

View Figure 16. In Figure 16 we have a drawing C of the Culprit and next to it we have a drawing C' of the result of part of Figure 3, where the under-crossing arc in C has been moved (making the complexity rise) until it has been drawn outside the rest of the diagram. Concentrate on C'. Notice that we can cut C' into two pieces, as shown in Figure 16. These two pieces, A and B, are rational tangles, and this cutting process shows that C' is the numerator closure of the tangle sum of A and B. This is written $C' = N(A + B)$. Each rational tangle T has a rational fraction $F(T)$ that tells all about it. In this case,

$$F(A) = \frac{1}{-1 + \frac{1}{-3}} = -3/4$$

and

$$F(B) = \frac{1}{1 + \frac{1}{2}} = 2/3.$$

We know that the numerator of $A + B$ is unknotted and we would like to understand why it is unknotted. We notice that the sum of the fractions of

A and B is

$$F(A) + F(B) = -3/4 + 2/3 = -1/12.$$

Thus the numerator of the sum of the fractions of A and B is -1. Does this -1 imply the unknottedness of the numerator of $A + B$? Well, the answer is that it does, and that will be the subject of much of the rest of the paper. See Theorem 6.

5. Collapsing to Unknots and Unlinks

In this section we first prove that the numerator of the sum of two rational tangles is a rational knot or link. We identify the knot or link, and use the result to characterize in Theorem 6 those unknots and unlinks that arise as numerators of the sum of two rational tangles.

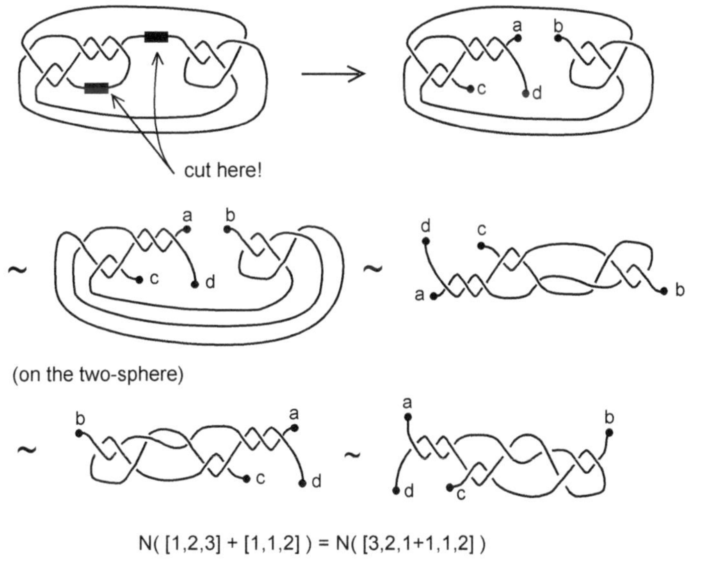

Fig. 17. The Numerator of a Sum of Rational Tangles is a Rational Link

Theorem 4 (Addition of Rational Tangles). *Let* $\frac{P}{Q} = [a_1, a_2, \ldots, a_n]$ *and* $\frac{R}{S} = [b_1, b_2, \ldots, b_m]$. *Let* $A = [\frac{P}{Q}]$ *and* $B = [\frac{R}{S}]$ *be the corresponding rational*

tangles. Then the knot or link $N(A+B)$ *is rational. In fact*

$$N(A+B) = N([a_n, a_{n-1}, \ldots, a_2, a_1 + b_1, b_2, \ldots, b_m]).$$

Proof. View Figure 17. In this figure we illustrate a special case of the Theorem. The general case is clear from this illustration. □

Definition 3. Given continued fractions $\frac{P}{Q} = [a_1, \ldots, a_n]$ and $\frac{R}{S} = [b_1, \ldots, b_m]$, let

$$[a_1, \ldots, a_n] \sharp [b_1, \ldots, b_m] := [a_n, \ldots, a_2, a_1 + b_1, b_2, \ldots, b_m].$$

If

$$\frac{F}{G} = [a_n, \ldots, a_2, a_1 + b_1, b_2, \ldots, b_m],$$

we shall write

$$\frac{P}{Q} \sharp \frac{R}{S} = \frac{F}{G}.$$

By Theorem 4, $N([\frac{P}{Q} \sharp \frac{R}{S}]) = N([\frac{P}{Q}] + [\frac{R}{S}])$.

Remark 2. The reader should note that $N([\frac{P}{Q}] + [\frac{R}{S}])$ is *not*, in general, isotopic to $N([\frac{P}{Q} + \frac{R}{S}]) = N([\frac{PS+QR}{QS}])$. This is the whole point of the previous Theorem. When we add two rational tangles we do not in general get a rational tangle, but the closure of this sum is a rational knot. This rational knot is the closure of a rational tangle distinct from $[\frac{P}{Q} + \frac{R}{S}]$. A good example is $P/Q = [3,2,1] = 10/3$ and $R/S = [2,2,1] = 7/3$. Then $P/Q + R/S = 51/9$, while $(P/Q)\sharp(R/S) = [1,2,3+2,2,1] = 51/35$. We have $N([10/3] + [7/3]) = N([51/35])$, which is not isotopic to $N([51/9])$ by the Schubert Theorem. Nevertheless, the numerator of $(P/Q)\sharp(R/S)$ is equal to $PS + QR$. This is the import of the next Theorem.

Theorem 5. *If* P/Q *has matrix*

$$M = M(\vec{a}) = M(a_1) \ldots M(a_n)$$

and R/S *has matrix*

$$N = M(\vec{b}) = M(b_1) \ldots M(b_m),$$

then $[a_1, \ldots, a_n] \sharp [b_1, \ldots, b_m]$ *has matrix*

$$M \sharp N := M^T N^E,$$

where N^E *denotes the matrix obtained by interchanging the rows of* N. *This gives an explicit formula for* $[a_1, \ldots, a_n] \sharp [b_1, \ldots, b_m]$. *This formula can be used to determine not only when* $N([\frac{P}{Q}] + [\frac{R}{S}])$ *is unknotted but also to find its knot type as a rational knot via Schubert's Theorem. In particular, we find that*

$$N([\frac{P}{Q}] + [\frac{R}{S}]) = N([\frac{PS + QR}{Q'S + UR}]) = N([\frac{Num(P/Q + R/S)}{Num(Q'/U + R/S)}])$$

where $|PU - QQ'| = 1$. *In other words, if we take the numerator closure of the sum of two rational tangles, this knot or link is equal to the numerator of a rational tangle whose arithmetical numerator is the artihmetical numerator of the sum of the fractions of the original two rational tangles.*

Proof. We know from Theorem 3 that

$$M(\vec{a}) = M(a_1) \ldots M(a_n) = \begin{pmatrix} P & Q' \\ Q & U \end{pmatrix},$$

and that

$$M(\vec{b}) = M(b_1) \ldots M(b_m) = \begin{pmatrix} R & S' \\ S & V \end{pmatrix}.$$

Let

$$F/G = [a_n, \ldots, a_2, a_1 + b_1, b_2, \ldots, b_m] = \frac{P}{Q} \sharp \frac{R}{S}.$$

Then we have, by Theorem 4, that

$$N([\frac{P}{Q}] + [\frac{R}{S}]) = N([\frac{F}{G}]),$$

and

$$M(\vec{c}) = M(a_n) \ldots M(a_2) M(a_1 + b_1) M(b_2) \ldots M(b_m) = \begin{pmatrix} F & G' \\ G & W \end{pmatrix}.$$

Now note the identity

$$\begin{pmatrix} a_1 & 1 \\ 1 & 0 \end{pmatrix} \begin{pmatrix} 0 & 1 \\ 1 & 0 \end{pmatrix} \begin{pmatrix} b_1 & 1 \\ 1 & 0 \end{pmatrix} = \begin{pmatrix} a_1 + b_1 & 1 \\ 1 & 0 \end{pmatrix}.$$

Thus

$$M(\vec{c}) = M(\vec{a})^T \begin{pmatrix} 0 & 1 \\ 1 & 0 \end{pmatrix} M(\vec{b}) = M(\vec{a})^T M(\vec{b})^E$$

where M^E denotes the matrix obtained from M by interchanging its two rows. In particular, this formula implies that

$$\begin{pmatrix} F & G' \\ G & W \end{pmatrix} = \begin{pmatrix} P & Q \\ Q' & U \end{pmatrix} \begin{pmatrix} S & V \\ R & S' \end{pmatrix} = \begin{pmatrix} PS+QR & PV+QS' \\ Q'S+UR & Q'V+US' \end{pmatrix}.$$

Thus

$$N([P/Q]+[R/S]) = N([(PS+QR)/(Q'S+UR)]) = N([\frac{Num(P/Q+R/S)}{Num(Q'/U+R/S)}])$$

where $|PU - QQ'| = 1$. This completes the proof of the Theorem. □

Theorem 6 (Unknot/Unlink Theorem). *Let* $\frac{P}{Q} = [a_1, \ldots, a_n]$ *and* $\frac{R}{S} = [b_1, \ldots, b_m]$ *be as in Theorem 5. Then* $N([\frac{P}{Q}] - [\frac{R}{S}])$ *is unknotted if and only if* $PS - QR = \pm 1$, *that is,* PS *and* QR *are consecutive integers. For links we have that* $N([\frac{P}{Q}] - [\frac{R}{S}])$ *is unlinked if and only if* $P/Q = R/S$.

Proof. It follows from Theorem 5 and the Schubert Theorem that $N([\frac{P}{Q}] - [\frac{R}{S}])$ is unknotted if and only if $PS - QR = \pm 1$. By the same token, for a link we have that $N([\frac{P}{Q}] - [\frac{R}{S}])$ is unlinked if and only if $PS - QR = 0$. Since Q and S are non-zero, it follows that $P/Q = R/S$. □

Remark 3. Suppose, as in Theorem 6, that $|PS - QR| = 1$ so that $N([\frac{P}{Q}] - [\frac{R}{S}])$ is an unknot. Then it follows from Theorem 6 that $N([\frac{Q}{P}] - [\frac{S}{R}])$ is also an unknot. Thus, the pair of fractions $(\frac{P}{Q}, \frac{R}{S})$ yields an unknot if and only if the pair of inverse fractions $(\frac{Q}{P}, \frac{S}{R})$ yields an unknot. In fact, the reader will enjoy proving the following identity for any rational tangles A and B :

$$N(\frac{1}{A} + \frac{1}{B}) = -N(A+B)$$

where the minus sign denotes taking the mirror image of the diagram $N(A + B)$, and equality is isotopy. Thus, for any pair of fractions $(\frac{P}{Q}, \frac{R}{S})$ (not necessarily forming an unknot)

$$N([\frac{P}{Q}] - [\frac{R}{S}]) = -N([\frac{Q}{P}] - [\frac{S}{R}]).$$

Remark 4. Our proof of this Theorem is elementary, based on rational tangle calculus. The conclusions of the Theorem can be reached by more sophisticated means as the reader will see in the papers [31, 32]. These papers are concerned with the question of the unknotting number of rational knots and links, and they solve the problem for unknotting number one.

We end this section with a way to produce infinitely many unknots. Consider a rational number, its corresponding continued fraction, and its matrix representation:

$$P/Q = [a_1, \ldots, a_n]$$

with

$$M = M(\vec{a}) = M(a_1)\ldots M(a_n) = \begin{pmatrix} P & Q' \\ Q & U \end{pmatrix}.$$

Note that since the determinant of this matrix is $(-1)^n$, we have the formula $PU - QQ' = (-1)^n$ from which it follows that

$$P/Q - Q'/U = (-1)^n/QU.$$

Hence, by Theorem 6, the rational knot

$$N([P/Q] - [Q'/U])$$

is unknotted and, as we shall see later, is a good candidate to produce a hard unknot.

6. Continued Fractions, Convergents and Lots of Unknots

In this section we'll show that we can do even better than the example above for producing infinitely many unknots. Remarkably, the fraction Q'/U (at the end of the last section) has an interpretation as the truncation of our continued fraction $[a_1, \ldots, a_n]$.

Lemma 1. *Let* $P/Q = [a_1, \ldots, a_n]$ *with corresponding matrix* $M = M(\vec{a}) = M(a_1)\ldots M(a_n) = \begin{pmatrix} P & Q' \\ Q & U \end{pmatrix}$. *Then* $Q'/U = [a_1, \ldots, a_{n-1}]$. *Conversely, if* $R/S = [a_1, \ldots, a_{n-1}]$ *is a reduced fraction, then* $|R| = |Q'|$ *and* $|S| = |U|$. *So, we can choose signs for* R *and* S *so that*

$$\begin{pmatrix} P & R \\ Q & S \end{pmatrix} = \begin{pmatrix} P & Q' \\ Q & U \end{pmatrix}.$$

Proof. To see this formula, let

$$N = M(a_1)\dots M(a_{n-1}) = \begin{pmatrix} R & S' \\ S & V \end{pmatrix},$$

so that

$$R/S = [a_1, \dots, a_{n-1}].$$

Then

$$\begin{pmatrix} P & Q' \\ Q & U \end{pmatrix} = M(a_1)\dots M(a_{n-1})M(a_n) = NM(a_n)$$

$$= \begin{pmatrix} R & S' \\ S & V \end{pmatrix} \begin{pmatrix} a_n & 1 \\ 1 & 0 \end{pmatrix}$$

$$= \begin{pmatrix} Ra_n + S' & R \\ Sa_n + V & S \end{pmatrix}.$$

This shows that $Q'/U = R/S = [a_1, \dots, a_{n-1}]$, as claimed. For the converse, note that the pairs $(P, Q), (Q', U), (R, S)$ are pairs of relatively prime integers. Note also that $Q'/U = R/S$. Thus $|R| = |Q'|$ and $|S| = |U|$. □

Definition 4. One says that $[a_1, \dots, a_{n-1}]$ is a *convergent* of $[a_1, \dots, a_{n-1}, a_n]$. We shall say that two fractions P/Q and R/S are *convergents* if the continued fraction of one of them is a convergent of the other.

We have, so far, proved the following result.

Proposition 2. *Let* $P/Q = [a_1, \cdots, a_{n-1}, a_n]$ *and let* $Q'/U = [a_1, \cdots, a_{n-1}]$, *a convergent of* P/Q. *Then*

$$N([P/Q] - [Q'/U]) = N([a_1, \cdots, a_{n-1}, a_n] - [a_1, \cdots, a_{n-1}])$$

is an unknot.

Remark 5. Another, more direct way to see this result is illustrated in Figure 18. In this figure we show $N([2,3,4]-[2,3])$ with the second tangle written as a $180°$ turn of the standard form for $[2,3]$ (this is isotopic to the standard form since we are working with rational tangles). In this form, we see clearly that the braidings of the two tangles cancel in the back of the 2-sphere to form an unknot.

The reader should note that

$$N([a_1,\dots,a_n]-[a_1,\dots,a_{n-1}]) = N([a_1,\dots,a_n]\sharp[-a_1,\dots,-a_{n-1}])$$

$$= N([a_n,\dots,a_2,a_1-a_1,\dots,-a_{n-1}]) = N([a_n,\dots,a_2,0,-a_2,\dots,-a_{n-1}])$$

$$= N([a_n,\dots,a_2-a_2,\dots,-a_{n-1}]) = \cdots$$

$$= N([a_n,a_{n-1}-a_{n-1}]) = N([a_n,0]) = U$$

where U is the unknotted circle. Thus we see directly that a continued fraction and the negative of its convergent will yield an unknot when we take the numerator of the sum of the corresponding tangles.

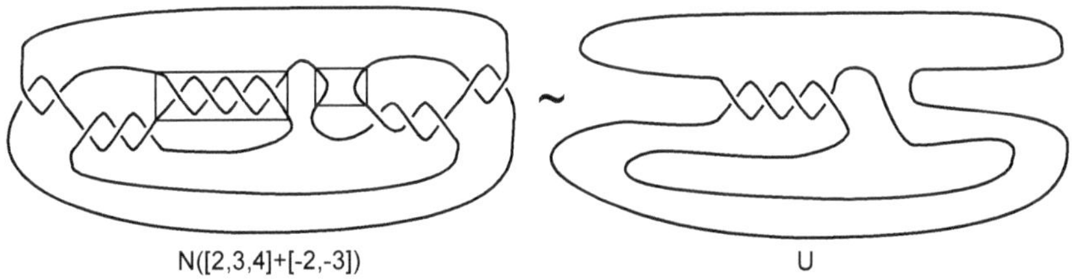

Fig. 18. Braid Cancellation to an Unknot

We see from the above that the two consecutive integers PU and QQ' produce two continued fractions $P/Q = [a_1,\dots,a_n]$ and $Q'/U = [a_1,\dots,a_{n-1}]$ so that the second fraction is a convergent of the first. *The property of one fraction being a convergent of the other is in fact, always a property of fraction pairs produced from consecutive integers.* For example, consider the integers 90 and 91. We have $90 = 10\times 9$ and $91 = 13\times 7$. Thus

$$13\times 7 - 10\times 9 = 1.$$

Hence

$$13/9 - 10/7 = 1/63,$$

so that $N([13/9] + [-10/7])$ is an unknot by Theorem 6. We see that

$$13/9 = [1, 2, 4] = [1, 2, 3, 1]$$

and

$$10/7 = [1, 2, 3]$$

so that 13/9 and 10/7 are convergents. We will now prove this statement about convergents. For the next Lemma see also [15].

Lemma 2. *Let P and Q be relatively prime integers and let s and r be a pair of integers such that $Ps - Qr = \pm 1$. Let $R = r + tP$ and $S = s + tQ$ where t is any integer. Then $\{R, S\}$ comprises the set of all solutions to the equation $PS - QR = \pm 1$. Similarly, all solutions of the equation $PS - QR = \mp 1$ are given in the form $R = -r + tP$ and $S = -s + tQ$.*

Proof. Without loss of generality we can assume that $Ps - Qr = 1$. We leave it to the reader to formulate the case where $Ps - Qr = -1$. Certainly R and S as given in the statement of the Lemma satisfy the equation $PS - QR = 1$. Suppose that R and S is some solution to this equation. Then it follows by taking the difference with the equation $Ps - Qr = 1$ that $P(S - s) = Q(r - R)$. Since P and Q are relatively prime, it follows at once from this equation and the uniqueness of prime factorization of integers that $R = r + tP$ and $S = s + tQ$ for some integer t. The last part of the Lemma follows by the same form of reasoning. This completes the proof.□

Lemma 3. *Let P and Q be relatively prime integers and let $P/Q = [a_1, \ldots, a_n]$ be a continued fraction expansion for P/Q. Let also $r/s = [a_1, \ldots, a_{n-1}]$ be a convergent for $[a_1, \ldots, a_n]$. Let $R = r + tP$ and $S = s + tQ$ where t is any integer. Then $R/S = [a_1, \ldots, a_n, t]$, and thus P/Q is a convergent of R/S.*

Proof. By Lemma 1, we have

$$M(\vec{a}) = M(a_1) \ldots M(a_n) = \begin{pmatrix} P & r \\ Q & s \end{pmatrix}.$$

Moreover, we have:

$$M(\vec{a})M(t) = \begin{pmatrix} P & r \\ Q & s \end{pmatrix} \begin{pmatrix} t & 1 \\ 1 & 0 \end{pmatrix} = \begin{pmatrix} Pt + r & P \\ Qt + s & Q \end{pmatrix}.$$

By Lemma 2 for $\{R = r + tP,\ S = s + tQ\}$ we have $PS - QR = (-1)^{n+1}$. So $R/S = [a_1, \ldots, a_n, t]$, and P/Q is a convergent of R/S. Thus all such solutions correspond to continued fractions of which P/Q is a convergent. □ This completes the proof.

Finally, we can sum up by pointing out that we have proved the following result.

Theorem 7 (Unknotting Convergents). *Let P/Q and R/S be reduced fractions. Then*

$$N([P/Q] - [R/S])$$

is an unknot if and only if P/Q and R/S are convergents.

Remark 6. In this sense Theorems 6 and 7 show us that the simplest method of collapse to the unknot is the only way that the collapse can occur when we take the numerator of the sum of two rational tangles. The reader will enjoy verifying that this class of unknots can be obtained from a closed loop of rope by consecutive twisting of adjacent strands of the rope, followed by a separation of the rope into the numerator sum of two rational tangles. Such processes occur naturally in situations involving DNA topology as described later in this paper.

Infinite Continued Fractions. Note that if we start with an infinite continued fraction

$$[a_1, a_2, a_3, \ldots]$$

and define the *truncations*

$$P_n/Q_n = [a_1, \ldots, a_n],$$

then we have

$$P_n/Q_n - P_{n-1}/Q_{n-1} = (-1)^n / Q_n Q_{n-1}.$$

This equation is usually used to analyze the convergence of the truncations of the continued fraction to a real number. Here we see a *single infinite continued fraction*, whether or not convergent, *producing an infinite family of unknot diagrams*:

$$K_n = N([P_n/Q_n] - [P_{n-1}/Q_{n-1}]).$$

Surely, the simplest instance of this phenomenon occurs with the infinite continued fraction for the Golden Ratio:

$$\frac{1+\sqrt{5}}{2} = [1, 1, 1, \ldots].$$

Here the convergents are $f_n/f_{n-1} = [1, 1, 1, \ldots, 1]$ (with n 1's) and f_n is the n-th Fibonacci number where $f_0 = 1, f_1 = 1$ and $f_{n+1} = f_n + f_{n-1}$ for $n \geq 1$. The matrices are

$$M(1)^n = \begin{pmatrix} f_n & f_{n-1} \\ f_{n-1} & f_{n-2} \end{pmatrix}.$$

Thus we have

$$f_n/f_{n-1} - f_{n-1}/f_{n-2} = (-1)^n/f_{n-1}f_{n-2}.$$

In this instance, this formula shows that the sequence of truncations is convergent (to the Golden Ratio).

In Figure 19 we illustrate a hard unknot based on the eighth and seventh truncations of the Golden Ratio continued fraction. This example is produced by using the standard closure of $N([f_8/f_7] - [f_7/f_6]) = N([34/21] - [21/13])$. By this same pattern, we can extract from the infinite continued fraction for the golden ratio an infinite family of hard unknots. Note the unavailability of Reidemeister moves on the first diagram. In Figure 19 the first diagram illustrates the numerator of the sum of the two tangles $[f_8/f_7] = [1, 1, 1, 1, 1, 1, 1, 1]$ and $-[f_7/f_6] = [-1, -1, -1, -1, -1, -1, -1]$ with each tangle in 3-strand braid form with the integer 1-twist on the left. In this form the diagram is a hard unknot, and this generalizes to infinitely many examples of the same type. The second example has one tangle turned by 180° so that the integer 1-twists face one another, giving an easy unknot. The third example has one tangle turned by 180° so that both integer 1-twists face into the outer region. This diagram is an easy unknot on the 2-sphere, since there is a Reidemeister II-move in the outer region. If we were to restrict moves to the plane, then this Reidemeister move would not be available. The third diagram is a hard unknot in the plane.

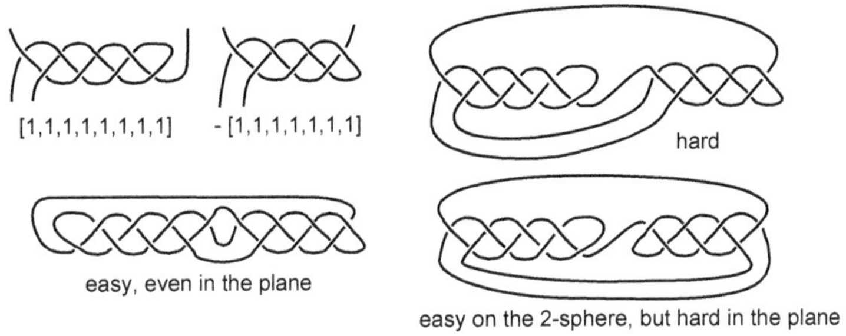

Fig. 19. $N([f_8/f_7] - [f_7/f_6]) = N([34/21] - [21/13])$

7. Constructing Hard Unknots

In this section we indicate how to construct hard unknots by using positive alternating tangles A and B such that $N(A - B)$ is unknotted. By our main results we know how to construct infinitely many such pairs of tangles by taking a continued fraction and its convergent, with the corresponding tangles in reduced (alternating) form.

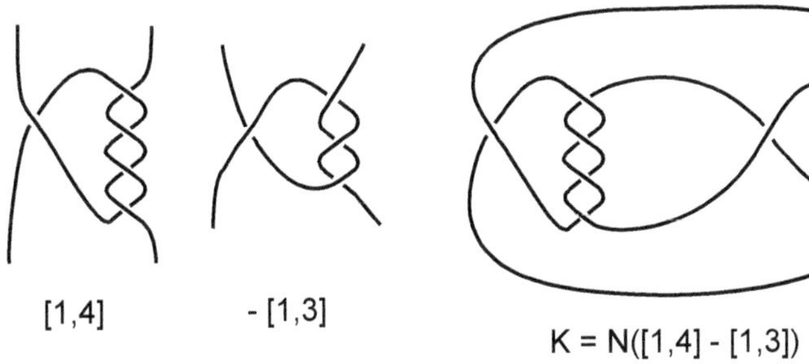

Fig. 20. $K = N([1, 4] - [1, 3])$

Let's begin with the case of $5/4 = [1, 4] = [1, 3, 1]$ and $4/3 = [1, 3]$. In Figure 20 we show the standard representations of $[1, 4]$ and $-[1, 3]$ as tangles, and $K = N([1, 4] - [1, 3])$. The reader will note that this diagram is a hard unknot with 9 crossings, one less than our original Culprit of Figure 16. In Figure 24 we give a diagram H that is isotopic to the mirror image of K. In Theorem 8 we show that H is one of a small collection of minimal hard unknot diagrams having the form $N(A-B)$ for reduced positive rational tangle diagrams A and B. In most cases, if one takes the standard representations of the tangles A and B, and forms the diagram for $N(A-B)$, the resulting unknot diagram will be hard. There are some exceptions however, and the next example illustrates this phenomenon.

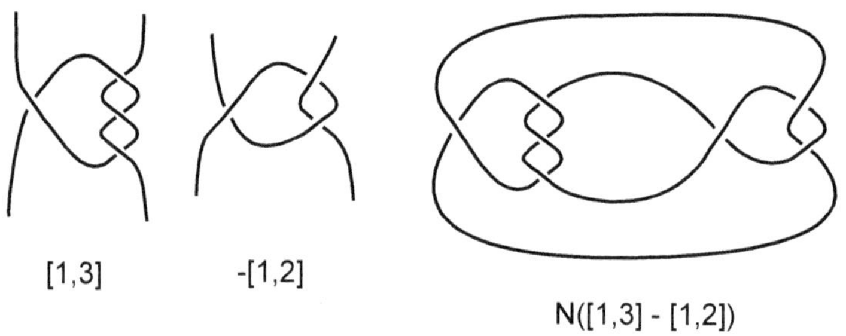

Fig. 21. $N([1,3]-[1,2])$

In Figure 21 we show the standard representations of $[1,3]$ and $[1,2]$ as tangles, and the diagram of $N([1,3]-[1,2])$. This diagram, while unknotted, is not a hard unknot diagram due to the three-sided outer region. This outer region allows a type III Reidemeister move on the surface of the 2-sphere.

The Tucking Construct. In Figure 22 we have replaced $[1,2]$ by $[1,2]^v$, the 180° turn of the tangle $[1,2]$ about the vertical direction in the page. Now we see that the literal diagram of $N([1,3]-[1,2]^v)$ is of course still unknotted and is also not a hard unknot diagram. However this diagram can be converted to a hard unknot diagram by tucking an arc as shown in the Figure. The resulting hard unknot is the same diagram of 10 crossings that we had in Figure 16 as our initial Culprit. Note that a direct tuck on Figure 21 does not work (there would be a type III move made available). Note also that the other possibilities of flipping both tangles in Figure 21 or flipping the first tangle do not lead to hard unknots. We call this strategem the *tucking construct.* The tucking prevents the appearance of a type II move in the outermost region of the diagram.

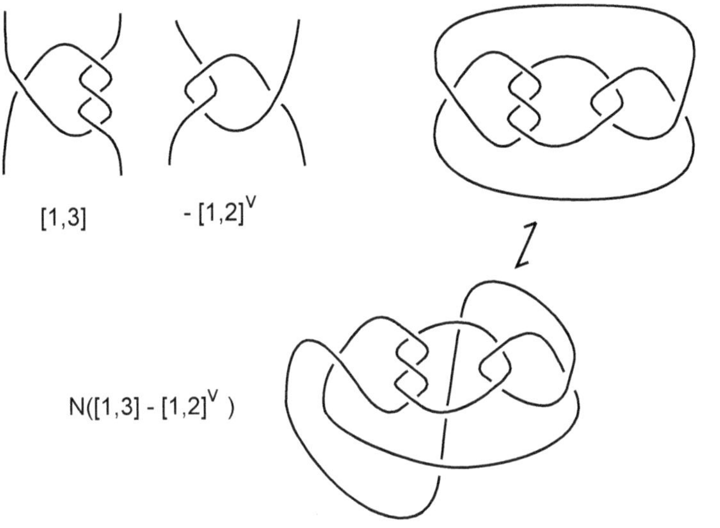

Fig. 22. $N([1,3]-[1,2]^v)$

The Culprit Revisited. Let's consider the example in Figure 16 again. Here we have $P/Q = -3/4$ and $R/S = 2/3$. We have $P/Q + R/S = -3/4 + 2/3 = -1/12$. Thus $N([-3/4] + [2/3])$ is an unknot by Theorem 6. This is exactly the unknot C' illustrated in Figure 16. Note that C is the tucking construct applied to C'.

We can make infinitely many examples of this type. View Figure 23. The pattern is as follows. Suppose that $T = [P/Q]$ and $T' = [R/S]$ are rational tangles such that $PS + QR = \pm 1$. Then we know that $N(T + (T')^v)$ is an unknot. Furthermore we can assume that each of the tangles T and T' are in alternating form. The two tangle fractions have opposite sign and hence the alternation of the weaves in each tangle will be of opposite type. We create a new diagram for $N(T + (T')^v)$ by putting an arc from the bottom of the closure entirely underneath the diagram as shown in Figure 23. This is the general form of the tucking construct.

Remark 7. View the bottom of Figure 23 and note that we have indicated that the hard unknot becomes an alternating and hence knotted diagram (this can be proved by applying the results in [25]) by switching one crossing

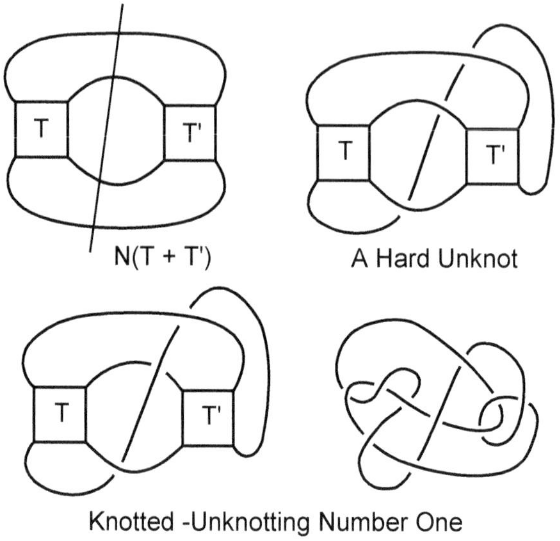

Fig. 23. The Tucking Construct

in the tucking construction. Thus the tucking construction is also an infinite source of knots of unknotting number one.

Hard Unknots Recipe. In a tangle $[a_1, \ldots, a_n]$, call a_1 the *integer part* of the tangle. Here we use tangles $T = [a_1, \ldots, a_n]$ and $T' = [-a_1, \ldots, -a_{n-1}]$ with all a_i positive. Except for very few cases with small number of total crossings, the following hold:

(1) If the integer parts both face in the same direction, then $N(T + T')$ is a hard unknot diagram.
(2) If the integer parts both face the outside region, thus making a type II move available on the 2-sphere, then the tucking construct will create a hard unknot.
(3) If the integer parts both face the inside region, then we do not get a hard unknot and we cannot remedy this fact using tucking.

The reader should compare these remarks to the contents of Figure 19.

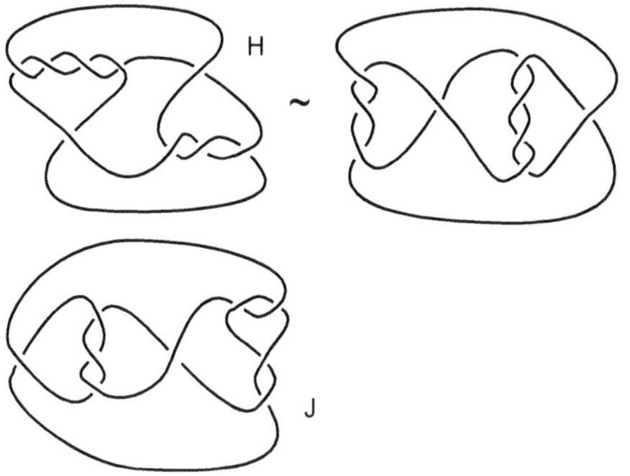

Fig. 24. Smallest hard unknots

8. The Smallest Hard Unknots

Figure 24 illustrates two hard unknot diagrams H and J with 9 crossings ($K = -H$ appears earlier in Figure 20).

Two equivalent versions of the diagram H appear in Figure 24. The right-hand version of H in this figure is of the form:

$$H = N([1,3] - [1,4]) = N([4/3] - [5/4]).$$

Note that $[1,3]$ and $[1,4] = [1,3,1]$ are convergents. Thus H and K are mirror images of each other. The diagram J in Figure 24 is of the form:

$$J = N([1,3] - [1,2,2]) = N([4/3] - [7/5]).$$

Note that $[1,3] = [1,2,1]$ and $[1,2,2] = [1,2,1,1]$ are convergents. Note also that the crossings in J corresponding to 1 in $[1,3]$ and -1 in $[-1,-2,-2]$ can be switched and we will obtain another diagram J', arising as sum of two alternating rational tangles, that is also a hard unknot. This diagram can be obtained from the diagram J without switching crossings, by performing flypes (Figure 12) on the subtangles $[1,3]$ and $[-1,-2,-2]$ of J, and then doing an isotopy of this new diagram on the 2-sphere. (We leave the verification of this statement to the reader.) Thus the diagram J' can be obtained from J by flyping. A similar remark applies to the diagram H, giving a corresponding diagram H', but in this case H' is easily seen to be equivalent to H by a planar isotopy that does not involve any Reidemeister moves. Thus, up to these sorts of modifications, we have produced essentially two hard unknot diagrams with 9 crossings.

We have the following result.

Theorem 8. *The diagrams H and J shown in Figure 24 are, up to flyping subtangle diagrams and taking mirror images, the smallest hard unknot diagrams in the form $N(A-B)$ where A and B are rational tangles in reduced positive alternating form.*

Proof. It is easy to see that we can assume that $A = [P/Q]$ where P and Q are positive, relatively prime and P is greater than Q. We leave the proof that one can choose P greater than Q to the reader.

We then know from Theorem 7 that $B = [-R/S]$ where one of P/Q and R/S is a convergent of the other. We can now enumerate small continued fractions. We know the total of all terms in A and B must be less than or equal to 9 since H and J each have nine crossings.

In order to make a 9-crossing unknot example of the form $N(A - B)$ where A and B are rational tangles in reduced positive alternating form, we must partition the number 9 into two parts corresponding to the number of crossings in each tangle. It is not hard to see that we need to use the partition $9 = 4+5$ in order to make a hard unknot of this form. Furthermore, 4 must correspond to the continued fraction $[1, 3]$, as $[2, 2]$ will not produce a hard unknot when combined with another tangle. Thus, for producing 9 crossing examples we must take $A = [1, 3]$. Then, in order that A and B be convergents, and B have 5 crossings, the only possibilities for B are $B = [1, 4]$ and $B = [1, 2, 2]$. These choices produce the diagrams H, H', J, J'. It is easy to see that no diagrams with less than 9 crossings will suffice to produce hard unknots, due to the appearance of Reidemeister moves related to the smaller partitions. This completes the proof. □

We made the following conjecture:

Conjecture 1. *Up to mirror images and flyping tangles in the diagrams, the hard unknot diagrams H and J of 9 crossings, shown in Figure 24, have the least number of crossings among all hard unknot diagrams.*

This conjecture has been verified for us by Slavik Jablan, using his program "Linknot" [23].

9. The Goeritz Unknot

The earliest appearance of a hard unknot is a 1934 paper of Goeritz [18]. In this paper Goeritz gives the hard unknot G shown in Figure 25. As the reader can see (for example by flyping vertically the tangle [-3] twice), this example is a variant on $N([4] + [-3])$ which is certainly unknotted. The Goeritz example has 11 crossings, due to the extra 2-twists that make it a hard unknot. It is part of an infinite family based on the pattern $N([n] + [-n+1])$.

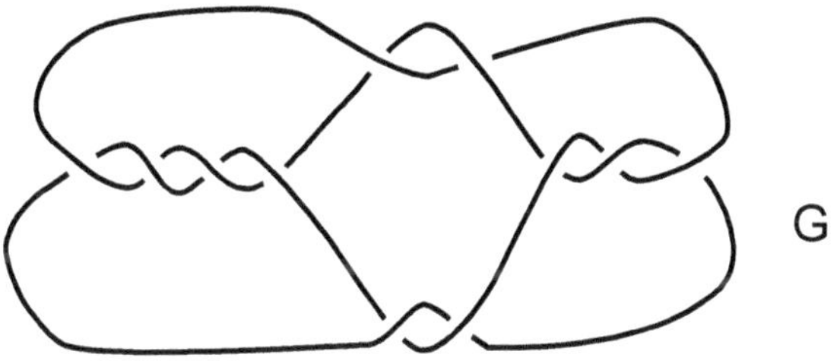

Fig. 25. The Goeritz Hard Unknot

In Figure 26 we illustrate the steps in another unknotting process for G. A move labeled *swing* consists in swinging the corresponding marked arc in the diagram. A swing move when realized by Reidemeister moves will complicate the diagram before taking it to the indicated simplification. After two swing moves G has been transformed into the diagram H of the previous section that is again a hard unknot diagram, but with 9 crossings rather than 11!

10. Recalcitrance Revisited

Suppose that $K[a]$ denotes the knot diagram shown in Figure 27, so that

$$K[a] = N(-[a] + S + [a] + T]).$$

We shall assume that $K[0] = N(S+T)$ is unknotted. We also assume that the diagram $K[a]$ is hard, a generalization of the form of the hard diagram

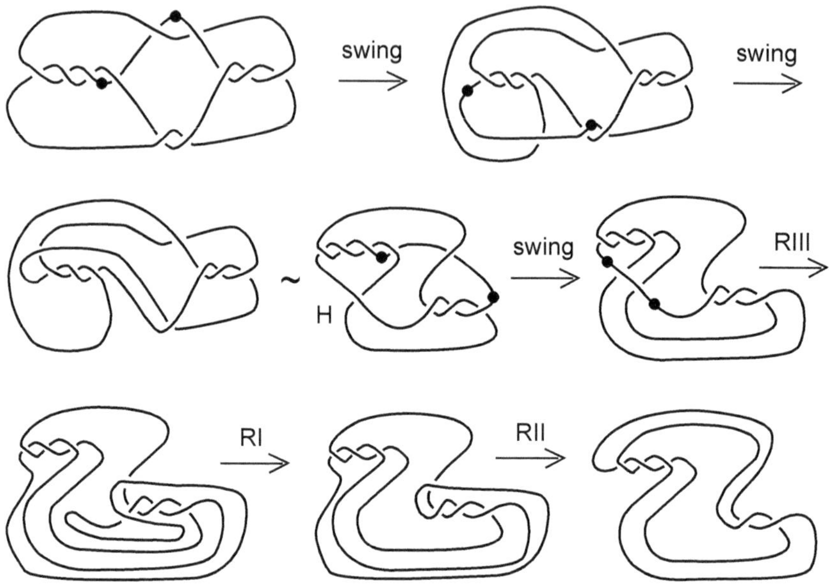

Fig. 26. Unknotting the Goeritz Unknot

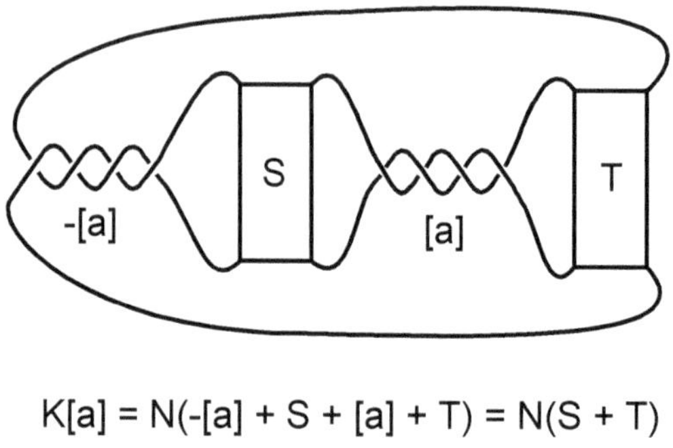

Fig. 27. Twisting Up the Recalcitrance

H in Figure 24. Let's also assume that it takes N Reidemeister moves to transform $[-1]+S+[1]$ to S. This transformation is easily accomplished in three dimensions by one full rotation, but may require many Reidemeister moves in the plane (keeping the ends of the tangle fixed). It is certainly possible to produce alternating rational tangles S and T with this property that no less than N Reidemeister moves can accomplish an isotopy from $[-1]+S+[+1]+T$ to $S+T$. Let $C = C(S)+C(T)$ denote the total number of crossings in the tangles S and T. Then the recalcitrance (Definition 2) of $K[0]$ is $R(0) = k/C$ for some k and the recalcitrance of $K[a]$ is generically given by the formula

$$R(a) = \frac{\overline{aN + k}}{2a + C}$$

since each extra turn of S will add N Reidemeister moves to the untying, and the number of crossings of $K[a]$ is equal to $2a + C$. We conclude from this that *for large a the recalcitrance $R(a)$ is as close as we like to $N/2$*. This shows that *the recalcitrance of an unknotted diagram can be arbitrarily large.* There is no upper limit for the ratio of the number of moves needed to undo the knot in relation to the number of crossings in the original knot diagram.

11. Collapsing to Knots and Links

In this section we point out that our results about unknotted closures of sums of rational tangles can be generalized to collapses to arbitrary knots and links. This leads to the concept of a *hard knot diagram*, generalizing our notion of a hard unknot diagram. We shall say that a knot or link diagram K is *hard* if it admits no type III moves and has no simplifying type I or type II moves. Any sequence of Reidemeister moves that takes this diagram to a smaller diagram for that knot or link must first increase the number of crossings of K. The concepts of this paper, such as recalcitrance, can be generalized to this domain. We can ask, given any diagram K for the least number of Reidemeister moves that will transform K to some minimal diagram for the knot type of K. Many questions can be asked at this juncture. In this section we formulate some collapse results and give examples. We encourage the reader to explore further!

It is worth putting these results in perspective. In this paper we are exploring the combinatorial space of all knot and link diagrams by using the Reidemeister moves. Hard unknots and hard knots are local minima in this combinatorial space. The problem with a local minimum is that it may not be a global minimum. Hard knots can often be made smaller (we want a minimal diagram for the knot), but again must be made larger before becoming smaller. Each time we find a local minimum we may have to face the complex possibilities involved in climbing up out of it before things can be simplified. These features are fundamental to the study of knots and links via diagrams.

Notation. Let R be *any* 2-tangle. We shall write

$$S = [a_1, \ldots, a_n, R]$$

to denote the tangle

$$[a_1] + \cfrac{1}{[a_2] + \cfrac{1}{\ldots + \cfrac{1}{[a_n] + \frac{1}{R}}}}.$$

Note that we will occasionally replace $[a]$ by a in the notation for continued fraction forms of tangles. Note that if R is an arbitrary tangle, then $(R^{rot})^{rot} = R^{\pi}$ where R^{π} denotes the result of turning R by 180° around an axis perpendicular to the page. In the case of rational tangles R, we know that $R^{\pi} = R$, but in the general case the tangle and its 180° rotate may be distinct topologically. This means that $1/(1/R) = R^{\pi}$ in the general case. Similar remarks apply to R^v, the result of rotating R about the vertical axis and R^h, the result of rotating R about the horizontal axis. Many of the isotopies that we use for rational tangles may replace R by rotations of R by 180° about horizontal or vertical axes or both. Note that $R^{\pi} = R^{hv}$.

Definition 5. Letting $P/Q = [a_1, a_2, \ldots, a_n]$ and $V/W = [b_1, b_2, \ldots, b_m]$, it is convenient to make the following definitions for any tangles T and S :

(1) $[P/Q]\sharp T = [a_n, a_{n-1}, \ldots, a_1 + T]$.
(2) $[a_1, a_2, \ldots, a_n, T]\sharp[b_1, b_2, \ldots, b_n, S] =$
$[T, a_n, a_{n-1}, \ldots, a_1 + b_1, b_2, \ldots, b_m, S]$

Note that this generalizes our previous definitions of the $\sharp$ operation. In fact, we have a generalization of Theorem 4, that the numerator closure of the sum of two rational tangles is a rational knot, to the following statement:

Theorem 9. *For any tangles T and S (not necessarily rational) the following equation expresses the numerator of the sum of the tangles indicated below. In this statement we have used the symbol $\hat{=}$ to mean that T and*

S on the right may differ from their counterparts on the left-hand side of the equation by 180° *rotations (as discussed above). In the case of S and T rational or invariant under such rotations, the sign* $\hat{=}$ *may be replaced by* $=$.

$$N([a_1, a_2, \ldots, a_n, T] + [b_1, b_2, \ldots, b_n, S])$$
$$\hat{=} N([T, a_n, a_{n-1}, \ldots, a_1 + b_1, b_2, \ldots, b_m, S]).$$

In particular,

$$N([a_1, a_2, \ldots, a_n, T] + [-a_1, -a_2, \ldots, -a_n, S]) \hat{=} N(T + S).$$

In the special case where $T = [\infty]$ *and* $S = [0]$ *this last equation becomes*

$$N([a_1, a_2, \ldots, a_n] + [-a_1, -a_2, \ldots, -a_{n-1}]) = U$$

where U *is unknotted, recovering Proposition* 2. *Similarly we have*

$$N([a_1, \ldots, a_n + T] + [-a_1, \ldots, -a_n]) \hat{=} N(T)$$

and

$$N([a_1, \ldots, a_n, T] + [-a_1, \ldots, -a_n]) \hat{=} D(T).$$

Note that since T and S are arbitrary tangles, we can also regard this Theorem as a statement about black boxes called T and S that are connected as in Figure 28 to form rationally extended tangles and closures. This is the same as considering tangles that contain 4-valent rigid graphical vertices. Such graphs can be modelled as embeddings in three-dimensional space by attaching strings to rigid bodies (such as geometrical balls) in place of the vertices.

Proof. The proof follows from the isotopies indicated in Figure 28 and the identities $[a_1, \ldots, a_n, 0] = [a_1, \ldots, a_{n-1}]$ and $[a_1, \ldots, a_n, \infty] = [a_1, \ldots, a_n]$. □

Remark 8. For T and S any 2-tangles, the cancellation to $N(T+S)$ in Theorem 9 can be stated more precisely as follows:

$$N([a_1, a_2, \ldots, a_n, T] + [-a_1, -a_2, \ldots, -a_n, S^v]^v) = N(T + S).$$

See Figure 29 for a direct illustration of this result. Note that this formula tells us that we have rational collapses to any knot or link, since we can always cut such a knot or link into one tangle or a sum of two tangles and embed these in cancelling rational patterns. Just as we have constructed hard unknot diagrams, we can construct hard *knot* diagrams.

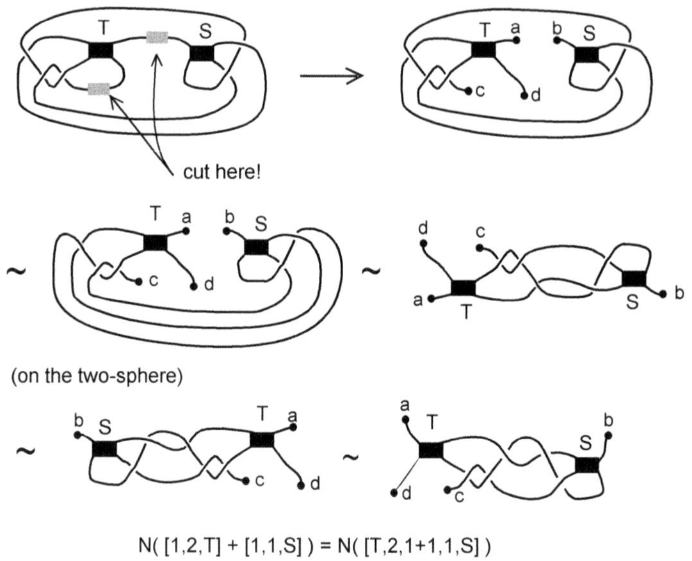

Fig. 28. The Black Boxes

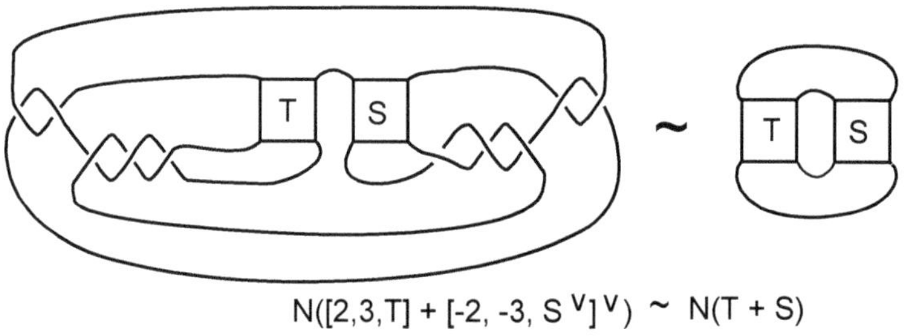

Fig. 29. Black Boxes and Braid Cancellation

The next Theorem is a direct consequence of Theorem 9. We leave its proof to the reader.

Theorem 10. *(i) Letting*

$$M = M(\vec{a}) = M(a_1)M(a_2)\ldots M(a_n) = \begin{pmatrix} P & Q' \\ Q & U \end{pmatrix},$$

and given F/G, a fraction in reduced form, then there exists a unique re-

duced fraction R/S such that

$$N([F/G]) = N([P/Q] + [R/S]),$$

and

$$\frac{F}{G} = \frac{Num(P/Q + R/S)}{Num(Q'/U + R/S)}$$

where $Num(a/b + c/d) = ad + bc$. Note that Q'/U is a convergent of P/Q, so we see that a combination of a fraction and its convergent appears in the decomposition of an arbitrary rational knot, fully generalizing our results about the unknot.

(ii) For any 2-tangle T,

$$N([P/Q] + [-P/Q']\sharp T) = N(T).$$

This is another generic form of the collapse result, and this one specializes directly to our unknot theorem when we take $T = [1/N]$ so that $N(T)$ is an unknot. In that case the fractions P/Q and $(P/Q')\sharp(1/N)$ are convergents.

(iii) If T and S are arbitrary 2-tangles and $[P/Q]$ denotes a specific rational tangle with a given continued fraction expansion, as above, then

$$N([P/Q]\sharp T + [-P/Q']\sharp S) = N(T + S).$$

Example 1. View Figure 30. This figure illustrates the identity

$$N(\frac{1}{[m] + \frac{1}{[n]+T}} - \frac{1}{[m]}) = D(T)$$

where $D(T)$ denotes the denominator of T, and T is any tangle. The example is due to DeWitt Sumners and is fashioned to illustrate that an equation

$$N(A + B) = K$$

for fixed K and variable A and B may have infinitely many solutions, as we have proved above. Let's analyse the class of examples in Figure 30 in our formalism. Note that for any tangle T,

$$N([T, 0]) = N(T + 1/[0]) = N(T + [\infty]) = D(T).$$

Then we have

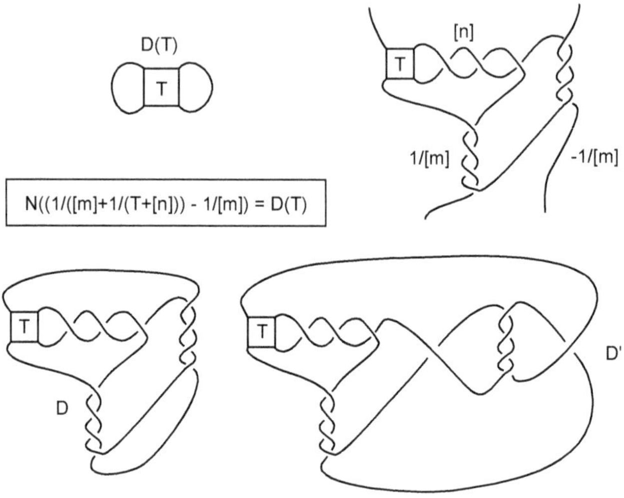

Fig. 30. Sumners' Example

$$N(\frac{1}{[m]+\frac{1}{[n]+T}}-\frac{1}{[m]}) = N([0,m,[n]+T]+[0,-m]) = N([[n]+T,m,0+0,-m])$$

$$= N([n+T,m,0,-m]) = N([[n]+T,0]) = D([n]+T) = D(T),$$

making this class a special case of the first part of the Proposition.

Example 2. View Figure 31. Here we have a knot $K = N(T+S)$ and have constructed three diagrams associated with K. The first diagram is $N([2,3,T]+[-2,-3,S])$, which by the results of this section is isotopic to K, and it is a hard diagram. The next diagram is made by applying a 180° turn and replacement to one of the subtangles of the previous diagram. It is hard in the plane but not on the 2-sphere. In the third diagram we remedy this situation by using the tucking construct, obtaining another hard diagram related to K.

Remark 9. Given a knot or link one can apply our methods of collapse in reverse, producing complex configurations that collapse back to that knot or link. This configuration can be allowed to undergo a crossing switch that locks it into a new knot or link that does not collapse. Thus we can create towers of knots and links such that each stage of the tower will collapse to

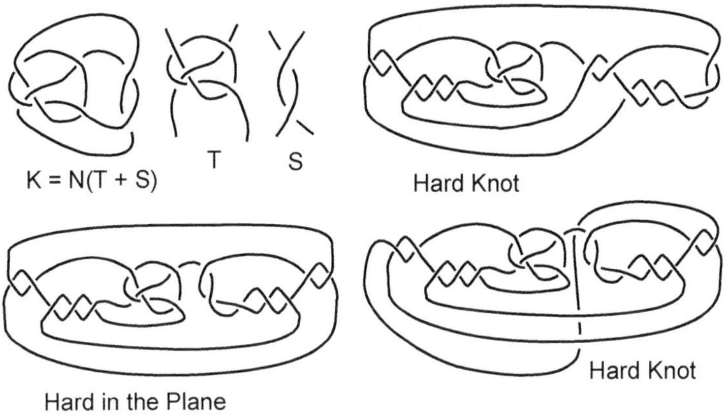

Fig. 31. A Hard Knot

the previous stage under a single crossing switch.

12. Stability in Processive DNA Recombination

In this section we use the techniques of this paper to study properties of the topology of processive DNA recombination. We follow the tangle model of DNA recombination [12, 13, 43] developed by C. Ernst and D.W. Sumners. In this model the DNA is divided into two regions corresponding to two tangles O and I and a recombination site that is associated with I. This division of the DNA is a model of how the enzyme that performs the recombination traps a part of the DNA, thereby effectively dividing it into the tangles O and I. The recombination site is represented by another tangle R. The entire arrangement is then a knot or link $K[R] = N(O+I+R)$. We then consider a single recombination in the form of starting with $R = [0]$, the zero tangle, and replacing R with a new tangle; for our purposes this will be the tangle $[1]$ or the tangle $[-1]$. Processive recombination consists in consecutively replacing again and again by $[1]$ or by $[-1]$ at the same site. Thus, in processive recombination we obtain the knots and links

$$K[n] = N(O + I + [n]).$$

The knot or link $K[0] = N(O + I)$ is called the *DNA substrate*, and the tangle $O + I$ is called the *substrate tangle*. It is of interest to obtain a uniform formula for knots and links $K[n]$ that result from the processive recombination.

In some cases the substrate tangle is quite simple and is represented as a single tangle $S = O + I$. For example, we illustrate processive recombination in Figure 32 with $R = [+1]$, $S = [-1/3] = [0, -3]$ and $I = [0]$ with $n = 0, 1, 2, 3, 4$. Note that by Proposition 1,

$$K[n] = N(S + [n]) = N([0, -3] + [n]) = N([-3, 0 + n]) = N([-3, n])$$

$$= N(-[3, -n]) = N(-[2, 1, n - 1]).$$

This formula gives the abstract form of all the knots and links that arise from this recombination process. This formula goes infinitely many steps further than Figure 32. Looking at Figure 32 we see that the sequence is: *Unknot, Hopf Link, Figure Eight Knot, Whitehead Link, Funny Twisty Diagram, and so on.* A single neat formula for the whole sequence is welcome.

We say that the formula

$$K[n] = N(-[2, 1, n - 1])$$

for $n > 1$ is *stabilized* in the sense that all the terms in the continued fraction have the same sign and the n is in *one single place* in the fraction. In general, a *stabilized fraction* will have the form

$$\pm[a_1, a_2, \dots a_{k-1}, a_k + n, a_{k+1}, \dots, a_n]$$

where all the terms a_i are positive for $i \neq k$ and a_k is non-negative. It is not quite obvious that the output of a recombination process will be stabilized in this sense. We shall prove below that this is always the case. For another example view Figure 33, where we show a process that stabilizes after the second term, with the transition from first to second term having its own idiosyncracy.

Let's see what the form of the processive recombination is for an arbitrary sequence of recombinations. We start with

$$O = [a_1, a_2, \dots, a_r]$$

$$I = [b_1, b_2, \dots, b_s].$$

Then

$$K[n] = N(O + (I + [n])) = N([a_1, a_2, \dots, a_r] + [n + b_1, b_2, \dots, b_s])$$

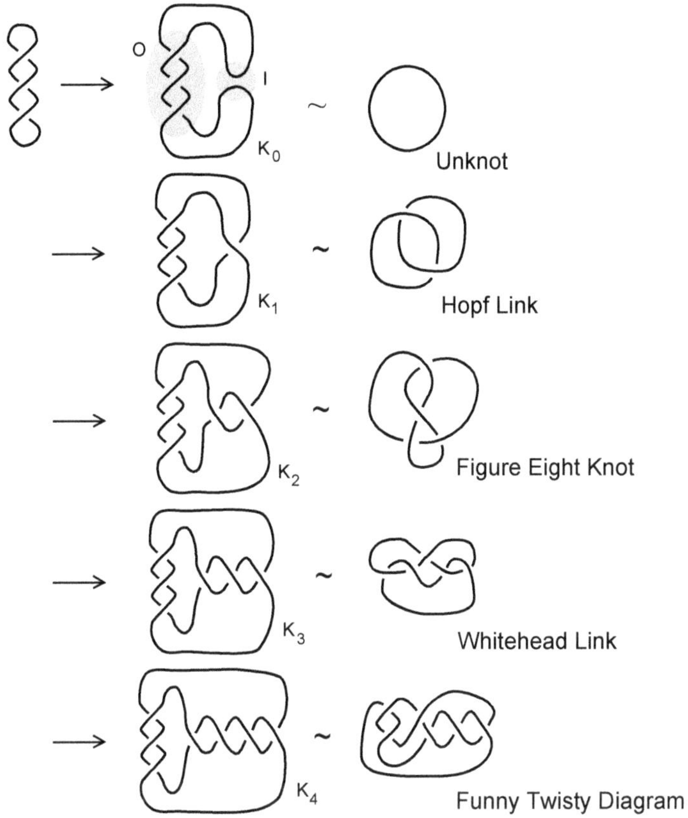

Fig. 32. Processive Recombination with $S = [-1/3]$

$$= N([a_r, a_{r-1}, \ldots, a_2, a_1 + n + b_1, b_2, \ldots, b_s]).$$

Proposition 3. *The formula*

$$K[n] = N([a_r, a_{r-1}, \ldots, a_2, a_1 + n + b_1, b_2, \ldots, b_{s-1}, b_s])$$

can be simplified to yield a stable formula (in the sense of the discussion above) for the processive recombination when n is sufficiently large.

Proof. Apply Proposition 1. □

Here is an example. Suppose we take $O = [1, 1, 1, 1]$ and $I = [-1, -1, -1]$ so that the DNA substrate is an (Fibonacci) unknot. Then by the above calculation:

$$K[n] = N([1, 1, 1, 1 + n + (-1), -1, -1]) = N([1, 1, 1, n, -1, -1]).$$

Suppose that n is positive. Applying Proposition 1, we get:

$$K[n] = N([1, 1, 1, n, -1, -1]) = N([1, 1, 1, n-1, 1, 0, 1]) = N([1, 1, 1, n-1, 2]),$$

and this is a stabilized form for the processive recombination.

More generally, suppose that $O = [a_1, \ldots, a_n]$ where all of the a_i are positive. Let $I = [-a_1, \ldots, -a_{n-1}]$. Then $K[0] = N(O+I)$ is a hard unknotted substrate by our result about convergents (Proposition 2 and Theorem 7). Consider $K[n]$ for positive n. We have

$$K[n] = N([a_n, a_{n-1}, \ldots, a_2, a_1 + n - a_1, -a_2, \ldots, -a_{n-1}])$$

$$= N([a_n, a_{n-1}, \ldots, a_2, n, -a_2, \ldots, -a_{n-1}])$$

$$= N([a_n, a_{n-1}, \ldots, a_2, (n-1), 1, (a_2 - 1), a_3, \ldots, a_{n-1}]).$$

If $a_2 - 1$ is not zero, the process terminates immediately. Otherwise there is one more step. In this way the knots and links resulting from the recombination process all have a uniform stabilized form. Further successive recombination just adds more twist in one entry in the continued fraction diagram whose closure is $K[n]$.

The reader may be interested in watching a visual demonstration of these properties of DNA recombination. For this we recommend the program *Ginterface* (TangleSolver) [44] of Mariel Vasquez. Her program can be downloaded from the internet as a Java applet, and it performs and displays DNA recombination. Figure 33 illustrates the form of display for this program. The reader should be warned that the program uses the reverse order from our convention when listing the terms in a continued fraction. Thus we say $[1, 1, 2]$ while the program uses $(2, 1, 1)$ for the same structure.

Remark 10. Along with this stabilization process, there is the general form of the recombination process itself that is related to our study of collapse. Our results apply to the process of successive recombination in DNA where a molecular configuration may undergo a crossing switch of

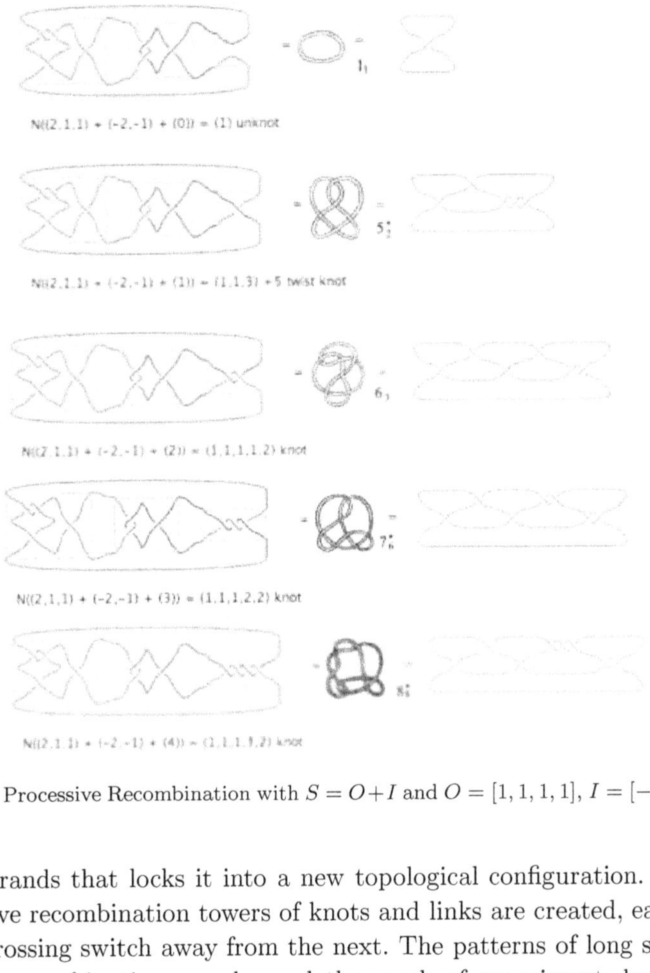

Fig. 33. Processive Recombination with $S = O + I$ and $O = [1, 1, 1, 1]$, $I = [-1, -1, -1]$.

DNA strands that locks it into a new topological configuration. Thus, in successive recombination towers of knots and links are created, each one a single crossing switch away from the next. The patterns of long sequences of such recombinations are beyond the reach of experiment, but can be understood mathematically. We hope that the results of this paper will apply to future experiments with DNA recombination.

13. Afterthoughts - Farey Series, Continued Fractions, Pick's Theorem and Ford Circles

The main theme of this paper has been our result (Theorem 6) that, given two fractions $\frac{p}{q}$ and $\frac{r}{s}$ such that $|ps - qr| = 1$, we can construct an unknot diagram $N([\frac{p}{q}] - [\frac{r}{s}])$ from the rational tangles $[\frac{p}{q}]$ and $[\frac{r}{s}]$. The arithmetic condition $|ps - qr| = 1$ has so many beautiful and surprising connections to other mathematics, that we must mention them in this last section of the paper! We shall touch on Farey series [11] and continued fractions, the Riemann Hypothesis, Pick's Theorem and Ford circles [6]. There is more, but we hope that this will give the reader a taste, and perhaps there will arise new connections with knots and unknots as well through this discussion.

To remind us to think of knots, we shall call fractions $\frac{p}{q}$ and $\frac{r}{s}$ such that $|ps - qr| = 1$ an *unknot pair*. Note that if $\frac{p}{q}, \frac{r}{s}$ is an unknot pair, then $\frac{q}{p}, \frac{s}{r}$ is also an unknot pair (see Remark 6 in Section 4). When $\frac{p}{q} < \frac{r}{s}$, then we have $ps - qr = -1$, the plus sign appearing when the fractions are in the other order. It is convenient to include the formal fractions $0/1$ and $1/0$ in these discussions just as we have done earlier in the paper with the tangles $[0]$ and $[\infty]$.

13.1. Farey Series and Continued Fractions

Given two fractions $\frac{a}{b}$ and $\frac{c}{d}$ such that $ad - bc = -1$, one can form the *mediant* fraction $\frac{e}{f} = \frac{a+c}{b+d}$. If $\frac{a}{b} < \frac{c}{d}$, then it is easy to see that $\frac{a}{b} < \frac{e}{f} < \frac{c}{d}$ and that $af - be = -1 = ed - fc$. This means that one can iterate the mediant construction, producing infinitely many unknot pairs from the given pair $(\frac{a}{b}, \frac{c}{d})$.

This iteration via the mediant fraction was discovered by John Farey [11] in 1816. It follows from Farey's construction that if one starts with the fractions $(\frac{0}{1}, \frac{1}{1})$ and iterates the mediant construction forever, then all rational numbers in the interval $[0, 1]$ will appear uniquely in their reduced forms.

In order to create any positive rational number greater than 1 we also consider the formal fraction $\frac{1}{0}$ and apply the mediant construction. Note that, then $\frac{1}{1} = \frac{0+1}{1+0}$ (one is the mediant of zero and infinity) so $\frac{0}{1}$ and $\frac{1}{0}$ can be seen as the "primal ancestors" of the rational numbers.

Given a real number $x > 1$, one can consider the rational numbers which, when expressed in lowest terms, have denominators less than x. The *Farey series* corresponding to x is the (ordered) set of positive rational numbers less than or equal to 1 which, in reduced form, have denominators less than

x. For example, the Farey series corresponding to 6 is (in order):

$$\{\frac{1}{5}, \frac{1}{4}, \frac{1}{3}, \frac{2}{5}, \frac{1}{2}, \frac{3}{5}, \frac{2}{3}, \frac{3}{4}, \frac{4}{5}, \frac{1}{1}\}.$$

The reader will notice that each adjacent pair of fractions in the Farey series above is an unknot pair. Each Farey series can be obtained by iterating the mediant construction starting with $\{\frac{0}{1}, \frac{1}{1}\}$, and omitting at each step the fractions whose denominators exceed x, until we reach a set for which every mediant is no longer permitted. For example, here is the generation of the Farey series corresponding to 6 (with $\frac{0}{1}$ and $\frac{1}{1}$ retained at the left and the right).

$$\{\frac{0}{1}, \frac{1}{1}\}$$

$$\{\frac{0}{1}, \frac{1}{2}, \frac{1}{1}\}$$

$$\{\frac{0}{1}, \frac{1}{3}, \frac{1}{2}, \frac{2}{3}, \frac{1}{1}\}$$

$$\{\frac{0}{1}, \frac{1}{4}, \frac{1}{3}, \frac{2}{5}, \frac{1}{2}, \frac{3}{5}, \frac{2}{3}, \frac{3}{4}, \frac{1}{1}\}$$

$$\{\frac{0}{1}, \frac{1}{5}, \frac{1}{4}, \frac{1}{3}, \frac{2}{5}, \frac{1}{2}, \frac{3}{5}, \frac{2}{3}, \frac{3}{4}, \frac{4}{5}, \frac{1}{1}\}$$

Adjacent pairs are unknot pairs because each fraction is the mediant of its neighbors.

Remark 11. Recall that the *Riemann Hypothesis* says that all the non-trival zeroes of the Riemann zeta function

$$\zeta(s) = \Sigma_{n=1}^{\infty} \frac{1}{n^s}$$

lie on the half-line $s = \frac{1}{2} + it$ where i is the square root of -1 and t is a real number. This famous unsolved problem is equivalent to a statement about the Farey series: Let $A(x)$ denote the number of terms in the Farey series corresponding to x. Let $\delta_j(x)$ denote the amount by which the j-th term of the Farey series for x differs from $j/A(x)$. The following conjecture is equivalent to the Riemann Hypothesis.

Conjecture 2 (Franel and Landau). *For each $\epsilon > 0$ there exists a constant K, depending upon ϵ, such that $\Sigma_{j=1}^{A(x)}|\delta_j(x)| < Kx^{\frac{1}{2}+\epsilon}$ as $x \longrightarrow \infty$.*

The equivalence is due to Franel and Landau [16]. See also [10]. Thus the Riemann Hypothesis is equivalent to a statement about how certain collections of unknots jostle one another as pairs of rational numbers in the real line.

In Figure 34 (top) we give an illustration of the Farey generating process for positive rational numbers. Here the tiers of numbers are connected in a graph, whose top nodes are the "ancestors" $\frac{0}{1}$ and $\frac{1}{0}$. A given fraction is the mediant of its two ancestors. In the graph each fraction has two lines upward to its immediate ancestors. The graph contains a natural binary tree starting from $\frac{1}{1}$ (see bottom illustration of Figure 34). We have indicated the genesis of this binary tree as a subgraph of this graph that shows all ancestors of the Farey fractions.

A given positive continued fraction $[a_1, a_2, ..., a_n]$ can be found (say n is odd) by starting at $\frac{1}{1}$ in the tree and heading downward by a_1 edges to the right, a_2 edges to the left, a_3 edges to the right, $\ldots$, a_{n-1} edges to the left, and finally $a_n - 1$ edges to the right. If n is even, the sequence will start with a_1 edges to the right and end with $a_n - 1$ edges to the left.

For example $\frac{7}{5} = [1, 2, 2]$ is related to the sequence of instructions $RLLR$ where R denotes "right" and L denotes "left". The reader will note that these instructions take one from $\frac{1}{1}$ to $\frac{7}{5}$ in the tree of Figure 34. For another example, take the instruction $RRLL$ and note that it takes one from $\frac{1}{1}$ to $\frac{7}{3}$. We have $\frac{7}{3} = [2, 3]$ which indeed is related to $RRLL$. On the other hand, we also have $\frac{7}{3} = [2, 2, 1]$ and this also is related to $RRLL$. Thus the truncation to $a_n - 1$ left or right steps in the last part of the path corresponds to our earlier discussion of the ambiguity of the last term in the sequence for the continued fraction.

The upshot of this structure is that one can use a Farey tree (also called the *Stern-Brocot tree*) to enumerate the possible unknots and their corresponding continued fraction pairs. It is easy to see from this prescription that a mediant and one of its ancestors are convergents. Compare these comments with our Theorem 8. Note also that since the rationals are dense in the real numbers, this tree provides a way to obtain for each

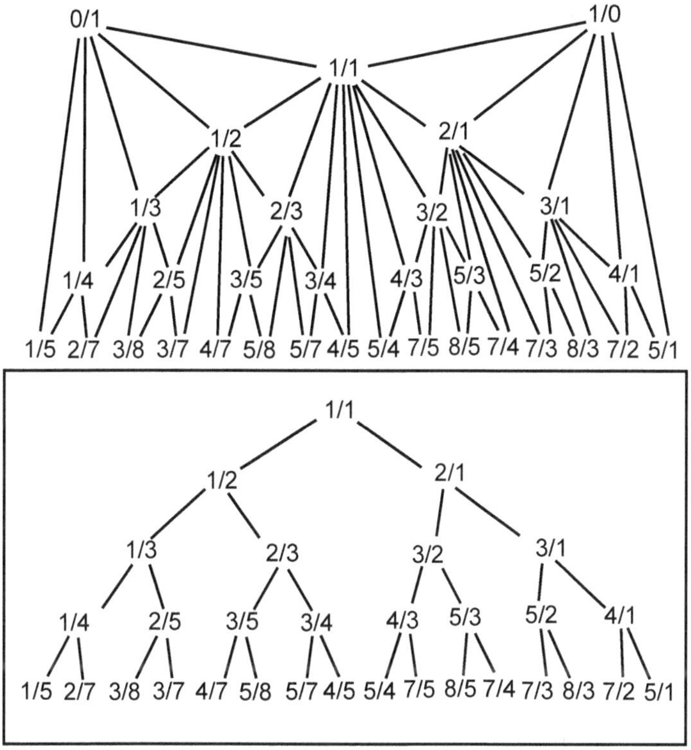

Fig. 34. The graph of Farey fractions and binary tree of ancestors

positive real number, a continued fraction that converges to it. Each real number is the limit of a path going down the binary tree. For example, the golden ratio $\phi = \frac{1+\sqrt{5}}{2} = [1, 1, 1, 1, \ldots]$ is the limit of the sequence $\{R, RL, RLR, RLRL, RLRLR, \ldots\}$. The inverse of the golden ratio, $\frac{1}{\phi} = \frac{\sqrt{5}-1}{2} = [0, 1, 1, 1, 1, \ldots]$, is the limit of the sequence $\{L, LR, LRL, LRLR, LRLRL, \ldots\}$. Each path down the infinite binary tree corresponds to the continued fraction expansion of a unique real number.

13.2. Pick's Theorem

Pick's Theorem [1], [2] states that the area of a polygon in the standard integral lattice in the Euclidean plane is given by the formula

$$A = I + B/2 - 1$$

where I denotes the number of lattice points in the interior of the polygon and B denotes the number of lattice points on the boundary. See Figure 35 for an example.

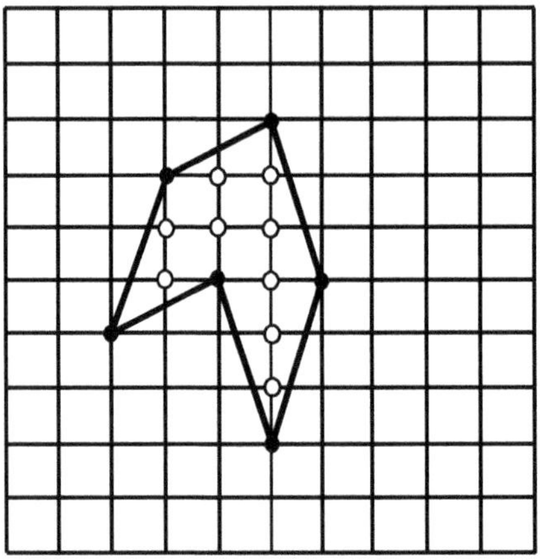

A = I + B/2 - 1 = 9 + 6/2 -1 = 11

Fig. 35. Area of a Lattice Polygon

The simplest case of Pick's Theorem is when there are no interior lattice points and the figure is a triangle with just three lattice points on the boundary. Call such a figure a *small triangle*. The area of a small triangle is, by Pick's Theorem, equal to $1/2$. If one node of the small triangle is at the origin, then the other two nodes can be viewed as integral vectors (a, b) and (c, d) in the plane. One can show that if a small triangle is formed by the origin and the tips of the two vectors, then the determinant $ad - bc$ has absolute value equal to 1. Hence the two vectors form an alternative basis for the integer lattice in the plane. They satisfy the formula $|ad - bc| = 1$ and yield unknot fraction pairs $\frac{a}{b}, \frac{c}{d}$ or $\frac{b}{a}, \frac{d}{c}$. This means that the fundamental case of Pick's Theorm is directly related to the structure of unknot pairs and to consecutive fractions in the Farey construction.

We can illustrate this relationship with the integer lattice by plotting vectors in a finite lattice. We will say that two vectors are *adjacent* in the integer lattice if together with the origin they form a triangle with no interior lattice points. Starting with a finite lattice, one can plot all lattice vectors with non-zero slopes less than or equal to one whose coordinates are relatively prime. This gives a finite collection of vectors and it is easy to see that two such vectors are adjacent, in the sense given above, if and only if one can be rotated about the origin into the other without encountering another vector in the collection. Thus adjacency becomes rotational adjacency in such a plot. In Figure 36 we illustrate this pattern by plotting all such vectors in the 5×5 lattice. This reproduces the Farey series for denominators no larger than 5 if we associate the vector (a, b) with the fraction b/a. The reader will note that the collection of fractions corresponding to the points in the figure are

$$\{1/5,\ 1/4,\ 1/3,\ 2/5,\ 1/2,\ 3/5,\ 3/4,\ 1/1\}.$$

The topic of Pick's Theorem brings us back to topology in that the general case of the Theorem follows from the special case of the triangle with no interior lattice points by the use of Euler's formula for plane graphs:

$$V - E + F = 2$$

when the connected graph in the plane has V nodes, E edges and F faces (including the outer, unbounded, face). One triangulates the polygon using the lattice points. All the faces are small triangles except for the outer face. From this it follows that $3(F-1) + B = 2E$, and we have that $V = I + B$. These two equations and the Euler formula give $(F-1)/2 = I + B/2 - 1$. The area of the polygon is $(F-1)/2$ since it is composed of $F-1$ triangles, each of area one-half. This gives Pick's Theorem. It is curious to think of the triangulation of the polygon giving us $F-1$ unknots to examine.

13.3. Ford Circles

Finally, there is an amazing geometric interpretation of pairs of fractions a/b and c/d such that $|ad - bc| = 1$. Associate to each reduced fraction a/b on the real line a circle tangent to the real line of diameter $1/b^2$. This is the *Ford circle* associated with the fraction (see, for example, [15], [6]). One can easily show that *two fractions form an unknot pair if and only if their Ford circles are tangent to one another.* This means that the mediant of an unknot pair produces a new circle that is tangent to both of its ancestral Ford circles. See Figure 37 for an illustration of this geometry.

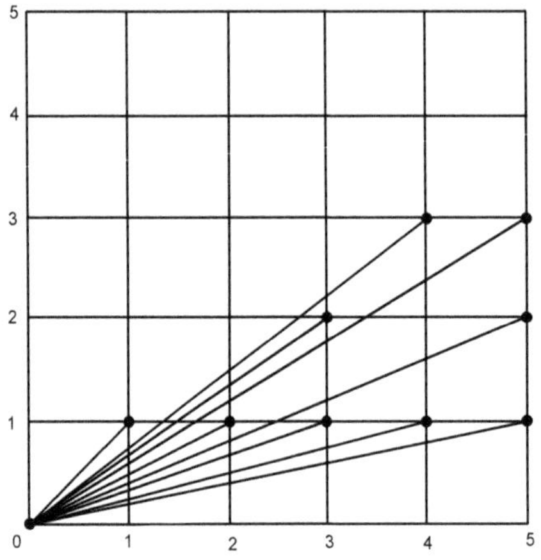

Fig. 36. Lattice Vectors and Farey Fractions

Ford circles can be regarded as curves in the complex plane. The set of Ford circles is invariant under the action of the modular group of transformations of the complex plane. By interpreting the upper half of the complex plane as a model of the hyperbolic plane (the Poincaré half-plane model) the set of Ford circles can be interpreted as a tiling of the hyperbolic plane. Thus, any two Ford circles are congruent in hyperbolic geometry. If C and C' are two tangent Ford circles, then the half-circle joining their respective fractions (on the real line) that is perpendicular to the real line is a hyperbolic line that also passes through the tangent point of the two circles. This half-circle is indicated with a dashed curve in Figure 37.

We have seen many aspects of the topology, number theory and geometry related to unknot pairs in the course of this paper. We hope that this last section stimulates the curiosity of the reader to make new explorations in this domain.

References

1. M. Aigner, G.M. Ziegler, "Proofs from The Book", Springer-Verlag (1991).
2. M. Bruckheimer, A. Arcavi, Farey series and Pick's area theorem, *The Mathematical Intelligencer*, **17**, No. 2 (1995), 64-67.
3. G. Burde, Verschlingungsinvarianten von Knoten und Verkettungen mit zwei Brücken, *Math. Zeitschrift*, **145** (1975), 235-242.

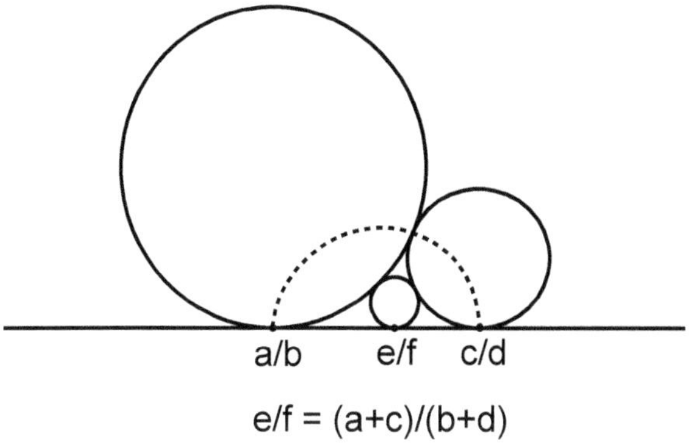

Fig. 37. Ford Circles

4. G. Burde, H. Zieschang, "Knots", de Gruyter Studies in Mathematics **5** (1985).
5. J.H. Conway, An enumeration of knots and links and some of their algebraic properties, *Proceedings of the conference on Computational problems in Abstract Algebra held at Oxford in 1967*, J. Leech ed., (First edition 1970), Pergamon Press, 329-358.
6. J.H. Conway, R.K. Guy, "The Book of Numbers", Springer Verlag (1996).
7. I. Darcy, Solving unoriented tangle equations involving 4-plats, *J. Knot Theory Remifications*, **14** (2005), no. 8, 993-1005.
8. I. Darcy, Solving oriented tangle equations involving 4-plats, *J. Knot Theory Remifications*, **14** (2005), no. 8, 1007-1027.
9. I.A. Dynnikov, Arc-presentations of links: Monotonic simplification, *Fund. Math*, **190** (2006), 29-76.
10. H.M. Edwards, "Riemann's Zeta Function", Academic Press (1974), Dover Publications (2001).
11. D. Epstein, C. Gunn, "Supplement to Not Knot", (booklet accompanying the video "Not Knot"),<http://www.geom.uiuc.edu/video/NotKnot/>, A. K. Peters, Ltd., Natick, MA 01760-4626, USA.
12. C. Ernst, D.W. Sumners, A calculus for rational tangles: Applications to DNA Recombination, *Math. Proc. Camb. Phil. Soc.*, **108** (1990), 489-515.
13. C. Ernst, D.W. Sumners, Solving tangle equations arising in a DNA recombination model. *Math. Proc. Cambridge Philos. Soc.*, **126**, No. 1 (1999), 23-36.
14. J. Farey, On a curious property of vulgar fractions, *Philosophical Magazine*, **47**, (1816), 385-386.
15. L.R. Ford, Fractions, *American Mathematical Monthly*, **45**, No. 9 (Nov., 1938), 586-601.
16. J. Franel, E. Landau, Les suites de Farey et le problème des nombres premiers, *Gottinger Nachr.*, (1924), 198-206.

17. J.S. Frame, Continued fractions and matrices, Classroom notes, C.B. Allendoerfer ed., *The Amer. Math. Monthly*, **56** (1949), 98-103.
18. L. Goeritz, Bemerkungen zur Knotentheorie, *Abh. Math. Sem. Univ. Hamburg*, **10** (1934), 201-210.
19. J.R. Goldman, L.H. Kauffman, Rational Tangles, *Advances in Applied Math.*, **18** (1997), 300-332.
20. J.H. Hass, C. Lagarias, The number of Reidemeister moves needed for unknotting, *J. Amer. Math. Soc.*, **14**, No. 2 (2001), 399-428.
21. G. Hemion, On the classification of the homeomorphisms of 2-manifolds and the classification of three-manifolds, *Acta Math.*, **142**, no. 1-2, (1979), pp. 123-155.
22. A. Henrich, L. H. Kauffman, On upper bounds for an unknotting sequence of Reidemeister moves. arXiv:1006.4176 (to appear).
23. S. Jablan (private conversation), $< http : //math.ict.edu.yu/ >$.
24. V.F.R. Jones, A polynomial invariant for links via von Neumann algebras, *Bull. Amer. Math. Soc.*, **129** (1985), 103-112.
25. L.H. Kauffman, State Models and the Jones Polynomial, *Topology*, 26 (1987), pp. 395-407.
26. L.H. Kauffman, S. Lambropoulou, On the classification of rational tangles, *Advances in Applied Math*, **33**, No. 2 (2004), 199237.
27. L.H. Kauffman, S. Lambropoulou, On the classification of rational knots. *L'Enseignement Mathématiques*, **49**, (2003), 357-410.
28. L. H. Kauffman, S. Lambropoulou, Unknots and molecular biology, *Milan. J. Math.*, **74** (2006), 227-263.
29. A. Kawauchi, "A Survey of Knot Theory", Birkhäuser Verlag (1996).
30. A.Ya. Khinchin, "Continued Fractions", Dover (1997) (republication of the 1964 edition of Chicago Univ. Press).
31. T. Kanenobu, H. Murakami, Two-bridge knots with unknotting number one, *Proc. Amer. Math. Soc.*, **98**, No. 3, Nov. 1986, 499-502.
32. P. Kohn, Two-bridge links with unlinking number one, *Proc. Amer. Math. Soc.*, **113**, No. 4, Dec. 1991, 1135-1147.
33. K. Kolden, Continued fractions and linear substitutions, *Archiv for Math. og Naturvidenskab*, **6** (1949), 141-196.
34. K. Millett (private conversation at Institutes Hautes Etudes Scientifiques, Burres Sur Yvette, France, circa 1988).
35. H. Morton, An irreducible 4-string braid with unknotted closure, *Proc. Camb. Phil. Soc.*, **93**, (1983), no. 2, pp. 259-261.
36. J.M. Montesinos, Revetêments ramifiés des noeuds, Espaces fibres de Seifert et scindements de Heegaard, *Publicaciones del Seminario Mathematico Garcia de Galdeano, Serie II, Seccion 3* (1984).
37. C.D. Olds, "Continued Fractions", New Mathematical Library, Math. Assoc. of Amerika, **9** (1963).
38. K. Reidemeister, "Knotentheorie" (Reprint), Chelsea, New York (1948).
39. K. Reidemeister, Knoten und Verkettungen, *Math. Zeitschrift*, **29** (1929), 713-729.
40. R. Scharein, KnotPlot (program available from the web), <http://wren.pims.math.ca/knotplot>.

41. H. SCHUBERT, Knoten mit zwei Brücken, *Math. Zeitschrift*, **65** (1956), 133-170.
42. L. SIEBENMANN, Lecture Notes on Rational Tangles, Orsay (1972) (unpublished).
43. D.W. SUMNERS, Untangling DNA, *Math.Intelligencer*, **12** (1990), 71-80.
44. M. VASQUEZ, TangleSolver (program available from the web),<http://math.berkeley.edu/ mariel/>.
45. H.S. WALL, "Analytic Theory of Continued Fractions", D. Van Nostrand Company, Inc. (1948).

L.H. KAUFFMAN: DEPARTMENT OF MATHEMATICS, STATISTICS AND COMPUTER SCIENCE, UNIVERSITY OF ILLINOIS AT CHICAGO, 851 SOUTH MORGAN ST., CHICAGO IL 60607-7045, U.S.A.

S. LAMBROPOULOU: DEPARTMENT OF MATHEMATICS, NATIONAL TECHNICAL UNIVERSITY OF ATHENS, ZOGRAFOU CAMPUS, GR-157 80 ATHENS, GREECE.

E-MAILS: `kauffman@math.uic.edu` `sofia@math.ntua.gr`

URLs: `http://www.math.uic.edu/~kauffman/`
`http://www.math.ntua.gr/~sofia`

KHOVANOV HOMOLOGY

Louis H. Kauffman

Department of Mathematics, Statistics
and Computer Science (m/c 249)
851 South Morgan Street
University of Illinois at Chicago
Chicago, Illinois 60607-7045
kauffman@uic.edu

Keywords: Bracket polynomial, Khovanov homology, cube category, simplicial category, tangle cobordism, chain complex, chain homotopy , unitary transformation, quantum computing, quantum information theory, link homology, categorif cation.

1. Introduction

This paper is an introduction to Khovanov homology. We start with a quick introduction to the bracket polynomial, reformulating it and the Jones polynomial so that the value of an unknotted loop is $q + q^{-1}$. We then introduce enhanced states for the bracket state sum so that, in terms of these enhanced states the bracket is a sum of monomials. Then in Section 3, we point out that the shape of the collection of bracket states for a given diagram is a cube and that this cube can be taken to be a category. It is an example of a *cube category*. We show that functors from a cube category to a category of modules naturally have homology theories associated with them. In Section 4 we show how to make a homology theory (Khovanov homology) from the states of the bracket so that the enhanced states are the generators of the chain complex. We show how a Frobenius algebra structure arises naturally from this adjacency structure for the enhanced states. Finally we show that the resulting homology is an example of homology related to a module functor on the cube category as described in Section 3. In Section 5, we give a short exposition of Dror BarNatan's tangle cobordism theory for Khovanov homology. This theory replaces Khovanov homology by an abstract chain homotopy class of a complex of surface cobordisms associated with the states of a knot or link diagram. In Section 6 we give a short exposition of the Rasmussen invariant and its application to f nding the four-ball genus of torus knots. In Section 7 we give a description of Khovanov homology as the homology of a simplicial module by following our description of the cube category in this context. I regard this section as a step in the direction of providing an abstract simplicial homotopy theory for Khovanov homology, but this project is not yet completed. In Section 8 we discuss a quantum context for Khovanov homology that is obtained by building a Hilbert space whose orthonormal basis is the set of enhanced states of a diagram K. Then there is a unitary transformation U_K of this Hilbert space so that the Jones polynomial J_K is the trace of $U_K : J_K = Trace(U_K)$. We discuss a generalization where the linear space of the Khovanov homology itself is taken to be the Hilbert space. In this case we can def ne a unitary transformation U'_K so that, for values of q and t on the unit circle, the Poincaré polynomial for the Khovanov homology is the trace of U'_K. Section 9 is is a discussion of other forms of link homology and categorif cation with selected references.

It gives the author great pleasure to thank the members of the Quantum Topology Seminar at the University of Illinois at Chicago for many useful conversations and to thank the Perimeter Institute in Waterloo, Canada for their hospitality while this paper was being completed.

2. Bracket Polynomial and Jones Polynomial

The bracket polynomial [15] model for the Jones polynomial [12–14, 52] is usually described by the expansion

$$\langle \times \rangle = A \langle \asymp \rangle + A^{-1} \langle) (\rangle$$

Here the small diagrams indicate parts of otherwise identical larger knot or link diagrams. The two types of smoothing (local diagram with no crossing) in this formula are said to be of type A (A above) and type B (A^{-1} above).

$$\langle \bigcirc \rangle = -A^2 - A^{-2}$$

$$\langle K \bigcirc \rangle = (-A^2 - A^{-2}) \langle K \rangle$$

$$\langle \gamma \rangle = (-A^3) \langle \smile \rangle$$

$$\langle \gamma' \rangle = (-A^{-3}) \langle \smile \rangle$$

One uses these equations to normalize the invariant and make a model of the Jones polynomial. In the normalized version we def ne

$$f_K(A) = (-A^3)^{-wr(K)} \langle K \rangle / \langle \bigcirc \rangle$$

where the writhe $wr(K)$ is the sum of the oriented crossing signs for a choice of orientation of the link K. Since we shall not use oriented links in this paper, we refer the reader to [15] for the details about the writhe. One then has that $f_K(A)$ is invariant under the Reidemeister moves (again see [15]) and the original Jones polynomial $V_K(t)$ is given by the formula

$$V_K(t) = f_K(t^{-1/4}).$$

The Jones polynomial has been of great interest since its discovery in 1983 due to its relationships with statistical mechanics, due to its ability to often detect the difference between a knot and its mirror image and due to the many open problems and relationships of this invariant with other aspects of low dimensional topology.

The State Summation. In order to obtain a closed formula for the bracket, we now describe it as a state summation. Let K be any unoriented link diagram. Def ne a *state*, S, of K to be the collection of planar loops resulting from a choice of smoothing for each crossing of K. There are two choices (A and B) for smoothing a given crossing, and thus there are $2^{c(K)}$ states of a diagram with $c(K)$ crossings. In a state we label each smoothing with A or A^{-1} according to the convention indicated by the expansion formula for the bracket. These labels are the *vertex weights* of the state. There are two evaluations related to a state. The f rst is the product of the vertex weights, denoted $\langle K|S\rangle$. The second is the number of loops in the state S, denoted $||S||$.
Def ne the *state summation*, $\langle K\rangle$, by the formula

$$\langle K\rangle = \sum_S < K|S > \delta^{||S||}$$

where $\delta = -A^2 - A^{-2}$. This is the state expansion of the bracket. It is possible to rewrite this expansion in other ways. For our purposes in this paper it is more convenient to think of the loop evaluation as a sum of *two* loop evaluations, one giving $-A^2$ and one giving $-A^{-2}$. This can be accomplished by letting each state curve carry an extra label of $+1$ or -1. We describe these *enhanced states* below. But before we do this, it will be useful for the reader to examine Figure 2. In Figure 2 we show all the states for the right-handed trefoil knot, labelling the sites with A or B where B denotes a smoothing that would receive A^{-1} in the state expansion.

Note that in the state enumeration in Figure 2 we have organized the states in tiers so that the state that has only A-smoothings is at the top and the state that has only B-smoothings is at the bottom.

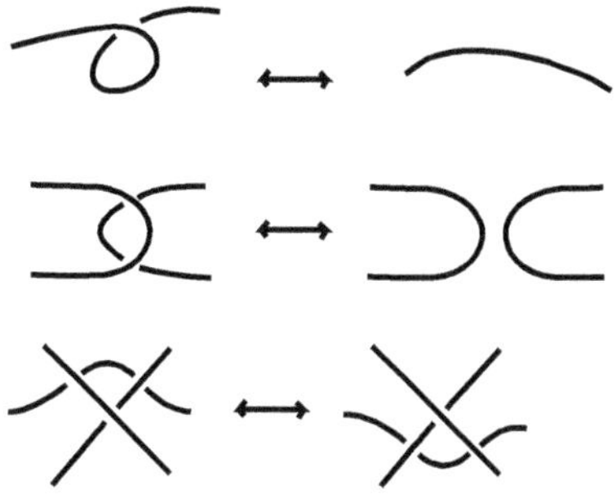

Fig. 1. **Reidemeister Moves**

Changing Variables. Letting $c(K)$ denote the number of crossings in the diagram K, if we replace $\langle K \rangle$ by $A^{-c(K)}\langle K \rangle$, and then replace A^2 by $-q^{-1}$, the bracket is then rewritten in the following form:

with $\langle \bigcirc \rangle = (q + q^{-1})$. It is useful to use this form of the bracket state sum for the sake of the grading in the Khovanov homology (to be described below). We shall continue to refer to the smoothings labeled q (or A^{-1} in the original bracket formulation) as *B-smoothings*.

We catalog here the resulting behaviour of this modifed bracket under the Reidemeister moves.

$$\langle \bigcirc \rangle = q + q^{-1}$$

$$\langle K \bigcirc \rangle = (q + q^{-1})\langle K \rangle$$

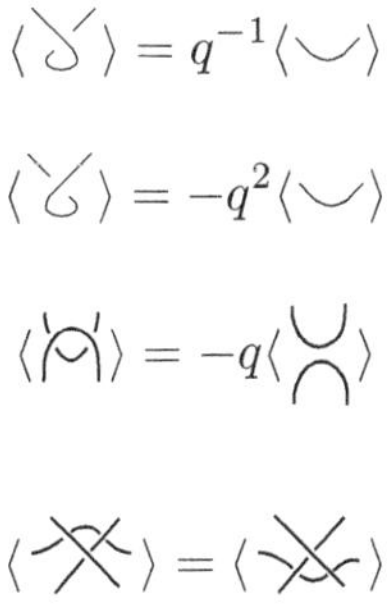

It follows that if we define

$$J_K = (-1)^{n_-} q^{n_+ - 2n_-} \langle K \rangle,$$

where n_- denotes the number of negative crossings in K and n_+ denotes the number of positive crossings in K, then J_K is invariant under all three Reidemeister moves. Thus J_K is a version of the Jones polynomial taking the value $q + q^{-1}$ on an unknotted circle.

Using Enhanced States. We now use the convention of *enhanced states* where an enhanced state has a label of 1 or -1 on each of its component loops. We then regard the value of the loop $q+q^{-1}$ as the sum of the value of a circle labeled with a 1 (the value is q) added to the value of a circle labeled with an -1 (the value is q^{-1}). We could have chosen the less neutral labels of $+1$ and X so that

$$q^{+1} \Longleftrightarrow +1 \Longleftrightarrow 1$$

and

$$q^{-1} \Longleftrightarrow -1 \Longleftrightarrow x,$$

since an algebra involving 1 and x naturally appears later in relation to Khovanov homology. It does no harm to take this form of labeling from the beginning. The use of enhanced states for formulating Khovanov homology was pointed out by Oleg Viro in [48].

Consider the form of the expansion of this version of the bracket polynonmial in enhanced states. We have the formula as a sum over enhanced states s :

$$\langle K \rangle = \sum_s (-1)^{i(s)} q^{j(s)}$$

where $i(s)$ is the number of B-type smoothings in s and $j(s) = i(s) + \lambda(s)$, with $\lambda(s)$ the number of loops labeled 1 minus the number of loops labeled -1 in the enhanced state s.

One advantage of the expression of the bracket polynomial via enhanced states is that it is now a sum of monomials. We shall make use of this property throughout the rest of the paper.

3. Khovanov Homology and the Cube Category

We are going to make a chain complex from the states of the bracket polynomial so that the homology of this chain complex is a knot invariant. One way to see how such a homology theory arises is to step back and note that the collection of states for a diagram K forms a category in the shape of a cube. A functor from such a category to a category of modules gives rise to a homology theory in a natural way, as we explain below.

Examine Figure 2 and Figure 3. In Figure 2 we show all the standard bracket states for the trefoil knot with arrows between them whenever the state at the output of the arrow is obtained from the state at the input of the arrow by a single re-smoothing of a site of type A to a site of type B. The abstract structure of this collection of states is a category with objects of the form $\langle ABA \rangle$ where this symbol denotes one of the states in the state diagram of Figure 2. In Figure 3 we illustrate this cube category (the states are arranged in the form of a cube) by replacing the states in Figure 2 by symbols $\langle XYZ \rangle$ where each literal is either an A or a B. A typical generating morphism in the 3-cube category is

$$\langle ABA \rangle \longrightarrow \langle BBA \rangle.$$

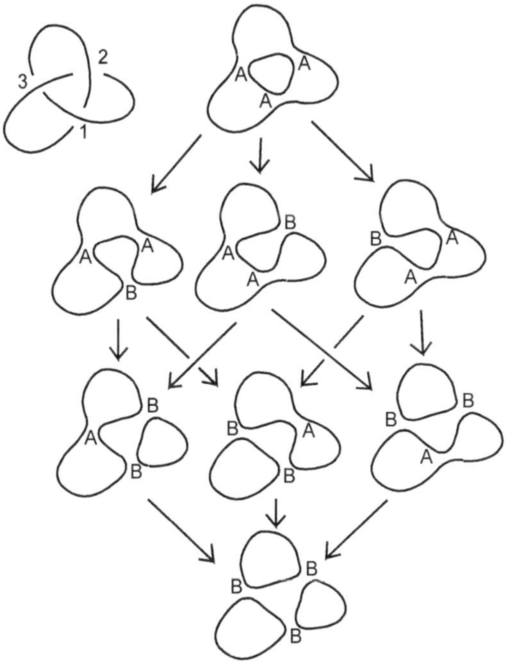

Fig. 2. **Bracket States and Khovanov Complex**

We formalize this way of looking at the bracket states as follows. Let $\mathcal{S}(K)$ denote a category associated with the states of the bracket for a diagram K whose objects are the states, with sites labeled A and B as in Figure 2. A morphism in

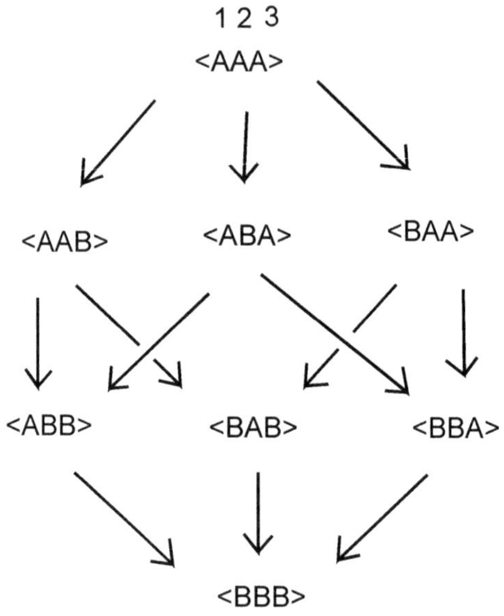

Fig. 3. **Cube Category**

this category is an arrow from a state with a given number of A's to a state with fewer A's.

Let $\mathcal{D}^n = \{A, B\}^n$ be the n-cube category whose objects are the n-sequences from the set $\{A, B\}$ and whose non-identity morphisms are arrows from sequences with greater numbers of A's to sequences with fewer numbers of A's. Thus $\mathcal{D}^n$ is equivalent to the poset category of subsets of $\{1, 2, \cdots n\}$. We make a functor $\mathcal{R} : \mathcal{D}^n \longrightarrow \mathcal{S}(K)$ for a diagram K with n crossings as follows. We map sequences in the cube category to bracket states by choosing to label the crossings of the diagram K from the set $\{1, 2, \cdots n\}$, and letting this functor take abstract A's and B's in the cube category to smoothings at those crossings of type A or type B. Thus each sequence in the cube category is associated with a unique state of K when K has n crossings. By the same token, we define a functor $\mathcal{S} : \mathcal{S}(K) \longrightarrow \mathcal{D}^n$ by associating a sequence to each state and morphisms between sequences corresponding to the state re-smoothings. With these conventions, the two compositions of these morphisms are the identity maps on their respective categories.

Let $\mathcal{M}$ be a pointed category with finite sums, and let $\mathcal{F} : \mathcal{D}^n \longrightarrow \mathcal{M}$ be a functor. In our case $\mathcal{M}$ will be a category of modules and $\mathcal{F}$ will carry n-sequences to certain tensor powers corresponding to the standard bracket states of a knot or link K. We postpone this construction for a moment, and point out that there is a natural structure of chain complex associated with the functor $\mathcal{F}$. First note that each object in $\mathcal{D}^n$ has the form

$$X = \langle X_0 \cdots X_{n-1} \rangle$$

where each X_i equals either A or B and we have morphisms

$$d_i : \langle X_0 \cdots X_i \cdots X_{n-1} \rangle \longrightarrow \langle X_0 \cdots \bar{X}_i \cdots X_{n-1} \rangle$$

whenever $X_i = A$ and (by definition) $\bar{X}_i = B$. We then define

$$\partial_i = \mathcal{F}(d_i) : \mathcal{F}\langle X_0 \cdots X_i \cdots X_{n-1} \rangle \longrightarrow \mathcal{F}\langle X_0 \cdots \bar{X}_i \cdots X_{n-1} \rangle$$

whenever d_i is defined. We then define the chain complex C by

$$C^k = \bigoplus_X \mathcal{F}\langle X_0 \cdots X_{n-1} \rangle$$

where each sequence $X = \langle X_0 \cdots X_{n-1} \rangle$ has k B's. With this we define

$$\partial : C^k \longrightarrow C^{k+1}$$

by the formula

$$\partial x = \Sigma_{i=0}^{n-1} (-1)^{c(X,i)} \partial_i(x)$$

for $x \in \mathcal{F}X = \mathcal{F}\langle X_0 \cdots X_{n-1} \rangle$ and $c(X,i)$ denotes the number of A's in the sequence X that precede X_i.

We want $\partial^2 = 0$ and it is easy to see that this is equivalent to the condition that $\partial_i \partial_j = \partial_j \partial_i$ for $i \neq j$ whenever these maps and compositions are defined. We can assume that the functor $\mathcal{F}$ has this property, or we can build it in axiomatically by adding the corresponding relations to the cube category in the form $d_i d_j = d_j d_i$ for $i \neq j$ whenever these maps are defined. In the next section we shall see that there is a natural way to define the maps in the state category so that this condition holds. Once we axiomatize this commutation relation at the level of the state category or the cube category, then the functor $\mathcal{F}$ will induce a chain complex and homology as above.

In this way, we see that a suitable functor from the cube category to a module category allows us to define homology that is modeled on the "shape" of the cube. The set of bracket states forms a natural functorial image of the cube category, and that makes it possible to define the Khovanov chain complex. In terms of the bracket states, we will map each state loop to a specific module V, and each state to a tensor power of V to the number of loops in the state. The details of this construction are in the next section.

We use a specif c construction for the Khovanov complex that is directly related to the enhanced states for the bracket polynomial, as we will see in the next section. In this construction we will use the enhanced states, regarding each loop as labeled with either 1 or x for a module $V = k[x]/(x^2)$ associated with the loop (where $k = Z/2Z$ or $k = Z$.) Thus the two labelings of the loop will corespond to the two generators of the module V. A state that is a collection of loops will be associated with $V^{\otimes r}$ where r is the number of loops in the state. In this way we will obtain a functor from the state category to a module category, and at the same time it will happen that any single enhanced state will correspond to a generator of the chain complex. In the next section we show how naturally this algebra appears in relation to the enhanced states. We then return to the categorical point of view and see how, surface cobordisms of circles provide an abstract category for the invariant.

4. Khovanov Homology

In this section, we describe Khovanov homology along the lines of [3, 31], and we tell the story so that the gradings and the structure of the differential emerge in a natural way. This approach to motivating the Khovanov homology uses elements of Khovanov's original approach, Viro's use of enhanced states for the bracket polynomial [48], and Bar-Natan's emphasis on tangle cobordisms [2, 3]. We use similar considerations in our paper [27].

Two key motivating ideas are involved in f nding the Khovanov invariant. First of all, one would like to *categorify* a link polynomial such as $\langle K \rangle$. There are many meanings to the term categorify, but here the quest is to f nd a way to express the link polynomial as a *graded Euler characteristic* $\langle K \rangle = \chi_q \langle \mathcal{H}(K) \rangle$ for some homology theory associated with $\langle K \rangle$.

We will use the bracket polynomial and its enhanced states as described in the previous sections of this paper. To see how the Khovanov grading arises, consider the form of the expansion of this version of the bracket polynomial in enhanced states. We have the formula as a sum over enhanced states s :

$$\langle K \rangle = \sum_{s} (-1)^{i(s)} q^{j(s)}$$

where $i(s)$ is the number of B-type smoothings in s, $\lambda(s)$ is the number of loops in s labeled 1 minus the number of loops labeled X, and $j(s) = i(s) + \lambda(s)$. This can be rewritten in the following form:

$$\langle K \rangle = \sum_{i,j} (-1)^i q^j dim(\mathcal{C}^{ij})$$

where we def ne $\mathcal{C}^{ij}$ to be the linear span (over the complex numbers for the purpose of this paper, but over the integers or the integers modulo two for other contexts) of the set of enhanced states with $i(s) = i$ and $j(s) = j$. Then the number of such states is the dimension $dim(\mathcal{C}^{ij})$.

We would like to have a bigraded complex composed of the $\mathcal{C}^{ij}$ with a differential

$$\partial : \mathcal{C}^{ij} \longrightarrow \mathcal{C}^{i+1\,j}.$$

The differential should increase the *homological grading* i by 1 and preserve the *quantum grading* j. Then we could write

$$\langle K \rangle = \sum_j q^j \sum_i (-1)^i dim(\mathcal{C}^{ij}) = \sum_j q^j \chi(\mathcal{C}^{\bullet\, j}),$$

where $\chi(\mathcal{C}^{\bullet\, j})$ is the Euler characteristic of the subcomplex $\mathcal{C}^{\bullet\, j}$ for a f xed value of j.

This formula would constitute a categorif cation of the bracket polynomial. Below, we shall see how *the original Khovanov differential ∂ is uniquely determined by the restriction that $j(\partial s) = j(s)$ for each enhanced state s.* Since j is preserved by the differential, these subcomplexes $\mathcal{C}^{\bullet\, j}$ have their own Euler characteristics and homology. We have

$$\chi(H(\mathcal{C}^{\bullet\, j})) = \chi(\mathcal{C}^{\bullet\, j})$$

where $H(\mathcal{C}^{\bullet\, j})$ denotes the homology of the complex $\mathcal{C}^{\bullet\, j}$. We can write

$$\langle K \rangle = \sum_j q^j \chi(H(\mathcal{C}^{\bullet\, j})).$$

The last formula expresses the bracket polynomial as a *graded Euler characteristic* of a homology theory associated with the enhanced states of the bracket state summation. This is the categorif cation of the bracket polynomial. Khovanov proves that this homology theory is an invariant of knots and links (via the Reidemeister moves of Figure 1), creating a new and stronger invariant than the original Jones polynomial.

We will construct the differential in this complex f rst for mod-2 coeff cients. The differential is based on regarding two states as *adjacent* if one differs from the other by a single re-smoothing at some site. Thus if (s, τ) denotes a pair consisting in an enhanced state s and site τ of that state with τ of type A, then we consider all enhanced states s' obtained from s by re-smoothing at τ and relabeling only those loops that are affected by the re-smoothing. Call this set of enhanced states

$S'[s,\tau]$. Then we shall def ne the *partial differential* $\partial_\tau(s)$ as a sum over certain elements in $S'[s,\tau]$, and the differential by the formula

$$\partial(s) = \sum_\tau \partial_\tau(s)$$

with the sum over all type A sites τ in s. It then remains to see what are the possibilities for $\partial_\tau(s)$ so that $j(s)$ is preserved.

Note that if $s' \in S'[s,\tau]$, then $i(s') = i(s) + 1$. Thus

$$j(s') = i(s') + \lambda(s') = 1 + i(s) + \lambda(s').$$

From this we conclude that $j(s) = j(s')$ if and only if $\lambda(s') = \lambda(s) - 1$. Recall that

$$\lambda(s) = [s:+] - [s:-]$$

where $[s:+]$ is the number of loops in s labeled $+1$, $[s:-]$ is the number of loops labeled -1 (same as labeled with x) and $j(s) = i(s) + \lambda(s)$.

Proposition. The partial differentials $\partial_\tau(s)$ are uniquely determined by the condition that $j(s') = j(s)$ for all s' involved in the action of the partial differential on the enhanced state s. This unique form of the partial differential can be described by the following structures of multiplication and comultiplication on the algebra $V = k[x]/(x^2)$ where $k = Z/2Z$ for mod-2 coeff cients, or $k = Z$ for integral coeff cients.

(1) The element 1 is a multiplicative unit and $x^2 = 0$.
(2) $\Delta(1) = 1 \otimes x + x \otimes 1$ and $\Delta(x) = x \otimes x$.

These rules describe the local relabeling process for loops in a state. Multiplication corresponds to the case where two loops merge to a single loop, while comultiplication corresponds to the case where one loop bifurcates into two loops.

Proof. Using the above description of the differential, suppose that there are two loops at τ that merge in the re-smoothing. If both loops are labeled 1 in s then the local contribution to $\lambda(s)$ is 2. Let s' denote a re-smoothing in $S[s,\tau]$. In order for the local λ contribution to become 1, we see that the merged loop must be labeled 1. Similarly if the two loops are labeled 1 and X, then the merged loop must be labeled X so that the local contribution for λ goes from 0 to -1. Finally, if the two loops are labeled X and X, then there is no label available for a single loop that will give -3, so we def ne ∂ to be zero in this case. We can summarize the result by saying that there is a multiplicative structure m such that $m(1,1) = 1, m(1,x) = m(x,1) = x, m(x,x) = 0$, and this multiplication describes the structure of the partial differential when two loops merge. Since this is the multiplicative structure of the algebra $V = k[x]/(x^2)$, we take this algebra as summarizing the differential.

Now consider the case where s has a single loop at the site τ. Re-smoothing produces two loops. If the single loop is labeled x, then we must label each of the two loops by x in order to make λ decrease by 1. If the single loop is labeled 1, then we can label the two loops by x and 1 in either order. In this second case we take the partial differential of s to be the sum of these two labeled states. This structure can be described by taking a coproduct structure with $\Delta(x) = x \otimes x$ and $\Delta(1) = 1 \otimes x + x \otimes 1$. We now have the algebra $V = k[x]/(x^2)$ with product $m : V \otimes V \longrightarrow V$ and coproduct $\Delta : V \longrightarrow V \otimes V$, describing the differential completely. This completes the proof. //

Partial differentials are def ned on each enhanced state s and a site τ of type A in that state. We consider states obtained from the given state by re-smoothing the given site τ. The result of re-smoothing τ is to produce a new state s' with one more site of type B than s. Forming s' from s we either amalgamate two loops to a single loop at τ, or we divide a loop at τ into two distinct loops. In the case of amalgamation, the new state s acquires the label on the amalgamated circle that is the product of the labels on the two circles that are its ancestors in s. This case of the partial differential is described by the multiplication in the algebra. If one circle becomes two circles, then we apply the coproduct. Thus if the circle is labeled X, then the resultant two circles are each labeled X corresponding to $\Delta(x) = x \otimes x$. If the original circle is labeled 1 then we take the partial boundary to be a sum of two enhanced states with labels 1 and x in one case, and labels x and 1 in the other case, on the respective circles. This corresponds to $\Delta(1) = 1 \otimes x + x \otimes 1$. Modulo two, the boundary of an enhanced state is the sum, over all sites of type A in the state, of the partial boundaries at these sites. It is not hard to verify directly that the square of the boundary mapping is zero (this is the identity of mixed partials!) and that it behaves as advertised, keeping $j(s)$ constant. There is more to say about the nature of this construction with respect to Frobenius algebras and tangle cobordisms. In Figures 4, 5 and 6 we illustrate how the partial boundaries can be conceptualized in terms of surface cobordisms. Figure 4 shows how the partial boundary corresponds to a saddle point and illustrates the two cases of fusion and f ssion of circles. The equality of mixed partials corresponds to topological equivalence of the corresponding surface cobordisms, and to the relationships between Frobenius algebras [23] and the surface cobordism category. In particular, in Figure 6 we show how in a key case of two sites (labeled 1 and 2 in that Figure) the two orders of partial boundary are

$$\partial_2\partial_1 = (1 \otimes m) \circ (\Delta \otimes 1)$$

and

$$\partial_1\partial_2 = \Delta \circ m.$$

In the Frobenius algebra $V = k[x]/(x^2)$ we have the identity

$$(1 \otimes m) \circ (\Delta \otimes 1) = \Delta \circ m.$$

Thus the Frobenius algebra implies the identity of the mixed partials. Furthermore, in Figure 5 we see that this identity corresponds to the topological equivalence of cobordisms under an exchange of saddle points.

In Figures 7 and 8 we show another aspect of this algebra. As Figure 7 illustrates, we can consider cup (minimum) and cap (maximum) cobordisms that go between the empty set and a single circle. With the categorical arrow going down the page, the cap is a mapping from the base ring k to the module V and we denote this mapping by $\eta : k \longrightarrow V$. It is the *unit* for the algebra V and is def ned by $\eta(1) = 1_V$, taking 1 in k to 1_V in V. The cup is a mapping from V to k and is denoted by $\epsilon : V \longrightarrow k$. This is the *counit*. As Figure 7 illustrates, we need a basic identity about the counit which reads

$$\Sigma\epsilon(a_1)a_2 = a$$

for any $a \in V$ where

$$\Delta(a) = \Sigma a_1 \otimes a_2.$$

The summation is over an appropriate set of elements in $V \otimes V$ as in our specif c formulas for the algebra $k[x]/(x^2)$. Of course we also demand

$$\Sigma a_1\epsilon(a_2) = a$$

for any $a \in V$. With these formulas about the counit and unit in place, we see that cobordisms will give equivalent algebra when one cancels a maximum or a minimum with a saddle point, again as shown in Figure 7.

Note that for our algebra $V = k[x]/(x^2)$, it follows from the counit identities of the last paragraph that

$$\epsilon(1) = 0$$

and

$$\epsilon(x) = 1.$$

In fact, Figure 8 shows a formula that holds in this special algebra. The formula reads

$$\epsilon(ab) = \epsilon(ax)\epsilon(b) + \epsilon(a)\epsilon(bx)$$

for any $a, b \in V$. As the rest of Figure 8 shows, this identity means that a single tube in any cobordism can be cut, replacing it by a cups and a caps in a linear combination of two terms. The tube-cutting relation is shown in its most useful form at the bottom of Figure 8. In Figure 8, the black dots are symbols standing for the special element x in the algebra.

It is important to note that we have a nonsingular pairing

$$\langle \mid \rangle : V \otimes V \longrightarrow k$$

def ned by the equation

$$\langle a|b\rangle = \epsilon(ab).$$

One can def ne a Frobenius algebra by starting with the existence of a non-singular bilinear pairing. In fact, a f nite dimensional associative algebra with unit def ned over a unital commutative ring k is said to be a *Frobenius algebra* if it is equipped with a non-degenerate bilinear form

$$\langle \mid \rangle : V \otimes V \longrightarrow k$$

such that

$$\langle ab|c\rangle = \langle a|bc\rangle$$

for all a, b, c in the algebra. The other mappings and the interpretation in terms of cobordisms can all be constructed from this def nition. See [23].

Remark on Grading and Invariance. In Section 2 we showed how the bracket, using the variable q, behaves under Reidemeister moves. These formulas correspond to how the invariance of the homology works in relation to the moves. We have that

$$J_K = (-1)^{n_-} q^{n_+ - 2n_-} \langle K\rangle,$$

where n_- denotes the number of negative crossings in K and n_+ denotes the number of positive crossings in K. $J(K)$ is invariant under all three Reidemeister moves. The corresponding formulas for Khovanov homology are as follows

$$\begin{aligned} J_K &= (-1)^{n_-} q^{n_+ - 2n_-} \langle K\rangle \\ &= (-1)^{n_-} q^{n_+ - 2n_-} \Sigma_{i,j} (-1)^i a^j dim(H^{i,j}(K) \\ &= \Sigma_{i,j} (-1)^{i+n_+} q^{j+n_+ - 2n_{-1}} dim(H^{i,j}(K)) \\ &= \Sigma_{i,j} (-1)^i q^j dim(H^{i-n_-, j-n_+ + 2n_-}(K)). \end{aligned}$$

It is often more convenient to def ne the *Poincaré polynomial* for Khovanov homology via

$$P_K(t,q) = \Sigma_{i,j} t^i q^j dim(H^{i-n_-,j-n_++2n_-}(K)).$$

The Poincaré polynomial is a two-variable polynomial invariant of knots and links, generalizing the Jones polynomial. Each coeff cient

$$dim(H^{i-n_-,j-n_++2n_-}(K))$$

is an invariant of the knot, invariant under all three Reidemeister moves. In fact, the homology groups

$$H^{i-n_-,j-n_++2n_-}(K)$$

are knot invariants. The grading compensations show how the grading of the homology can change from diagram to diagram for diagrams that represent the same knot.

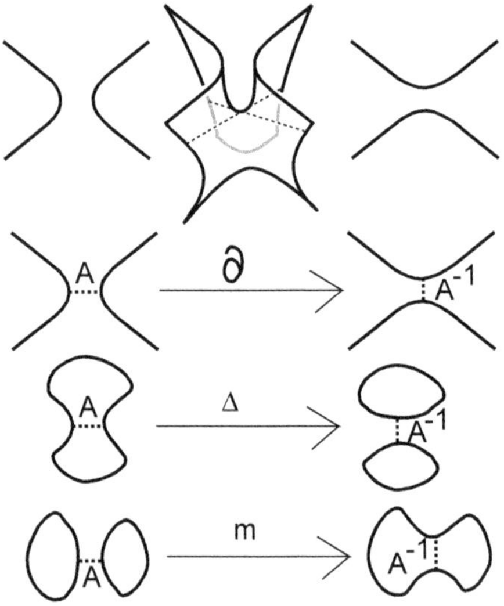

Fig. 4. **SaddlePoints and State Smoothings**

Remark on Integral Differentials. Choose an ordering for the crossings in the link diagram K and denote them by $1, 2, \cdots n$. Let s be any enhanced state of K and let $\partial_i(s)$ denote the chain obtained from s by applying a partial boundary at the i-th site of s. If the i-th site is a smoothing of type A^{-1}, then $\partial_i(s) = 0$. If the

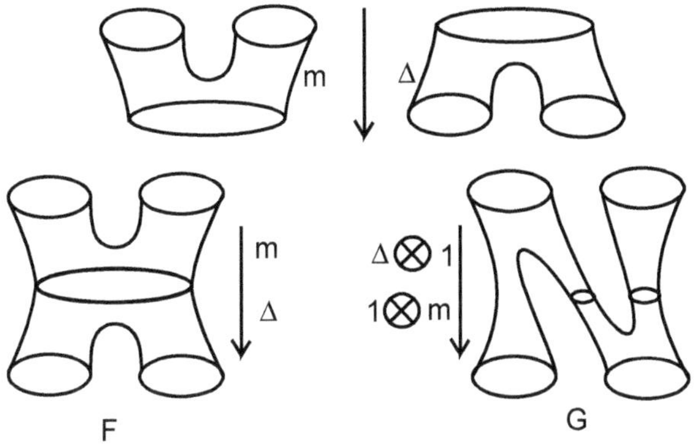

Fig. 5. **Surface Cobordisms**

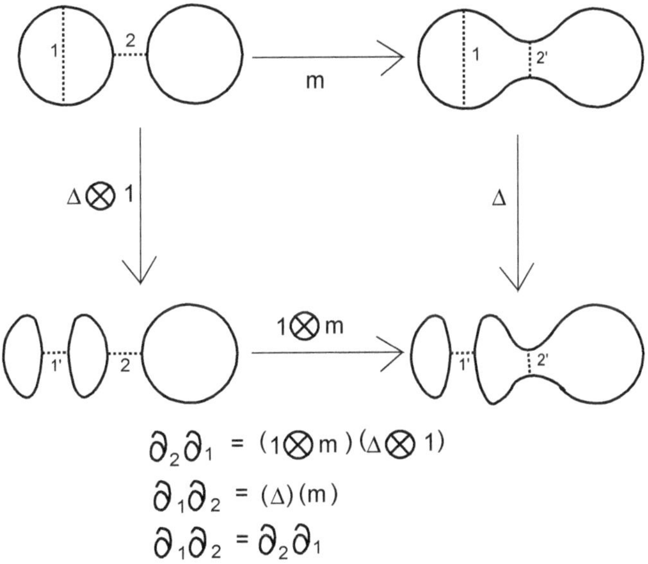

Fig. 6. **Local Boundaries Commute**

i-th site is a smoothing of type A, then $\partial_i(s)$ is given by the rules discussed above (with the same signs). The compatibility conditions that we have discussed show that partials commute in the sense that $\partial_i(\partial_j(s)) = \partial_j(\partial_i(s))$ for all i and j. One then def nes signed boundary formulas in the usual way of algebraic topology. One way to think of this regards the complex as the analogue of a complex in de

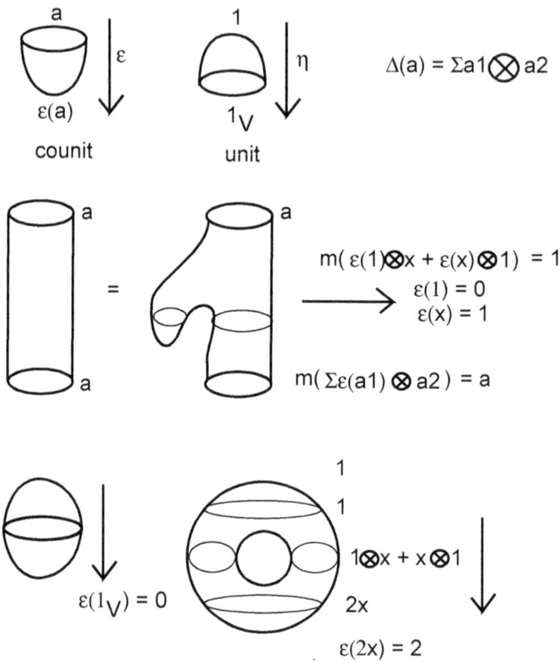

Fig. 7. **Unit and Counit as Cobordisms**

Rham cohomology. Let $\{dx_1, dx_2, \cdots, dx_n\}$ be a formal basis for a Grassmann algebra so that $dx_i \wedge dx_j = -dx_j \wedge dx_i$ Starting with enhanced states s in $C^0(K)$ (that is, states with all A-type smoothings) def ne formally, $d_i(s) = \partial_i(s)dx_i$ and regard $d_i(s)$ as identical with $\partial_i(s)$ as we have previously regarded it in $C^1(K)$. In general, given an enhanced state s in $C^k(K)$ with B-smoothings at locations $i_1 < i_2 < \cdots < i_k$, we represent this chain as $s\,dx_{i_1} \wedge \cdots \wedge dx_{i_k}$ and def ne

$$\partial(s\,dx_{i_1} \wedge \cdots \wedge dx_{i_k}) = \sum_{j=1}^{n} \partial_j(s)\,dx_j \wedge dx_{i_1} \wedge \cdots \wedge dx_{i_k},$$

just as in a de Rham complex. The Grassmann algebra automatically computes the correct signs in the chain complex, and this boundary formula gives the original boundary formula when we take coeff cients modulo two. Note, that in this formalism, partial differentials ∂_i of enhanced states with a B-smoothing at the site i are zero due to the fact that $dx_i \wedge dx_i = 0$ in the Grassmann algebra. There is more to discuss about the use of Grassmann algebra in this context. For example, this approach clarif es parts of the construction in [28].

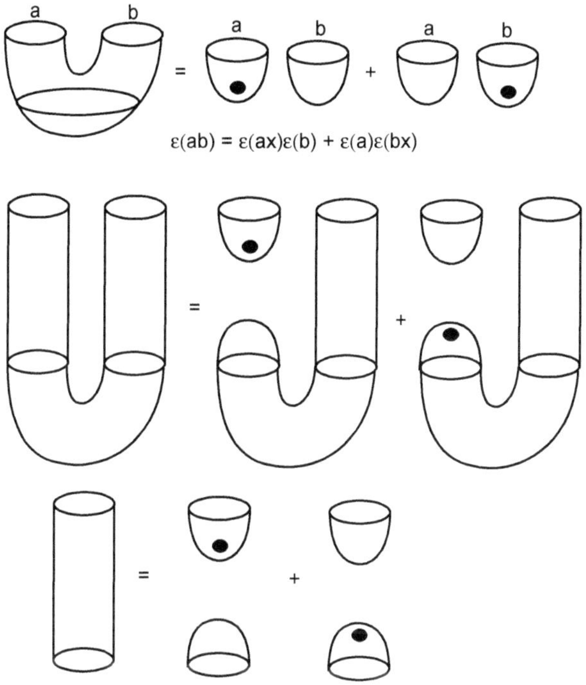

Fig. 8. **The Tube Cutting Relation**

It is of interest to examine this analogy between the Khovanov (co)homology and de Rham cohomology. In that analogy the enhanced states correspond to the differentiable functions on a manifold. The Khovanov complex $C^k(K)$ is generated by elements of the form $s\,dx_{i_1} \wedge \cdots \wedge dx_{i_k}$ where the enhanced state s has B-smoothings at exactly the sites $i_1, \cdots, i_k$. If we were to follow the analogy with de Rham cohomology literally, we would define a new complex $DR(K)$ where $DR^k(K)$ is generated by elements $s\,dx_{i_1} \wedge \cdots \wedge dx_{i_k}$ where s is *any* enhanced state of the link K. The partial boundaries are defined in the same way as before and the global boundary formula is just as we have written it above. This gives a *new* chain complex associated with the link K. Whether its homology contains new topological information about the link K will be the subject of a subsequent paper.

In the case of de Rham cohomology, we can also look for compatible unitary transformations. Let M be a differentiable manifold and $\mathcal{C}(M)$ denote the DeRham complex of M over the complex numbers. Then for a differential form of the type $f(x)\omega$ in local coordinates $x_1, \cdots, x_n$ and ω a wedge product of a subset of $dx_1 \cdots dx_n$, we have

$$d(f\omega) = \sum_{i=1}^{n} (\partial f/\partial x_i) dx_i \wedge \omega.$$

Here d is the differential for the DeRham complex. Then $\mathcal{C}(M)$ has as basis the set of $|f(x)\omega\rangle$ where $\omega = dx_{i_1} \wedge \cdots \wedge dx_{i_k}$ with $i_1 < \cdots < i_k$. We could achieve $Ud + dU = 0$ if U is a very simple unitary operator (e.g. multiplication by phases that do not depend on the coordinates x_i) but in general it will be an interesting problem to determine all unitary operators U with this property.

A further remark on de Rham cohomology. There is another relation with the de Rham complex: In [41] it was observed that Khovanov homology is related to Hochschild homology and Hochschild homology is thought to be an algebraic version of de Rham chain complex (cyclic cohomology corresponds to de Rham cohomology), compare [45].

5. The Cube Category and the Tangle Cobordism Structure of Khovanov Homology

We can now connect the constructions of the last section with the homology construction via the cube category. Here it will be convenient to think of the state category $\mathcal{S}(K)$ as a cube category with extra structure. Thus we will denote the bracket states by sequences of A's and B's as in Figures 2 and 3. And we shall regard the maps such as $d_2 : \langle AABA\rangle \longrightarrow \langle ABBA\rangle$ as corresponding to re-smoothings of bracket states that either join or separate state loops. We take $V = k[x]/(x^2)$ with the coproduct structure as given in the previous section. The maps from $m : V \otimes V \longrightarrow V$ and $\Delta : V \longrightarrow V \otimes V$ allow us to def ne the images of the re-smoothing maps d_i under a functor $\mathcal{F} : \mathcal{S}(K) \longrightarrow \mathcal{M}$ where $\mathcal{M}$ is the category generated by V by taking tensor powers of V and direct sums of these tensor powers. It then follows that the homology we have described in the previous section is exactly the homology associated with this functor.

The material in the previous section also suggests a modif cation of the state category $\mathcal{S}(K)$. Instead of taking the maps in this category to be simply the abstract arrows generated by elementary re-smoothings of states from A to B, we can regard each such smoothing as a surface cobordism from the set of circles comprising the domain state to the set of circles comprising the codomain state.

With this, in mind, two such cobordisms represent equivalent morphisms whenever the corresponding surfaces are homeomorphic relative to their boundaries. Call this category $\mathcal{C}obS(K)$. We then easily generalize the observations of the previous section, particularly Figures 4, 5 and 6, to see that we have the desired commuting relations $d_i d_j = d_j d_i$ (for $i \neq j$) in $\mathcal{C}obS(K)$ so that any functor from $\mathcal{C}obS(K)$ to a module category will have a well-def ned chain complex and associated homology. This applies, in particular to the functor we have constructed, using the Frobenius algebra $V = k[x]/(x^2)$.

In [3] BarNatan takes the approach using surface cobordisms a step further by making a categorical analog of the chain complex. For this purpose we let $\mathcal{C}obS(K)$ become an additive category. Maps between specif c objects X and Y added formally and the set $Maps(X, Y)$ is a module over the integers. More generally, let $\mathcal{C}$ be an additive category. In order to create the analog of a chain complex, let $\mathcal{M}at(C)$ denote the *Matrix Category of C* whose objects are n-tuples (vectors) of objects of $\mathcal{C}$ (n can be any natural number) and whose morphisms are in the form of a matrix $m = (m_{ij})$ of morphisms in C where we write

$$m : O \longrightarrow O'$$

and

$$m_{ij} : O_i \longrightarrow O'_j$$

for

$$O = (O_1, \cdots, O_n),$$

$$O' = (O'_1, \cdots, O'_m).$$

Here O_i and O'_j are objects in $\mathcal{C}$ while O and O' are objects in $\mathcal{M}at(C)$. Composition of morphisms in $\mathcal{M}at(C)$ follows the pattern of matrix multiplication. If

$$n : O' \longrightarrow O''$$

then

$$n \circ m : O \longrightarrow O''$$

and

$$(n \circ m)_{i,j} = \Sigma_k n_{i,k} \circ m_{k,j}$$

where the compositions in the summation occur in the category C.

We then def ne the *category of complexes over* $\mathcal{C}$, denoted $\mathcal{K}om(\mathcal{M}at(C))$ to consist of sequences of objects of $\mathcal{M}at(C)$ and maps between them so that consecutively composed maps are equal to zero.

$$\cdots \longrightarrow O^k \longrightarrow O^{k+1} \longrightarrow O^{k+2} \longrightarrow \cdots .$$

Here we let $\partial_k : O^k \longrightarrow O'^{k+1}$ denote the differential in the complex and we assme that $\partial_{k+1}\partial_k = 0$. A morphism between complexes O^* and O'^* consists in a family of maps $f_k : O^k \longrightarrow O'^k$ such that $\partial'_k f_k = f_{k+1}\partial_k$. Such morphisms will be called *chain maps*.

At this abstract level, we cannot calculate homology since kernels and cokernels are not available, but we can def ne the *homotopy type* of a complex in $\mathcal{K}om(\mathcal{M}at(C))$. We say that two chain maps $f : O \longrightarrow O'$ and $g : O \longrightarrow O'$ are *homotopic* if there is a sequence of mappings $H_k : O^k \longrightarrow O^{k-1}$ such that

$$f - g = H\partial + \partial H.$$

Specif cally, this means that

$$f_k - g_k = H_{k+1}\partial_k + \partial_{k+1}H_k.$$

Note that if $\phi = H\partial + \partial H$, then

$$\partial\phi = \partial H\partial = \phi\partial.$$

Thus any such ϕ is a chain map. We call two complexes O and O' *homotopy equivalent* if there are chain maps $F : O \longrightarrow O'$ and $G : O' \longrightarrow O$ such that both FG and GF are homotopic to the identity map of O and O' respectively. The homotopy type of a complex is an abstract substitute for the homology since, in an abelian category (where one can compute homology) homology is an invariant of homotopy type.

We are now in a position to work with the category $\mathcal{K}om(\mathcal{M}at(\mathcal{C}obS(K)))$ where K is a link diagram. The question is, what extra equivalence relation on the category $\mathcal{C}obS(K)$ will ensure that the homotopy types in $\mathcal{K}om(\mathcal{M}at(\mathcal{C}obS(K)))$ will be invariant under Reidemeister moves on the diagram K.

BarNatan [3] gives an elegant answer to this question. His answer is illustrated in Figure 9 where we show the *4Tu Relation*, the *Sphere Relation* and the *Torus Relation*. The key relation is the the $4Tu$ relation, which serves a number of purposes, including being a basic homotopy in the category $\mathcal{K}om(\mathcal{M}at(\mathcal{C}obS(K)))$. Note that the $4Tu$ relation can be described as follows: There are four local bits of surface, call them S_1, S_2, S_3, S_4. Let $C_{i,j}$ denote this conf guration with a tube connecting S_i and S_j. Then in the cobordism category we take the identity

$$C_{1,2} + C_{3,4} = C_{1,3} + C_{2,4}.$$

It is a good exercise for the reader to show that the $4Tu$ relation follows from the tube cutting relation of Figure 8. The Sphere and Torus relations assert that the 2-sphere has value 0 and that the torus has value 2, just as we have seen by using the Frobenius algebra in Figure 7.

To illustrate how things work once we factor by these relations, we show in Figures 10 and 11 how one sees the homotopy equivalance of the complexes for a diagram before and after the second Reidemeister move. In Figure 10 we show the complexes and indicate chain maps F and G between them and homotopies in the complex for the diagram before it is simplif ed by the Reidemeister move. In Figure 11 we show these maps and their compositions in the form of a four-term identity. The reader should f rst just look at the identity of involving these four cobordisms and recognize that it is exactly the $4Tu$ relation. On the other hand, each of the terms in the relation is factored into mappings involving F_1, G_1 and the homotopies H_1 and H_2 and the boundary mappings in the complex. Study of Figure 11 will convince the reader that the complexes before and after the second Reidemeister move are homotopy equivalent. A number of details are left to the reader. For example, note that in Figure 10 we have indicated the categorical chain complexes Z and W by showing only how they differ locally near the change corresponding to a Reidemeister two move. We give, via Figures 10 and 11, chain maps $F : W \longrightarrow Z$ and $G : Z \longrightarrow W$. These maps consist in a particular cobordism on one part of the complex and an identity map on the other part of the complex. We have specif cally labeled parts of these mappings by F_1 and by G_1. Using the implicit def nitions of F_1 and G_1 given in Figure 11, the reader will easily see that $G_1 F_1 = 0$ since this composition includes a 2-sphere. From this it follows that GF is the identity mapping on the complex W. We also leave to the reader to check that the mappings F and G commute with the boundary mappings so that they are mappings of complexes. The part of the homotopy indicated shows that FG is homotopic to the identity (up to sign) and so shows that the complexes Z and W are homotopy equivalent. One needs the value of the torus equal to 2 for homotopy invariance under the f rst Reidemeister move. This is a fragment of the full derivation of homotopy equivalences corresponding to all three Reidemeister moves that is given in the paper [3].

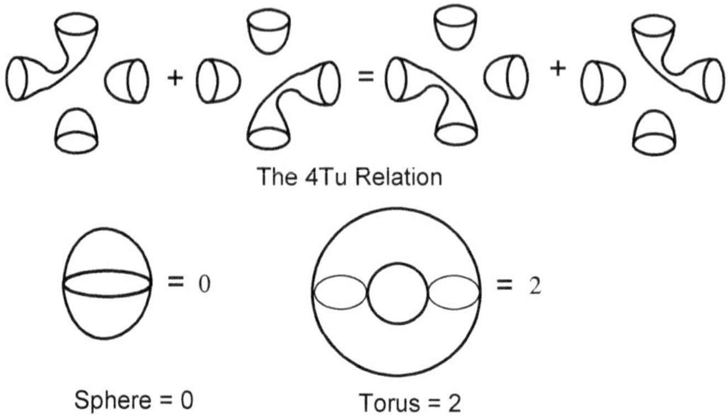

Fig. 9. **The 4Tu Relation, Sphere and Torus Relations**

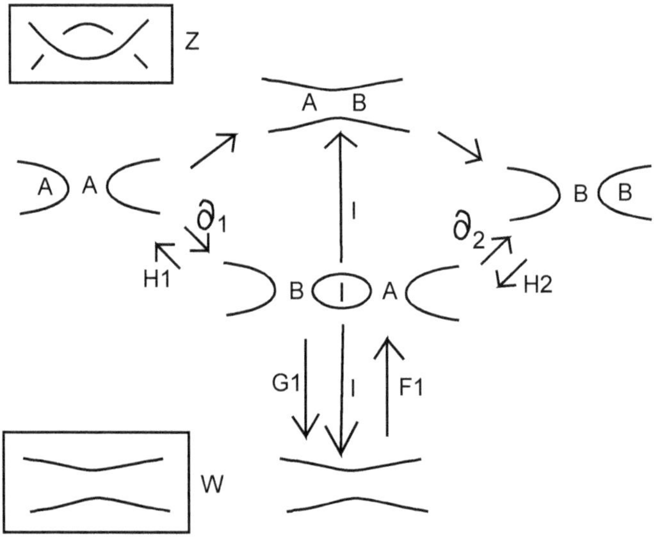

Fig. 10. **Complexes for Second Reidemeister Move**

6. Other Frobenius Algebras and Rasmussen's Theorem

Lee [25] makes another homological invariant of knots and links by using a different Frobenius algebra. She takes the algebra $\mathcal{A} = k[x]/(x^2 - 1)$ with

$$x^2 = 1,$$

$$\Delta(1) = 1 \otimes x + x \otimes 1,$$

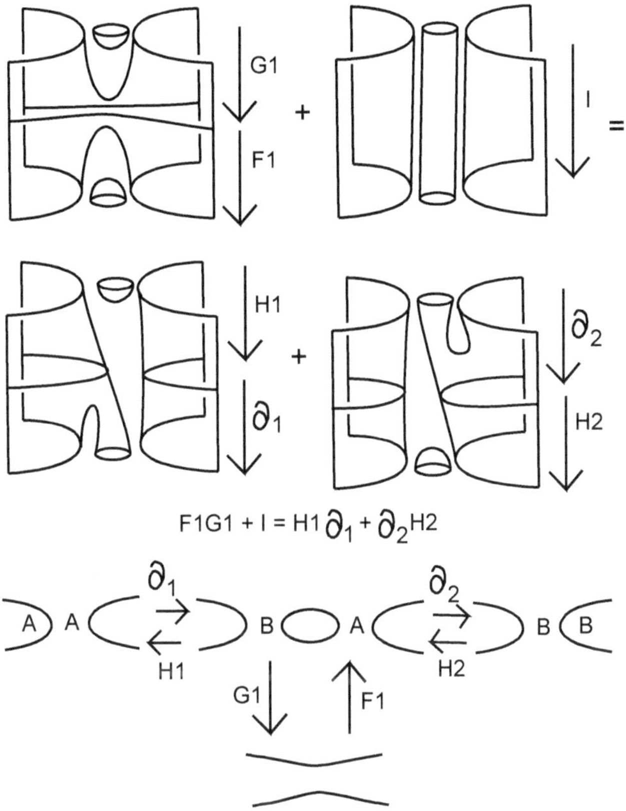

Fig. 11. **Homotopy for Second Reidemeister Move**

$$\Delta(x) = x \otimes x + 1 \otimes 1,$$

$$\epsilon(x) = 1,$$

$$\epsilon(1) = 0.$$

This gives a link homology theory that is distinct from Khovanov homology. In this theory, the quantum grading j is not preseved, but we do have that

$$j(\partial(\alpha)) \geq j(\alpha)$$

for each chain α in the complex. This means that *one can use j to f lter the chain complex for the Lee homology.* The result is a spectral sequence that starts from Khovanov homology and converges to Lee homology.

Lee homology is simple. One has that the dimension of the Lee homology is equal to $2^{comp(L)}$ where $comp(L)$ denotes the number of components of the link L. Up to homotopy, Lee's homology has a vanishing differential, and the complex behaves well under link concondance. In his paper [4] Dror BarNatan remarks "In a beautiful article Eun Soo Lee introduced a second differential on the Khovanov complex of a knot (or link) and showed that the resulting (double) complex has non-interesting homology. This is a very interesting result." Rasmussen [44] uses Lee's result to def ne invariants of links that give lower bounds for the four-ball genus, and determine it for torus knots. This gives an (elementary) proof of a conjecture of Milnor that had been previously shown using gauge theory by Kronheimer and Mrowka [24].

Rasmussen's result uses the Lee spectral sequence. We have the quantum (j) grading for a diagram K and the fact that for Lee's algebra $j(\partial(s)) \geq j(s)$. Rasmussen uses a normalized version of this grading denoted by $g(s)$. Then one makes a f ltration $F^k C^*(K) = \{v \in C^*(K) | g(v) \geq k\}$ and given $\alpha \in Lee^*(K)$ def ne

$$S(\alpha) := max\{g(v) | [v] = \alpha\}$$

$$s_{min}(K) := min\{S(\alpha) | \alpha \in Lee^*(K), \alpha \neq 0\}$$

$$s_{max}(K) := max\{S(\alpha) | \alpha \in Lee^*(K), \alpha \neq 0\}$$

and

$$s(K) := (1/2)(s_{min}(K) + s_{max}(K)).$$

This last average of s_{min} and s_{max} is the Rasmussen invariant.

We now enter the following sequence of facts:

(1) $s(K) \in Z$.
(2) $s(K)$ is additive under connected sum.
(3) If K^* denotes the mirror image of the diagram K, then

$$s(K^*) = -s(K).$$

(4) If K is a positive knot diagram (all positive crossings), then

$$s(K) = -r + n + 1$$

where r denotes the number of loops in the canonical oriented smoothing (this is the same as the number of Seifert circuits in the diagram K) and n denotes the number of crossings in K.

(5) For a torus knot $K_{a,b}$ of type (a, b), $s(K_{a,b}) = (a-1)(b-1)$.
(6) $|s(K)| \leq 2g^*(K)$ where $g^*(K)$ is the least genus spanning surface for K in the four ball.
(7) $g^*(K_{a,b}) = (a-1)(b-1)/2$. This is Milnor's conjecture.

This completes a very skeletal sketch of the construction and use of Rasmussen's invariant.

7. The Simplicial Structure of Khovanov Homology

Let $\mathcal{S}$ denote the set of (standard) bracket states for a link diagram K. One way to describe the Khovanov complex is to associate to each state loop λ a module V isomorphic to the algebra $k[x]/(x^2)$ with coproduct as we have described in the previous sections. The generators 1 and x of this algebra can then be regarded as the two possible enhancements of the loop λ. In the same vein we associate to a state S the tensor product of copies of V, one copy for each loop in the state. The local boundaries are defned exactly as before, and the Khovanov complex is the direct sum of the modules associated with the states of the link diagram. We will use this point of view in the present section, and we shall describe Khovanov homology in terms of the n-cube category and an associated simplicial object. The purpose of this section is to move towards, albeit in an abstract manner, a description of Khovanov homology as the homology of a topological space whose homotopy type is an invariant of the knot of the underlying knot or link. We do not accomplish this aim, but the constructions given herein may move toward that goal. An intermediate possibility would be to replace the Khovanov homology by an abstract space or simplicial object whose generalized homotopy type was an invariant of the knot or link.

Let $\mathcal{D}^n = \{A, B\}^n$ be the n-cube category whose objects are the n-sequences from the set $\{A, B\}$ and whose non-identity morphisms are arrows from sequences with greater numbers of A's to sequences with fewer numbers of A's. Thus $\mathcal{D}^n$ is equivalent to the poset category of subsets of $\{1, 2, \cdots n\}$. Let $\mathcal{M}$ be a pointed category with fnite sums, and let $\mathcal{F} : \mathcal{D}^n \longrightarrow \mathcal{M}$ be a functor. In our case $\mathcal{M}$ is a category of modules (as described above) and $\mathcal{F}$ carries n-sequences to certain tensor powers corresponding to the standard bracket states of a knot or link K. We map sequences to states by choosing to label the crossings of the diagram K from the set $\{1, 2, \cdots n\}$, and letting the functor take abstract A's and B's in the cube category to smoothings at those crossings of type A or type B. Thus each sequence in the cube category is associated with a unique state of K when K has n crossings. Nevertheless, we shall describe the construction more generally.

For the functor $\mathcal{F}$ we f rst construct a semisimplicial object $\mathcal{C}(F)$ over $\mathcal{M}$, where a semisimplicial object is a simplicial object without degeneracies. This means that it has partial boundaries analogous to the partial boundaries that we have discussed before but none of the degeneracy maps that are common to simplicial theory (see [46] Chapter 1). For $k \geq 0$ we set

$$\mathcal{C}(F)_k = \oplus_{v \in \mathcal{D}_k^n} \mathcal{F}(v)$$

where $\mathcal{D}_k^n$ denotes those sequences in the cube category with k A's. Note that we are indexing dually to the upper indexing in the Khovanov homology sections of this paper where we counted the number of B's in the states.

We introduce face operators (partial boundaries in our previous terminology)

$$d_i : \mathcal{C}(F)_k \longrightarrow \mathcal{C}(F)_{k-1}$$

for $0 \leq i \leq k$ with $k \geq 1$ as follows: d_i is trivial for $i = 0$ and otherwise d_i acts on $\mathcal{F}(v)$ by the map $\mathcal{F}(v) \longrightarrow \mathcal{F}(v')$ where v' is the sequence resulting from replacing the i-th A by B. The operators d_i satisfy the usual face relations of simplicial theory:

$$d_i d_j = d_{j-1} d_i$$

for $i < j$.

We now expand $\mathcal{C}(F)$ to a simplicial object $\mathcal{S}(F)$ over $\mathcal{M}$ by applying freely degeneracies to the $\mathcal{F}(v)$'s. Thus

$$\mathcal{S}(F)_m = \oplus_{v \in \mathcal{D}_k^n, k+t=m}\ s_{i_1} \cdots s_{i_t} \mathcal{F}(v)$$

where $m > i_1 > \cdots > i_t \geq 0$ and these degeneracy operators are applied freely modulo the usual (axiomatic) relations among themselves and with the face operators. Then $\mathcal{S}(F)$ has degeneracies via formal application of degeneracy operators to these forms, and has face operators extending those of $\mathcal{C}(\mathcal{F})$. It is at this point we should remark that in our knot theoretic construction there is only at this point an opportunity for formal extension of degeneracy operators above the number of crossings in the given knot or link diagram since to make specif c degeneracies would involve the creation of new diagrammatic sites. There may be a natural construction of this sort and if there is, such a construction could lead to a signif cant homotopy theory for Khovanov homology.

When the functor $\mathcal{F} : \mathcal{D}^n \longrightarrow \mathcal{M}$ goes to an abelian category $\mathcal{M}$, as in our knot theoretic case, we can recover the homology groups via

$$H_\star N\mathcal{S}(\mathcal{F}) \cong H_\star \mathcal{C}(\mathcal{F})$$

where $NS(\mathcal{F})$ is the normalized chain complex of $\mathcal{S}(\mathcal{F})$. This completes the abstract simplicial description of this homology.

8. Quantum Comments

States of a quantum system are represented by unit vectors in a Hilbert space. Quantum processes are unitary transformations applied to these state vectors. In an appropriate basis for the HIlbert space, each basis vector represesents a possible measurement. If $|\psi\rangle$ is a unit vector, then, upon measurement, one of the basis vectors will appear with probability, the absolute square of its coeff cient in $|\psi\rangle$. One can, in principle, f nd the trace of a given unitary transformation by instantiating it in a certain quantum system and making repeated measurements on that system. Such a scheme, in the abstract, is called a quantum algorithm, and in the concrete is called a quantum computer. One well-known quantum algorithm for determining the trace of a unitary matrix is called the "Hadamard Test" [47].

In [19] we consider the Jones polynomial and Khovanov homology in a quantum context. In this section we give a sketch of these ideas. Recall from Section 2 that we have the following formula for the Jones polynomial.

$$J_K = (-1)^{n_-} q^{n_+ - 2n_-} \langle K \rangle.$$

Using the enhanced states formulation of Section 2, we form a Hilbert space $\mathcal{H}(K)$ with orthonormal basis the set of enhanced states of K. For the Hilbert space we denote a basis element by $|s\rangle$ where s is an enhanced state of the diagram K. Now using q as in Section 2, let q be any point on the unit circle in the complex plane. Def ne $U_K : \mathcal{H}(K) \longrightarrow \mathcal{H}(K)$ by the formula

$$U_K|s\rangle = (-1)^{i(s)+n_-} q^{j(s)+n_+ - 2n_-} |s\rangle.$$

Then U_K def nes a unitary transformation of the Hilbert space and we have that

$$J_K = Trace(U_K).$$

The Hadamard Test applied to this unitary transformation gives a quantum algorithm for the Jones polynomial. This is not the most eff cient quantum algorithm for the Jones polynomial. Unitary braid group represention can do better [1, 20, 21]. But this algorithm has the conceptual advantage of being directly related to Khovanov homology. In particular, let $C^{i,j}$ be the subspace of $\mathcal{H}(K)$ with basis the set of enhanced states $|s\rangle$ with $i(s) = i$ and $j(s) = j$. Then $\mathcal{H}(K)$ is the direct sum of these subspaces and we see that $\mathcal{H}(K)$ is identical with the Khovanov complex for K with coeff cients in the complex numbers. Furthermore, letting $\partial : \mathcal{H}(K) \longrightarrow \mathcal{H}(K)$ be the boundary mapping that we have def ned for the Khovanov complex, we have

$$\partial \circ U_K + U_K \circ \partial = 0.$$

Thus U_K induces a mapping on the Khovanov homology of K. As a linear space, the Khovanov homology of K,

$$\mathcal{H}omology(\mathcal{H}(K)) = Kernel(\partial)/Image(\partial)$$

is also a Hilbert space on which U_K acts and for which the trace yields the Jones polynomial.

If we are given more information about the Khovanov homology as a space, for example if we are given a basis for $H^{i-n_-,j-n_++2n_-}(K)$ for each i and j, then we can extend U to act on $H^{i-n_-,j-n_++2n_-}(K)$ as an eigenspace with eigenvalue $t^i q^j$ where q and t are chosen unit complex numbers. Then we have an extended U'_K with

$$U'_K|\alpha\rangle = t^i q^j |\alpha\rangle$$

for each $\alpha \in H^{i-n_-,j-n_++2n_-}(K)$. With this extension we have that the trace of U'_K recovers a specialization of the Poincaré polynomial (Section 4) for the Khovanov homology.

$$Trace(U'_K) = \Sigma_{i,j} t^i q^j dim(H^{i-n_-,j-n_++2n_-}(K)) = P_K(t,q).$$

Thus, in principle, we formulate a quantum algorithm for specializations of the Poincarè polynomial for Khovanov homolgy.

Placing Khovanov homology in an appropriate quantum mechanical, quantum information theoretic, or quantum f eld theory context is a fundamental question that has been considered by a number of people, including Sergei Gukov [10, 11] and Edward Witten [52–54]. The constructions discussed here are elementary in nature but we would like to know how they interface with other points of view. In particular, if one thinks of the states in the state expansion of the bracket polynomial as analogs of the states of a physical system such as the Potts model in statistical mechanics, then the loop conf guration of a given state corresponds to a decomposition of the underlying graph of the statistical mechanics model into regions of constant spin (where spin designates the local variable in the model). Working with a boundary operator, as we did with the Khovanov chain complex, means taking into account adjacency relations among these types of physical states.

9. Discussion

The subject of Khovanov homology is part of the larger subject of categorif cation in general and other link homologies in particular. The term *categorif cation* was coined by Crane and Frenkel in their paper [9] speculating on the possibility for invariants of four-manifolds via a categorical generalization of Hopf algebras where all structures are moved up one categorical level. Just such a shift is seen in the Khovanov homology where loops that were once scalars become modules and the original Jones polynomial is seen as a graded Euler characteristic of a homology theory. There is now a complex literature on categorif cations of quantum groups (aka Hopf algebras) and relationships of this new form of representation theory with the construction of link homology. For this we refer the reader to the following references [7, 8, 32–34, 43, 49–51]. It is possible that the vision of Crane and Frenkel for the construction of invariants of four dimensional manifolds will come true.

Other link homology theories are worth mentioning. In [35–37] Khovnaov and Rozansky construct a link homology theory for specializations of the Homf ypt polynomial. Their theory extends integrally to a Khovanov homology theory for virtual knots, but no calculations are known at this writing. Khovanov homology does extend integrally to virtual knot theory as shown by Manturov in [28]. The relationship of the Manturov construction to that of Khovanov and Rozansky is not known at this time. In [27] Dye, Kauffman and Maturov show how to modify mod-2 Khovanov homology to categorify the arrow polynomial for virtual knots. This leads to many new calculations and examples [29, 30]. In [38, 39] Ozsvath, Szabo, Sarkar and Thurston construct combinatorial link homology based on Floer homology that categorif es the Alexander polynomial. Their techniques are quite different from those explained here for Khovanov homology. The combinatorial def nition should be compared with that of Khovanov homology, but it has a f avor that is different, probably due to the fact that it categorif es a determinant that calculates the Alexander polynomial. This Knot Floer Homology theory is very powerful and can detect the three-dimensional genus of a knot (the least genus of an orientable spanning surface for the knot in three dimensional space). Caprau in [5] has a useful version of the tangle cobordism approach to Khovanov homology and Clark, Morisson and Walker [6] have an oriented tangle cobordism theory that is used to sort out the functoriality of Khovanov homology for knot cobordisms. There is another signif cant variant of Khovanov homology termed *odd Khovanov homology* [40]. Attempts to f nd other global interpretations of Khovanov homology have led to very signif cant lines of research [7, 8, 43], and attempts to f nd general constructions for link homology corresponding to the quantum link invariants coming from quantum groups have led to research such as that of Webster [50, 51] where we now have theories for such constructions that use the categorif cations of quantum groups for classical Lie algebras..

There have been three applications of Khovanov homology that are particularly worth mentioning. One, we have discussed in Section 6, is Rasmussen's use of Khovanov homology [44] to determine the slice genus of torus knots without using gauge theory. Another is the proof by Kronheimer and Mrowka [24] that Khovanov homology detects the unknot. The work of Kronheimer and Mrowka interrelates Khovanov homology with their theory of knot instanton homology and allows them to apply their gauge theoretic results to obtain this striking result. A proof that Khovanov homolgy detects the unknot by purely combinatorial topological means is unknown at this writing. By the same token, it is still unknown whether the Jones polynomial detects classical knots. Finally, we mention the work of Shumakovitch [42] where, by calculating Khovanov homology, he shows many examples of knots that are topologically slice but are not slice in the differentiable category. Here Khovanov homology circumvents a previous use of gauge theory.

References

1. D. Aharonov, V. Jones, Z. Landau, A polynomial quantum algorithm for approximating the Jones polynomial, quant-ph/0511096.
2. D. Bar-Natan, On Khovanov's categorif cation of the Jones polynomial, Algebra Geometry Topology Vol 2. (2002) 337-370, math QA/0201043.
3. D. Bar-Natan (2005), Khovanov's homology for tangles and cobordisms, *Geometry and Topology*, **9-33**, pp. 1465-1499. `arXiv:mat.GT/0410495`
4. D. Bar-Natan and S. Morrison, The Karoubi envelope and Lee's degeneration of Khovanov homology. Algebr. Geom. Topol. 6 (2006), 14591469. math.GT/0606542.
5. C. Caprau, The universal sl(2) cohomology via webs and foams, *Topology and its Applications*, Volume 156 (2009), 1684-1702. math.GT. arXiv:0802.2848
6. D. Clark, S. Morrison, K. Walker, Fixing the functoriality of Khovanov homology. *Geom. Topol.* 13 (2009), no. 3, 14991582.
7. S. Cautis, J. Kamnitzer, Knot homology via derived categories of coherent sheaves. I. The sl(2)-case. *Duke Math. J.* 142 (2008), no. 3, 511588.
8. S. Cautis, J. Kamnitzer, Knot homology via derived categories of coherent sheaves. II. slm case. *Invent. Math.* 174 (2008), no. 1, 165232.
9. L. Crane, I. Frenkel, Four-dimensional topological quantum f eld theory, Hopf categories, and the canonical bases. *Topology and physics. J. Math. Phys.* 35 (1994), no. 10, 51365154.
10. S. Gukov, Gauge theory and knot homologies. Fortschr. Phys. 55 (2007), no. 5-7, 473490.

11. S. Gukov, Surface operators and knot homologies, ArXiv:0706.2369.
12. V.F.R. Jones, A polynomial invariant for links via von Neumann algebras, *Bull. Amer. Math. Soc.*, **129** (1985), 103–112.
13. V.F.R. Jones. Hecke algebra representations of braid groups and link polynomials, *Ann. of Math.* **126**, (1987), pp. 335-338.
14. V.F.R. Jones, On knot invariants related to some statistical mechanics models, *Pacifc J. Math.*, **137**, no. 2 (1989), pp. 311-334.
15. L.H. Kauffman, State models and the Jones polynomial, *Topology*, **26** (1987), 395–407.
bibitemKaAM L. H. Kauffman, New invariants in knot theory, Amer. Math. Monthly (1987).
16. L.H. Kauffman, Statistical mechanics and the Jones polynomial, *AMS Contemp. Math. Series*, **78** (1989), 263–297.
17. L.H. Kauffman, Knots and Physics, World Scientifc Publishers (1991), Second Edition (1993), Third Edition (2002).
18. L H. Kauffman, Topological quantum information, Khovanov homology and the Jones polynomial. Topology of algebraic varieties and singularities, 245264, *Contemp. Math.*, 538, Amer. Math. Soc., Providence, RI, 2011 arXiv:1001.0354
19. L. H. Kauffman. A quantum model for the Jones polynomial, Khovanov Homology and Generalized Simplicial Homology, *AMS Contemp Math*, **536**, Cross Disciplinary Advances in Quantum Computing, ed. by Mahdavi et al., pp. 75 - 94.
20. L.H. Kauffman, Quantum computing and the Jones polynomial, in *Quantum Computation and Information*, S. Lomonaco, Jr. (ed.), AMS CONM/305, 2002, pp. 101–137. math.QA/0105255
21. L. H. Kauffman and S. Lomonaco Jr., A Three-stranded quantum algorithm for the Jones polynonmial, in *Quantum Information and Quantum Computation V*, Proceedings of Spie, April 2007, edited by E.J. Donkor, A.R. Pirich and H.E. Brandt, pp. 65730T1-17, Intl Soc. Opt. Eng.
22. M. Khovanov, A categorifcation of the Jones polynomial, Duke J. Math. 101(200), no. 3, 359-426, mathQA/9908171.
23. J. Kock, "Frobenius Algebras and 2D Topological Quantum Field Theories", London Mathematical Society, Student Texts No. 59 (2004).
24. P. B. Kronheimer, T. S. Mrowka, Khovanov homology is an unknot-detector. math.GT.arXiv:1005.4346.
25. E. S. Lee, An endomorphism of the Khovanov invariant, *Adv. Math.* 197 (2), 554-586 (2005). math.GT/0210213.
26. S. J. Lomonaco Jr. and L. H. Kauffman, Quantum Knots and Mosaics, *Journal of Quantum Information Processing*, Vol. 7, Nos. 2-3, (2008), pp. 85 - 115. arxiv.org/abs/0805.0339
27. H.A. Dye, L.H. Kauffman, V.O. Manturov, On two categorifcations of the arrow polynomial for virtual knots, arXiv:0906.3408.
28. V.O. Manturov, Khovanov homology for virtual links with arbitrary coeffcients, math.GT/0601152. (Russian) Izv. Ross. Akad. Nauk Ser. Mat. 71 (2007), no. 5, 111–148. J. Knot Theory Ramifcations 16 (2007), no. 3, 345377.
29. A. Kaestner Ph.D. Thesis, University of Illinois at Chicago (2011).
30. A. Kasetner and L. H. Kauffman, On Categorifcations of the Arrow Polynomial (to appear).

31. M. Khovanov,(1997), A categorif cation of the Jones polynomial, *Duke Math. J*,**101** (3), pp.359-426.
32. M. Khovanov, A. Lauda, A categorif cation of quantum sl(n). *Quantum Topol.* 1 (2010), no. 1, 192.
33. M. Khovanov, A. Lauda, A diagrammatic approach to categorif cation of quantum groups II.*Trans. Amer. Math. Soc.* 363 (2011), no. 5, 26852700.
34. M. Khovanov, Categorif cations from planar diagrammatics. *Jpn. J. Math.* 5 (2010), no. 2, 153181.
35. M. Khovanov, L. Rozansky, Virtual crossings, convolutions and a categorif cation of the SO(2N) Kauffman polynomial. *J. Gkova Geom. Topol. GGT* 1 (2007), 116214.
36. M. Khovanov, L. Rozansky, Matrix factorizations and link homology. *Fund. Math.* 199 (2008), no. 1, 191.
37. M. Khovanov, L. Rozansky, Matrix factorizations and link homology. II. *Geom. Topol.* 12 (2008), no. 3, 13871425.
38. C. Manolescu, P. Ozsvth, S. Sarkar, A combinatorial description of knot Floer homology. *Ann. of Math.* (2) 169 (2009), no. 2, 633660.
39. C. Manolescu, P. Ozsvth, Z. Szab, D. Thurston, On combinatorial link Floer homology. *Geom. Topol.* 11 (2007), 23392412.
40. P. Ozsvth, J. Rasmussen, and Z. Szab, Odd Khovanov homology. Preprint (2007), http://arxiv.org/abs/0710.4300.
41. J. H. Przytycki, When the theories meet: Khovanov homology as Hochschild homology of links, `arXiv:math.GT/0509334`
42. A. N. Shumakovitch, Rasmussen invariant, slice-Bennequin inequality, and sliceness of knots, *J. Knot Theory Ramif cations* 16 (2007), no. 10, 14031412.
43. P. Seidel, I. A. Smith, A link invariant from the symplectic geometry of nilpotent slices. *Duke Math. J.* 134 (2006), no. 3, 453514.
44. J. Rasmussen, Khovanov homology and the slice genus, math.GT/0402131.
45. J-L. Loday, Cyclic Homology, Grund. Math. Wissen. Band 301, Springer-Verlag, Berlin, 1992 (second edition, 1998).
46. J. P. May, Simplicial Objects in Algebraic Topology, University of Chicago Press (1967).
47. M. A. Nielsen and I. L. Chuang, Quantum Computation and Quantum Information, Cambridge University Press (2000).
48. O. Viro (2004), Khovanov homology, its def nitions and ramif cations, *Fund. Math.*, **184** (2004), pp. 317-342.
49. B. Webster, Khovanov-Rozansky homology via a canopolis formalism. *Algebr. Geom. Topol.* 7 (2007), 673699.
50. B. Webster, Knot invariants and higher representation theory I: diagrammatic and geometric categorif cation of tensor products. math.GT.arXiv:1001.2020.
51. B. Webster, Knot invariants and higher representation theory II: the categorif cation of quantum knot invariants. math.GT. arXiv:1005.4559
52. E. Witten, Quantum Field Theory and the Jones Polynomial, *Comm. in Math. Phys.*, **121**, (1989), 351-399.
53. E. Witten, Knot Invariants from Four-Dimensional Gauge Theory. Davide Gaiotto, Edward Witten. physics.hep-th. arXiv:1106.4789
54. E. Witten, Fivebranes and Knots. Edward Witten. physics.hep-th. arXiv:1101.3216

BRAID EQUIVALENCES AND THE L–MOVES

Sofia Lambropoulou

ABSTRACT. In this survey paper we present the L–moves between braids and how they can adapt and serve for establishing and proving braid equivalence theorems for various diagrammatic settings, such as for classical knots, for knots in knot complements, in c.c.o. 3–manifolds and in handlebodies, as well as for virtual knots, for flat virtuals, for welded knots and for singular knots. The L–moves are local and they provide a uniform ground for formulating and proving braid equivalence theorems for any diagrammatic setting where the notion of braid and diagrammatic isotopy is defined, the statements being first geometric and then algebraic.

INTRODUCTION

The central problem in classical knot theory is the complete classification of knots and links up to ambient isotopy in three-space. Reidemeister [Rd] translated isotopy in space into a diagrammatic equivalence relation. Further, the braid group provides a fundamental algebraic structure associated with knots. The Alexander theorem tells us that every knot or link can be isotoped to braid form. The capstone of this relationship is the Markov theorem, giving necessary and sufficient conditions for two braids to close to isotopic links. Since the discovery of the Jones polynomial [Jo] –the first invariant for knots and links that was constructed using braids– the Markov theorem received renewed attention.

2000 *Mathematics Subject Classification.* 57M25, 57M27, 57N10.

Key words and phrases. Braiding, Markov theorem, L–moves, conjugation, commuting, stabilization, links in knot complements, links in 3–manifolds, mixed links, mixed braids, braid equivalence, twisted loop conjugation, twisted band move, links in handlebodies, virtual braids, welded braids, singular braids.

The author was partially supported by the Mathematisches Forschungsinstitut Oberwolfach.

Among other proofs of the Markov theorem, in [LR1] we introduced a new type of braid move, the L–move, and we proved an *one-move* Markov theorem. The L–move is a very simple uniform geometric move that can be applied anywhere in a braid to produce a braid with the same closure. More precisely, an *L–move* consists in cutting a strand of the braid and taking the top end to the bottom of the braid and the bottom end to the top of the braid, both entirely over or both entirely under the braid, creating a new pair of corresponding braid strands. By small braid isotopies, an L–move is equivalent to adding an in-box crossing in the braid, aquiring thus an algebraic expression. See Fig. 5 for illustrations. Consequently, in [LR1, LR2, HL] the L–move approach was used for proving braid equivalence theorems (analogues of the Markov theorem) for classical knots and links in knot complements, in closed, connected, oriented (c.c.o.) 3–manifolds and in handlebodies. Further, using the L–move methods, we proved in [KL2] braid equivalence theorems for the virtual braid group, for the welded braid group and for some analogues of these structures. Finally, in [La3] an L–move singular braid equivalence is given, using the algebraic singular braid equivalence of [Ge].

Given a diagrammatic knot theory there are deep interrelations between the diagrammatic knot isotopy in this theory, the braid structures and the corresponding braid equivalence. On the other hand, braid equivalence theorems are important for understanding the structure and classification of knots and links in various settings. Also, for constructing invariants of knots and links using algebraic means.

In this survey paper we present the L–move methods of the author and collaborators and how they can serve for establishing and proving braid equivalence theorems for the diagrammatic settings mentioned above. We have omitted all technicalities and most proofs. For details, the interested reader is referred to the papers [LR1], [LR2], [La2], [HL], [KL2] and [La3]. In Section 1 we discuss braiding and braid equivalence for knots in S^3. We describe a generic braiding method that can adapt to any diagrammatic setting, we define the L–moves, and we give the one-move Markov theorem (Theorem 3). In Section 2 we present the L–move methods for deriving braid equivalences, geometric and algebraic, for knots in knot complements, in c.c.o. 3–manifolds and in handlebodies. Finally, in Section 3 we describe how the L–moves adapt for deriving braid equivalences for virtual knots, welded knots, flat knots and for singular knots.

More precisely: the diagrammatic equivalence in a specific topological setting, but also moves that are not allowed in the setting, define the corresponding braid isotopy, the way the closure of a braid is realized, and also the types of L–moves needed for the corresponding braid equivalence.

In the case of knots and links in knot complements, in c.c.o. 3–manifolds and in handlebodies, the basic idea is to represent the manifold by a braid in S^3 and then links and braids in the manifold by mixed links and mixed braids in S^3. Then to translate isotopy in these spaces in terms of mixed link isotopy in S^3, according to the hosting manifold each time. The main difference among the knot theories in these spaces is that in a handlebody a knot may not pass beyond the boundary of the handlebody from either end, and this is reflected both in the definition of the closure of a braid (see Figs. 27, 28) as well as in the corresponding braid equivalence. Further, in the case of c.c.o. 3–manifolds we have an extra isotopy move (from the case of knot complements) coming from the handle sliding moves related to the surgery description of the manifold. All these are explained in Section 2.

Regarding the L–move braid equivalence corresponding to the knot isotopy in a given setting, there are two conceptual steps: first create the L–move braid equivalence on a geometric level. Then take into account the algebraic structures of the braids and the algebraic expressions of the L–moves (using the interpretation of an L–move as introducing an in-box crossing, see Fig. 6 for braids in S^3) and turn the geometric braid equivalence into an algebraic braid equivalence. The interpretation of an L–move as introducing an in-box crossing is also very crucial in the search of the types of L–moves needed in a specific diagrammatic setting, as they are related to the types of kinks allowed in the given diagrammatic isotopy. For example, in the virtual domain we have L–moves introducing a real or a virtual crossing facing to the right or to the left of the braid. Moreover, the presence of the two forbidden moves in the theory leads to the fact that the strands of an L–move cross the other strands of the braid only virtually, and also to a type of virtual L–move coming from a 'trapped' virtual kink (see Definitions 10, 11 and 12). As another example, in the singular domain we do not have L–moves introducing a singular crossing, as the closure of such a move would contract to a kink with a singular crossing, and this is not an isotopy move in the theory. All these are discussed in Section 3.

Regarding the algebraic braid equivalences, surprisingly conjugation by a braid generator may sometimes be achieved simply by L–moves within the braid. Then, such conjugations will not appear in the geometric statement of the braid equivalence. But there are also situations where a braid generator, even if it has an inverse, cannot be conjugated. For example, the loop generators a_i in a handlebody of genus $m \geq 2$ (see Subsection 2.4), which can be conjugated in the case of knot complements. Then, such conjugations will not appear even in the algebraic braid equivalence of the theory. Finally, there are braid structures where some generators do not have inverses (e.g. the singular crossings τ_i in singular knot theory), so conjugation by these generators may not appear in the braid statement. Yet, a cutting line is permitted to cut a closed braid before or after a braid generator x_i. For this reason we will adopt the following more general move, which is equivalent to conjugation if x_i has an inverse.

- *Commuting in* $\mathcal{B}_n$: $\alpha x_i \sim x_i \alpha$,

where $\mathcal{B}_n$ denotes the set of braids with n strands in the given diagrammatic knot theory. There may be also situations where the cutting line of the closure cannot cut anywhere (for example, singular knot theory in a handlebody). For this reason we will adopt the more general stabilization move:

- *Stabilization in* $\cup_n \mathcal{B}_n$: $\alpha_1 \alpha_2 \sim \alpha_1 x_n^{\pm 1} \alpha_2$,

where $\alpha_1, \alpha_2 \in \mathcal{B}_n$ and $x_n \in \mathcal{B}_{n+1}$. Stabilization by a real crossing may occur in all the above-mentioned knot theories. This is because the isotopy move RI is permitted in all these theories. Yet, stabilization by a singular crossing is not permitted in singular braids.

The L–moves between braids are very fundamental and they provide a flexible conceptual center from which to deduce many results. They may be adapted to any diagrammatic setting where the notion of braid and diagrammatic isotopy is defined, and they provide a uniform ground for formulating and proving braid equivalence theorems, first geometric and then algebraic. Moreover, the local algebraic versions of the braid equivalences in the various settings, as they arise using the L–moves, promise to be useful for constructing invariants of knots and links in these settings.

As we said earlier, the differences in the various settings lie in the knot isotopies allowed or not allowed in the theory, and these will be reflected in the way the closure of a braid is defined and in the corresponding braid equivalences. For instance, one could work out braid equivalences for the virtual or the welded or the singular knot theory in knot complements, in c.c.o. 3–manifolds or in handlebodies. In future work we intend to apply our methods to transverse knots and to knots in thickened surfaces.

All the issues treated in this paper are expanded, discussed thoroughly and proved in detail in the author's monograph [La4].

1. Braid equivalence in S^3

A *link* of k components is the homeomorphic image of k copies of the circle into three-space and a knot is a link of one component. The central problem in classical Knot Theory is the complete classification of knots and links in S^3 up to ambient isotopy. Reidemeister [Rd] translated isotopy in space into a discrete diagrammatic equivalence relation on knot diagrams in the plane, generated by the three well–known Reidemeister moves RI, RII, RIII together with planar isotopy, see Fig. 1. Further, isotopy of oriented

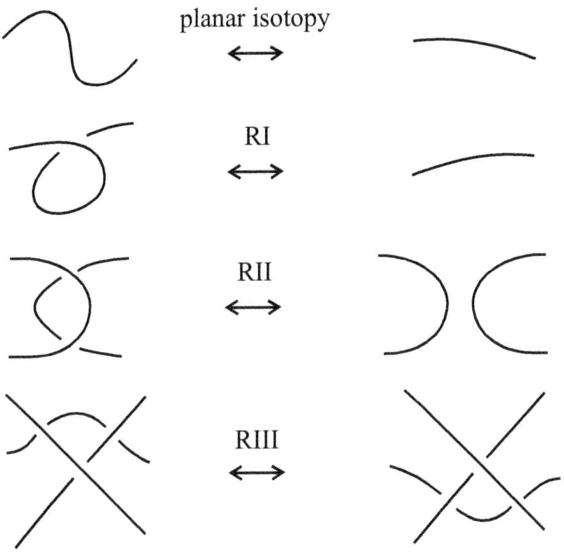

FIGURE 1. The diagrammatic isotopy moves

links uses all variations of the Reidemeister moves resulting in from all different orientations of the arcs involved. In the sequel, whenever we say 'knots' we shall mean both knots and links.

Braids, as topological objects, are similar to links, but they have an algebraic structure. A *braid* on n strands consists in n arcs embedded in a thickened square $[0,1] \times [0,\epsilon] \times [0,1]$, such that the ends are arranged into n collinear top endpoints in $[0,1] \times [0,\epsilon] \times \{0\}$ and n collinear bottom endpoints in $[0,1] \times [0,\epsilon] \times \{1\}$ and such that the embedding is monotonous with respect to the height function, that is, there are no local maxima or minima. Two braids are isotopic if one can be deformed to the other via an isotopy that preserves the braid structure. Braids shall be identified with braid diagrams (that is, regular projections on $[0,1] \times \{\epsilon\} \times [0,1]$) with no risk of confusion. Algebraically, the set of braids on n strands forms the classical braid group B_n [Ar1, Ar2], [Ch], whose generators $\sigma_1, \ldots, \sigma_{n-1}$ are the elementary positive crossings between consecutive strands and they satisfy the defining *braid relations*:

$$(1) \quad \sigma_i \sigma_{i+1} \sigma_i = \sigma_{i+1} \sigma_i \sigma_{i+1} \qquad \text{and} \qquad \sigma_i \sigma_j = \sigma_j \sigma_i \quad \text{for} \quad |i-j| > 1.$$

The first relation is the most characteristic, and it reflects the braid isotopy move RIII (all arrows down). Also, the fact that the generators σ_i are invertible is reflected into the braid isotopy move RII. An excellent reference on braid groups is [KT].

By definition a braid has a natural orientation (from top to bottom). Taking the closure of a braid β, that is, joining its corresponding endpoints with simple unlinked arcs to the side of the braid yields an oriented link, denoted $\widehat{\beta}$. Conversely we have:

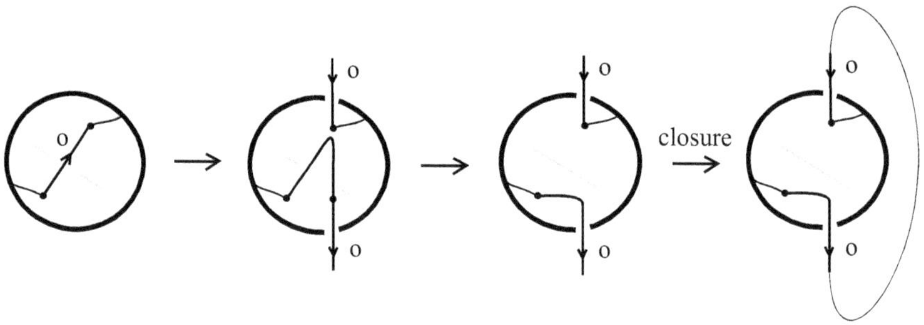

FIGURE 2. The basic braiding move

Theorem 1 (J. W. Alexander, 1923). *Any oriented link in S^3 is isotopic to the closure of a braid (not unique).*

Brunn [Br] proved that any link has a projection with a single multiple point; from which it follows immediately (by appropriate perturbations) that we can braid any link diagram. Apart from Alexander [Al], other proofs of Theorem 1 have been given by Morton [Mo], Yamada [Ya], Vogel [Vo], by the author with Rourke [La1, LR1], and by the author with Kauffman [KL1], where the emphasis is given on braiding virtual knots.

We shall present here two braiding algorithms, that of [La1, LR1] and that of [KL1], both conceptually very simple and both based on the same idea. The context of both braidings is that of [La1, LR1]: Any link diagram can be arranged to be in general position with respect to the standard height function on the plane. This means that it does not contain any horizontal arcs and it can be seen as a composition of horizontal stripes, each containing either a local minimum or a local maximum or a crossing. The idea of the braiding is on the one hand to keep the *down–arcs* of the diagram that are oriented downward and, on the other hand, to eliminate the *up–arcs* that go upward and produce instead braid strands. First consider up–arcs that contain no crossings. Call such an arc in the diagram a *free up–arc*.

Definition 1. A *braiding move* on a free up–arc is the following elimination operation of it: We cut the arc at any point. We then pull the two ends the upper upward and the lower downward, keeping them aligned, and so as to obtain a pair of corresponding braid strands, both running entirely *over* the previously constructed tangle or entirely *under* it. View Fig. 2 for an abstract illustration. In Fig. 2 the up–arc is cut at the topmost point and the two ends are both pulled over the rest of the diagram, which is represented by the bold circle. The closure of the resulting tangle is a link diagram, obviously isotopic to the original one. Indeed, from the up–arc we created a stretched loop, which by move RI is isotopic to the original free up–arc. Note that in Fig. 2 a faint arc is also illustrated inside the braid. It should be ignored in this definition; we shall refer to it later.

In [KL1], before performing the braiding moves we prepare the diagram by rotating all crossings that contain up-arcs, so that the arcs that pass through any crossing are directed downward. There are two types of rotation: If both arcs of the crossing go up, then we rotate the crossing by 180 degrees. If only one arc in the crossing goes up, then we rotate it by 90 degrees. See Figs. 3 and 4. These rotations may produce new free up–arcs. After adjusting all the crossings we then braid all the free up–arcs. The resulting tangle is the desired braid, the closure of which is isotopic to the original diagram.

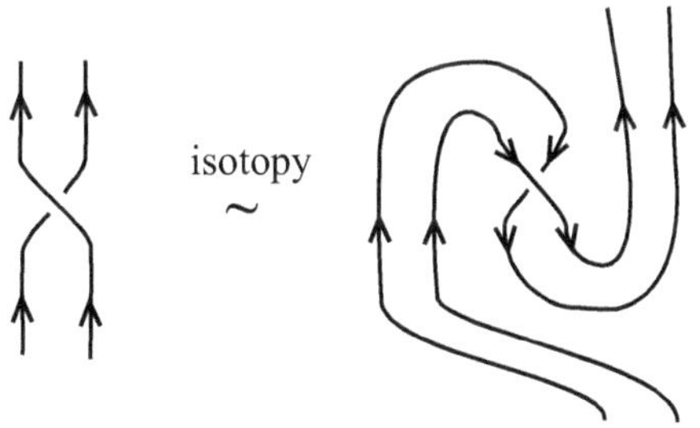

FIGURE 3. Full twist

In [La1, LR1] we do not rotate the crossings. Instead, we apply the braiding move also to the up–arcs going through crossings. The difference here is that we have to label each up–arc as *over* or *under* according to its position in the crossing, and then the resulting strands from the braiding move will be both over or under the rest of the diagram accordingly, see [LR1] for details. For example, in Fig. 2 consider the up–arc creating a

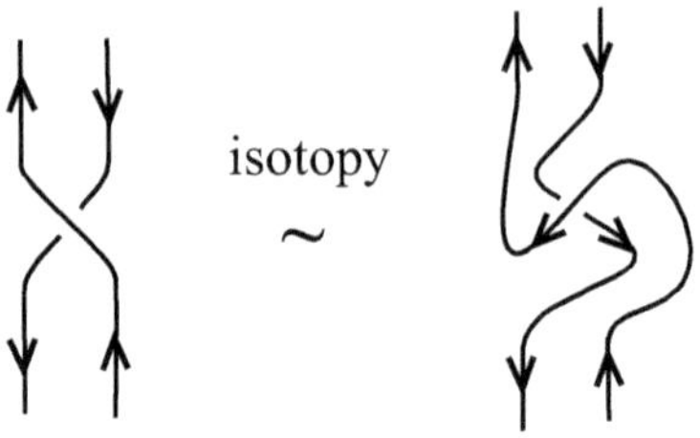

FIGURE 4. Half twist

crossing with the faint arc. Then the up–arc is the over arc and it is labelled "o", so the two new strands are stretched over the rest of the diagram.

Remark 1. The advantage of the braiding technique in [KL1] is that it applies equally well to the virtual category, to flat virtuals, to welded knots and to singular knots, see [KL1] for details. It applies, in fact, to all the categories in which braids are constructed. The braiding technique in [La1, LR1] is preferred in situations where we need to analyze how the isotopy moves on diagrams affect the final braids, see [LR1, HL, KL1] for details.

Trying now to use braids for studying knots we need to know when two braids correspond to isotopic knots. The Markov theorem gives an equivalence relation in the set of all braids $B_\infty := \cup_{n=1}^{\infty} B_n$ reflecting isotopy of knots:

Theorem 2 (A. A. Markov, 1936). *Two oriented links in S^3 are isotopic if and only if any two corresponding braids in B_∞ differ by braid relations and the moves:*
(i) Conjugation in B_n: $\sigma_i^{-1}\alpha\sigma_i \sim \alpha$,
(ii) Bottom stabilization in B_∞: $\alpha \sim \alpha\sigma_n^{\pm 1}$, where $\alpha \in B_n$.

Conjugation can be generalized to the braid equivalence move *commuting* described in the Introduction. Indeed, conjugation is clearly related to isotopy move RII occurring in the back side of a closed braid $\widehat{\alpha}$. Opening $\widehat{\alpha}$ with a cutting line, this line will either place both crossings of the move at the same side of the braid (and then the initial RII move becomes the braid cancellation $\sigma_i^{-1}\sigma_i$) or it will separate the two crossings, one at the top and one at the bottom of the braid, yielding a conjugate of α. Since the σ_i's are invertible, commuting in this case is equivalent to conjugation. Regarding the bottom stabilization, we just note that the more general stabilization move described in the Introduction together with move (i) implies bottom stabilization. Stabilization is related to the isotopy move RI, since the closure of a braid with a stabilization move is a knot with a kink at that place.

The Markov theorem is not easy to prove. Its proof is very technical and it depends on the braiding algorithm used. Markov [Ma] used Alexander's braiding algorithm. The Markov theorem was originally stated by Markov [Ma] with *three* braid moves and then Weinberg [We] reduced them to the *two* braid moves of Theorem 2. Birman [Bi1] gave a more rigorous proof, filling in all details of Markov's original proof, using a more rigorous version of Alexander's braiding algorithm. Bennequin [Be] gave another proof using contact topology. After the discovery of the Jones polynomial [Jo] –the first knot invariant that was constructed using braids and the Markov theorem– other proofs of Theorem 2 have been given by: Morton [Mo] using his threading algorithm, by the author with Rourke [La1, LR1] using the braiding algorithm of [La1, LR1] and the new type of braid move, the L–move (see below), by Traczyk [Tr] using Vogel's braiding algorithm, and by Birman and Menasco [BM] using Bennequin's ideas.

In [LR1] using the L–moves we proved, in fact, the following sharper version of Theorem 2 (cf. [LR1], Thm. 2.3).

Theorem 3 (One-move Markov theorem). *Two oriented links in S^3 are isotopic if and only if any two corresponding braids differ by braid relations and the L-moves.*

Definition 2. An *L–move* consists in cutting a strand of the braid at a point, then pulling the two ends of the cut, the upper downward and the lower upward, both running entirely *over* the braid or entirely *under* it, and keeping them aligned (on the vertical line of the cutpoint) so as to obtain a new pair of corresponding braid strands. Thus there are two types of L–moves, an *under L–move*, or *L_u–move*, and an *over L–move*, or *L_o–move*. Fig. 5 illustrates an example of the two L–moves. Further, by a small braid isotopy that does not change the relative positions of endpoints, an L–move is equivalent to adding an in-box crossing (positive or negative) facing the right-hand side or the left-hand side of the braid box.

So, the L–move generalizes the stabilization move. View last instance of Fig. 5, where a positive in-box crossing is formed and two dots are placed to indicate the starting points of the L_o–move. See also Fig. 6 for an abstract illustration. Closing the new pair of strands created from an L–move results in a kink in the tangle diagram, so it corresponds to the isotopy move RI.

Remark 2. The version of an L–move introducing an in-box crossing gives rise to the following algebraic expression for an L_o–move and an L_u–move respectively:

$$\alpha = \alpha_1\alpha_2 \overset{L_o}{\sim} \sigma_i^{-1}\dots\sigma_n^{-1}\alpha_1'\sigma_{i-1}^{-1}\dots\sigma_{n-1}^{-1}\sigma_n^{\pm 1}\sigma_{n-1}\dots\sigma_i\alpha_2'\sigma_n\dots\sigma_i \tag{2}$$

$$\alpha = \alpha_1\alpha_2 \overset{L_u}{\sim} \sigma_i\dots\sigma_n\alpha_1'\sigma_{i-1}\dots\sigma_{n-1}\sigma_n^{\pm 1}\sigma_{n-1}^{-1}\dots\sigma_i^{-1}\alpha_2'\sigma_n^{-1}\dots\sigma_i^{-1} \tag{3}$$

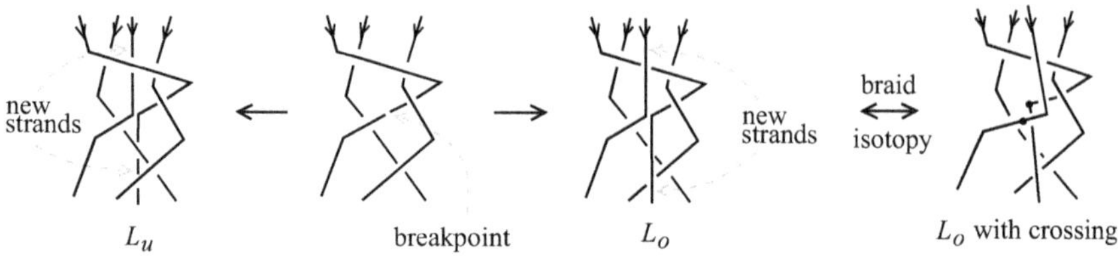

FIGURE 5. The L–moves

where α_1, α_2 are elements of B_n and α'_1, $\alpha'_2 \in B_{n+1}$ are obtained from α_1, α_2 by replacing each σ_j by σ_{j+1} for $j = i, \ldots, n-1$.

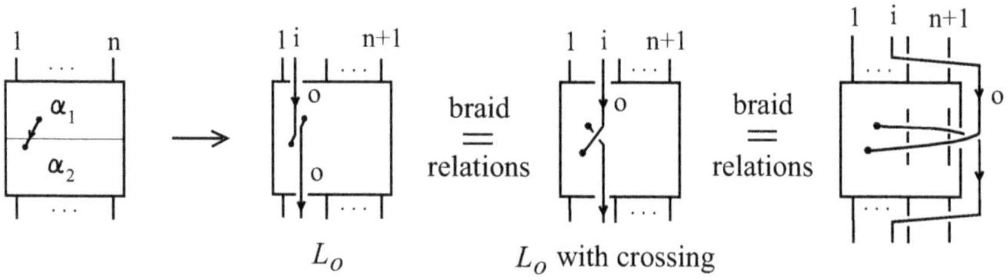

FIGURE 6. L_o–move, L_o–move with crossing and its algebraic expression

Theorems 2 and 3 imply that braid conjugation follows from the L–moves. This fact is also directly illustrated in Fig. 7.

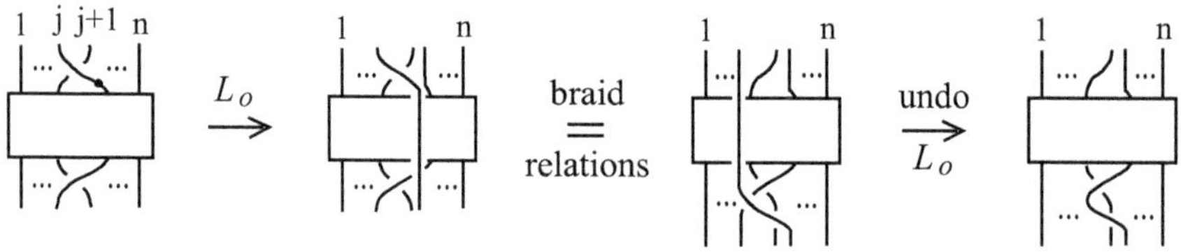

FIGURE 7. Braid conjugation follows from the L–moves

Remark 3. Theorem 3 has three main advantages: First, there is only one type of braid equivalence move. Second, the L–moves are very local and fundamental. Thus, adapted each time to a given diagrammatic setting (where diagrammatic isotopy is given and the notion of a braid can be defined), they may provide a uniform ground for formulating and proving geometric analogues of the Markov theorem. Third, due to the fact that the L–moves have algebraic expressions, these geometric statements can be turned into algebraic statements, according to the braid structures of the given situation (which, in turn, depend on the diagrammatic equivalence and the manifold where the diagrams live).

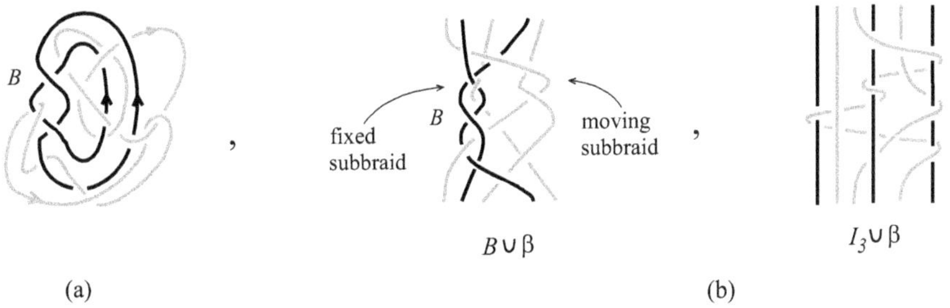

FIGURE 8. A mixed link and two mixed braids related to knot complements and c.c.o. 3-manifolds

2. Braid equivalence in knot complements, c.c.o. 3-manifolds and handlebodies

It is a well–known result in Topology that any c.c.o. 3-manifold can be obtained via surgery on an unoriented framed link in S^3 with integral framings. The intermediate stage of surgery is the construction of the complement of the surgery link. By the term 'knot complement' we shall be referring throughout to both knot and link complements. Another category of bounded 3-dimensional manifolds that give rise to c.c.o. 3-manifolds are the handlebodies, via the Heegaard decomposition. The special case of the solid torus is the only manifold common in both categories, and its knot theory has been studied quite extensively from various viewpoints.

2.1. **The common diagrammatic setting.** As we shall establish in this section, knots and braids in all three categories of 3-dimensional manifolds may be studied via very similar diagrammatic representations in S^3 and very similar techniques. The differences lie in the knot isotopies allowed or not allowed in each manifold, and these will be reflected in the way we define the closure of a braid and in the corresponding braid equivalences. The underlying braid moves that lead to algebraic braid equivalences in all these manifolds are the L–moves.

Let $S^3 \backslash K$ be the complement of the oriented link K in S^3. By the Alexander theorem and the definition of ambient isotopy, $S^3 \backslash K$ is homeomorphic to $S^3 \backslash \widehat{B}$, where $\widehat{B}$ is the closure of some braid B and it is isotopic to K. Throughout this section the braid B will remain fixed. Let, further, L be an oriented link in $S^3 \backslash \widehat{B}$. Fixing $\widehat{B}$ pointwise on its projection plane we may represent L unambiguously by the *mixed link* $\widehat{B} \bigcup L$ in S^3, that is, a link in S^3 consisting of the *fixed part* $\widehat{B}$ and the *moving part* L that links with $\widehat{B}$, view Fig. 8a for an example. A *mixed link diagram* is a diagram of $\widehat{B} \bigcup L$ projected on the plane of $\widehat{B}$ which is equipped with the top-to-bottom direction of B.

Let now M be a c.c.o. 3-manifold obtained by surgery on an unoriented framed link in S^3. Without loss of generality this link is the closure $\widehat{B}$ of some braid B. We shall refer to $\widehat{B}$ as the *surgery link* and we shall be writing $M = \chi(S^3, \widehat{B})$. So M may be represented in S^3 by the framed braid $\widehat{B}$. Further, fixing $\widehat{B}$ pointwise, we have that links in M can be unambiguously represented by mixed links in S^3 –exactly as in the case of knot complements. So, we follow the same diagrammatic setting.

Trying to find an analogue of the Markov theorem for links in $S^3 \backslash \widehat{B}$ or $\chi(S^3, \widehat{B})$ we notice first that $\widehat{B}$ must remain fixed. So, we need a braiding process for mixed link diagrams that maps $\widehat{B}$ to B and not to an L–equivalent braid or even to a conjugate of B. Such braids are called geometric mixed braids. A *geometric mixed braid* is a braid $B \bigcup \beta$ on $m + n$ strands that contains the subbraid B as a fixed subbraid. View Fig. 8b for two examples. The subbraid B represents the manifold where the knot theory lives in and it shall be called *the fixed subbraid* in the case of knot complements and *the surgery braid* in the case of c.c.o. 3-manifolds, while the subbraid β that complements B shall be called *the moving subbraid* and it represents the link in the manifold. These braids were introduced in [La1, LR1]. The closure of a geometric mixed braid is defined to be the standard closure of the mixed braid seen as a classical braid on $m + n$ strands. As we shall see below, every mixed link may be braided to a geometric mixed braid.

Let H_m denote the handlebody of genus m. The *handlebody* of genus m is usually defined as (*a closed disc* $\setminus$ *m open discs*) $\times I$, where I is the unit interval. Equivalently, H_m can be defined as ($S^3 \setminus$ *an open tubular neighbourhood of* I_m), where I_m denotes here the identity braid on m infinitely extended strands, all meeting at the point at infinity, see Fig. 9a. Thus H_m may be represented in S^3 by the usual identity braid I_m. Let now L be an oriented link in H_m. Then, fixing I_m pointwise on its projection plane, L may be represented unambiguously by the mixed (m, m)-tangle $I_m \bigcup L$ in S^3 which, by abuse of language, we shall call *mixed link* (view

Fig. 9b). The subbraid I_m shall be called *the fixed part* and L *the moving part* of the mixed link. A *mixed link diagram* is then a diagram of $I_m \bigcup L$ projected on the plane of I_m, which is equipped with the top-to-bottom direction.

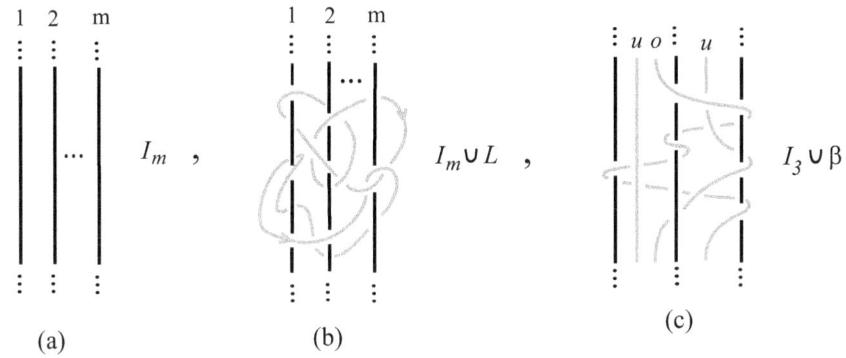

FIGURE 9. Representing H_m - a mixed link and a mixed braid related to a handlebody

Now, applying the braiding algorithm of [La1, LR1], by which all down–arcs remain fixed, we can easily braid an oriented mixed link $I_m \bigcup L$ in S^3 to a geometric mixed braid $I_m \bigcup \beta$, as defined above. In the case of a handlebody, though, the strands of the fixed subbraid I_m are thought infinitely extended. This forces the strands of the moving subbraid to be labelled with labels 'u' and 'o' for 'under' and 'over' (and this is related to the definition of the closure, see details in the subsection on handlebodies). These braids were introduced in [HL]. View Fig. 9c for an example.

We note that, in all settings, if we remove the fixed braid from a mixed link or a geometric mixed braid we are left with an oriented link resp. a braid in S^3.

2.2. **The common algebraic setting.** An *algebraic mixed braid* on n moving strands is a braid on $m+n$ strands such that the first m strands form the identity braid I_m. View Fig. 10 for an example. These braids were introduced in [La2]. We denote the set of all such braids $B_{m,n}$. Clearly, if in the definition of a geometric mixed braid B stands for the identity braid I_m, then the set of algebraic mixed braids $B_{m,n}$ is a subset of the corresponding set of geometric mixed braids.

The set $B_{m,n}$ is a subset of B_{m+n} closed under concatenation and under inverses, so it has a group structure, and shall be called *the mixed braid*

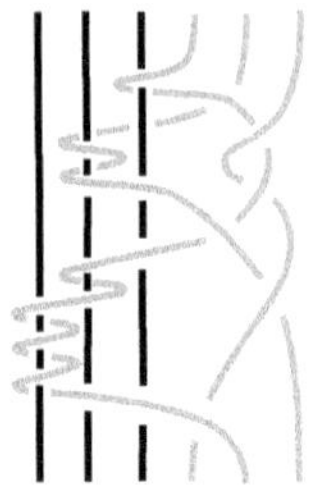

$\in B_{3,3}$

FIGURE 10. An algebraic mixed braid

group. The mixed braid group is generated by the classical crossings σ_j and by the 'loops' a_i as illustrated in Fig. 11. Moreover, $B_{m,n}$ has the following presentation (cf. [La2]):

(4)

$$B_{m,n} = \left\langle \begin{array}{c} a_1, \ldots, a_m, \\ \sigma_1, \ldots, \sigma_{n-1} \end{array} \,\middle|\, \begin{array}{l} \sigma_k \sigma_j = \sigma_j \sigma_k, \quad |k-j| > 1 \\ \sigma_k \sigma_{k+1} \sigma_k = \sigma_{k+1} \sigma_k \sigma_{k+1}, \quad 1 \leq k \leq n-1 \\ a_i \sigma_k = \sigma_k a_i, \quad k \geq 2, \ 1 \leq i \leq m \\ a_i \sigma_1 a_i \sigma_1 = \sigma_1 a_i \sigma_1 a_i, \quad 1 \leq i \leq m \\ a_i (\sigma_1 a_r \sigma_1^{-1}) = (\sigma_1 a_r \sigma_1^{-1}) a_i, \quad r < i. \end{array} \right\rangle$$

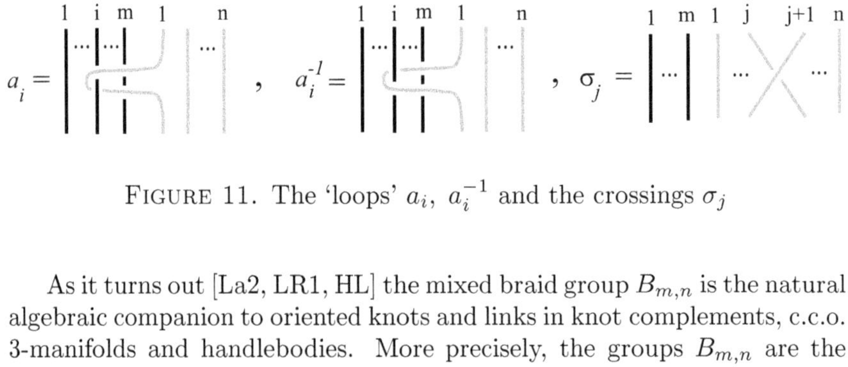

FIGURE 11. The 'loops' a_i, a_i^{-1} and the crossings σ_j

As it turns out [La2, LR1, HL] the mixed braid group $B_{m,n}$ is the natural algebraic companion to oriented knots and links in knot complements, c.c.o. 3-manifolds and handlebodies. More precisely, the groups $B_{m,n}$ are the appropriate braid structures for studying knots and links in the complement of the m–unlink or a connected sum of m lens spaces or in a handlebody of genus m. For arbitrary knot complements and c.c.o. 3-manifolds we have cosets $C_{m,n}$ of these groups in the groups B_{m+n}, which are determined by the specific hosting manifold, represented in S^3 by the braid B [La2]. Then an element of the related coset $C_{m,n}$ consists in an elements of the group $B_{m,n}$ followed by the natural embedding of B in $B_{m,n}$. Yet, as we shall explain below, by parting and combing mixed braids, the braid equivalences in all these spaces can be formulated only in terms of the mixed braid groups $B_{m,n}$. The capstone for studying braid equivalence in all above-mentioned spaces is the following result ([LR1], Thm. 4.7).

Theorem 4 (Relative version of the L–move braid equivalence). *Let L_1, L_2 be oriented link diagrams which both contain a common braided portion B. Suppose that there is an isotopy of L_1 to L_2 which finishes with a homeomorphism fixed on B. Suppose further that B_1 and B_2 are braids obtained from our braiding process applied to L_1 and L_2 respectively. Then B_1 and B_2 are equivalent by braid isotopy and by L–moves that do not affect the common braided portion B.*

Using Theorem 4, geometric versions of the Markov theorem were proved for links in knot complements and for links in arbitrary c.c.o. 3-manifolds by means of L–move analogues (cf. [LR1], Thms. 5.5 and 5.10). These results were consequently turned into algebraic statements in [LR2] (Thms. 4 and 5), via the braid groups $B_{m,n}$. Theorem 4 led also to an L–move analogue of the Markov theorem for knots in a handlebody as well as to algebraic analogues of it using the braid groups $B_{m,n}$ (cf. [HL], Thms. 3, 4 and 5). We shall now explain all this in more detail.

2.3. **Braid equivalence in knot complements.** Two oriented links L_1, L_2 are isotopic in $S^3\backslash\widehat{B}$ if and only if the mixed links $\widehat{B}\bigcup L_1$ and $\widehat{B}\bigcup L_2$ are isotopic in S^3 by an ambient isotopy which keeps $\widehat{B}$ pointwise fixed. In terms of diagrams, mixed link isotopy is generated by the ordinary Reidemeister moves for the moving part and by the mixed moves RII, RIII of Fig. 12 involving also the fixed part (see [LR1] for details).

FIGURE 12. The mixed Reidemeister moves

Further, in [La1, LR1] we proved the following (cf. Thm. 5.3 [LR1]):

Theorem 5 (Braiding theorem for $S^3\backslash\widehat{B}$ and $\chi(S^3,\widehat{B})$). *Any oriented link L in $S^3\backslash\widehat{B}$ can be represented in S^3 by some geometric mixed braid $B\bigcup\beta$, the closure of which is isotopic in S^3 to the mixed link $\widehat{B}\bigcup L$.*

Proof. The fixed part $\widehat{B}$ may be viewed in S^3 as the braid B union an arc, k say, at infinity. This arc is the identification of the two horizontal arcs containing the endpoints of B and thus it realizes the closure of B, view

Fig. 13. In Fig. 13a the braid B is drawn curved but this is just for the purpose of picturing it. Let L be a link in $S^3\backslash\widehat{B}$. By general position L misses a small regular neighbourhood of k, $N(k)$, and therefore it can be isotoped into the complement of $N(k)$ in S^3. By expanding $N(k)$ we can view its complement as the cylinder $T = D^2 \times I$ that contains B (Fig. 13b). We then apply the braiding of [LR1] in T. This will leave B untouched but it will braid L in T. Finally we cut $\widehat{B}$ open along k in order to obtain a geometric mixed braid. □

Let $B\bigcup\beta$ be a braid in $S^3\backslash\widehat{B}$. A *geometric L–move* consists in cutting a strand of the subbraid β at some point and then performing an L–move as described in Definition 2. Using, now, Theorem 4 we have the following L–move version of the Markov theorem for knot complements (cf. Thm. 5.5 [LR1]).

Theorem 6 (Geometric mixed braid equivalence for $S^3\backslash\widehat{B}$). *Two oriented links L_1, L_2 in $S^3\backslash\widehat{B}$ are isotopic in $S^3\backslash\widehat{B}$ if and only if any two corresponding mixed braids $B\bigcup\beta_1$ and $B\bigcup\beta_2$ in S^3 differ by braid isotopies and by L–moves that do not touch the fixed subbraid B.*

The passage from Theorem 4 to Theorem 6 is by showing that, if during the isotopy the hypothetical closing arc k is crossed (see Fig. 13c), then the corresponding mixed braids differ by L–moves applied only on the moving subbraids.

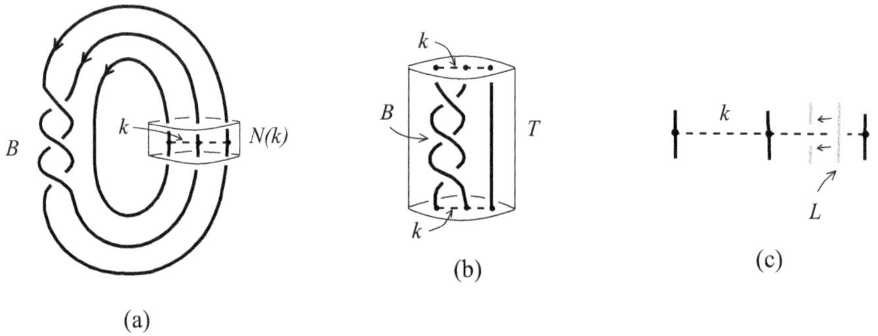

FIGURE 13. Crossing the closing arc k

We would like now to pass from the geometric braid equivalence to an algebraic statement. Our strategy is the following. We first part the mixed braids and we translate the L–equivalence of Theorem 6 to an equivalence of parted mixed braids. *Parting* a geometric mixed braid $B\bigcup\beta$ on $m+n$

strands means to separate its endpoints into two different sets, the first m belonging to the subbraid B and the last n to β, and so that the resulting braids have isotopic closures. The generators of the groups $B_{m,n}$ become already apparent in the parted mixed braid equivalence. We then comb the parted mixed braids and we translate the parted mixed braid equivalence to an equivalence of algebraic mixed braids. *Combing* a parted mixed braid means to separate the fixed subbraid from the moving subbraid using mixed braid isotopy.

Lemma 1. *Every mixed braid may be represented by a parted mixed braid with isotopic closure (cf. Lemma 1* [LR2]*).*

Proof. Let $B \bigcup \beta$ be a geometric mixed braid. We attach arbitrarily arrays of labels 'o' or 'u' to each pair of corresponding strands of the moving subbraid β, with as many entries as the number of fixed strands on their right. We then pull each pair of corresponding moving strands to the right, *over* or *under* each strand of B lying on their right, according to the label in the array of the pair. We start from the rightmost pair, respecting the position of the endpoints. See the first two illustrations of Fig. 14 for a parting of an abstract geometric mixed braid. Obviously, the closures of the initial and of the parted mixed braids are isotopic (they differ by planar isotopy and by mixed RII moves). □

Fig. 14 illustrates two different partings of an abstract geometric mixed braid, an arbitrary one and the standard parting, whereby all strands are pulled *over*. We shall use the same notation for parted mixed braids as for geometric mixed braids.

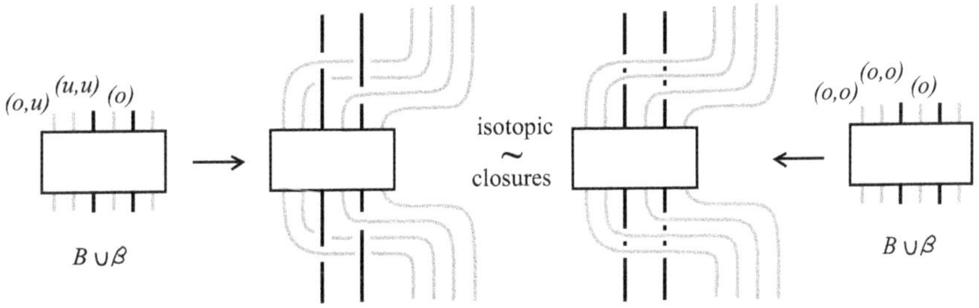

FIGURE 14. Parting a mixed braid – the standard parting

Let now $C_{m,n}$ denote the set of parted mixed braids on n moving strands, related to $S^3 \backslash \widehat{B}$. Note that for different fixed braids B the corresponding sets of parted mixed braids will be also different. Yet, we will use

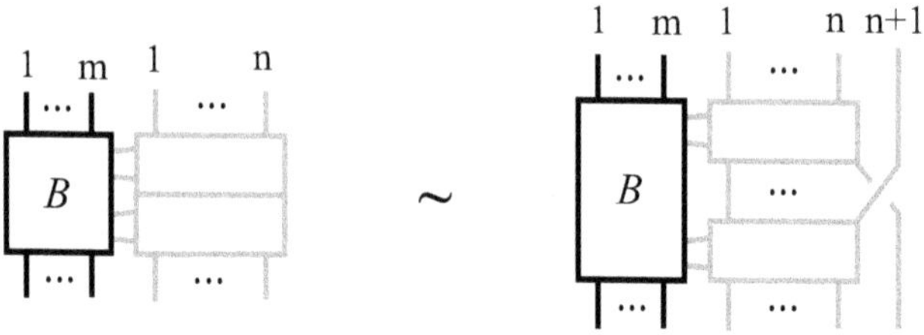

FIGURE 15. The stabilization move

the same notation for them with no risk of confusion. By adding an extra moving strand on the right of a parted mixed braid on n moving strands, $C_{m,n}$ embeds naturally into $C_{m,n+1}$. Let $C_{m,\infty} := \bigcup_{n=1}^{\infty} C_{m,n}$ denote the disjoint union of all sets $C_{m,n}$. We define below some moves in $C_{m,\infty}$.

Definition 3. *(1) Loop conjugation* of a parted mixed braid in $C_{m,n}$ is its concatenation from above by a loop a_i (or by a_i^{-1}) and from below by a_i^{-1} (corr. a_i).

(2) Markov conjugation of a parted mixed braid in $C_{m,n}$ is its concatenation from above by a crossings σ_j (or by σ_j^{-1}) and from below by σ_j^{-1} (corr. σ_j).

(3) A *parted L–move* is defined to be an L–move between parted mixed braids. As for links in S^3, a parted L–move may be considered to introduce an in–box crossing (view left-hand illustration of Fig. 16).

(4) A *stabilization move* is the insertion of an extra strand with a crossing $\sigma_n^{\pm 1}$ at the right hand side of a parted mixed braid on n moving strands or the reverse operation. View Fig. 15.

The relations among moves (2), (3) and (4) are analogous to links in S^3 and are illustrated in Figs. 16 and 17.

The lemma below says that, up to loop conjugations, we can choose any parting for a geometric mixed braid. Indeed, Fig. 18 illustrates that the change of only one label of one moving strand is realized by a loop conjugation.

Lemma 2. *Two partings of a geometric mixed braid differ by loop conjugations (cf. Lemma 2* [LR2]*).*

Finally, we need the following lemma:

Lemma 3. *A mixed braid with an L–move performed can be parted to a parted mixed braid with a parted L–move performed (cf. Lemma 3* [LR2]*.)*

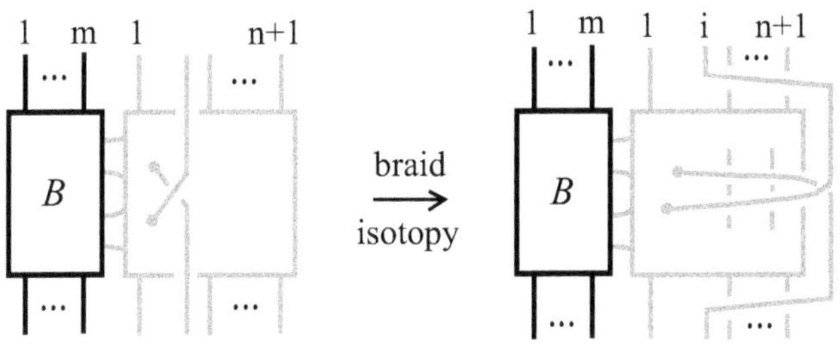

FIGURE 16. Parted L–move = stabilization move and Markov conjugation

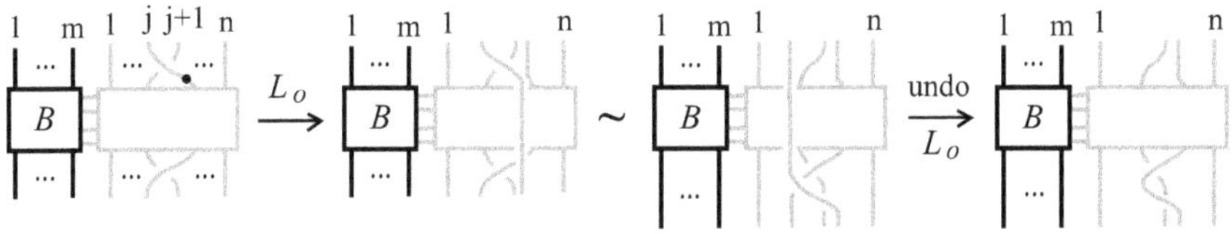

FIGURE 17. Markov conjugation is a composition of L–moves

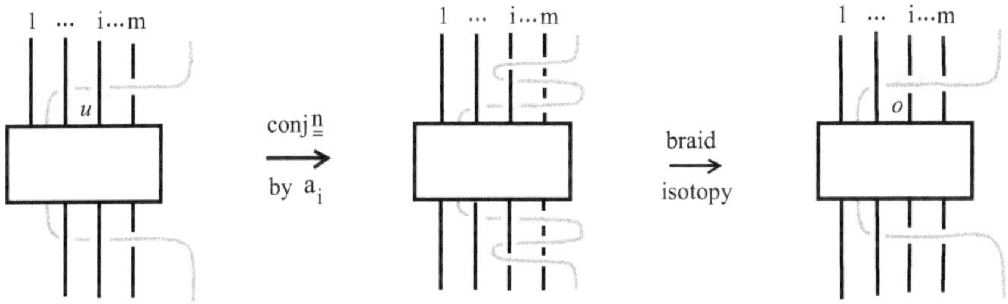

FIGURE 18. Change of parting labels $\longleftrightarrow$ conjugation by a_i

Proof. If the L–move is an L_o–move we choose to part its strands by pulling them to the right and over all other strands in between. Then the crossing of the L–move slides over to the right by a braid isotopy. See Fig. 19. If the L–move is an L_u–move we pull the two strands under the fixed strands in between. □

We are now in a position to state two versions of the braid equivalence in $S^3\backslash\widehat{B}$ for parted mixed braids.

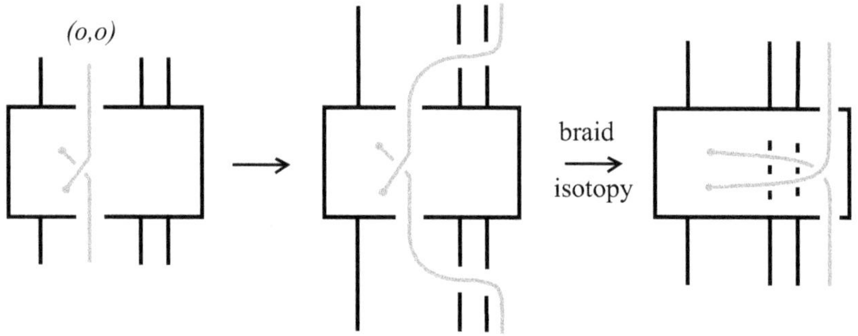

FIGURE 19. An L_o–move + parting = a parted L_o–move

Theorem 7 (Parted mixed braid equivalence for $S^3\backslash\widehat{B}$). *Two oriented links in $S^3\backslash\widehat{B}$ are isotopic if and only if any two corresponding parted mixed braids in $C_{m,\infty}$ differ by a finite sequence of braid isotopies, parted L–moves and loop conjugations.*

Equivalently, by a finite sequence of braid isotopies, stabilization moves, Markov conjugations and loop conjugations.

Proof. Clear by Lemmas 2 and 3 and by Figs. 16 and 17. □

We shall now move to a completely algebraic statement for the braid equivalence in $S^3\backslash\widehat{B}$. Unless the fixed subbraid B represents the complement of the m–unlink or a connected sum of m lens spaces or a handlebody of genus m, B is not the identity braid on m strands and, so, concatenating two elements of $C_{m,n}$ is not a closed operation, since it alters the braid description of the manifold. So, the set $C_{m,n}$ of parted mixed braids is not a subgroup of B_{m+n}. Yet, we have the following result.

Lemma 4. *For the fixed subbraid B on m strands the set $C_{m,n}$ is a coset of $B_{m,n}$ in B_{m+n} (cf. Proposition 1* [La2].*)*

Proof. Let $A \in C_{m,n}$. We shall show that A can be written as a product $\alpha\, B$, where α is an algebraic mixed braid in $B_{m,n}$ followed by the fixed subbraid B embedded naturally in B_{m+n}. Indeed, we notice first that, by symmetry, Artin's combing for pure braids can be also applied starting from the bottom of the braid: first the pure braiding of the 2nd strand with the 1st and the others remaining fixed, then the pure braiding of the 3rd strand with the 1st and the 2nd one and the others remaining fixed, and so on. So, we multiply A from the top with a braid p on the last n strands and with a braid σ on the first m strands, such that $pA\sigma$ is a pure braid in B_{m+n}. Then we apply to it Artin's combing as above. This will separate A into

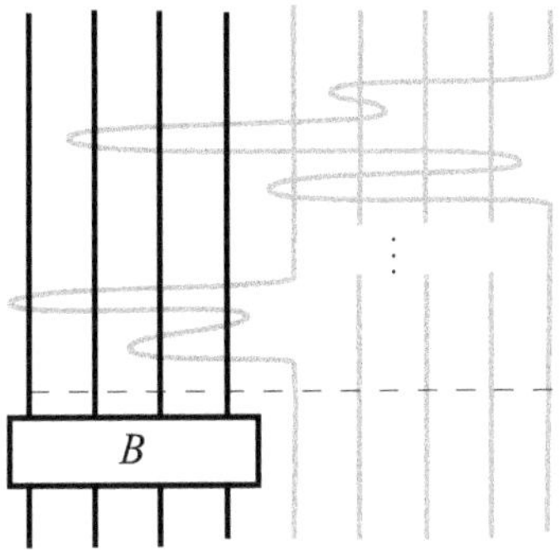

FIGURE 20. A combed mixed braid

two parts: the top one (after multiplying from the top with p^{-1}), an element of $B_{m,n}$, representing the link in $S^3\backslash\widehat{B}$, followed by the fixed braid B embedded in B_{m+n} (after multiplying from the bottom with σ^{-1}). Thus, the set $C_{m,n}$ of combed mixed braids is a coset of $B_{m,n}$ in B_{m+n}. □

A parted mixed braid in the form $\alpha\, B$, where $\alpha \in B_{m,n}$, shall be called a *combed mixed braid.* View Fig. 20 for an abstract illustration of a combed mixed braid.

In view of the above result we would like to restate the braid equivalence of Theorem 7 for parted mixed braids into an equivalence relation between their corresponding algebraic representatives after combing. For this we need to understand how exactly the combing is done and how it affects the parted braid equivalence moves. Indeed, if we regard a parted mixed braid as an element of the classical braid group B_{m+n} and if Σ_k denotes the crossing between the kth and the $(k+1)$st strand of the fixed subbraid, then the crossings σ_j of the moving part commute with the crossings of the fixed part, so they are not affected by the combing. Thus the only generating elements of the moving part that are affected by the combing are the loops a_i. Fig. 21 illustrates the relations between Σ_1 and the loops a_i. For arbitrary Σ_k the relations are completely analogous. Cf. [LR2] for details.

The group $B_{m,n}$ of algebraic mixed braids embedds naturally into the group $B_{m,n+1}$ and we shall denote $B_{m,\infty} := \bigcup_{n=1}^{\infty} B_{m,n}$ the disjoint union of all algebraic mixed braid groups. In $B_{m,\infty}$ we define the following moves.

Definition 4. *(1) Algebraic Markov conjugation* is a Markov conjugation between algebraic mixed braids, and it has the algebraic expression:

$$\alpha \sim \sigma_j^{\pm 1}\alpha\sigma_j^{\mp 1}$$

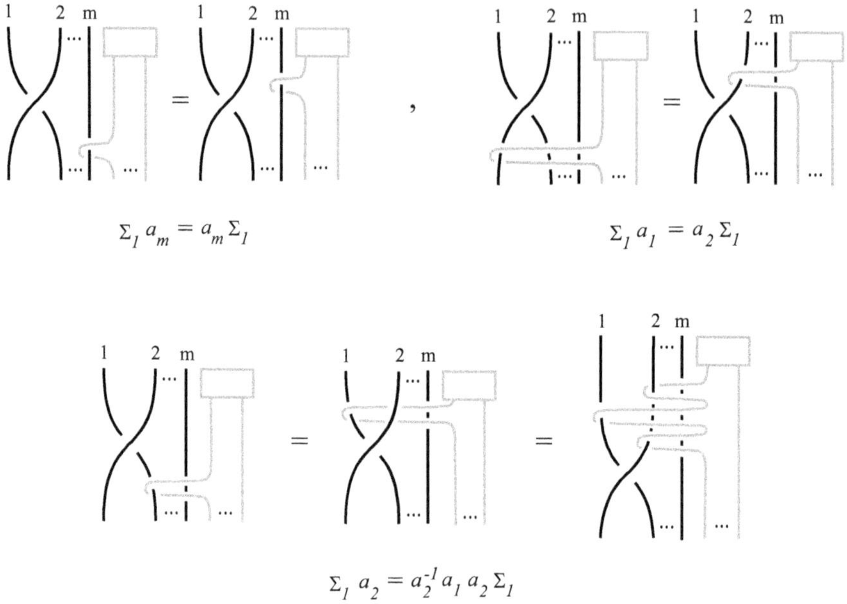

FIGURE 21. Combing the a_i's to the top

where $\alpha, \sigma_j \in B_{m,n}$.

(2) An *algebraic stabilization move* is a stabilization move between algebraic mixed braids, and it has the algebraic expression:

$$\alpha_1\alpha_2 \sim \alpha_1\sigma_n^{\pm 1}\alpha_2$$

where $\alpha_1, \alpha_2 \in B_{m,n}$.

(3) An *algebraic L–move* is a parted L–move between algebraic mixed braids. From Fig. 16 one can easily derive analogous algebraic expressions for an algebraic L_o–move and an algebraic L_u–move as in Eqs. 2 and 3 for the classical situation, where now $\alpha_1, \alpha_2 \in B_{m,n}$ and α_1', $\alpha_2' \in B_{m,n+1}$ are obtained from α_1, α_2 by replacing each σ_j by σ_{j+1} for all $j = i, \ldots, n-1$.

(4) *Twisted loop conjugation* is defined to be the multiplication of an algebraic mixed braid by a loop a_i or a_i^{-1} from the top and by its combing through the fixed braid B from the bottom, and it has the algebraic expression:

$$\beta \sim a_i^{\mp 1}\beta\rho_i^{\pm 1}$$

for $\beta, a_i, \rho_i \in B_{m,n}$, where ρ_i is the combing of the loop a_i through B. (Note that the combing of the loop a_i^{-1} through B is ρ_i^{-1}.)

We stress that the twisted loop conjugation is a move between algebraic mixed braids. For the first three moves view Figs. 16 and 17, where the braid B is now the identity braid. We are now in a position to restate Theorem 7 in terms of algebraic mixed braids.

Theorem 8 (Algebraic mixed braid equivalence for $S^3\backslash\widehat{B}$). *Two oriented links in $S^3\backslash\widehat{B}$ are isotopic if and only if any two corresponding algebraic mixed braid representatives in $B_{m,\infty}$ differ by a finite sequence of algebraic braid relations and the following moves:*
(1) algebraic stabilization moves,
(2) algebraic Markov conjugations,
(3) twisted loop conjugations.

Equivalently, by a finite sequence of algebraic braid relations and the following moves:
(1′) algebraic L–moves,
(2′) twisted loop conjugations.

2.4. **Braid equivalence in c.c.o. 3-manifolds.** Let $M = \chi(S^3, \widehat{B})$. A link L in M may be seen as a link in $S^3\backslash\widehat{B}$ with the extra freedom to slide across the 2–discs bounded in M by the specified longitudes of the components of the surgery link $\widehat{B}$.

Definition 5. Let b be the oriented boundary of a ribbon and let $L_1 \bigcup \widehat{B}$ and $L_2 \bigcup \widehat{B}$ be two oriented mixed links, so that $L_2 \bigcup \widehat{B}$ is the band connected sum (over b) of a component, c, of L_1 and the specified (from the framing) longitude of a surgery component of $\widehat{B}$. This is not an isotopy between the mixed links in S^3 but it reflects isotopy between the links L_1 and L_2 in M and we shall call it a *band move.*

A band move can be, thus, performed in two steps: firstly, one of the small edges of b is glued to a part of c so that the orientation of the band agrees with the orientation of c, while the other small edge of b, which we shall call *little band*, approaches a surgery component of $\widehat{B}$ in an arbitrary way. Secondly, the little band is replaced by an arc running in parallel with the specified longitude of the surgery component, such that the orientation of the arc agrees with the orientation of b and the resulting link is $L_2 \bigcup \widehat{B}$.

The first step of a band move is an isotopy move between the links L_1 and L_2 in $S^3\backslash\widehat{B}$. So, from now on whenever we say *band move* we shall always be referring to the second step. Since $\widehat{B}$ is oriented in our setting, there are two types of band moves according as in the second step the orientation of the arc replacing the little band agrees (type α) or disagrees (type β) with the orientation of the surgery component –and implicitly of its specified longitude. View Fig. 22, where p is the integral framing of the

FIGURE 22. The two types of band moves

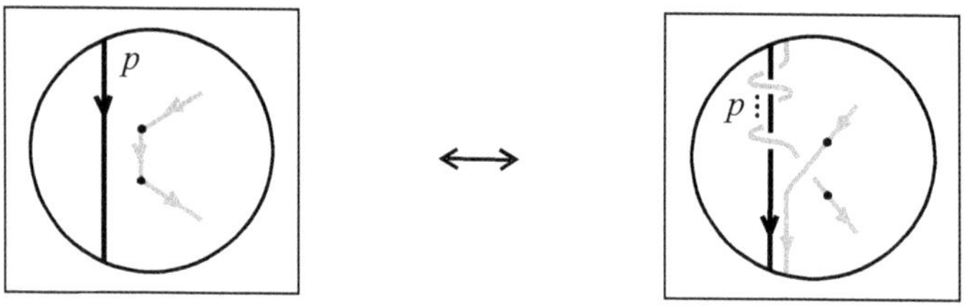

FIGURE 23. The braid band move

surgery component. The discussion can be summarized by saying that two oriented links L_1, L_2 are isotopic in M if and only if the mixed links $\widehat{B} \bigcup L_1$ and $\widehat{B} \bigcup L_2$ differ in S^3 by ambient isotopies that keep $\widehat{B}$ pointwise fixed and by the band moves (see [LR1] for details).

Note now that neither of the two types of band moves can appear as a move between braids; so, in order to extend our geometric mixed braid equivalence to $\chi(S^3, \widehat{B})$, we modify appropriately the band move of type α, by twisting the little band before performing the move resulting in a single braid band move that captures on the braid level both types of band moves. Namely:

Definition 6. A *braid band move* is a move between mixed braids that upon closure is a band move. View Fig. 23, where only the move has been focused upon. The replacement of the little band links only with one of the strands of the same surgery component and runs in parallel to all remaining strands of that surgery component. A braid band move can be positive or negative depending on the type of crossing we choose for performing it.

We can now extend Theorem 6 to the following result (cf. Thm. 5.10 [LR1]).

Theorem 9 (Geometric mixed braid equivalence for $\chi(S^3, \widehat{B})$)**.** *Let L_1, L_2 be two oriented links in $\chi(S^3, \widehat{B})$ and let $B_1 \bigcup B$, $B_2 \bigcup B$ be two corresponding geometric mixed braids in S^3. Then L_1 is isotopic to L_2 if and*

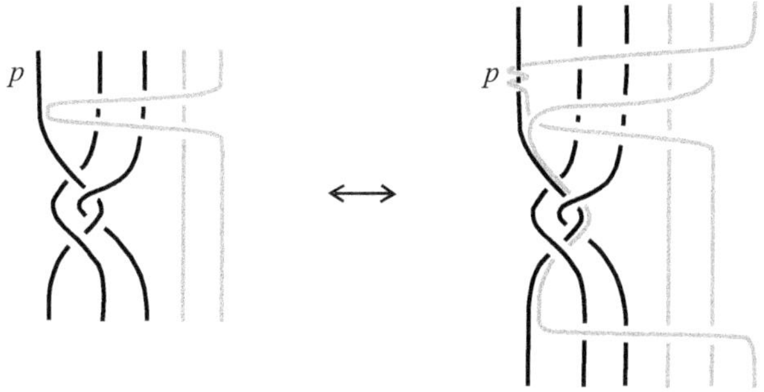

FIGURE 24. A parted band move

only if $B_1 \bigcup B$ is equivalent to $B_2 \bigcup B$ in S^3 by a finite sequence of braid isotopies, the braid band moves and L–moves that do not affect B.

We would like next to extend Theorem 7 to parted mixed braids related to c.c.o. 3–manifolds. We first define:

Definition 7. *A parted band move* is a braid band move between parted mixed braids, such that: it takes place at the top of the braid (before any crossings of the surgery braid are encountered), and the little band starts from the last strand of the moving subbraid and moves over each strand of the parted mixed braid until it reaches the specified surgery strand from the right. Then, after the band move is performed, we apply to the resulting mixed braid the standard parting, bringing the new strands over to the last position of the moving subbraid. View Fig. 24 for an example, where the moving subbraid has been simplified to the identity braid on two strands.

Lemma 5 (cf. Lemma 5 [LR2]). *A braid band move may be always assumed, up to L–moves that do not affect B and mixed braid isotopy, to take place at the top part of a mixed braid and on the right of the specific surgery strand. Moreover, performing a braid band move on a parted mixed braid and then parting, the result is equivalent, up to L–moves and loop conjugation, to performing a parted band move.*

Lemma 5 enables us to extend Theorem 7 to parted mixed braids related to c.c.o. 3–manifolds (cf. Thm. 3 [LR2]):

Theorem 10 (Parted mixed braid equivalence for $\chi(S^3, \widehat{B})$). *Two oriented links in $\chi(S^3, \widehat{B})$ are isotopic if and only if any two corresponding parted mixed braids in $C_{m,\infty}$ differ by a finite sequence of parted mixed braid isotopies, parted L–moves, loop conjugations and parted band moves.*

Equivalently, by a finite sequence of parted mixed braid isotopies, stabilization moves, Markov conjugations, loop conjugations and parted band moves.

In order to move toward a completely algebraic statement we need to understand how a parted band move is combed through the surgery braid B and to give algebraic expressions for parted band moves between algebraic mixed braids.

Definition 8. An *algebraic band move* is defined to be a parted band move between elements of $B_{m,\infty}$. View Fig. 25 for an abstract example. Note that the isotopy of the little band in the dotted box is treated as 'invisible', that is, as identity in the braid group. Setting:

$$\lambda_{n-1} := \sigma_{n-1} \dots \sigma_1 \quad \text{and} \quad t_{k,n} := \sigma_n \dots \sigma_1 a_k \sigma_1^{-1} \dots \sigma_n^{-1},$$

an algebraic band move has the following algebraic expression:

$$\beta_1 \beta_2 \sim \beta_1' \, t_{k,n}^{p_k} \, \sigma_n^{\pm 1} \, \beta_2',$$

where $\beta_1, \beta_2 \in B_{m,n}$ and $\beta_1', \beta_2' \in B_{m,n+1}$ are the words β_1, β_2 respectively with the substitutions:

$$\begin{array}{lcl} a_k^{\pm 1} & \longleftrightarrow & [(\lambda_{n-1}^{-1} \sigma_n^2 \lambda_{n-1}) \, a_k]^{\pm 1} \\ a_i^{\pm 1} & \longleftrightarrow & (\lambda_{n-1}^{-1} \sigma_n^2 \lambda_{n-1}) \, a_i^{\pm 1} \, (\lambda_{n-1}^{-1} \sigma_n^2 \lambda_{n-1})^{-1}, \quad \text{if } \ i < k \\ a_i^{\pm 1} & \longleftrightarrow & a_i^{\pm 1}, \quad \text{if } \ i > k. \end{array}$$

Further, a *twisted algebraic band move* is defined to be a parted band move that has been combed through B, so it is the composition of an algebraic band move with the combing of the parallel strand, view Fig. 26. Moreover, it has the following algebraic expression:

$$\beta_1 \beta_2 \sim \beta_1' \, t_{k,n}^{p_k} \, \sigma_n^{\pm 1} \, \beta_2' \, r_k$$

where r_k is the combing of the parted parallel strand to the kth surgery strand through the surgery braid.

We stress that a twisted algebraic band move is a move between algebraic mixed braids.

The above definitions lead immediately to the following (cf. Lemma 8[LR2]):

Lemma 6. *Performing a parted band move on a parted mixed braid and then combing, the result is the same as combing the mixed braid and then performing an algebraic band move.*

We are, finally, in the position to state the following result.

Theorem 11 (Algebraic mixed braid equivalence for $\chi(S^3, \widehat{B})$)**.** *Two oriented links in $\chi(S^3, \widehat{B})$ are isotopic if and only if any two corresponding*

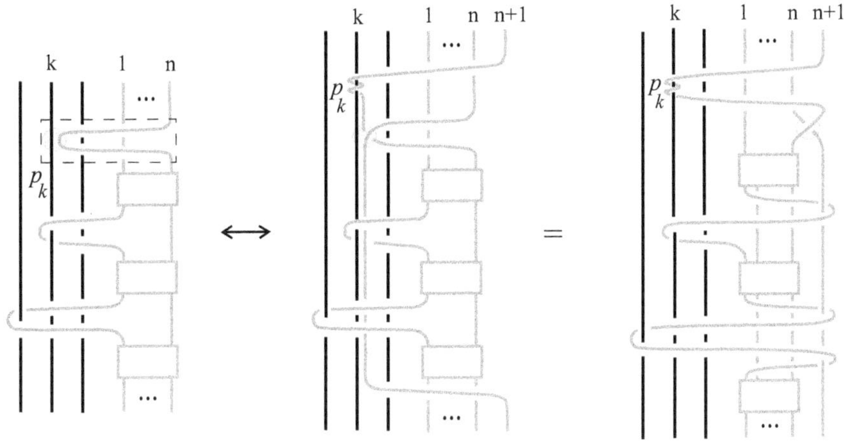

FIGURE 25. An algebraic band move and its algebraic expression

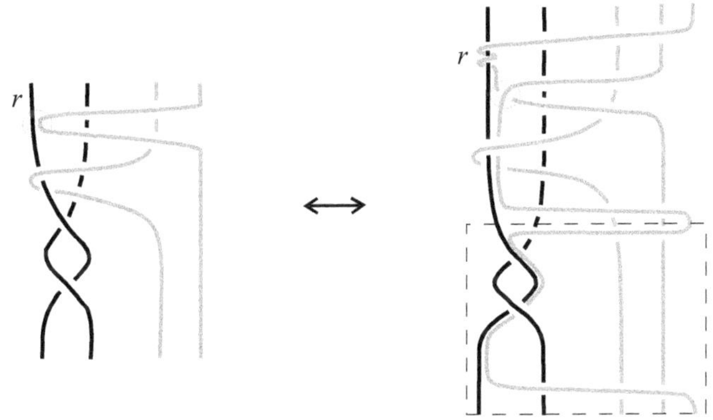

FIGURE 26. Parted band move = algebraic band move + combing

algebraic mixed braid representatives in $B_{m,\infty}$ differ by a finite sequence of algebraic mixed braid relations and the following moves:
(1) algebraic stabilization moves (Definition 4),
(2) algebraic Markov conjugations (Definition 4),
(3) twisted loop conjugations (Definition 4),
(4) twisted algebraic band moves (Definition 8).

Equivalently, by a finite sequence of algebraic mixed braid relations and the following moves:

(1′) algebraic L–moves (Definition 4),
(2′) twisted loop conjugations (Definition 4),
(3′) twisted algebraic band moves (Definition 8).

Proof. By Theorem 8 we only have to consider the case when a parted band move takes place, and by Theorem 10 we only have to consider the behaviour of a parted band move with respect to combing, and this is Lemma 6. □

In [LR2] we give the precise algebraic braid equivalence moves for knot complements and c.c.o. 3–manifolds defined by: the m–unlink, the Hopf link, a daisy chain, the right-handed trefoil.

Remark 4. The element $t_{k,n}^{p_k}$ in Definition 8 of an algebraic band move is just a Markov conjugate of the loop $a_k^{p_k}$ and these are the appropriate words for defining inductive Markov traces on quotient algebras of group algebras of $B_{m,n}$. Note also that the words in the parentheses of the substitutions of the loops get significantly simplified if we apply a quadratic relation on the σ_i's. Moreover, in Theorem 11 we obtain the most local description and the best possible control over the band moves of links in closed 3–manifolds, and this is very useful for the study of skein modules of closed 3–manifolds[Pr].

2.5. **Braid equivalence in handlebodies.** A link inside a handlebody H_m may not pass beyond the boundary of the handlebody from either end. Representing H_m by the identity braid I_m with infinitely extended strands, this means that a link inside H_m may not cross the closing arc k of Fig. 13. So, two oriented links L_1, L_2 in a handlebody H_m are isotopic if and only if the mixed links $I_m \bigcup L_1$ and $I_m \bigcup L_2$ are isotopic in S^3 by an ambient isotopy which keeps I_m (with infinitely extended strands) pointwise fixed.

The fact that crossing the closing arc k is *forbidden* in the category of handlebodies affects the definition of the closure of a geometric mixed braid in this category. Indeed, let L be an oriented link in H_m giving rise to the oriented mixed link $I_m \bigcup L$ in S^3, which gets braided to the geometric mixed braid $I_m \bigcup \beta$ (recall the braiding discussion for handlebodies in Subsection 2.1). In order to ensure that the closure of $I_m \bigcup \beta$ will be isotopic to $I_m \bigcup L$, the *closure* is defined by joining with simple closing arcs the endpoints of the corresponding strands of the moving part only, each closing arc passing *over* or *under* the mixed braid, according to the instruction 'u' and 'o' on the top endpoint of the geometric mixed braid, recall Fig. 9c. View Fig. 27 for an abstract illustration of the closure. To emphasize that different choices of closing labels will give rise, in general, to non-isotopic links in H_m view the example in Fig. 28.

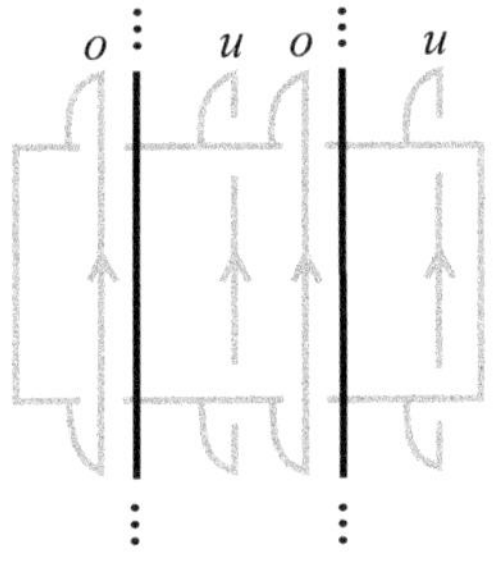

FIGURE 27. The closure of a geometric mixed braid related to a handlebody

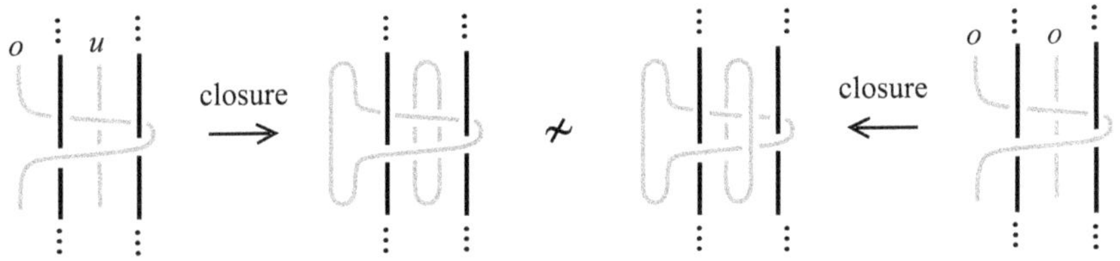

FIGURE 28. Different closing labels yield non-isotopic links

The isotopy restriction for handlebodies, as supposed to knot complements, leads to the following definition of L–moves (cf. [HL]):

Definition 9 (Geometric L–moves in H_m). Let $I_m \bigcup \beta$ be a braid in H_m. A *geometric L–move* consists in cutting a strand of the subbraid β at a point, then pulling the two ends of the cut, the upper downward and the lower upward, both running entirely *over* the braid (*geometric L_o–move*) or entirely *under* it (*geometric L_u–move*), and keeping them aligned so as to obtain a new pair of corresponding braid strands, which are labelled 'o' and 'u' *according to the type of the L–move.* View Fig. 29 for an example. Further, by small braid isotopies, an L–move is equivalent to adding an in-box crossing (positive or negative) facing the right-hand side (or, equally, the left-hand side) of the braid.

Clearly, closing the new pair of strands created from an L–move results in a kink in the tangle diagram, so it corresponds to the isotopy move RI.

We are now in a position to state the following (cf. Thm. 3 [HL]).

Theorem 12 (Geometric braid equivalence for H_m). *Two oriented links in H_m are isotopic if and only if any two corresponding geometric mixed braids differ by a finite sequence of geometric mixed braid isotopies and geometric L–moves.*

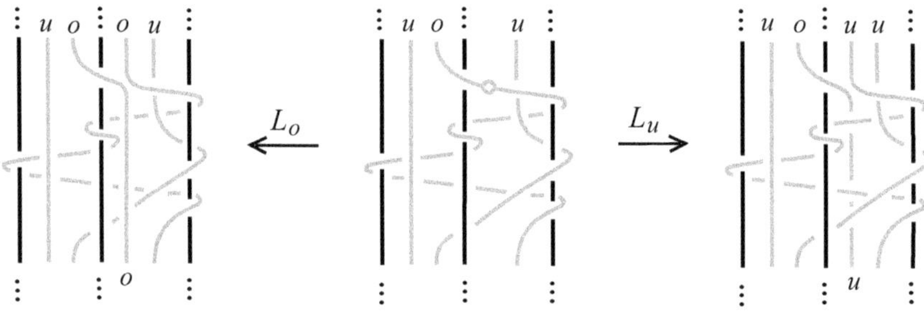

FIGURE 29. The two types of L–moves in H_m

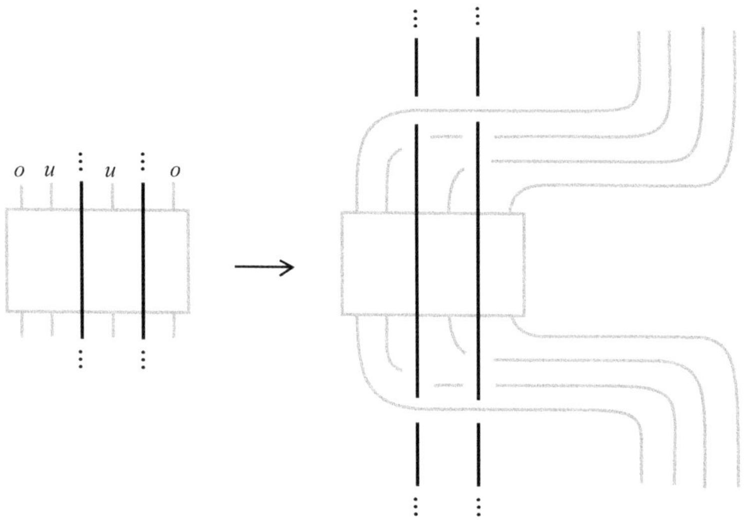

FIGURE 30. The parting of a geometric mixed braid related to H_m

Moving toward an algebraic statement, we define the *parting* of a geometric mixed braid as described in Lemma 1 and as illustrated in Fig. 14, only, here, the pulling of each pair of corresponding moving strands to the right is done according to its label 'o' or 'u'. Note that label 'o' resp. 'u' means to pull the pair of corresponding strands over *all* resp. under *all* fixed strands to the right. So, if we would like to relate to the proof of Lemma 1, an array of labels here would contain the same symbol 'o' or 'u'. View Fig. 30 for an example. A parted mixed braid related to a handlebody does not have labels 'o' or 'u' attached to its moving strands.

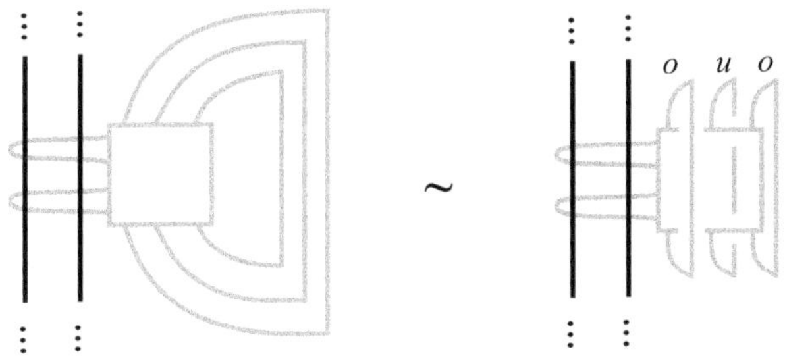

FIGURE 31. The closure of a parted mixed braid

Consequently, the *closure* of a parted mixed braid is realized by joining the pairs of corresponding endpoints of the moving part only with simple unlinked arcs, as in the standard closure of classical braids (view left-hand illustration of Fig. 31). This is not in contradiction with the fact that the parted mixed braids form a subset of the geometric mixed braids. Indeed, putting arbitrarily labels 'o' and 'u' to the endpoints of the moving strands of a parted mixed braid and applying the definition of closure of a geometric mixed braid will result in isotopic mixed links, as no surpassing of the fixed strands is involved (view right-hand illustration of Fig. 31).

From the above, a parted mixed braid is the same as an algebraic mixed braid. So, we will be working with the braid groups $B_{m,n}$ and the algebraic moves of Definition 4. Note that, the fact that the parting cannot assume arbitrary labels (as it was the case for knot complements) forbids loop conjugation as a braid equivalence move for handlebodies. Finally, the agreement of closing labels, L–moves and parting labels yields the following:

Lemma 7. *Parting a geometric mixed braid containing a geometric L–move gives rise to an algebraic mixed braid containing an algebraic L–move.*

The above together with Thm. 12 lead to the following (cf. Thms. 4 and 5[HL]):

Theorem 13 (Algebraic mixed braid equivalence for H_m). *Two oriented links in H_m are isotopic if and only if any two corresponding algebraic mixed braids in $B_{m,\infty}$ differ by a finite sequence of algebraic mixed braid relations and the following moves:*
(1) algebraic stabilization moves (Definition 4),
(2) algebraic Markov conjugations (Definition 4).

Equivalently, by a finite sequence of algebraic mixed braid relations and the following moves:
(1′) algebraic L–moves (Definition 4).

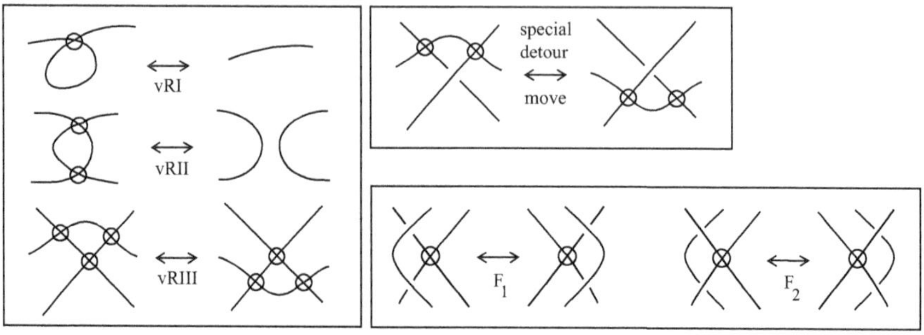

FIGURE 32. Virtual isotopy moves and the forbidden moves

3. OTHER DIAGRAMMATIC SETTINGS

In this section we discuss braid equivalences for virtual knots, for flat virtuals, for welded knots and for singular knots. In all these settings the closure of the corresponding braids is realized as the ordinary closure of classical braids.

3.1. **Virtual, flat virtual and welded braid equivalence.** Virtual knot theory was introduced by Kauffman [Kau2] and it is an extension of classical diagrammatic knot theory. In this extension one adds a *virtual crossing* that is neither an over-crossing nor an under-crossing. Virtual isotopy generalizes the ordinary Reidemeister moves for classical links and is generated by planar isotopy and the moves RI, RII, RIII of Fig. 1 and the moves vRI, vRII, vRIII and the special detour move of Fig. 32, the last one being the key move in the theory. Moves like F1 and F2 with two real crossings and one virtual are forbidden in virtual knot theory. The virtual braid group VB_n is generated by the classical crossings and the virtual crossings v_i, satisfying $v_i^2 = 1$, the braid relations in Eq. 1 for the classical crossings and for the virtual crossings, and the mixed relations

$$v_i \sigma_{i+1} v_i = v_{i+1} \sigma_i v_{i+1}$$

which are not symmetric. Cf. [KL1] and references therein.

A *flat virtual knot* is like a virtual knot but without the extra under/over structure at the real crossings. Instead we have shadow crossings [Tu], the flat crossings. The study of flat virtual knots and links was initiated in [Kau2]. The isotopy moves and the two forbidden moves in virtual knot theory are completely analogous for the flat virtual setting. In the flat virtual braid group on n strands the generators are the virtual crossings v_i and the flat crossings c_i, such that $c_i^2 = 1$. The mixed relations

$$v_i c_{i+1} v_i = v_{i+1} c_i v_{i+1}$$

are here also not symmetric. Cf. [KL1] and references therein.

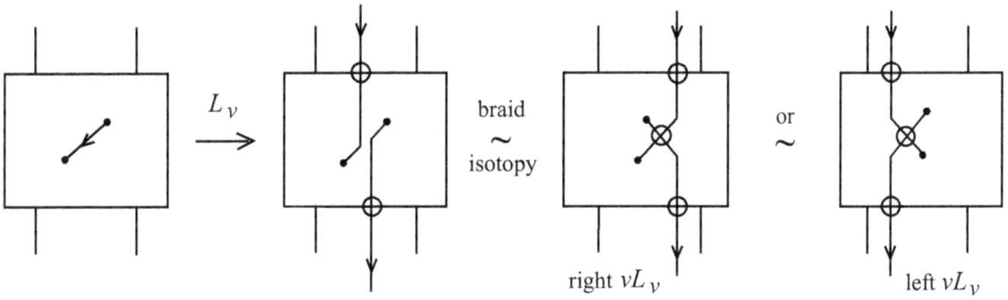

FIGURE 33. A basic L_v–move and the two vL_v–moves

Welded braids were introduced in [FRR]. *Welded knots* satisfy the same isotopy relations as the virtuals, but for welded knots one of the two forbidden moves of Fig. 32 is allowed, the move F1 which contains an over arc and one virtual crossing. The welded braid group on n strands is a quotient of the virtual braid group, so it can be presented with the same generators and relations, with the additional relations:

$$v_i \sigma_{i+1} \sigma_i = \sigma_{i+1} \sigma_i v_{i+1} \text{ (F1)}.$$

Braid equivalence theorems for the above categories have been established in [Ka] and [KL2]. We will present here the L–move approach of [KL2].

Definition 10. A *basic* L_v*–move* on a virtual braid, consists in cutting an arc of the braid open and pulling the upper cutpoint downward and the lower upward, so as to create a new pair of braid strands with corresponding endpoints, and such that both strands cross entirely *virtually* with the rest of the braid. See Fig. 33. In abstract illustrations this is indicated by placing virtual crossings on the border of the braid box. See also Fig. 37 for a concrete example. Further, by a small braid isotopy that does not change the relative positions of endpoints, a basic L_v–move can be equivalently seen as introducing an in-box virtual crossing to a virtual braid which faces either the *right* or the *left* side of the braid. If we want to emphasize the introduction of a virtual crossing, we will say *virtual* L_v*–move*, abbreviated to vL_v*–move*. View Fig. 33 for abstract illustrations.

In the closure of a basic L_v–move or a vL_v–move the detoured loop contracts to a virtual kink. But a kink could be also created by a real crossing, positive or negative. So we define:

Definition 11. A *real* L_v*–move*, abbreviated to $+L_v$*–move* or $-L_v$*–move*, is a virtual L–move that introduces a real in-box crossing (positive or negative) on a virtual braid, and it can face either the *right* or the *left* side of the braid. View Fig. 34 for abstract illustrations.

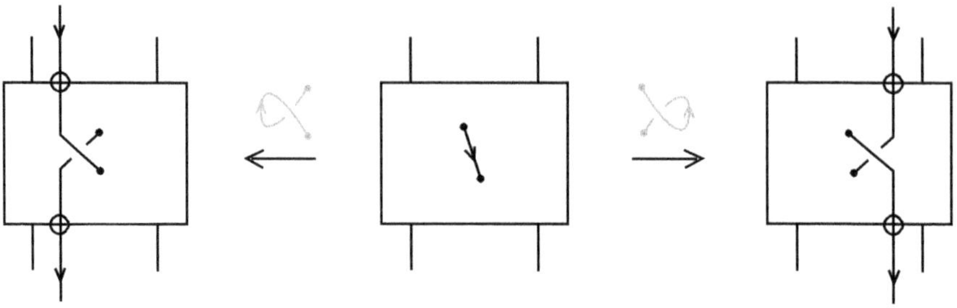

FIGURE 34. Left and right real L_v–moves

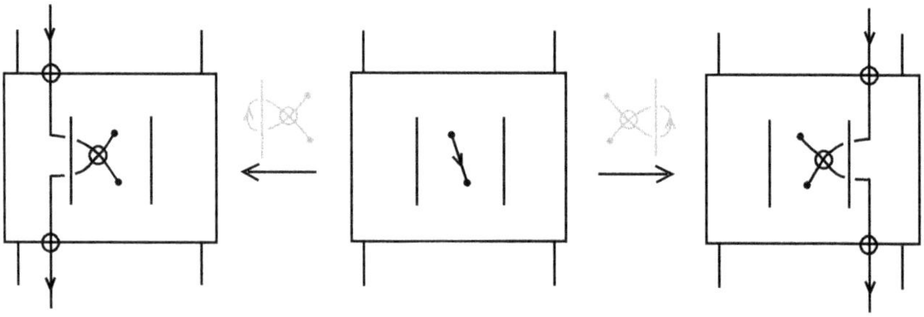

FIGURE 35. Left and right under–threaded L_v–moves

Further, if the crossing of the kink is virtual, then, in the presence of the forbidden moves, there is another possibility for an L_v–move on the braid level. Namely, we have:

Definition 12. A *threaded L_v–move* on a virtual braid is a virtual L–move with a virtual crossing in which, before pulling open the little up-arc of the kink, we perform a Reidemeister II move with real crossings, using another arc of the braid, the *thread.* See Fig. 35. There are two possibilities: an *over-threaded L_v–move* and an *under-threaded L_v–move*, depending on whether we pull the kink over or under the thread, both with the variants *right* and *left.*

A threaded L_v–move cannot be simplified in the braid. If the crossing of the kink were real, then, using a braided RIII move with the thread, the move would reduce to a real L_v–move. Similarly, if the forbidden moves were allowed, a threaded L_v–move would reduce to a vL_v–move.

The effect of a virtual L–move, basic, real or threaded, is to stretch (and cut open) an arc of the braid around the braid axis using the detour move, after twisting it and possibly after threading it. Conversely, such a

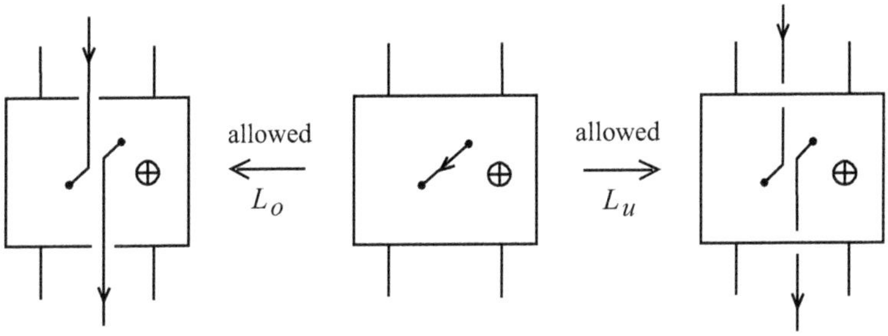

FIGURE 36. The allowed classical L–moves

move between virtual braids gives rise to isotopic closures, since the virtual L–moves shrink locally to kinks (grey diagrams in Figs. 34 and 35). We may finally introduce the notion of a classical L–move.

Definition 13. An *allowed classical L_o–move* resp. *L_u–move* on a virtual braid is like an L–move between classical braids (recall Definition 2), such that the closures of the virtual braids before and after the move are isotopic. See Fig. 36 for abstract illustrations. An allowed classical L–move (over or under) may also introduce an in-box crossing (positive, negative or virtual) or it may even involve a thread.

In order that a classical L–move between virtual braids is allowed, in the sense that it gives rise to isotopic virtual links upon closure, it is required that the virtual braid has no virtual crossings on the entire vertical zone either to the left or to the right of the new strands of the L–move, so that we can perform the closure of the new strands on that side. Otherwise, the presence of forbidden moves preclude such moves. As it turns out, the allowed classical L–moves can be expressed in terms of L_v–moves and real conjugation [KL2].

In Fig. 37 we illustrate an example of various types of L–moves taking place at the same point of a virtual braid. We are now in a position to state the following results (cf. Thms. 2 and 3 [KL2]).

Theorem 14 (Geometric equivalence for virtual braids). *Two oriented virtual links are isotopic if and only if any two corresponding virtual braids differ by virtual braid isotopy and a finite sequence of the following moves:*
(i) real conjugation,
(ii) right virtual L_v-moves,
(iii) right real L_v-moves,
(iv) right and left under-threaded L_v-moves.

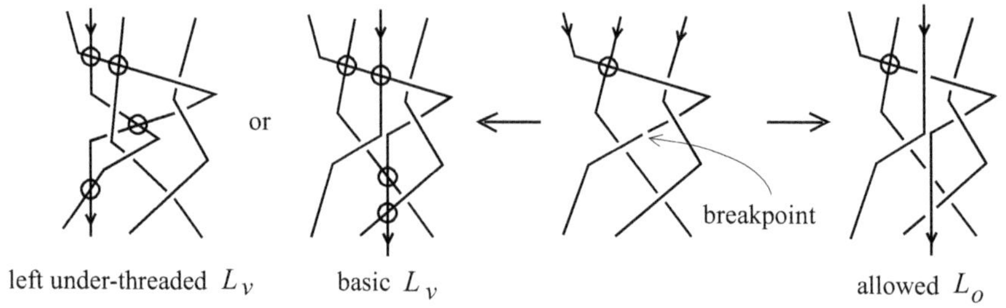

FIGURE 37. A concrete example of introducing L–moves in a virtual braid

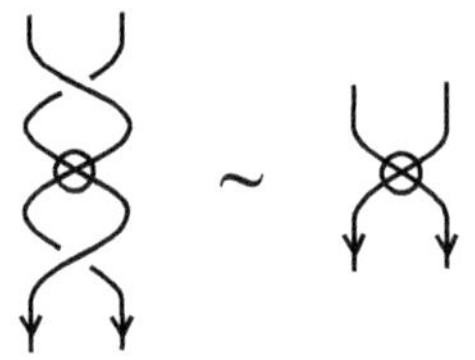

FIGURE 38. The simplest pair of real conjugates

In [KL2] we stated the following conjecture.

Conjecture 1. *Real conjugation is not a consequence of the L_v-moves.* In other words, it should be possible to construct a virtual braid invariant that will not distinguish L_v-move equivalent virtual braids, but will distinguish virtual braids that differ by real conjugation. As the simplest possible puzzle, try to show that there is no sequence of L_v-moves connecting the pair of equivalent braids shown in Fig. 38.

Theorem 15 (Algebraic equivalence for virtual braids). *Two oriented virtual links are isotopic if and only if any two corresponding virtual braids in $\cup_n VB_n$ differ by a finite sequence of braid relations and the following moves:*
(i) virtual and real conjugation: $v_i \alpha v_i \sim \alpha \sim \sigma_i{}^{-1} \alpha \sigma_i$,
(ii) right virtual and real stabilization: $\alpha v_n \sim \alpha \sim \alpha \sigma_n^{\pm 1}$,
(iii) algebraic right under-threading (Fig. 39): $\alpha \sim \alpha \sigma_n^{-1} v_{n-1} \sigma_n^{+1}$,
(iv) algebraic left under-threading (Fig. 39): $\alpha \sim \alpha v_n v_{n-1} \sigma_{n-1}^{+1} v_n \sigma_{n-1}^{-1} v_{n-1} v_n$,
where $\alpha, v_i, \sigma_i \in VB_n$ *and* $v_n, \sigma_n \in VB_{n+1}$.

The statements for flat virtual braid equivalence are completely analogous. The only difference is that real crossings are substituted by flat

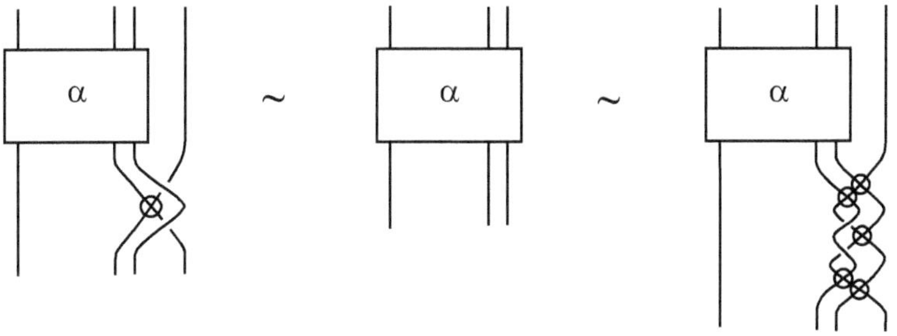

FIGURE 39. Algebraic right and left under-threading

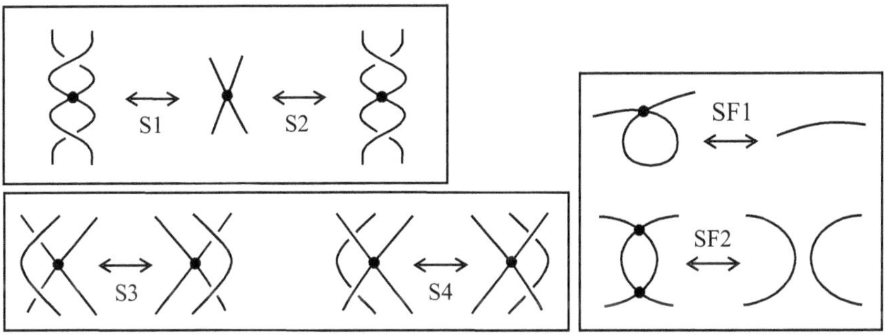

FIGURE 40. Diagrammatic moves for singular knots

crossings, cf. Thms. 4 and 5 [KL2]. For the welded braid equivalence, move (iv) of Theorem 14 is not needed and moves (iii) and (iv) of Theorem 15 are not needed, cf. Thms. 6 and 7 [KL2].

We close by pointing out that in [KL2] it would be quite difficult to compare our local algebraic formulation of the braid equivalence for virtuals with that of Kamada [Ka] without the fundamental L–move context.

3.2. **Singular braid equivalence.** Singular knots are related to Vassiliev's theory of knot invariants. In singular knot theory isotopy is generated by planar isotopy, by the moves RI, RII, RIII of Fig. 1 and by the moves S1, S2, S3, S4 of Fig. 40. Moves SF1 and SF2 of Fig. 40 are forbidden in this theory, see [Kau1].

The singular crossings τ_i generate, together with the real crossings σ_i and their inverses, the singular braid monoid SB_n (introduced in different contexts by Baez [Ba], Birman [Bi2] and Smolin [Sm]) and they satisfy the braid relations Eq. 1 and the relations below.

$$
(5)\qquad
\begin{array}{rclcl}
\sigma_i\sigma_i^{-1} & = & \sigma_i^{-1}\sigma_i & = 1 & \text{for all } i \\
\sigma_{i+1}\sigma_i\tau_{i+1} & = & \tau_i\sigma_{i+1}\sigma_i & & \text{for } |i-j|=1 \\
\sigma_i\tau_j & = & \tau_j\sigma_i & & \text{for } |i-j|>1 \\
\sigma_i\tau_i & = & \tau_i\sigma_i & & \text{for all } i \\
\tau_i\tau_j & = & \tau_j\tau_i & & \text{for } |i-j|>1
\end{array}
$$

The algebraic braid equivalence for singular braids was established by Gemein [Ge]:

Theorem 16 (Algebraic equivalence for singular braids [Ge]). *Two singular braids in $\cup_n SB_n$ have isotopic closures if and only if they differ by a finite sequence of braid relations in $\cup_n SB_n$ and the following moves:*
(i) singular commuting: $\tau_i\omega \sim \omega\tau_i, \quad \omega, \tau_i \in SB_n$
(ii) real conjugation: $\sigma_i\omega \sim \omega\sigma_i, \quad \omega, \sigma_i \in SB_n$
(iii) real stabilization: $\omega \sim \omega\sigma_n^{\pm 1}, \quad \omega \in SB_n$.

Using the above result, in [La3] the geometric braid equivalence for singular braids was easily established:

Theorem 17 (Geometric equivalence for singular braids). *Two oriented singular links are isotopic if and only if any two corresponding singular braids differ by a finite sequence of braid relations in $\cup_n SB_n$ and the following moves:*
(i) singular commuting: $\tau_i\omega \sim \omega\tau_i, \quad \omega, \tau_i \in SB_n$
(ii) classical L-moves (Definition 2).

References

[Al] J. W. Alexander, A lemma on systems of knotted curves, *Proc. Nat. Acad. Sci. U.S.A.* **9** (1923) 93–95.
[Ar1] E. Artin, Theorie der Zöpfe, *Abh. Math. Sem. Hamburg Univ.* **4** (1925) 47–72.
[Ar2] E. Artin, Theory of braids, *Ann. of Math.* (2) **48** (1947) 101–126.
[Be] D. Bennequin, Entrlacements et équations de Pfaffe, *Asterisque* **107-108** (1983) 87–161.
[Ba] J. Baez, Link invariants and perturbation theory, *Lett. Math. Phys.* **2** (1992) 43–51.
[Bi1] J. S. Birman, *Braids, links and mapping class groups*, Annals of Mathematics Studies, Vol. 82 (Princeton University Press, Princeton, 1974).
[Bi2] J. Birman, New points of view in knot theory, *Bull. Amer. Math. Soc. (N.S.)* **28**(2) (1993) 253–287.
[BM] J. S. Birman, W. W. Menasco, On Markov's Theorem, *J. Knot Theory Ramifications* **11** no.3 (2002), 295–310.
[Br] H. Brunn, Über verknotete Curven, *Verh. des intern. Math. Congr.* **1**, 256–259 (1897).
[Ch] W.-L. Chow, On the algebraical braid group, *Ann. of Math.* (2) **49** (1948) 654–658.
[FRR] R. Fenn, R. Rimanyi, C.P. Rourke, The braid permutation group, *Topology* **36** (1997) 123–135.

[Ge] B. Gemein, Singular braids and Markov's theorem, *J. Knot Theory Ramifications* **6**(4) (1997) 441–454.

[HL] R. Häring-Oldenburg, S. Lambropoulou, Knot theory in handlebodies, *J. Knot Theory Ramifications* **11**(6) (2002) 921–943.

[Jo] V. F. R. Jones, Hecke algebra representations of braid groups and link polynomials, *Ann. Math.* **126** (1987) 335–388.

[Ka] S. Kamada, Braid representation of virtual knots and welded knots, *Osaka J. Math.* **44** (2007), no. 2, 441–458. See also *arXiv:math.GT/0008092*.

[KT] C. Kassel and V. Turaev, *Braid groups*, Graduate Texts in Mathematics, Vol. 247 (Springer, 2008).

[Kau1] L. H. Kauffman, Invariants of graphs in three-space, *Trans. Amer. Math. Soc.* **311**(2) (1989) 697–710.

[Kau2] L. H. Kauffman, Virtual Knot Theory, *European J. Comb.* **20** (1999) 663–690.

[KL1] L. H. Kauffman and S. Lambropoulou, Virtual Braids, *Fundamenta Mathematicae* **184** (2004) 159–186.

[KL2] L. H. Kauffman and S. Lambropoulou, Virtual Braids and the L-Move, *J. Knot Theory Ramifications* **15**(6) (2006) 773–811.

[La1] S. Lambropoulou, "A study of braids in 3–manifolds", Ph.D. thesis, Warwick, 1993.

[La2] S. Lambropoulou, Braid structures in handlebodies, knot complements and 3-manifolds, in *Proceedings of Knots in Hellas '98*, Series on Knots and Everything Vol. 24 (World Scientific, 2000), pp. 274–289.

[La3] S. Lambropoulou, L-moves and Markov theorems, *J. Knot Theory Ramifications* **16** no. 10 (2007), 1–10.

[La4] S. Lambropoulou, "Braid structures and braid equivalence in different manifolds and in different settings", World Scientific, Series on Knots and Everything, in preparation.

[LR1] S. Lambropoulou and C. P. Rourke, Markov's theorem in three–manifolds, *Topology and its Applications* **78** (1997) 95–122.

[LR2] S. Lambropoulou and C. P. Rourke, Algebraic Markov equivalence for links in 3-manifolds, *Compositio Math.* **142** (2006) 1039–1062.

[Ma] A.A. Markov, Über die freie Äquivalenz der geschlossenen Zöpfe, *Recueil Mathématique Moscou* **1**(43) (1936) 73–78.

[Mo] H.R. Morton, Threading knot diagrams, *Math. Proc. Cambridge Philos. Soc.* **99** (1986) 247–260.

[Pr] J.H. Przytycki, Skein Modules, eprint: math.GT/0602264; Chapter IX of the book "KNOTS: From combinatorics of knot diagrams to the combinatorial topology based on knots", Cambridge University Press,to appear 2014, pp. 600.

[Rd] K. Reidemeister, Elementare Begründung der Knotentheorie, *Abh. Math. Sem. Hamburg Univ.* **5** (1927) 24–32.

[Sk] R. Skora, Closed braids in 3–manifolds, *Math. Zeitschrift* **211** (1992) 173–187.

[Sm] L. Smolin, *Knot theory, loop space and the diffeomorphism group*, New perspectives in canonical gravity, 245–266, Monogr. Textbooks Phys. Sci. Lecture Notes, **5**, Bibliopolis, Naples, 1988.

[Su] P.A. Sundheim, Reidemeister's theorem for 3–manifolds, *Math. Proc. Camb. Phil. Soc.* **110** (1991) 281–292.

[Tu] V. Turaev, Virtual strings, *Ann. Inst. Fourier (Grenoble)* **54**(7) (2004) 2455–2525.

[Tr] P. Traczyk, A new proof of Markov's braid theorem, preprint (1992), Banach Center Publications, Vol.42, Institute of Mathematics Polish Academy of Sciences, Warszawa (1998).

[Vo] P. Vogel, Representation of links by braids: A new algorithm, *Comment. Math. Helvetici* **65** (1990) 104–113.

[We] N. Weinberg, Sur l' equivalence libre des tresses fermée, *Comptes Rendus (Doklady) de l' Académie des Sciences de l' URSS* **23**(3) (1939) 215–216.

[Ya] S. Yamada, The minimal number of Seifert circles equals the braid index of a link, *Invent. Math.* **89** (1987) 347–356.

Department of Mathematics, National Technical University of Athens, Zografou Campus, GR-157 80 Athens, Greece.

E-mail address: `sofia@math.ntua.gr`

URL: `http://www.math.ntua.gr/~sofia`

FREE KNOTS AND PARITY

V. O. Manturov*

Peoples' Friendship University of Russia,
Faculty of Sciences, 3 Ordjonikidze St., Moscow, 117923
and
Mechanical and Mathematical Dept., GSP-1 Leninskie Gory,
1 MSU (main building) Moscow, 119991
E-mail: vomanturov@yandex.ru

We consider knot theories possessing *parity*: each crossing is decreed *odd* or *even* according to some universal rule. If this rule satisfies some simple axioms concerning the behaviour under Reidemeister moves, then this leads to a possibility of constructing new invariants and proving minimality and non-triviality theorems for knots from these classes, and constructing maps from knots to knots.

Our main example is virtual knot theory and its simplification, *free knot theory*. By using Gauss diagrams, we show the existence of non-trivial free knots (counterexample to Turaev's conjecture) and construct simple yet deep invariants that rely on parity. Some invariants are valued in graph-like objects and some other are valued in groups.

We discuss applications of parity to virtual knots and ways of extending well-known invariants.

The existence of a non-trivial parity for classical knots remains an open problem.

Keywords: Knot, link, free knot, parity.

1. Introduction

The present paper describes the state-of-art of the theory of free knots and links, a drastic simplification of virtual knot theory. The main tool in our investigation is the concept of *parity*. In some knot theories, we can distinguish crossings based on their parity allowing us to construct new and powerful invariants as well as improve some well-known invariants such as the Kauffman bracket polynomial.

The first example of parity, called *Gaussian*, comes from the chord diagram of a link: a chord (and, hence, the corresponding crossing) is even

*Partially supported by RFBR, No. 07–01–00648

whenever the number of chords linked with it, is even. Note that this parity is trivial for Gauss diagrams of classical knots: this fact was familiar to C.F.Gauss. Based on the properties of the Gaussian parity, one can axiomatize the notion of parity for crossings. Summing up the properties of the Gaussian parity, one can axiomatize the notion of parity for crossings. Most of theorems we present here, work for *different* parities in different situations. Here we would like to emphasize the point of view that if there is a possibility to distinguish between *different types* (not necessarily two types) of crossings when looking at a diagram, this information can be used to *improve many known invariants.*[1,17–19] To get an example of such *property* (which is not a parity) one can take the class of long knots (with two ends pulled to the infinity) and then distinguish between *early overcrossings*, where an overcrossing occurs first when walking along the knot diagram, and the remaining crossings, called *early undercrossings*. This approach was useful for studying *long virtual knots*, see Refs. 16, 21 and 22.

The present paper deals with various ways of applying different parities. Some key features are highlighted, but proofs of many theorems are omitted here; proofs can be found in the series of preprints,[18–20] which constitute the paper.[17] For further developments, see also Refs. 1, 3 and 23.

While the paper has been refereed, the author and his colleagues have classified parities for virtual knots, proved that there are no non-trivial parities for classical knots and constructed a projection from knots in thickened surfaces to classical knots, see Refs. 11 and 22.

The concepts of the present paper have some far-reaching generalizations in other situations. In particular, parity is very helpful in the theory of graph-links, see review[8] in the present volume.

The present paper is organized as follows. In the next section we describe the notions of virtual knots and free knots. Section 3 is devoted to the first examples and applications of parity; we construct "forgetting mappings" and reprove the theorem on "closedness" of the class of even diagrams. Then in section 4 we redefine the concept of parity in terms of homology of graphs. Section 5 deals with invariants of free knots and virtual knots and minimality results. We conclude the section by a very simply constructed invariant of free knots valued in a certain group, see Ref. 23; this invariant gives an obstruction for a free knot to be slice.

1.1. *Acknowledgement*

I am grateful to O.V.Manturov, D.P.Ilyutko and the referee of the paper for useful comments concerning this work.

2. Virtual knots. Free knots

Virtual knots were introduced by Kauffman[13] as an extension of classical knots to knots in thickened surfaces considered up to isotopy and stabilizations/destabilizations. Virtual knots have a simple diagrammatic descriptions by planar diagrams and Reidemeister moves. Another description of virtual knots comes from Gauss diagrams[7]: one admits all possible Gauss diagrams (not necessarily realizing planar Gauss codes) and factorizes them by formal Reidemeister moves.

By a *framed 4-graph* we mean a 4-valent graph where for each vertex we fix a way of splitting of the four emanating half-edges into two pairs of edges called *(formally) opposite.*

A *virtual diagram* is a generic immersion of a framed 4-graph in $\mathbb{R}^2$ with each vertex endowed with a classical crossing structure, i.e. one pair of opposite half edges is defined to form an overcrossing; the other two half-edges form an undercrossing; when drawing a diagram on the plane, intersections of different edges of the framed 4-graphs in interior points are called *virtual crossings*, and they are encircled.

By abusing notation in accordance with the fact that an unknot has a crossingless diagram, we shall allow a 4-graph (four-valent graph) to have connected components homeomorphic to a circle. We shall denote the one-component crossingless diagram by Γ_0.

A *virtual link* is an equivalence class of virtual diagrams by usual Reidemeister moves applied to classical crossings and the *detour move.* The detour move changes the immersion of an edge of the graph: one takes an edge fragment drawn on the plane (which has only virtual crossings and self-crossings inside) and redraws it arbitrarily in a generic way with all new crossings specified as virtual.

So, every virtual knot or link has a "frame" being a framed 4-graph, whence virtual crossings "are not there". As shown in Ref. 7, classical links embed into virtual links, i.e., *every two classical knots which are equivalent as virtual knots are equivalent as classical knots (isotopic).*

In fact, embedding theorems of such sort saying that the equivalence relation on a subset coincides with the induced equivalence relation coming from a set, can be proved by constructing projection "maps" from the whole set to a subset.

Consider now a simplification of virtual knots obtained by "forgetting" all structures except framing at classical crossings. *Free knots and links* (also known as homotopy classes of Gauss words, resp., Gauss phrases, following Turaev[27]) are defined as follows. Framed 4-graphs naturally arise from knot (or virtual knot) projections or from generically immersed curves in 2-surfaces.

A *free link* is an equivalence class of framed 4-graphs modulo the following *Reidemeister moves*:

1) addition/removal of a loop formed by two half-edges of the same edge approaching a vertex from non-opposite sides; 2) addition/removal of a bigon with two edges being non-opposite at both vertices; 3) the triangular move pulling a piece of unicursal curve through a crossing formed by two other pieces of unicursal curves, see Fig. 1.

By a *unicursal component* of a framed 4-graph we mean either a connected component homeomorphic to the circle or an equivalence class of edges (of a non-circular component) where the equivalence is generated by the opposite relation at vertices. The pair of opposite half edges allows one to define the notion of *component* and to count the number of *unicursal components* of a framed graph, which turns out to be invariant under Reidemeister moves. A *free knot* is a single-component free link.

Analogously one defines *long free knot.* To this end, one uses *long 4-graphs* instead of free graphs; here we allow two vertices a, b to have valency one (the others having valency four) in such a way that the edges x, y incident to a and b should belong to the same unicursal component in the above sense. One may also think of these two edges of the long 4-graph to be *noncompact*, i.e., we may think that the vertices of valency one are removed and the ends of the edges are taken to infinity. For long 4-graphs, one uses Reidemeister moves applied to regions away from infinity.

There are obvious "forgetting maps" from virtual links to free links and from homotopy classes of curves on 2-surfaces to free links; classical knots (or planar curves) project to free unknots. Such functorial mappings allow one to create a filtration for virtual knots; one filtration for any given parity.

Parities exist for free knots and links; free knots and links naturally appear as simplifications of virtual knots or homotopy classes of curves on surfaces; thus, the study of invariants of free links may enlarge our knowledge about virtual knots or knots in surfaces. For two-component free links there is an obvious invariant: the parity of number of the intersection points; for an n-component free links there is a collection of mutual parities which can be encoded by a graph on n vertices.

Several years ago, V.G. Turaev[25] conjectured that all *free knots* were trivial. This conjecture was recently disproved by myself[17] and (independently, just some days later) by A.Gibson.[5] To do that, I used the concept of *parity.* The existence of a non-trivial parity for classical knots remains an open problem. For different examples of parity, see Ref. 17.

3. The concept of parity

We shall encode knots and virtual knots[13] by Gauss diagrams.[7] We say that two chords of a chord diagram are *linked* if both ends of one of them lie in different connected components of the two-component set obtained from the circle by deleting the ends of the second components. We say that a chord c of a Gauss diagram is *even* if the number of chords it is linked with, is even, and *odd*, otherwise. We denote the set of chords linked with c by E_c; any chord is assumed not to be linked with itself; by $E_a + E_b$ we shall mean $E_a \cup E_b \backslash (E_a \cap E_b)$. For two chords a, b, we write $\langle a, b\rangle = 1$ if a and b are linked, and $\langle a, b\rangle = 0$, otherwise. This notion has a simple geometrical meaning: when drawing a chord diagram as a round circle with straight intervals (chords) inside, two chords are linked whenever the corresponding intervals intersect inside the disc.

We say that a vertex of the four-valent framed graph corresponding to the Gauss diagram is *odd* whenever the chord corresponding to this vertex is odd.

We call the parity of vertices of the framed 4-graph (and corresponding chords of the chord diagram) the *Gaussian parity.*

Let us summarize the properties of *even and odd chords* with respect to their behaviour under Reidemeister moves. It is easy to see that

(1) The parity of a vertex participating in the first Reidemeister move, is even.
The first Reidemeister move corresponds to an addition/removal of an even vertex, and the pairwise incidence relation of the remanining vertices does not change.
(2) The second Reidemeister move adds (removes) two vertices a, b of the same parity.
Indeed, for such vertices a, b we see that $E_a + E_b$ is either $\emptyset$ or coincides with $a + b$; after applying a Reidemeister move the parity of the remaining vertices does not change. Neither does the pairwise incidence of the remaining vertices.
(3) The sum of parities (modulo 2) of the three vertices participating in a third Reidemeister move is even. The parity of a vertex participating in a third Reidemeister move does not change after this move is applied.
Indeed, when performing the third Reidemeister move, we have three vertices (chords) a, b, c, for which $E_a + E_b + E_c \subset \{a, b, c\}, |E_a + E_b + E_c| = 0$ or 2. After performing the move, we get instead of a, b, c three vertices a', b', c' with pairwise switched incidences (w.r.t. a, b, c) the

remaining incidences are unchanged: for $f, g \notin \{a, b, c\} : \langle f, g\rangle$ remains unchanged: a) $\langle f, a\rangle = \langle f, a'\rangle; \langle f, b\rangle = \langle f, b'\rangle; \langle f, c\rangle = \langle f, c'\rangle$ and b) the number of *odd* vertices amongst a, b, c is *even* (is equal to zero or two).

Definition 3.1. By a *parity* we mean a function on the set of all four-valent framed graphs which satisfies the conditions above (assuming all vertices not taking part in a Reidemeister move preserve their parity).

The collection of axioms given above will be further referred to as *parity axiomatics.*

Remark 3.1. Here, the vertices of a 4-valent graph are not enumerated, so the parity function on should be symmetric with respect to the action of the group of symmetries of the graph respecting the framing.

In the case when we deal not merely with free knots or links, but with knots having some decorations at vertices, by *parity* we shall mean the property of crossings satisfying the same axioms with respect to the Reidemeister moves. Thus, for instance, the second parity axiom for the second Reidemeister move requires that the parity is the same for both vertices of any "empty bigon" in the case of free knots, whence for virtual knots, we shall require this only for those pairs of crossings which can participate in a second Reidemeister move, that is, for two crossings *of opposite signs*(, 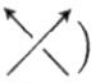) forming a bigon.

Certainly, each parity defined for free links induces a parity for virtual knots, but not vice versa. In the sequel, we shall state some theorems using parity. Unless specified otherwise, such theorems will hold *for any parity.*

As said above, the Gaussian parity is trivial for classical knots. The existence of non-trivial parity for classical knots remains an open problem.

3.1. *The functorial map f*

It turns out that the parity axioms lead to a simple and powerful map on the set of free knots and, more generally, on the set of virtual knots.

Let K be a virtual knot diagram. Let f be a diagram obtained from K by making all *odd* crossings virtual. In other words, we remove all odd chords. The following theorem follows from definitions.

Theorem 3.1. *The map f is a well-defined map on the set of all virtual knots. For a virtual knot diagram K, $f(K) = K$ if and only if all crossings of K are even. Otherwise, the number of classical crossings of $f(K)$ is strictly less than the number of classical crossings of K.*

3.2. *Other examples of parities. The index and the parity hierarchy*

The Gaussian parity (defined via intersection as above) is not the only parity for knots. We shall give definitions of two additional types of parity which can be applied in different situations.

The first example deals with two-component free links. We call a crossing of a 2-component free link *even* if it is formed by two branches of the same component; otherwise it is called *odd.*

The other example deals with one-component virtual knots. Consider the set of virtual knots with all crossings having *even* Gaussian parity. It follows from Theorem 3.1 that this set is closed with respect to the Reidemeister moves, i.e., whenever two diagrams having even crossings are equivalent, they are equivalent by a sequence of Reidemeister moves where all intermediate diagrams have all even crossings. Denote this class of knots by $\mathfrak{V}^1$; evidently, it includes all classical knots.

Let D be a virtual diagram, and let $G(D)$ be the corresponding Gauss diagram. We shall endow the diagram G with signs and arrows in the usual way (+ corresponds to  and − corresponds to ); an arrow is directed from the vertex of the Gauss diagram corresponding to the arc forming an overpass to the vertex of the Gauss diagram corresponding to the arc forming an underpass).

With each classical crossing we associate its *index* (cf. Ref. 25) which will be a non-negative integer. Let X be a classical crossing of D, and let $c(X)$ be the corresponding (oriented) chord of $G(D)$. Without loss of generality assume this chord is oriented from the bottom to the top. Then all the chords intersecting it will be directed either from the left to the right, or from the right to the left. Let us calculate the sum of signs of all those chords intersecting $c(X)$ form the left to the right and subtract the sum of signs of those chords intersecting $c(X)$ from the right to the left.

The *index of c* is the absolute value of the above quantity; we shall denote it by $ind(X) \equiv ind(c(X))$.

Clearly, a chord is even whenever its index is even.

We shall show that there exists a natural *filtration* on the set of all virtual knots, where $\mathfrak{V}^j$ consists of all those diagrams with all chords having index divisible by 2^j:

$$\mathfrak{V}^0 \supset \mathfrak{V}^1 \supset \mathfrak{V}^2 \supset \cdots \supset \mathfrak{V}^n \supset \cdots,$$

which starts with the set of $\mathfrak{V}^0$ of all virtual knots and has as a limit some set $\mathfrak{V}^\infty$ of "knots of index zero" including all classical knots.

Let us collect some facts concerning the index; the proof can be obtained by a simple check:

Proposition 3.1.

(1) Any chord taking part in a Reidemeister-1 move has index 0.
(2) In the Reidemeister-2 move indices of the two chords are the same.
(3) The index is invariant under the third Reidemeister move $ind(a) = ind(a'), ind(b) = ind(b'), ind(c')$*; moreover, if some three chords* a, b, c *take part in a third Reidemeister move then* $ind(a) \pm ind(b) \pm ind(c) = 0$, *see Fig. 1.*
(4) The index remains unchanged under any Reidemeister move for those chords not participating in this move.
(5) All crossings of a classical diagram have index zero.

The last statement easily follows from the fact that the index of a chord does not change when we remove a bigon consisting formed by two consecutive classical crossings (no matter alternating or non-alternating), and this move can undo every classical diagram.

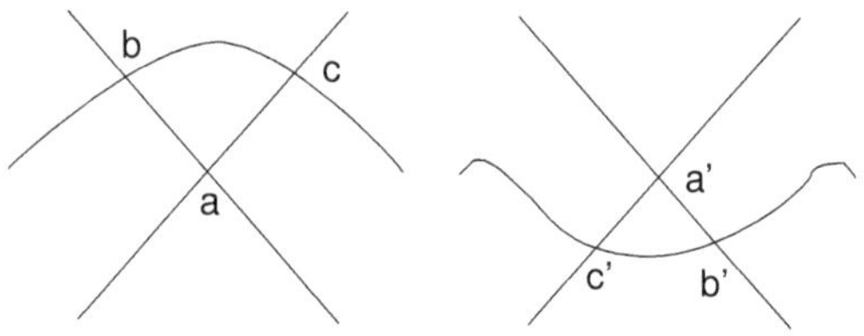

Fig. 1. Correspondence of crossings for the third Reidemeister move

The properties described above show that the index can be used for defining parities.

Indeed, from Statement 3.1 we get the following

Prop 3.1. There is a well defined parity on diagrams from $\mathfrak{V}^i$: let K be a knot diagram from $\mathfrak{V}^i$; we define those crossings of K having index divisible by 2^{i+1}, to be even, and those having index congruent to 2^i modulo four, to be odd.

From Proposition 3.1, it follows that the parity defined in this way, satisfies all the parity axioms (see page 326).

Note that the corresponding map f defined on $\mathfrak{V}^i$ takes $\mathfrak{V}^i$ to $\mathfrak{V}^0$: indeed, if we take a diagram having all chords with index divisible by, say, 16, and remove all those whose index is not divisible by 32, we may get a chord diagram with *odd* chords.

Applying the functorial map f (see page 326) to knots from $\mathfrak{V}^1$ with respect to the parity described above (Gaussian parity) we get the set $\mathfrak{V}^2$ of diagrams with all indices divisible by four.

Arguing as above, we define the sets $\mathfrak{V}^k$ of diagrams with indices divisible by $2^k, k \in \mathbb{N}$. Let $\mathfrak{V}^\infty$ be the set of diagrams with all crossings having index zero.

Theorem 3.2. *Let K, K' be two diagrams of virtual knots from $\mathfrak{V}^k$ (where k is a positive integer or ∞), representing equivalent virtual knots. Then there is a chain of diagrams $K = K_0 \to K_1 \cdots \to K_n = K'$ from $\mathfrak{V}^k$ where every two adjacent diagrams differ by a Reidemeister move (or a detour move).*

The proof of Theorem 3.2 is obtained as follows. Take an arbitrary chain connecting K to K'. Now we define a map f as in page 326 for every diagram as follows: if L belongs to $\mathfrak{V}^j$ and not to $\mathfrak{V}^{j+1}$ then $f(L)$ is obtained from L by killing all chords whose indices are not divisible by 2^{j+1}; if L has all chords of index 0, then $f(L) = L$. Analogously, define f_k to be the map which is identical on $\mathfrak{V}^k$ and behaves as f on all sets $\mathfrak{V}^j$, $j < k$.

Now, if there is a diagram of non-zero index (resp., of index not divisible by 2^k in a chain connecting K to K', one can apply f (resp., f_k) to some diagrams from the chain to get a new chain with the total number of chords smaller than that in the initial one.

Repeating the process, we stabilize at a chain of diagrams of index 0 (resp., of index divisible by 2^k).

Note that the class $\mathfrak{V}^\infty$ is quite interesting: it is an "approximation" of classical knots in the set of virtual knots. All invariants defined on the set $\mathfrak{V}^\infty$, can be translated to virtual knots by using f (in virtue of Theorem 3.2). Obviously, all classical diagrams are diagrams belong to $\mathfrak{V}^\infty$. A non-classical knot diagram D from $\mathfrak{V}^\infty$ is given in Fig. 2.

4. Parity as homology

Consider a framed 4-graph Γ with one unicursal component. The homology group $H_1(\Gamma, \mathbb{Z}_2)$ is generated by "halves" corresponding to vertices: for every vertex v we have two halves of the graph $\Gamma_{v,1}$ and $\Gamma_{v,2}$, obtained by

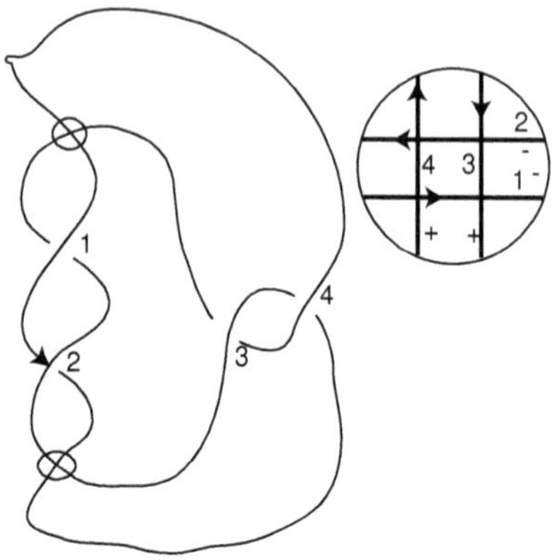

Fig. 2. A non-classical diagram from $\mathfrak{V}^\infty$

smoothing at this vertex, see Fig. 3. If the set of framed 4-graphs (possibly, with some further decorations at vertices) is endowed with a *parity*, we may assume that we are given the following cohomology class h: for each of the halves $\Gamma_{v,1}, \Gamma_{v,2}$ we set $h(\Gamma_{v,1}) = h(\Gamma_{v,2}) = p(v)$, where $p(v)$ is the parity of the vertex v. Taking into account that every two halves sum up to give the cycle generated by the whole graph, we have defined a "characteristic" cohomology class h from $H_1(\Gamma, \mathbb{Z}_2)$.

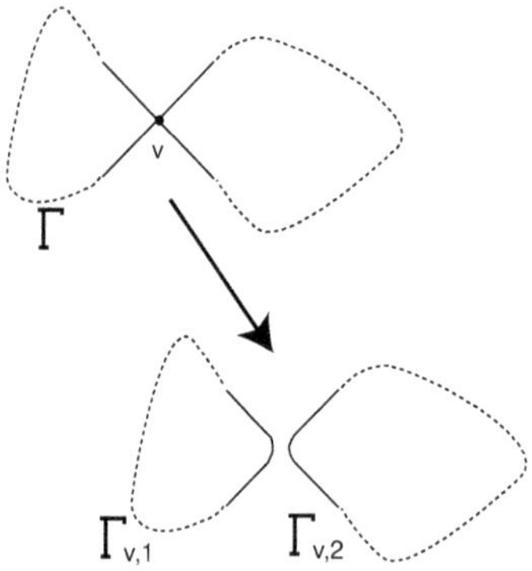

Fig. 3. The graphs $\Gamma_{v,1}$ and $\Gamma_{v,2}$

Collecting the properties of this cohomology class we see that

(1) For every framed 4-graph Γ we have $h(\Gamma) = 0$.

(2) If Γ' is obtained from Γ by a first Reidemeister move adding a loop then for every basis $\{\alpha_i\}$ of $H_1(\Gamma, \mathbb{Z}_2)$ there exists a basis of the group $H_1(\Gamma, \mathbb{Z}_2)$ consisting of one element β corresponding to the loop and a set of elements α_i' naturally corresponding to α_i.
Then we have $h(\beta) = 0$ and $h(\alpha_i) = h(\alpha_i')$.

(3) Let Γ' be obtained from Γ by a Reidemeister move II increasing the number of crossings by two. Then for every basis $\{\alpha_i\}$ of $H_1(\Gamma, \mathbb{Z}_2)$ there exists a basis in $H_1(\Gamma', \mathbb{Z}_2)$ consisting of one "bigon" γ, the elements α_i' naturally corresponding to α_i and one additional element δ, see Fig. 4, left.

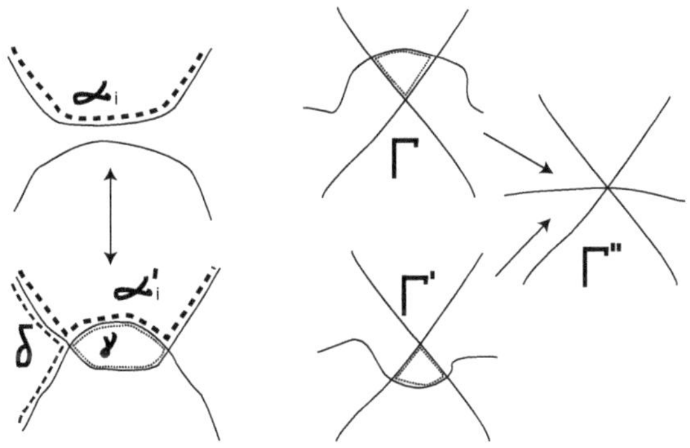

Fig. 4. The cohomology condition for Reidemeister moves

Then the following holds: $h(\alpha_i) = h(\alpha_i')$, $h(\gamma) = 0$.

(4) Let Γ' be obtained from Γ by a third Reidemeister move. Then there exists a graph Γ'' with one vertex of valency 6 and the other vertices of valency 4 which is obtained from either of Γ or Γ' by contracting the "small" triangle to the point. This generates the mappings $i : H_1(\Gamma, \mathbb{Z}_2) \to H_1(\Gamma'', \mathbb{Z}_2)$ and $i' : H_1(\Gamma', \mathbb{Z}_2) \to H_1(\Gamma'', \mathbb{Z}_2)$, see Fig. 4, right.
Then the following holds: the cocycle h is equal to zero for small triangles, besides that if for $a \in H_1(\Gamma, \mathbb{Z}_2), a' \in H_1(\Gamma', \mathbb{Z}_2)$ we have $i(a) = i'(a')$, then $h(a) = h(a')$.

Note that in (3) no condition on $h(\delta)$ is imposed.

Thus, every parity for free knots generates some $\mathbb{Z}_2$-cohomology class for all framed 4-graphs with one unicursal component, and this class behaves nicely under Reidemeister moves.

The converse is true as well. Assume we are given a certain "universal" $\mathbb{Z}_2$-cohomology class for all four-valent framed graphs satisfying the conditions 1)-4) described above. Then it originates from some *parity.* Indeed, it is sufficient to define the parity of every vertex to be the parity of the corresponding half. The choice of a particular half does not matter, since the value of the cohomology class on the whole graph is zero. One can easily check that parity axioms follow.

This point of view allows to find parities for those knots lying in $\mathbb{Z}_2$-homologically nontrivial manifolds. For more details, see.[17]

5. Invariants and minimality examples

By *smoothing* of Γ at v we mean any of the two framed 4-graphs obtained by removing v and repasting the edges as $a-b$, $c-d$ or as $a-d, b-c$, see Fig. 5.

As an example showing the power of the notion of parity we present the following theorem.

Theorem 5.1 ([17]). *Let K be a four-valent framed graph with one unicursal component, s.t. all chords of K are odd and no decreasing Reidemeister move II can be applied to K. Then K is a minimal diagram of the corresponding free link in the following strong sense: for any diagram K' equivalent to K there is a smoothing of K' isomorphic to the graph K.*

We shall call four-valent framed graphs from the formulation of Theorem 5.1 *irreducibly odd.* In fact, we shall see that irreducibly odd diagrams are mininmal in a stronger sense, see Corollary 5.1 ahead.

To prove this theorem, we shall introduce an invariant $[\cdot]$ valued in linear combinations of some graphs (more precisely, equivalence classes of graphs), which allows to reduce all the Reidemeister moves to the second Reidemeister move only.

Let $\mathfrak{G}$ be the set of all equivalence classes of framed graphs with one unicursal component modulo second Reidemeister moves. Consider the linear space $\mathbb{Z}_2\mathfrak{G}$.

Let Γ be a framed graph, let v be a vertex of Γ with four incident half-edges a, b, c, d, s.t. a is opposite to c and b is opposite to d at v.

The rest of the graph (together with all framings at vertices except v) remains unchanged. We may then consider further smoothings of Γ at *several* vertices.

Consider the following sum

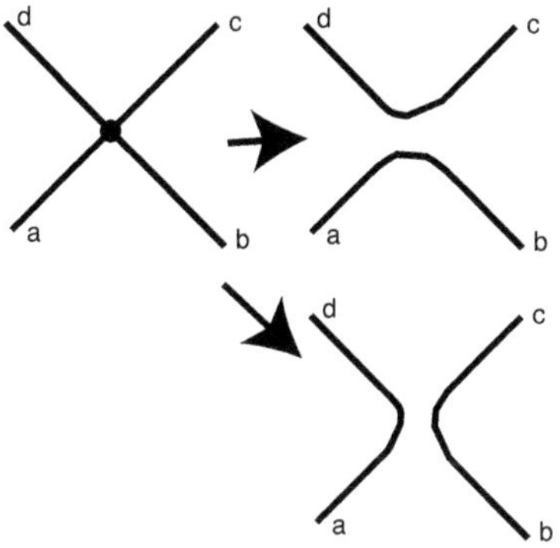

Fig. 5. Two smoothings of a vertex of for a framed graph

$$[\Gamma] = \sum_{s\ even.,1\ comp} \Gamma_s, \tag{1}$$

which is taken over all smoothings in all *even* vertices, and only those summands are taken into account where Γ_s has one unicursal component.

Thus, if Γ has k even vertices, then $[\Gamma]$ contains at most 2^k summands, and if all vertices of Γ are odd, then we shall have exactly one summand, the graph Γ itself. Consider $[\Gamma]$ as an element of $\mathbb{Z}_2\mathfrak{G}$. In this case it is evident that if all vertices of Γ are even then $[\Gamma] = [\Gamma_0]$: by construction, all summands in the definition of $[\Gamma]$ are equal to $[\Gamma_0]$, it can be easily checked that the number of such summands is odd.

Now, we are ready to formulate the main result of the present section:

Theorem 5.2 ([17]). *If Γ and Γ' represent the same free knot then in $\mathbb{Z}_2\mathfrak{G}$ the following equality holds: $[\Gamma] = [\Gamma']$.*

Theorem 5.2 yields the following

Corollary 5.1. *Let Γ be an irreducibly odd framed 4-graph with one unicursal component. Then any representative Γ' of the free knot K_Γ, generated by Γ, has a smoothing $\tilde{\Gamma}$ equivalent to Γ as a framed 4-graph. In particular, Γ is a minimal representative of the free knot K_Γ with respect to the number of vertices.*

Proof of the Corollary. By definition of $[\Gamma]$ we have $[\Gamma] = \Gamma$. Thus if Γ' generates the same free knot as Γ we have $[\Gamma'] = \Gamma$ in $\mathbb{Z}_2\mathfrak{G}$.

Consequently, the sum representing $[\Gamma']$ in $\mathfrak{G}$ contains at least one summand which is equivalent to Γ in $\mathbb{Z}_2\mathfrak{G}$. Thus Γ' has at least as many vertices as Γ does.

Moreover, the corresponding smoothing of Γ' is a diagram, which is equivalent to Γ by second Reidemeister moves. The irreducibility condition yields that one of smoothings of Γ' coincides with Γ. □

The invariant $[\cdot]$ is constructive. To see that, one should just look at the structure of the set $\mathbb{Z}_2\mathfrak{G}$.

Example 5.1. The simplest example of an irreducibly odd graph and its chord diagram (which is minimal according to Theorem 5.1 is depicted in Fig. 6.

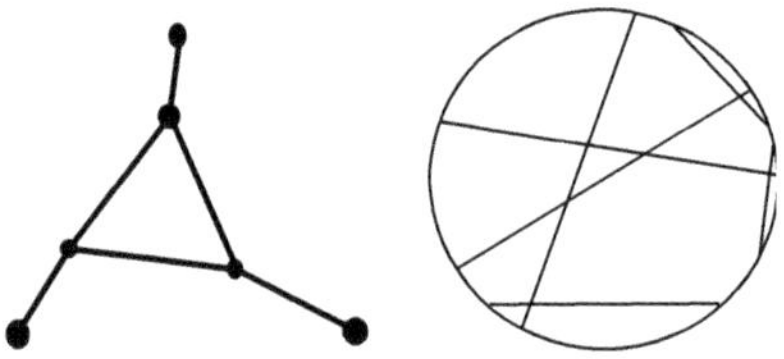

Fig. 6. An irreducibly odd graph and its chord diagram

Example 5.2. A free link whose minimality can be established analogously by using $\{\cdot\}$ is shown in Fig. 7.

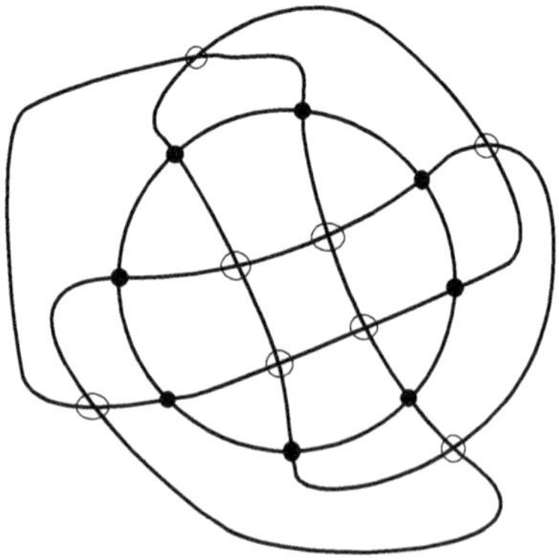

Fig. 7. Minimal free two-component link

Remark 5.1. In the case of free links with a parity, the bracket $[\cdot]$ does not work verbatim: by definition, for any free link L with all odd crossings we have $[L]=0$ since the only even smoothing of L gives more than one component. Nevertheless, there is a simple modification of $[\cdot]$, denoted by $\{\cdot\}$, which is the sum over all even smoothings, which yield links with no split components. $\{\cdot\}$ allows to prove minimality theorems for links with arbitrarily many components, for more details, see Ref. 17.

5.1. *The set* $\mathbb{Z}_2\mathfrak{G}$

Having a framed 4-graph, one can consider it as an element of $\mathbb{Z}_2\mathfrak{G}$. It is natural to try simplifying it: we call a graph in $\mathbb{Z}_2\mathfrak{G}$ *irreducible* if no decreasing second Reidemeister move can be applied to it.

Theorem 5.3. *Every 4-valent framed graph G with one unicursal component considered as an element of $\mathbb{Z}_2\mathfrak{G}$ has a unique irreducible representative, which can be obtained from G by consequtive application of second decreasing Reidemeister moves.*

This allows to recognize elements $\mathbb{Z}_2\mathfrak{G}$ easily, which makes the invariants constructed in the previous subsection digestable. In particular, the minimality of a framed 4-graph in $\mathbb{Z}_2\mathfrak{G}$ is easily detectable: one should just check all pairs of vertices and see whether any of them can be cancelled by a second Reidemeister move (or in $\mathbb{Z}_2\mathfrak{G}$ one should also look for free loops).

5.2. *The "even" Kauffman bracket*

We have shown that the parity consideration allows to construct new invariants. In fact, parity considerations allow to improve some existing invariants, as well. Below we show how to strengthen the Kauffman bracket by using parity, for details see Ref. 17. A construction of a generalized Alexander polynomial and a generalized quandle are given in D.M. Afanasiev's paper.[1]

We shall explicitly write down all formulae for virtual knot theory; note that these generalizations work for the case of graph-links.[8–10]

We recall the definition of the Kauffman bracket. For a virtual diagram K, we set

$$\langle K\rangle = \sum_s a^{\alpha(s)-\beta(s)}(-a^2-a^{-2})^{\gamma(s)-1}, \tag{2}$$

where $\alpha(s)$ and $\beta(s)$ are the numbers of smoothings of type A : 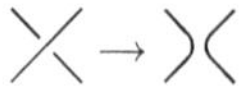
and of type B : 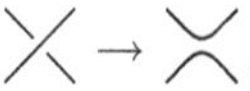, respectively, and $\gamma(s)$ denotes the number of circles (unicursal components) in the state s (after smoothing). Note that in the definition of the Kauffman bracket we *smooth all crossings*. Thus, the value of the Kauffman bracket is just a Laurent polynomial in a.

Consider the free module $\mathcal{F}$ over $\mathbb{Z}[a, a^{-1}]$ generated by all framed 4-graphs. Let $\tilde{\mathcal{F}}$ be the quotient of $\mathcal{F}$ modulo

(1) The second Reidemeister move,
(2) The relation $L \sqcup \bigcirc = (-a^2 - a^{-2})L$, where L stays for any framed 4-graph and $L \sqcup \bigcirc$ denote the split sum of L with a circle.

The algorithmic recognizability of elements from $\tilde{\mathcal{F}}$ is similar to that of $\mathbb{Z}_2\mathfrak{G}$.

Let us use the Gaussian parity for virtual knots, and let us construct the *even Kauffman bracket* taking values in the module $\mathcal{F}$, as follows.

By *states* we shall mean those states considered by smoothings at *even crossings*. Set

$$\langle K \rangle_{even} = \sum_{s_{even}} a^{\alpha(s)-\beta(s)} K_s, \tag{3}$$

where K_s is the free link obtained from the diagram K by smoothing with respect to the state s. Here K_s is considered as an element from $\tilde{\mathcal{F}}$.

It is easy to see that the invariant $[\cdot]$ (as well as $\{\cdot\}$) is a specification of the bracket defined above. Then the following theorem takes place

Theorem 5.4.[17] *The bracket $\langle \cdot \rangle even$ is invariant under the second and the third classical Reidemeister moves (and the detour move). When applying the first Reidemeister move, the bracket $\langle \cdot \rangle_{even}$ gets multiplied by $(-a)^{\pm 3}$. The following normalization for $\langle \cdot \rangle_{even}$ is invariant under all Reidemeister moves: $X_{even}(K) = (-a)^{-3w(K)} \langle K \rangle even$, where $w(K)$ stays for the writhe number of the oriented diagram K*

We call $X_{even}(K)$ the *even Jones polynomial* of the virtual knot K.

Remark 5.2. The even Jones polynomial is a generalization of the invariant $\{\cdot\}$ for any knot theory possessing a parity and a distinguished rule for A- and B-type smoothing respecting the Reidemeister moves. Moreover, for the Gaussian parity, the polynomial $X_{\mathrm{even}}(K)$ is a generalization of the usual Jones polynomial for classical knots and for knots with all crossings being even. In this case, in the definition of $\langle \cdot \rangle_{\mathrm{even}}$, all elements K_s

are framed 4-graphs with 2 unicursal components, which in the module $\tilde{\mathcal{F}}$ are multiples of the unknot with coefficients being powers of the polynomial $(-a^2 - a^{-2})$. Considering the generator of the module $\tilde{\mathcal{F}}$, corresponding to the unknot, as 1, we get the standard Jones polynomial.

5.3. *Atoms and parity*

Theorem 5.1 and Corollary 5.1 deal with only those free knots having odd chords. However, the class of knots having with all even chords is very important. These are knots corresponding to *orientable atoms.* We shall describe the notion of atom and its connection to the Gaussian parity, and also construct an example of a trivial link with all even chords.

Definition 5.1. An *atom* (originally introduced by Fomenko,[4]) is a pair (M, Γ) consisting of a 2-manifold M and a graph Γ embedded in M together with a colouring of $M \backslash \Gamma$ in a checkerboard manner. An atom is called *orientable* if the surface M is orientable; Γ is called the *frame* of the atom, by *genus* (*Euler characteristic, orientation*) of the atom we mean that of the surface M (atoms and their genera were also studied by Turaev,[26] and atom genus is also called the Turaev genus[26]).

Having an atom V, one can construct a virtual link diagram from it as follows. Take a generic immersion of atom's frame into $\mathbb{R}^2$, for which the formally opposite structure of edges coincides with the opposite structure induced from the plane. Put virtual crossings at the intersection points of images of different edges and restore classical crossings at images of vertices in such a way that the two angles corresponding to the A-smoothing of the Kauffman bracket correspond to the white angles of the atom. Since we disreagard virtual crossings, at most we can expect that these diagrams are well defined up to detours. However, this allows us to get different virtual link types from the same atom, since for every vertex V of the atom with four emanating half-edges a, b, c, d (ordered cyclically on the atom) we may get two different clockwise-orderings on the plane of embedding, (a, b, c, d) and (a, d, c, b). This corresponds to a move called *virtualisation.*

Definition 5.2. By a *virtualisation* of a classical crossing of a virtual diagram we mean a local transformation shown in Fig. 8.

The above statements summarise as

Proposition 5.1. *(see, e.g.,[16]). Let L_1 and L_2 be two virtual links obtained from the same atom by using different immersions of its frame. Then L_1*

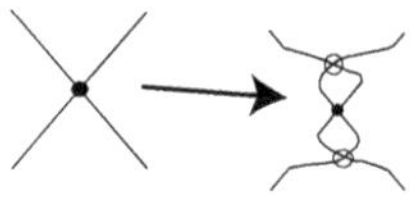

Fig. 8. Virtualisation

differs from L_2 by a sequence of detours and virtualisations.

At the level of Gauss diagrams, virtualisation does not change the sign crossings, but inverts the arrow directions. So, atoms retain the information about signs of Gauss diagrams, but not of their arrows.

A further simplification comes when we want to forget about the signs and pass to flat virtual links (see also Ref. 29): in this case we don't want to know which branch forms an overpass at a classical crossing, and which one forms an underpass. We need only remember is its frame with opposite edge structure of vertices (the *A-structure*). Having that, we take any atom with this frame and restore a virtual knot up to virtualisation and crossing change.

The A-structure of an atom's frame is exactly a 4-valent framed graph.

This perfectly agrees with the fact that **free links are virtual links modulo virtualization and crossing changes.** Having a framed 4-graph, one can consider *all atoms* which can be obtained from it by attaching black and white cells to it. In fact, it turns out that for a given framed 4-graph either all such surfaces are orientable or they are all non-orientable.

To see this, one should introduce the *source-sink* orientation. By a *source-sink* orientation of a 4-valent framed graph we mean an orientation of all edges of this graph in such a way that for each vertex some two opposite edges are outgoing, whence the remaining two edges are incoming.

The following statement is left to the reader as an exercise

Example 5.3. Let G be a framed 4-graph. Then the following conditions are equivalent:

1. G admits a source-sink orientation

2. At least one atom obtained from G by attaching black and white cells is orientable.

3. All atoms obtained from G by attaching black and white cells are orientable.

Moreover, if G has one unicursal component, then each of the above conditions is equivalent to the following: every chord of the corresponding

Gauss diagram $C(D)$ is even.

Theorem 3.1 leads to a new proof of a partial case the following

Theorem 5.5. *(First proved by O.Ya. Viro and V.O.Manturov, 2005, first published in Ref. 8) The set of virtual links with orientable atoms is closed. In other words, if two virtual diagrams K and K' have orientable atoms and they are equivalent, then there is a sequence of diagrams $K = K_0 \to K_1 \cdots \to K_n = K'$ all having orientable atoms where K_i is obtained from K_i by a Reidemeister move.*

Indeed, for the case of knots, this is a direct corollary from Theorem 3.1. Indeed, assume atoms corresponding to K and K' are orientable. Then all classical crossings of K and all crossings of K' have even Gaussian parity. Now take any sequence of diagrams and Reidemeister moves $K = K_0 \to K_1 \cdots \to K_n = K'$, and consequently apply the map killing even crossings to this sequence until all intermediate diagrams become even. As a result we get a chain of diagrams connected by Reidemeister moves and detour moves taking K to K' where all intermediate diagrams have orientable atoms.

In the general case (of links), this theorem can be proved by a modification of parity, called *relative parity*, investigated by D.Yu.Krylov and the author.[15]

We give two examples: for a planar 4-valent framed graph we present a source-sink orientation (left picture, Fig. 9), and for a non-orientable 4-valent framed graph (right picture, Fig. 9, the artefact of immersion is depicted by a virtual crossing) we see that the source-sink orientation taken from the left crossing leads to a contradiction for the right crossing.

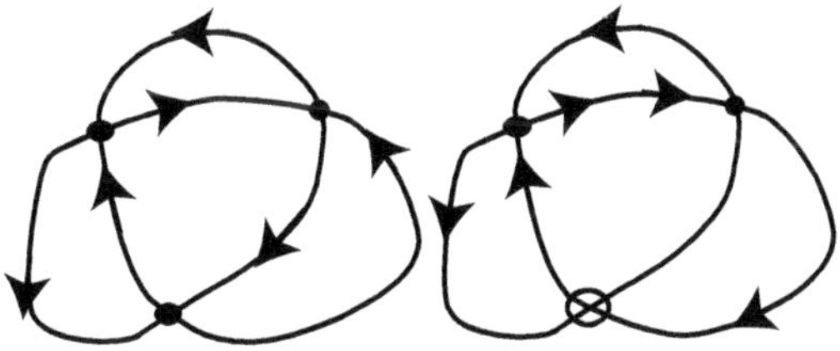

Fig. 9. The source-sink condition: satisfied (left); violated (right)

5.4. *Turaev's cobracket and its even analogue*

In many situations, it is easier to find *links* rather than *knots* with desired non-triviality properties. So, we shall first define a map from free 1-component links to $\mathbb{Z}_2$-linear combinations of 2-component links, and then we shall study the latter by an invariant similar to that constructed in the paper.[17]

Ideologically, the first map is a simplified version of Turaev's cobracket[24] which establishes a structure of Lie coalgera on the set of curves immersed in 2-surfaces (up to some equivalence, the Lie *algebra* structure was introduced by Goldman in a similar way). We shall use a simplification of Turaev's construction adopted to the case of free knots and links.

The second map takes a certain state sum for a 2-component free link, where we distinguish between two types of crossings, and smooth only crossings of the first type. What should these "two types" mean, will be discussed later.

In some sense, the invariant $[\cdot]$ of free knots constructed in Ref. 17 is a diagrammatic extension of a terrifically simplified Alexander polynomial (we forget about the variable and signs taking $\mathbb{Z}_2$-coefficients). The invariant $\{\cdot\}$ suggested in the present paper is in the same sense an extension of the terrifically simplified Kauffman bracket, but again we use diagrams as coefficients.

Altogether, these two constructions (the bracket and Turaev's Δ) provide an example of non-trivial and minimal diagrams of free knots with orientable atoms.

There is a simple and fertile idea due to Goldman[6] and Turaev[24] of transforming two-component curves into one-component curves and vice versa.

Here we simplify Turaev's idea for our purposes and we shall call the map we are going to construct "Turaev's Δ". Let $\mathfrak{G}_l$ is be the set of all equivalence classes of framed graphs with l unicursal components by the second Reidemeister move and the relation that takes every framed 4-graph with a split component to 0. We shall construct a map from $\mathbb{Z}_2\mathfrak{G}$ to $\mathbb{Z}_2\mathfrak{G}_2$ as follows.

Given a framed 4-graph G. We shall construct an element $\Delta(G)$ from $\mathbb{Z}_2\mathfrak{G}_2$ as follows. For each crossing c of G, there are two ways of smoothing it. One way gives a knot, and the other smoothing gives a 2-component link G_c. We take the one giving a 2-component link and write

$$\Delta(G) = \sum_c G_c \in \mathbb{Z}_2\mathfrak{G}_2 \tag{4}$$

Theorem 5.6. *$\Delta(G)$ is a well defined mapping from $\mathbb{Z}_2\mathfrak{G}$ to $\tilde{\mathbb{Z}}_2\mathfrak{G}_2$.*

The proof is standard, see Ref. 17 one should consider the three Reidemeister move. The first move adds a new summand which has a free loop (the latter assumed to be trivial in $\tilde{\mathbb{Z}}_2\mathfrak{G}_2$); for the second Reidemeister move we get two new identical summands, which cancel each other because we are dealing with $\mathbb{Z}_2$ coefficients. For the third Reidemeister moves the LHS and the RHS will lead to the summands identical up to second Reidemeister moves.

Proposition 5.2. *The free knot K_1 shown in Fig. 10 is minimal.*

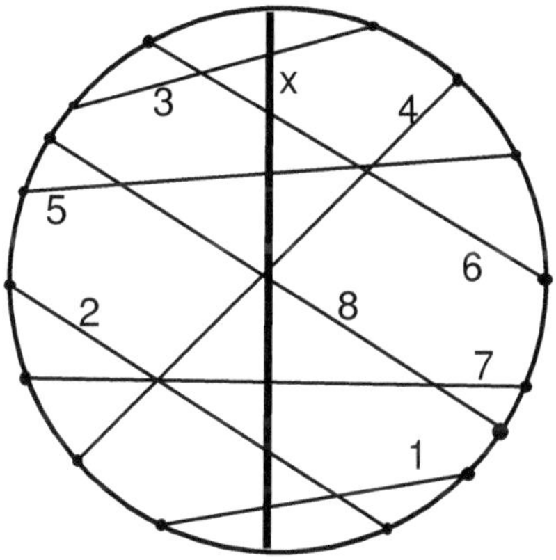

Fig. 10. A Non-trivial free knot with all even crossings

To see the minimality, one takes $\Delta(K_1)$ which, as an element of $\mathbb{Z}_2\mathfrak{G}_2$, coincides with the link shown in Fig. 7; there is exactly one summand obtained by smoothing at x; all other summands cancel by symmetry over $\mathbb{Z}_2$. Now, considering $\{\Delta(K_1)\}$, we get the desired claim.

Analogously one defines the maps Δ_{even} and Δ_{odd} corresponding to smoothing at even (resp., odd) crossings.[17]

6. Parity and groups

It turns out that the parity argument can be used to construct some extremely simple invariants of free knots; moreover, these invariants turn out to be obstructions for a free knot to be slice.

The idea goes as follows. Consider the group $G = \langle a, b, b' | a^2 = b^2 = b'^2 = 1, ab = b'a \rangle$. For a Gauss diagram D, we shall construct a word $\gamma(D)$ in the alphabet a, b, b' such that the conjugacy class of this word in G is an invariant of the corresponding free knot. Let D be an oriented chord diagram on a circle C, and let X be a fixed point on C distinct from chord ends.

We say that an odd chord c of D is *of the first type* if it is linked with an odd number of *odd chords*, and *of the second type* if it is linked with an odd number of *even* chords.

With the pointed diagram D we associate a word in the alphabet a, b, b' in the following way. We start walking along the orientation of the circle C from the point X. Every time when we meet a chord end, we write a letter a, b, or b' depending on whether the corresponding chord is *even*, *first type odd* or *second type odd.*

Denote the obtained word by $\gamma(D)$; this word generates an element $[\gamma(D)]$ in the group G.

Theorem 6.1. *If pointed chord diagrams* (D, X) *and* (D', X') *generate equivalent (oriented) free knots then* $[\gamma(D)] = [\gamma(D')]$ *in* G.

The proof of this theorem is a simple check of the invariance under Reidemeister moves. Theorem 6.1 has an obvious

Corollary 6.1. *The conjugacy class of the element* $[\gamma(D, X)]$ *in* G *is an invariant of free knots.*

Indeed, moving the fixed point through a chord end, we change the word cyclically, which means conjugation in G.

The group G admits a simple combinatorial description. Its Cayley graph looks like a vertical strip on a squared paper between $x = 0$ and $x = 1$: we choose the point $(0, 0)$ to be the unit in the group; the multiplication by a on the right is corresponds to one step in the horizontal direction (to the right if the first coordinate of the point is equal to zero, and to the left if this first coordinate is equal to one), the multiplication by b is one step upwards if the sum of coordinates is even and one step downwards if this sum is odd, and the multiplication by b' is one step downwards if the sum of coordinates is even and one step upwards if the sum of coordinates is even, see Fig. 11.

With each pointed chord diagram one associates an element from G having coordinates $(0, 4l)$. Moreover, the conjugacy class of the element

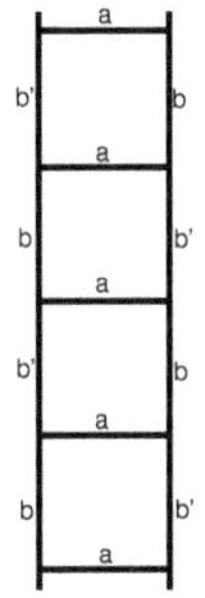

Fig. 11. The Cayley graph of the group G

$(0, 4l)$ for $l \neq 0$ consists of the two elements: $(0, 4l)$ and $(0, -4l)$. Thus, for each *long free knot* one gets an integer-valued invariant, equal to l; we shall denote this invariant by $l(K)$; each compact free knot has, in turn, the invariant equal to $|l|$; we shall denote the latter by $L(K)$.

In Fig. 12 we depict a free knot K_1 with $l(K_1) = 4$.

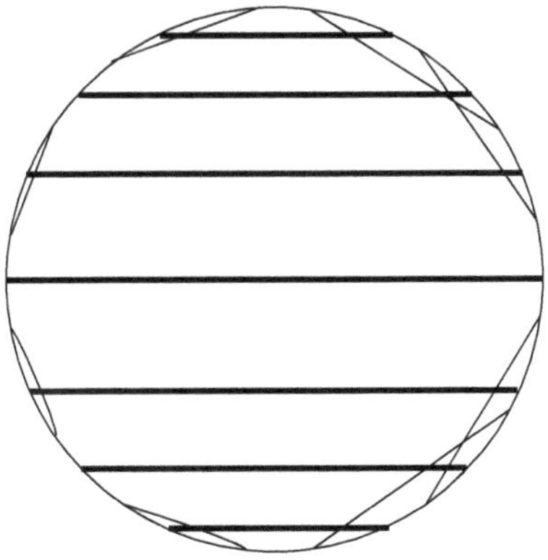

Fig. 12. A non-slice free knot

So, it is possible to prove non-triviality of some free knots by using parity arguments in a much simpler ways than in the initial sections of the present paper.

In some consequent paper we shall prove that $L(K)$ is a *cobordism invariant of free knots*, i.e., if $L(K)$ is non-zero then the free knot K is not null-cobordant. The theorem about sliceness will be proved elsewhere. This invariant was strengthened in Ref. 23.

References

1. D.M.Afanasiev, *On strengthening tnvariants of virtual knots by using parity*, *Sbornik Mathematics*, to appear.
2. J.S.Carter (1991), *Closed Curves that never extend to proper maps of disks*, *Proc. Amer. Math. Soc.*, **113**, No. 3, pp. 879-888.
3. M.Chrisman, V.O.Manturov (2010), *Combinatorial Formulae for Finite-Type Invariants via Parities*, Arxiv:Math.GT/1002.0539, v.2.
4. Fomenko A. T. (1991), The theory of multidimensional integrable hamiltonian systems (with arbitrary many degrees of freedom). Molecular table of all integrable systems with two degrees of freedom, *Adv. Sov. Math*, *6*, pp. 1-35.
5. A.Gibson (2009) *Homotopy Invariants of Gauss words*, Arxiv:Math.GT./0902.0062
6. Goldman, W. M., Invariant functions on Lie groups and Hamiltonian flows of surface group representations. *Invent. Math.* 85 , no. 2,(1986) 263-302
7. Goussarov M., Polyak M., and Viro O, Finite type invariants of classical and virtual knots// Topology. 2000. V. 39. P. 1045–1068.
8. D.P.Ilyutko, V.O.Manturov, *Graph-links*, ibid.
9. D. P. Ilyutko, V. O. Manturov (2009), Introduction to graph-link theory, *Journal of Knot Theory and Its Ramifications*, **18** (6), pp. 791-823.
10. D.P.Ilyutko, V.O.Manturov (2009), *Doklady Mathematics*, v. 428, **5**, pp. 591-594.
11. Ilyutko, D.P., Manturov, V.O.,Nikonov I.M, Virtual Knot Invariants arising from parities, arxiv.math.GT/1102.5081
12. L. H. Kauffman, Virtual knot theory, Eur. J. Combinatorics. 1999. V. 20, N. 7, pp. 662–690.
13. L. H. Kauffman, State Models and the Jones Polynomial (1987), *Topology*, **26**, pp. 395–407.
14. D.Yu. Krylov, V.O.Manturov, Relative parity, *Work in Progress*
15. V. O. Manturov (2004), *Knot Theory*, Champan and Hall/CRC, Boca Raton, 416 pp.
16. V.O.Manturov, *Parity in Knot Theory*, *Sbornik Mathematics*, to appear.
17. V.O.Manturov, *On Free Knots*, ArXiv:Math.GT/0901.2214 v2
18. V.O.Manturov, *On Free Knots and Links*, ArXiv: Math.GT/0902.0127
19. V.O.Manturov, *Free Knots are Not Invertible*, ArXiv:Math.GT/0909.2230
20. V.O.Manturov (2004), *Long Virtual Knots and Their Invariants*, *J. Knot Theory Ramif.*,**13** (8), pp. 1029-1039
21. V.O.Manturov (2005), *O Dlinnykh virtual'nykh uzlakh*, On long virtual knots, *Doklady Mathematics*, **71**, No. 2, pp. 253-255.
22. A Functorial Map from Knots in Thickened Surfaces to Classical Knots and Generalisations of Parity, arxiv.math.GT/ 1011.4640
23. V.O.Manturov, O.V.Manturov, *Free Knots and Groups*, Arxiv:Math.GT/09122694, v.2.
24. V.G.Turaev, Skein quantization of Poisson algebras of loops on surfaces. *Ann. Sci. Ecole Norm. Sup.* (4) 24, no. 6, (1991) 635704.

25. V.G.Turaev,Topology of words, *Proc. Lond. Math. Soc.* (3) 95 (2007), no. 2, 360412.
26. V.G.Turaev, A simple proof of the Murasugi and Kauffman theorems on alternating links, *L'Enseignement Mathématique.* **33** (1987), pp. 203- 225.
27. V.G.Turaev, *Cobordisms of Words*, Arxiv:Math.CO/0511513, v.2.
28. V.G.Turaev, *Cobordisms of Knots on Surfaces*, Arxiv:Math.GT/0703055, v.1.
29. V.G. Turaev, *Virtual open strings and their cobordisms*, preprint, 2004, arXiv: math.GT/ 0311185 v5

PHYSICAL KNOT THEORY: AN INTRODUCTION TO THE STUDY OF THE INFLUENCE OF KNOTTING ON THE SPATIAL CHARACTERISTICS OF POLYMERS

Kenneth C. Millett*

Department of Mathematics, University of California
Santa Barbara, CA 93106, USA
**E-mail: millett@math.ucsb.edu*

This paper contains selected topics from four lectures given at the Abdus Salam International Centre for Theoretical Physics in May 2009. We introduce the study of the influence of knotting and linking on the spatial characteristics of linear and ring polymer chains with examples of scientific interest. We describe a few basic concepts of the geometry and topology of knots and measures of the spatial shape of open and closed polymer chains. We then present some fundamental mathematical results concerning them. Next we discuss random sampling methods of collections of open and closed chains that are employed to provide estimates of the spatial properties of the chains. Finally, we discuss implementations of the sampling algorithms, survey consequences of theoretical and experimental results, and discuss some interesting problems deserving further research.

Keywords: Knot, polymer, macromolecule, shape, spatial characteristics, open and closed chains.

1. Introduction

Humans have long exploited the special properties of polymer chains to manufacture materials used in their everyday life. Polymers are very large molecules existing in a wide variety of topological types, both singly or in collections. While they exhibit substantial structural regularity, e.g. DNA Figure 1, their large scale topology can be linear or circular in nature (the cases of principal interest here) or resemble quite complex graphs, for example star or wheel graphs. They can be natural, e.g. silk, cotton, shellac, rubber, amber, and the shells of crabs and lobsters or synthetic, e.g. bakelite, nylon, neoprene, polystyrene, and polypropylene to name just a few. They are found everywhere, including in the DNA, proteins, and other molecular structures that make up our bodies.

A common mathematical model for polymers is a polygonal chain, i.e. a collection of vertices and connecting edges. The polygons can be open, i.e. topologically equivalent to a closed interval, or closed, i.e. topologically equivalent to a circle. They are studied individually or in large collections. From the topological perspective, a polygonal ring in 3-space defines a knot while a collection determines a link. The key feature that distinguishes physical knot theory from the classical knot theory is that there are constraints on the chains arising from the physical properties of the polymers being studied. For example, the heterogeneity of the polymer may lead one to study equilateral polygons. The material properties may impose constraints on the bending or twisting of the polygon. The thickness of the material may impose a requirement that the collection of small tubular neighborhoods of the polygons consist of mutually disjoint tori. The imposition of these constraints gives birth to a whole new class of mathematical problems that are not simply of a topological nature but include geometric features in fundamental ways.

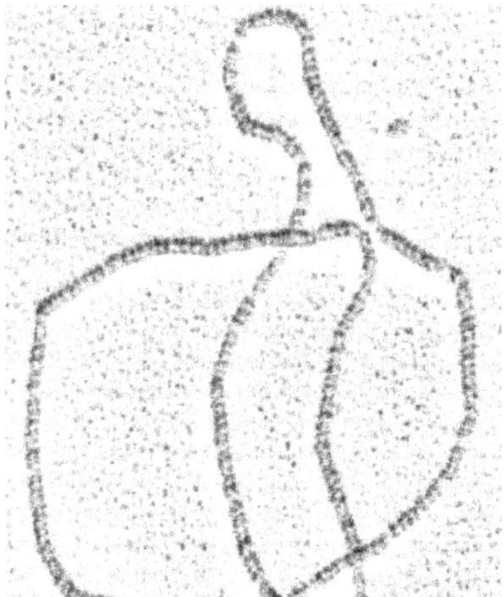

Fig. 1. Electron micrograph of a RecA protein coated DNA trefoil knot generated by E. coli DNA topoisomerase acting on nicked circular DNA. Micrograph courtesy of Andrzej Stasiak, University of Lausanne

What is the relationship between knotting and linking, the average topological structure or geometric structure of configurations, and the biological or physical properties or the functionality of the polymers they model? What sorts of measures does one employ to characterize the shape of polymers? We will discuss a few of these questions, beginning with those more frequently studied: the identification of the knot type, the radius of gyration, and the average crossing number. The identification of knotting in ring polymers and the extension to knotting in open chains due to applications to protein structures is central. In addition to the specific type of knotting, we will be interested in the average scale of the knotting and the average shape of the knots. Furthermore, we will look at slipknots, i.e. knotted segments of a polymer chain which are contained in a larger unknotted chain, and their properties.

While there are some key results for which we have rigorous proofs, much of the information comes from the random sampling of the space of all possible configurations. As a consequence, we will describe two standard strategies, the polygonal fold method and the crankshaft rotation method, and the rigorous proofs that they properly sample the set of configurations. Implementations of these methods have provided a rich set of data whose analysis provides estimates of the shape parameters discussed earlier. A discussion of the consequences of this information and their implications for future research will complete this introduction to physical knot theory.

2. Models of biological and physical polymers

In this section, we will focus attention on structures that are locally linear such as found in models for DNA and proteins, see Figure 1, where we replace the sugar-phosphate backbone or related structures by a linear chain of vertices and edges. We will require that there is no "branching" or "crosslinking" in the structure.

2.1. *Intrinsic structure*

The molecular chains can be modeled as unions of linear segments (polygonal arcs) whose lengths, vertex angles, and torsion statistically reflect the structural properties of the polymer, e.g. Kuhn length. We will assume that the polymers are homogeneous so that, in our polygonal models, the edge lengths will be equal to unity. The individual chains can be either open, i.e. topologically equivalent to closed intervals, or closed, i.e. topologically equivalent to circles. A chain, K, is described by its vertex set $\{v_1, v_2, \ldots, v_n\}$ with edges connecting a vertex with it successor and, if the chain is closed, the last vertex with the first. Thus, if $\{v_1, v_2, v_3, \ldots, v_n\}$ are vertices, the edge vectors of a closed chain $\{e_1 = v_2 - v_1, e_2 = v_3 - v_2, \ldots, e_n = v_1 - v_n\}$. Note that an open chain of

n edges has one more vertex than a closed chain of n edges. Examples of 50 edge open and closed chains with initial vertex at $\{0, 0, 0\}$ are shown in Figure 2.

Fig. 2. Linear and circular 50 edge chains

2.2. *Extrinsic structure of closed chains*

The conformations of closed polygonal chains in 3-space can be different in many distinct ways. Indeed, this is also the case for open chains if one focuses on individual "frozen" conformations rather than allowing deformations without self intersections. By allowing for changes in lengths and bending of edges, one arrives at the topological notion of knotting and linking.

A closed (circular) chain is knotted if it can not be deformed, in a structure preserving manner, to the standard planar equilateral polygon of that number of edges, Figure 3. Two closed chains represent the same *knot type* if one chain can be deformed to the other in a structure preserving manner. If it is the topological structure that is to be respected, one studies the notion of topological knotting. If, however, one must preserve the length of the edges, a geometric stucture has been imposed and the result is the distinctly different concept of geometric knotting. A particularly elusive example was the 8_{19} eight edge equilateral polygonal knot, Figure 8, whose existence I proved via numerical methods. Note that, in these figures, care is paid to insure that the 2-dimensional images of the knot accurately capture the structure of the 3-dimensional polygonal chain by insuring that image is unchanged under small changes of perspective and that the under/over crossing structure is clearly discernible, see Figure 3. Such images are called knot presentations.

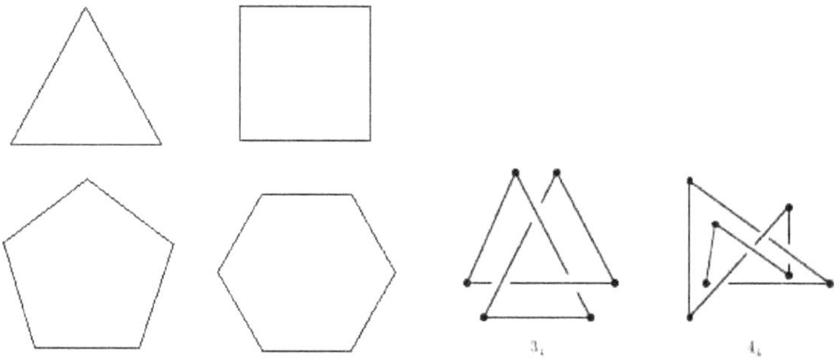

Fig. 3. Standard regular polygons, right, and the six edge equilateral trefoil and the seven edge equilateral figure-eight knots

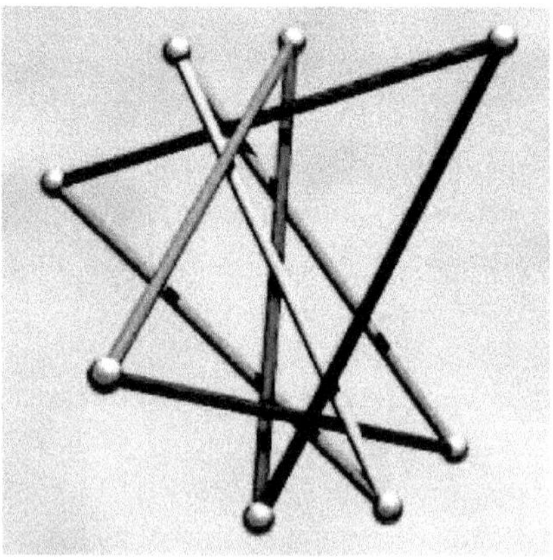

Fig. 4. An octagonal equilateral 8_{19}, Image provided by Peterson Tretheway

2.3. *Extrinsic structure of open chains*

In our study of the spatial structure of open chains, we emphasize that this concerns the extrinsic properties of the "*frozen*" configuration. Knotting or linking of open chains is *not* invariant under topological deformations. For chains in which the edges are of unequal length, there are examples of knotting and linking,[1–6] under deformations that respect the linear structure and perserve lengths of edges. Whether or not this is true for open equilateral chains is unknown.

Problem 2.1. *Are there knotted open equilateral chains?*

Thus, our open equilateral chains will be frozen in position unless explicitly stated otherwise. To detect knotting in such open chains, I prefer to consider the statistics of closures of the chain to the "sphere-at-infinity".[7–9] There are many other strategies that have been employed.[10–19] These include some which "shorten" the open chain by means of an algorithm that often includes choices that lead to some level of indeterminacy as to the final result. Such uncertainty leads to concerns that must be addressed in applications of these methods. One such example is shown in Figure 5. Here, if one applies a popular shortening method,[14] the outcome depends upon the end at which one starts. Another strategy is to simply connect the end vertices to a fixed point, directly to each other, or to "infinity". Another looks for solid balls capturing the "knotted" portion in its interior. These methods can lead to an indentification of knotting that is not consistent with what might expect considering the large scale structure of the conformation or are of such limited application that they are difficult to use in these contexts. While it is possible that the extent of indeterminacy that arises in some of these methods is insufficiently large as to undermine the analysis, this is difficult to assess. As a consequence, I prefer the statistically robust method connecting the ends of the chain to points on the "sphere-at-infinity." For all practical purposes, one may consider this to be a sphere of very large radius, compared to the diameter of the conformation, centered at the centroid of the chain, see Figure 5. By applying this method to subchains of closed chains, one can identify the site and scale of knotting in a large chain.

Thus, to identify knotting in open chains, one collects a random sample of closures of appropriate size, computes a knot invariant such as the HOMFLYPT polynomial,[20] Figure 6, to identify the topological knot type of each closure, and analyzes the resulting collection of knot types (actually their polynomial surrogates). These statistics give the *spectrum of knotting*

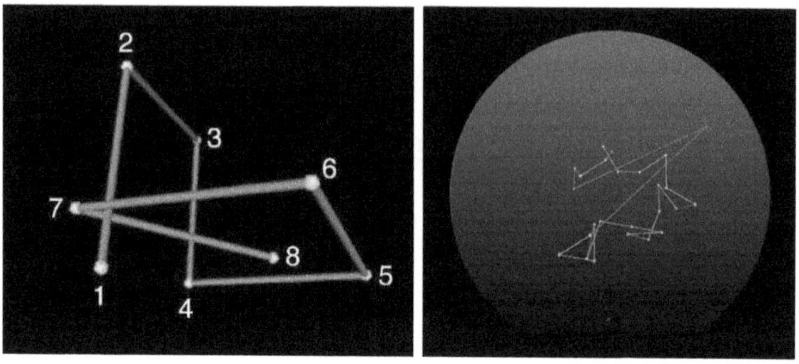

Fig. 5. Curve shortening example and a "sphere-at-infinity" closure example[7]

of the chain. The selection an appropriate threshold to establish the knot type of the open chain, for example 0.6, results in the identification of a trefoil knot for the spectrum shown in Figure 7.

Fig. 6. In Berkeley, summer 1985, Vaughan Jones with James **H**oste, Adrian **O**ceanu, Kenneth **M**illett, Peter **F**reyd, W. B. Raymond **L**ickorish, and David **Y**etter (HOMFLY) Absent were J. **P**rzytycki and P. **T**raczyk

3. Elements of physical knot theory

In applications of polygonal models, both open and closed chains, to the study of physical or biological polymers, many aspects of the topological

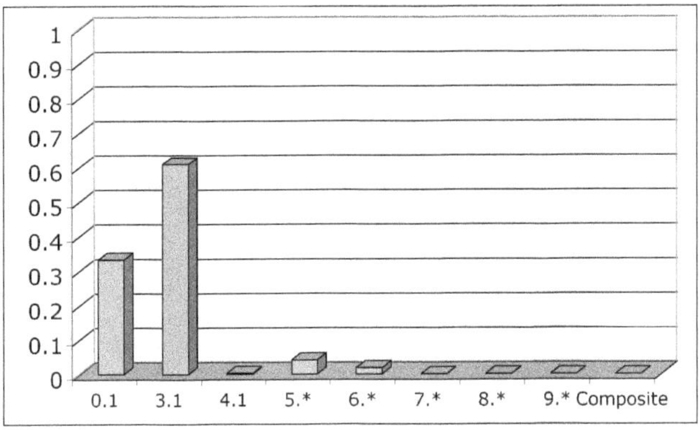

Fig. 7. A sample spectrum of the knotting of an open chain.[8] The trivial knot and trefoil account for more than 90% of the sample with the remainder being, principally, five and six crossing knots.

and geometrical structures have been employed. Some of these arose in the topological study of knots while others have developed in response to the goal of modelling physical and biological properties.

3.1. *Average crossing number*

The $ACN(K)$ is understood as the average of the crossing number of a configuration over all orthogonal planar projections. These planes are parametrized by their normal unit vectors, i.e. points on the unit sphere.

$$ACN(K) = \frac{1}{4\pi}\int_{S^2} cr(v)dA \tag{1}$$

If the curve is parameterized by $\gamma(t)$, the ACN is given by the following integral:

$$ACN(K) = \frac{1}{2\pi}\int_{S^1}\int_{S^1} \frac{|(\gamma'(t),\gamma'(s),\gamma(t)-\gamma(s))|\ dt\ ds}{\|\gamma(t)-\gamma(s)\|^3} \tag{2}$$

in which $|(\gamma'(t),\gamma'(s),\gamma(t)-\gamma(s))|$ denotes the absolute value of the triple vector product and $\|\ \|$ denotes the vector norm in $\mathbf{R}^3$.

3.2. *Curvature and torsion*

If $\gamma(s)$ is an arc length parameterization of a curve,[21,22] $T(s) = \frac{d\gamma(s)}{ds}$. The curvature is defined by $\kappa(s) = \|\frac{dT(s)}{ds}\|$. The nomal vector is $N(s) = \frac{\frac{dT(s)}{ds}}{\|\frac{dT(s)}{ds}\|}$.

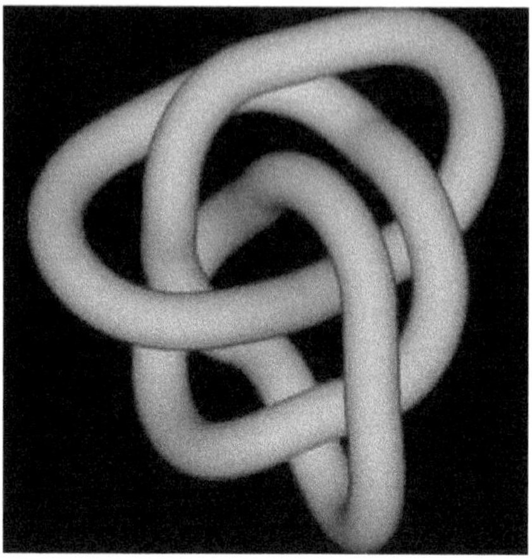

Fig. 8. An 8_{19} knot with ACN 11.2271

The binormal vector is $B(s) = T(s) \times N(s)$. The torsion is defined by $\tau(s) = -\|\frac{dB(s)}{ds}\|$.

3.3. *Total curvature and total torsion*

For spatial curves we define the *total curvature* and the *total torsion* to be the integrals of the curvature and torsion as follows:

$$TCU(K) = \int_{S^1} \kappa(s)ds \tag{3}$$

$$TTO(K) = \int_{S^1} \tau(s)ds \tag{4}$$

A closed curve must have total curvature at least 2π. The total curvature and total torsion of the knot in Figure 8 are 26.4085 and 118.602, respectively.

Theorem 3.1 (Fary-Milnor[23]). *If K is a smooth non-trivial knot, its total curvature is greater than 4π.*

As our models are based on polygonal chains, we translate these concepts to the polygonal setting. The $\{v_1, v_2, \dots, v_{n+1}\}$ are the $n+1$ vertices of an n edge polygonal chain. These give the edge vectors $\{e_1 = v_2 - v_1, e_2 = v_2 - v_2, , e_n = v_{n+1} - v_n\}$ for an open chain. If the chain is closed, we set v_1

equal to v_{n+1}. The angle, in radians, between e_i and e_{i+1} defines the local curvature at v_{i+1}. The total curvature is defined to be the sum of these angles (measured in radians). Adjacent vectors, e_i and e_{i+1}, determine a plane. The angle between adjacent planes defines the torsion at each edge. The total torsion is the sum of these angles.

3.4. *Radius of gyration*

The *center of mass* and *squared radius of gyration* of a chain, K are defined by:

$$CM(K) = \frac{\Sigma v_i}{n} \tag{5}$$

$$RGN(K)^2 = \frac{\Sigma \|v_i - CM(K)\|^2}{n} \tag{6}$$

The RGN is one of the most frequently employed measures of the spatial extent of a configuration. The radius of gyration of the 8_{19} knot in Figure 8 is 3.165.

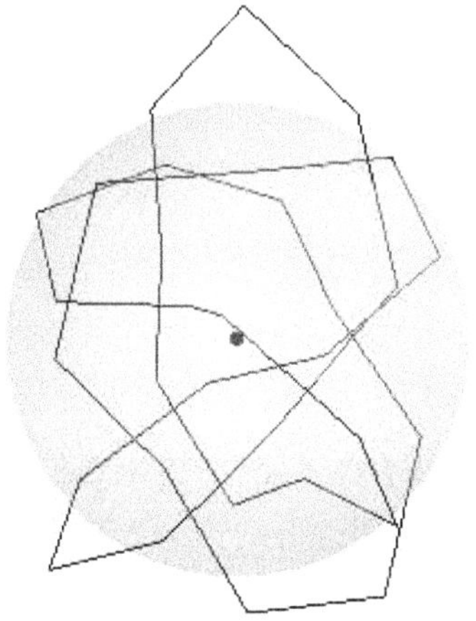

Fig. 9. The center of mass and the radius of gyration sphere associated to a polygonal trefoil

3.5. *Injectivity radius of a knot in 3-space*

The injectivity radius of a smooth knot is 3-space is the upperbound of the radii of embedded normal tubes centered on the knot.[24] This radius is determined by limits imposed by the curvature and by contact points of distance portions of the normal tube 11. Gonzalez and Maddocks[25] give a very general definition as follows: For any triple of distinct points in 3-space, let $r(x, y, z)$ be the radius of the unique circle they determine. If $T(x)$ is a line tangent to the curve at x, let $r(T(x), y)$ be the radius of the unique circle tangent to $T(x)$ at x and containing y. $\rho_x(L)$ is the infimum of $r(x, y, z)$ over distinct points x, y, z with x fixed. For a smooth link, the function $\rho_x(L)$ is continuous and $0 < \rho_x(L) < \frac{1}{\kappa(s(x))} = \rho(s(x))$, where $s(x)$ is the arc length parameter of the point x and $\rho(s(x))$ is defined to be the *radius of curvature of* L *at* x. The *thickness*, $\rho(L)$, is the infimum of $\rho_x(L)$ over all $x \epsilon L$.

The images of thick knots often do not display the maximal tubes due to the difficulty of distinguishing features of the structure, see Figure 10.

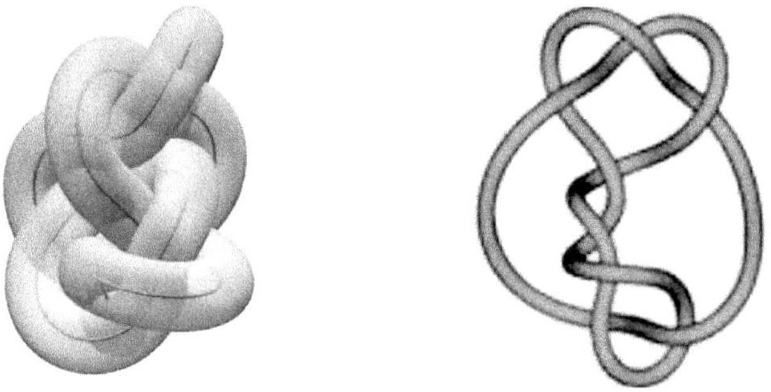

Fig. 10. Two images of thick 8_6 knots. Images provided, respectively, by Cantarella and Scharein

For polygonal knots, Eric Rawdon,[26–28] has defined a polygonal injectivity radius reminiscent of the smooth category in terms of the vertex angles and the minimal distal distances.

4. Theoretical results on the random sampling of equilateral open and closed chains

Due to biological, physical, geometric, and topological interest, the collections of open or closed (or ring) equilateral polygons have attracted much

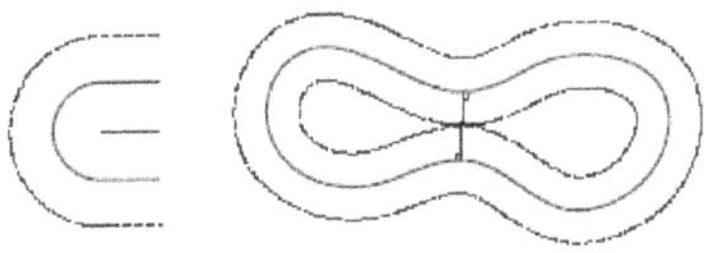

Fig. 11. Injectivity radius constraints[26]

attention. The results of Randell,[29,30] Calvo,[31–35] among others, tell us exactly which knots can be represented by closed chains with eight or fewer edges. Beyond eight edges, however, there are only estimates for the number of knots representable with exactly n edges. The frequency of a specific knot type, for a fixed number of edges, is another quantity that has not been determined rigorously. It has only been estimated via numerical studies. Generating and analyzing large random samples is, therefore, a fundamental component of the modeling of macromolecules. The theoretical results presented here should be considered as a first step toward models that take curvature, torsion, and injective radius into consideration.

The space of open embedded equilateral chains is the same as the space of embedded random walks. In the following, we will first consider open chains and then those which are closed, i.e. in which the final vertex equals the first vertex.

4.1. *Open chains (or random walks)*

An open chain, based at the origin, is determined by a set of n vertices, $v_0 = \mathbf{0}, v_1, ..., v_n$. The associated edge vectors are determined by $\mathbf{e}_i = v_i - v_{i-1}$ and, in turn, determine the vertices. The edge vectors have unit length and, thereby, correspond to points on the unit radius 2-dimensional sphere.

$$\begin{aligned}
v_0 &= \mathbf{0}, \\
v_1 &= \mathbf{e}_1, \\
v_2 &= \mathbf{e}_1 + \mathbf{e}_2, \\
&\vdots \\
v_n &= \mathbf{e}_1 + \mathbf{e}_2 + \mathbf{e}_3 + \ldots + \mathbf{e}_n.
\end{aligned}$$

Thus, an n-edge equilateral chain can be considered as a point in the product of n 2-dimensional spheres, i.e.

$$(\mathbf{e}_1, \mathbf{e}_2, \ldots, \mathbf{e}_n) \in \mathbf{S}^2 \times \mathbf{S}^2 \times \ldots \times \mathbf{S}^2$$

The requirement that the edges meet only at common vertices imposes a collection of constraints that insures that embedded chains define an open subset, $\mathbf{C}_n$, of $\mathbf{S}^2 \times \mathbf{S}^2 \times \ldots \times \mathbf{S}^2$. An example of such a chain is shown in Figure 12. The meaning of "random" is determined by the natural measure on the space of configurations. We will use the product measure on $\mathbf{S}^2 \times \mathbf{S}^2 \times \ldots \times \mathbf{S}^2$. A random edge vector $\mathbf{e}$ is determined as follows:

Random Chain Generation Method

(1) With respect to the uniform measure on $[0,1]$, select two numbers, a and b.
(2) Define $\alpha = 2\pi a$ and $\beta = ArcCos[2b-1]$
(3) Define $\mathbf{e} = (Cos[\alpha]Sin[\beta], Sin[\alpha]Sin[\beta], Cos[\beta])$.
(4) Check that the addition of $\mathbf{e}$ does not introduce any intersections in the chain defined to that point. If so, reject the vector and try again.

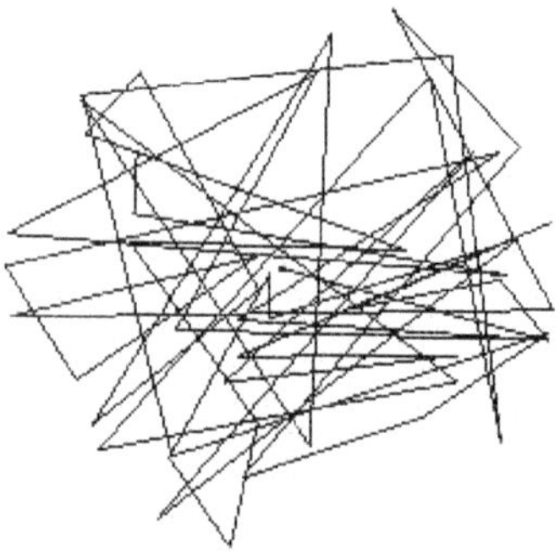

Fig. 12. A "random" 50 edge chain in 3-space

4.2. *Closed chains (or polygons) in 3-space*

Closed chains are determined by a list of unit vectors $\mathbf{e}_i$, $(\mathbf{e}_1, \mathbf{e}_2, \ldots, \mathbf{e}_n) \in \mathbf{S}^2 \times \mathbf{S}^2 \times \ldots \times \mathbf{S}^2$. The requirement that $v_n = \mathbf{0}$ is equivalent to $\mathbf{e}_1 + \mathbf{e}_2 + \mathbf{e}_3 + \ldots + \mathbf{e}_n = \mathbf{0} = (0,0,0)$. This imposes three relations on the set of edges thereby decreasing the dimension to $2n-3$. The requirement that the edges meet only at common vertices imposes a collection of constraints that insures that the set of embedded closed chains defines an open subset, $\mathbf{P}_n$, of the $2n-3$ dimensional subset of $\mathbf{S}^2 \times \mathbf{S}^2 \times \ldots \times \mathbf{S}^2$. An example of a "random" twenty-edge equilateral polygon is show in Figure 13.

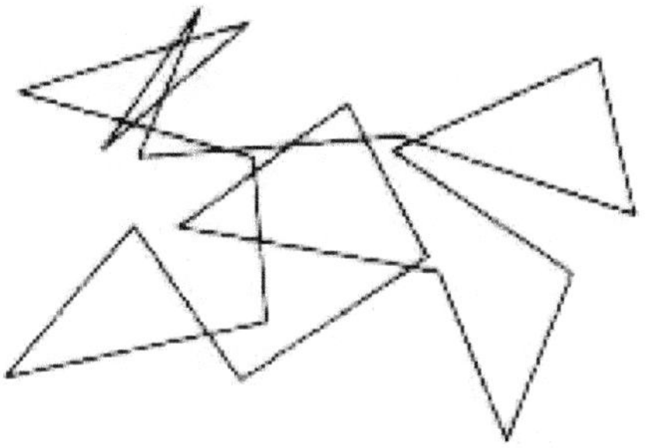

Fig. 13. A "random" 20 edge equilateral polygon in 3-space

4.3. *An important property of these spaces*

These spaces can be considered as subsets of the product of n copies of $\mathbf{R}^3$. As sampling occurs via numerical procedures which are not exact, small numerical "errors" naturally arise. Such "errors" represent "approximations" of the coordinates of the vertices of the polygonal chain. They can also be considered as corresponding to perturbations of exact data. These errors can impact the conclusions that one may wish to infer from an analysis of the data. Thus, a fundamental question that one must ask is "How close to having an equilateral polygon with specified properties must one be in order to insure the mathematical existence of a nearby equilateral polygon with exactly these topological or geometric properties?" Suppose that P is an n vertex polygon, $v_0, v_1, ..., v_n$ and P' is another with vertices $v'_0, v'_1, ..., v'_n$. We measure the distance between P and P' by $d(P, P') = max\{\|v_i - v'_i\|\}$, This metric on $\mathbf{R}^{3n}$ gives a measure of how close two polygons are. Similarly $max\{|\|e_i\| - 1|\}$ is a measure of how close P is to being an equilateral polygon with unit length edges. Answering our question requires the conjunction of several considerations that we shall discuss next.

4.3.1. *Chains and tubes*

We look closely at the injective radius, see Sec. 3.5, to develop a reply for polygonal models.[27] Let $\{\mathbf{v}_1, \mathbf{v}_2, \ldots, \mathbf{v}_n\}$ be the vertices of a 3-space polygon, P. Let $\|\mathbf{e}\|$ denote the norm of the edge vector $\mathbf{e}$. Given a vertex $\mathbf{v}_j$, let $angle(\mathbf{v}_j)$ be the turning angle from prior edge, $\mathbf{e}_{j-1}$, to the next edge, $\mathbf{e}_j$. Define

$$Rad(\mathbf{v}_j) = \frac{min\{\|\mathbf{e}_{j-1}\|, \|\mathbf{e}_j\|\}}{2\ Tan(\frac{angle(\mathbf{v}_j)}{2})} \tag{7}$$

and

$$MinRad(P) = min\{Rad(\mathbf{v}_j)\}. \tag{8}$$

For $x \in P$, let $\mathbf{d}_x : P \to \mathbf{R}$ be defined by $\mathbf{d}_x(y) = \|x - y\|$. A point, y, is a turning point if $\mathbf{d}_x$ changes from increasing to decreasing or vice-versa at y. Let $DC \in P \times P$ denote the collection of points, (x,y) with $x \neq y$, that are turning points of $\mathbf{d}_y$ and $\mathbf{d}_x$, respectively. Let

$$dscd(P) = min\{\|x - y\| : (x, y) \in DC\}. \tag{9}$$

Rawdon[26] defines the polygonal injective radius as

$$R(P) = min\{MinRad(P), \frac{dscd(P)}{2}\}. \tag{10}$$

The following theorem provides a way to analyze the effect of pertrubations or numerical approximations on the structure of a closed equilateral polygonal chain. The analogous result for open chains easily follows by the same methods.

Theorem 4.1 (The Closed Chain Tube Theorem[27]). *If $r < R(P)$, the union of solid balls of radius r centered at points on the closed polygon P is an embedded solid torus.*

Theorem 4.2 (The Open Chain Tube Theorem). *If $r < R(C)$, the union of solid balls of radius r centered at points on the open polygonal chain C is an embedded ball.*

These theorems help provide one step in answering our question.

Theorem 4.3 (The Perturbation Theorem[27]). *If P is equilateral polygonal chain (open or closed) of constant edge length, E, and the scale of a perturbation, $r = d(P, P')$, is smaller than minimum of the polygonal injective radius of $R(P)$ and $\frac{12ER(P)^2 - E^3}{8R(P)^2 + 2E^2}$, then all perturbed polygonal chains, P', are geometrically equivalent P.*

One application of the theorem in response to our questions concerns the existence of an equilateral eight edge 8_{19} knot. Random searches or perturbations of examples, without requiring equal length edges, lead to the impression that this might be an example of a knot whose polygonal stick number was different from its equilateral stick number, i.e. it could be represented as an eight edge polygon but not an equilateral one. An early example was discovered to fail to satisfy the conditions of the Perturbation Theorem, i.e. the edges lengths were not quite equal and, simultaneously, the edges almost crossed so that its injective radius was to small to insure that it could be approximated by an equilateral polygon of the same topological knot type. Later, in an unrelated effort, I developed a new random search strategy to test the existence of this knot. It appeared to be successful and, this time, the Perturbation Theorem confirmed the existence of the configuration shown in Figure 4, whose coordinates are (approximately) the following:

$$v_1 = (0.0, 0.0, 0.0)$$
$$v_2 = (4.244988698233, -5.29806224161222, 4.23209255993696)$$
$$v_3 = (1.621389440625, -2.446558361029, -2.76688254749867)$$
$$v_4 = (6.9553834171687, -3.1449123567692, 3.1543343844196)$$
$$v_5 = (-0.58394895866445, -5.81008379802231, 2.919114241909)$$
$$v_6 = (3.9278286450686, -2.446558361029, -2.76688254749867)$$
$$v_7 = (2.0591187707439, -5.1915142515365, 4.5113823944207)$$
$$v_8 = (5.3795054883023, -5.2350054125733, -2.76688254749867)$$

4.4. *Markov Chain Monte Carlo (MCMC) sampling of the space of closed equilateral chains*

A discrete stochastic process (or random process) is a sequence of random states, K_i. Markov Chain Monte Carlo sampling refers to discrete methods of sampling a probability distribution, a sample space with a specific unit measure, using a discrete stochastic process whose equilibrium distribution gives the measure on the space. This means that, given any measurable subset, the limiting ratio of the number of samples in the subset and the total number of samples approaches the proportional measure of the subset as the number of samples goes to infinity. A Markov chain is a stochastic process for which future samples depend only on the present case or, perhaps, a finite number of previous states within a time interval. The fundamental property of MCMC sampling is that, independent of the initial state, this ratio converges to the measure. A key problem is to determine how many samples are required in order to estimate the desired measure within a predetermined error bound. A good MCMC will require a smaller number of samples, independent of the initial state. Balanced against this is the time required to achieve statistical independence of sample choices. Thus we have **The Random Sampling Objective:** *Provide a MCMC algorithm whose sampling of P_n converges to the measure rapidly and with a high degree of accuracy.* We will describe several of the most commonly used algorithms for sampling the space of closed equilateral chains.

Algorithm 4.1 (The Basic Algorithm).

(i) Select an initial configuration, e.g. the standard planar regular polygon.
(ii) Perform a sequence of successive random ***modification steps*** *giving closed equilateral chains until the sampling criteria are met. Add the configuration to the sampling sequence.*
(iii) Replace the configuration with this one and return to the previous step until the desired number of samples is achieved.

Variation: An initial configuration is randomly generated to determine the sample.

The two principal methods differ in the specific choice of the **modification step**:

Algorithm 4.2 (The Polygonal Fold Modification (PFM) Step).

(i) Randomly select a pair of non-adjacent vertices of the polygon and select one of the segments they bound.
(ii) Using these vertices to define an axis, randomly rotate the selected segment to define a new polygon.
(iii) Verify that the polygon is embedded to complete the polygonal fold construction.

Algorithm 4.3 (The Crankshaft Rotation Modification (CRM) Step).

(i) Randomly select a pair of edge vectors, $\mathbf{e}_i$ *and* $\mathbf{e}_j$.

(*ii*) *The random rotation of* $\mathbf{e}_i$ *about axis defined by* $\mathbf{0}$ *and* $\mathbf{e}_i + \mathbf{e}_j$ *defines a new pair of edge vectors,* $\mathbf{e}'_i$ *and* $\mathbf{e}'_j$*, that replace* $\mathbf{e}_i$ *and* $\mathbf{e}_j$*, respectively, to define a new polygon.*

(*iii*) *Verify that the polygon is embedded to complete the crankshaft rotation step.*

Fundamental Properties.

- The PFM and CRM steps preserve the lengths of the edges.
- The inverse of a PFM, resp CRM, step is a PFM, resp CRM, step.
- The PFM and CRM algorithms are aperiodic, i.e. the probability that a configuration occurs a second time is zero.
- The PFM and CRM algorithms are positive recurrent, i.e. given any two configuration, there is a finite number of CRM (or PFM) modifications taking the initial configuration to the second. This implies that, given any neighborhood of the second, there is a positive probability of starting at the first and ending in the neighborhood in a finite number of steps.

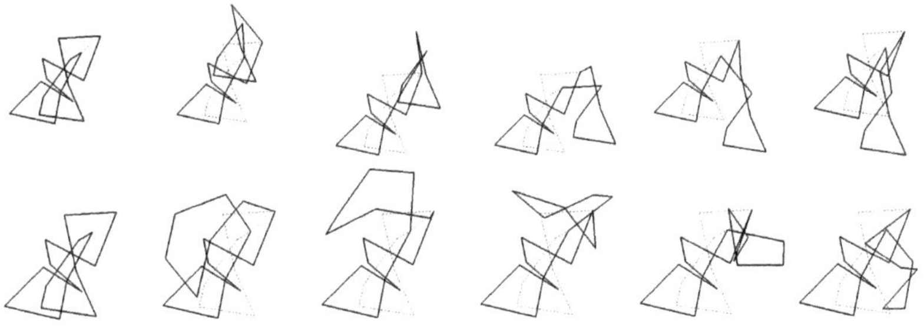

Fig. 14. CRM moves (above) versus PFM moves (below) demonstrating their fundamentally different nature

Figure 14 compares the effects of a crankshaft rotation with those of the polygonal fold. In each vertical pair of frames of this figure, the dashed lines indicate the original position of the polygon. In addition, a pair of dotted lines is drawn joining the endpoints of the two edges selected for the rotation. An animated version of this figure can be viewed at `www.avemaria.edu/calvopubs`. The modifications of a polygon produced by each method are quite distinct geometric operations. The PFM requires, on average, $\frac{n}{4}$ times as many calculations as CRM.[36] Each of these methods will asymptotically produce the required sample of the space of equilateral polygons as a consequence the following two theorems.

Theorem 4.4. *(Millett,[37] Kapovic & Milson,[38] Toussaint[39]) Every configuration can be reached from every other configuration by a finite sequence of PFM steps.*

Theorem 4.5. *(Alvarado, Calvo & Millett[36]) Every configuration can be reached from every other configuration by a finite sequence of CRM steps.*

5. Applications of sampling spaces of open and closed equilateral chains

We next focus on the application of these random sampling strategies to the acquisition of data illuminating the questions and quantities described earlier, in Section 3. All the methods we have discussed have been employed during one or more of these studies. In particular, the regular polygon, the hedgehog and random triangles, see Figure 15, have been used as starting configurations.

Problem 5.1. *How many randomization steps are required before sampling?*

We observe that, on average, the PFM requires more calculations per step than the CRM. In our implementation, both PFM and CRM use current sample as the initial polygon for each step and, as a consequence, require some number of modifications before sampling again.

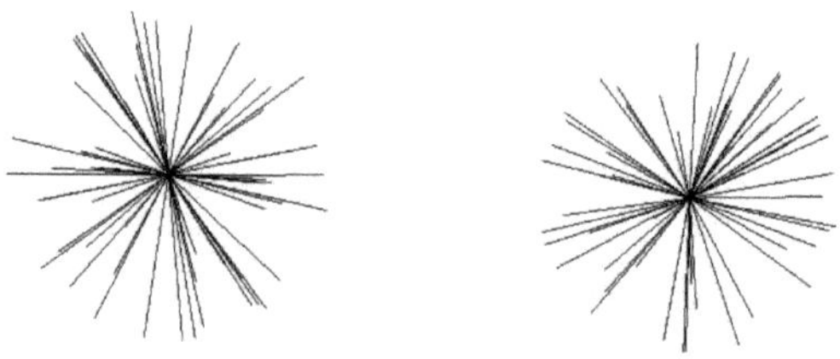

Fig. 15. Initial hedgehog (left) and triangle (right) sets of 60 unit vectors. Although appearing quite similar, the antipodal symmetry of the initial hedgehog data distinguishes the left from the right sets of unit vectors.

5.1. *Measures of randomness*

One measure of the extent to which an algorithm is random is given by using the maximum *Pearson Correlation Index*, PCI, to analyze the collection of edge vectors that is generated. We can, thereby, estimate the number of steps needed to achieve a PCI close to that of a completely random process. The PCI (also known as the Pearson Product-Moment Correlation) provides a measure of the degree to which two sets of data are linearly related.[40] It is defined as the covariance between the two sets of data divided by their standard deviations and, therefore, takes on values in the closed interval $[-1, 1]$. PCI values near ± 1 indicate a strong linear relationship between the data sets, while values near 0 indicate a weak relationship. In our case, we determine the correlation between two n-sided polygons by comparing the sequences of $3n$ coordinates obtained from the n edge vectors. We calculate the PCI for these sequences n times, once for each choice of starting vertex, taking as our final result the maximum absolute value observed. We use the absolute value of the Pearson Correlation in order to estimate its average value over a large populations of pairs of configurations. As the standard of comparison, we report the average value for a collection of random vectors equalling the number of edges in the polygons.

Table 1. Edge vector correlations (PCI) for the Crankshaft Rotation Method using various sampling rates.[36]

edges	crankshaft rotations					random	
$n = 12k$	n	$2n$	$3n$	$4n$	$5n$	mean	std dev
12	0.3457	0.3371	0.3354	0.3377	**0.3330**	0.3280	0.0828
24	0.2737	0.2665	0.2648	0.2673	**0.2679**	0.2695	0.0605
36	0.2326	**0.2346**	0.2339	0.2307	0.2302	0.2352	0.0484
48	**0.2106**	0.2086	**0.2106**	0.2089	0.2093	0.2127	0.0413
60	**0.1946**	0.1931	0.1919	0.1932	0.1928	0.1972	0.0360
72	**0.1811**	0.1808	0.1786	0.1797	0.1808	0.1836	0.0319
84	0.1696	0.1692	0.1692	**0.1699**	0.1694	0.1732	0.0289
96	0.1623	0.1614	**0.1628**	0.1609	0.1612	0.1650	0.0269
108	0.1546	0.1542	**0.1554**	0.1535	0.1550	0.1574	0.0247
120	**0.1498**	0.1459	0.1480	0.1490	0.1497	0.1510	0.0236

Alvarado et. al.[36] have shown that a conservative sampling rate of $4n$ crankshaft rotationsis adequate for polygons through 120 edges. In fact, a sampling rate of n crankshaft rotations is sufficient to achieve PCI values within one standard deviation of the random distribution without the closure condition.

5.2. *Estimation of knot populations*

One of the more challenging problems of statistical knot theory is the estimation of the population of knots that can be realized as an equilateral n-gon. As important, for physical knot theory, is the determination of the probability distribution function, PDF, of the knot types that are realized for a fixed number of edges and, for a fixed knot type, as a function of the number of edges. To estimate the PDF, we first use one of the random knot generation methods. Next we employ topological invariants which are sufficiently effective discriminators of knots and that can be calculated sufficiently easily within the target population. With Bruce Ewing, I developed a computer code in 1985 to calculate the HOMFLYPT polynomial,[20,41–44] see Figure 6. This continues to be a very effective tool. A very clear picture has emerged from an analysis of the resulting data, see Figure 16.

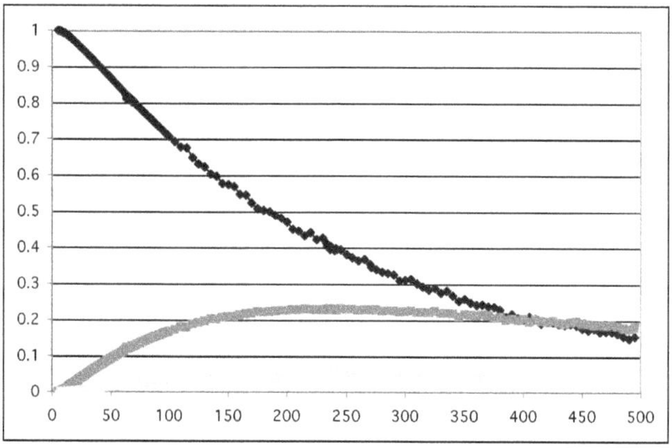

Fig. 16. The probability of a knot type (distinct HOMFLYPT polynomials) as a function of the number of edges: $0_1, 3_1, and\ 4_1$. Sample size: 2.51×10^9

Problem 5.2. *As a function of the number of edges, how many distinct knot types are represented by equilateral polygons of length n?*

The estimation of the number of distinct knot types and their relative probability is also difficult.[45,46] Although we know that there are only finitely many knot types, there is no useful estimate of the number of topological knot types as a function of the number of edges. Analysis of knot populations, see Figure 17, shows that even with a relatively large sample size, here 2.51×10^9, there are many knot types (or HOMFLYPT polynomials) that are observed only once. Despite knowing that this number is bounded, an analysis of the number of knots as a function of the sample size, shown in Figure 18, does not yet suggest asymptotic convergence to a finite population much less provide the means to estimate the size of the population.

Problem 5.3. *What is the probability distribution of the knot types occurring in equilateral n-gons?*

Another objective is the estimation of the probability of a fixed knot type as a function of the number of edges. One can prove that the probability of each fixed knot type goes to zero exponentially as the number of edges goes to infinity. There are numerical estimates of the exponential constant for some models and for some of the more likely knot types, e.g. $0_1, 3_1$, and 4_1. Although one might conjecture that there is an elementary functional model for the probability distribution, Figure 16, this is apparently not the case due to differences in behaviour between the finite range and the asymptotic range.[47]

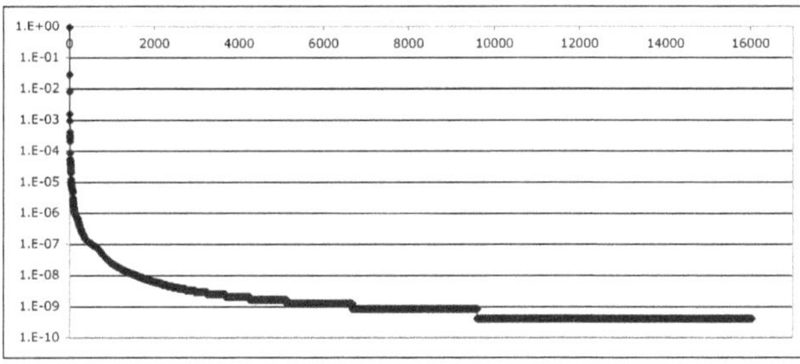

Fig. 17. The observed probability distribution function for distinct HOMFLYPT polynomials of polygons with 32 edges. Total knots 16,497 Sample size 2.51×10^9

5.2.1. *The probability of knotting in 3-space random walks.*

We may also use the *HOMFLYPT* invariant, as indicated earlier, study knots and slipknots in random walks through the knot spectrum. Just as in the case of random polygons, we know that the probability that random walk contains a knot (slipknot) goes to one as its length goes to infinity.[9,10,48,49]

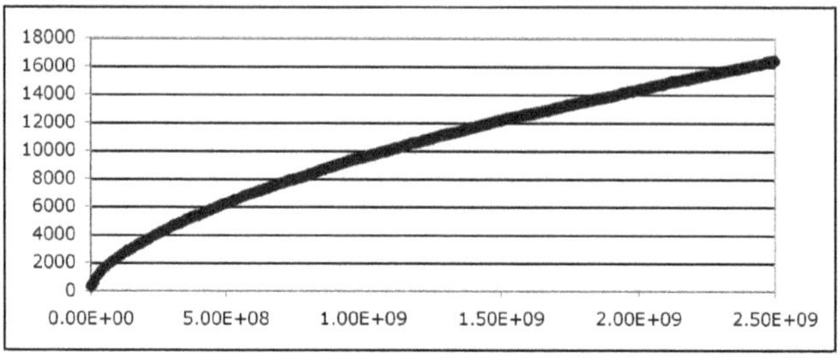

Fig. 18. The number of observed knots as a function of the sample size. Total knots 16,457 Sample size 2.51×10^9

Problem 5.4. *How does knotting in random walks compare to that in random polygons?*

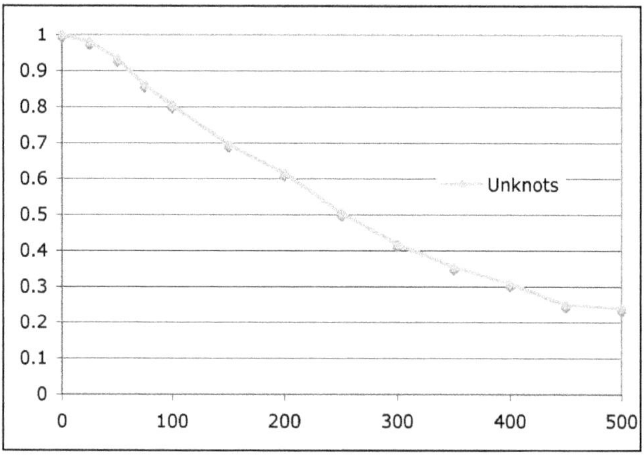

Fig. 19. Knotting in random walks[7,8]

5.3. *Knot and slipknot localization*

Problem 5.5. *What is the average size of the knotted (or slipknotted) segments of a random chain or polygon?*

One can also employ the HOMFLYPT polynomial to identify which specific segments in an open or closed chain support a knot and to specify the knot type by systematically checking all subsegments of increasing size until knotted segments are identified. By analyzing larger segments containing the knotted segment, we can also determine if it is a slipknot, i.e. is it contained in a larger unknotted segment. This provides a method by which we may determine the average lengths of knots and slipknots of specific types that occur. This allows us to determine the extent to which knotting is a local or a global matter. We will say that the average knotting is *global* if the average length of knots is proportional to length of the chain, An, for large n. The average knotting will be said to be *weakly local* if its extent is bounded by An^{σ}, with $\sigma < 1$, for large n. And, finally, we will say that the average knotting is *strongly local* if its extent is bounded by a constant A.

Conjecture 5.1. *On average, knotting and slipknotting is strongly local in random walks.*

Recent simulation data, see Figure 20, suggests that this conjecture is true with the average length of knots bounded by 25 and the average length of slipknots bounded by 50. This is likely to be a challenging theoretical problem as we know that both local and global knots occur with probability one as the length of the walk goes to infinity.[10,13,48,49]

For random polygons in the simple cubic lattice, there is strong evidence that the knot size scales as the $\frac{3}{4}$ power[15,18,50] and is, therefore, weakly local.

Problem 5.6. *How do knots and slipknots in equilateral* 3*-space polygons scale?*

5.3.1. *Implications of knot and slipknot localization on the average shape*

If knot localization in equilateral polygons is sufficiently strong, the characteristics of the random unknotted polygon would determine the average asymptotic spatial properties for any fixed knot type. On the other hand,when one considers the entire populaltion of n-gons, the contributions from the exponential growth in numbers of increasingly complex compact knots appears to have a measurable influence on the average properties. For example, this accounts for the existence of equilibrium lengths of polygons having fixed knot types.[51–54] These powerful phenemona appear to account for the principle influences of knotting on the polygonal models of macromolecules.

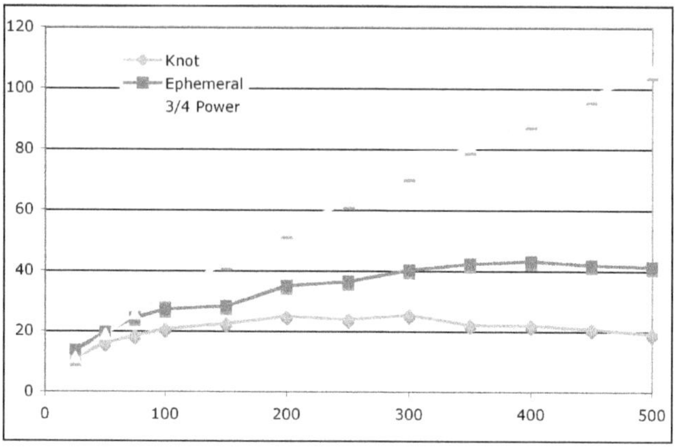

Fig. 20. Average knot localization in random walks

Conjecture 5.2. *As the length of a walk or polygon of a fixed knot type increases, the average spatial properties converge to those of the unknot.*

5.3.2. *Influence of knotting on physical properties: squared radius of gyration*

The squared radius of gyration, the average squared distance from the center of mass, is the principal physical measure of the size of a polymer. In the case of polygonal models, the mass is concentrated at the vertices and the squared radius of gyration is the average of the squared distance of the vertices from the center of mass. Polymers at thermodynamic equilibrium are modeled by random walks of two types. In good solvents, they are modeled by self-avoiding random walks as segments do not approach each other due to the repulsive forces between non-adjacent segments. Polymers chains under the θ condition (where segments neither repel nor attract) and for polymers in the dense melt phase (where individual chains are entangled with many others) are modeled by ideal random walks. For non-self-avoiding random walks, e.g. those in 3-space, $< {R_G}^2 >$, is proportional to $N^{2\nu}$, where N is the number of steps or segments, with $\nu \approx 0.500$. For self-avoiding random walks, $< {R_G}^2 >$, is proportional to $N^{2\nu}$, with $\nu \approx 0.588$. A. Grosberg[55] argued that, in the long chain regime, polymers with effective diameter zero of a fixed knot type should have scaling exponent ν of 0.588 and that the length required for this behavior should increase with the complexity of the knot. Dobay et. al.[51] showed that random walks, without excluded volume, have radius of gyration that scales at 0.500 but individual knot types (including the unknot) have radius of gyration scaling at 0.588.

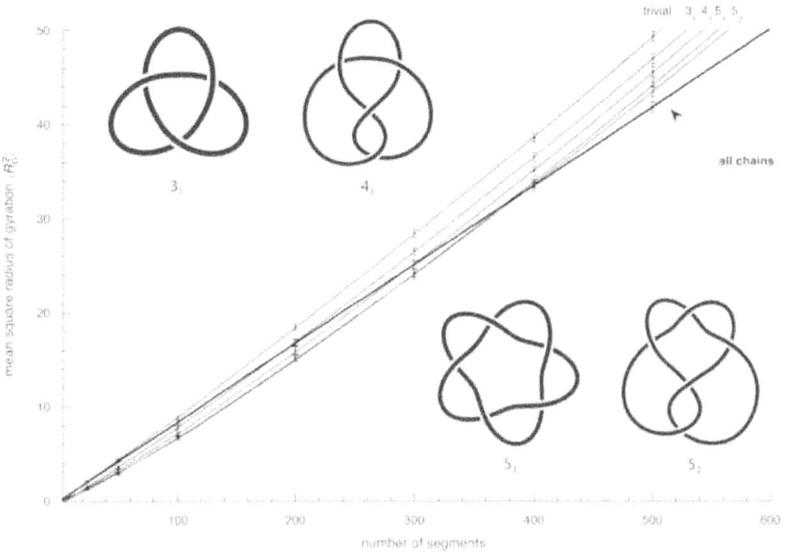

Fig. 21. Closed polygons: average squared radius of gyration for all types and specified knot types showing the difference in scaling.[51]

The *equilibrium length* is the number of edges at which the average property, e.g. the radius of gyration, equals the average of the entire population. In Figure 21, the crossing behaviour that defines the radius of gyration equilibrium lengths for the simplest knot types are shown.

One can make the same analysis of for random walks in which the knot types are identified using the MDS method.[7] The scaling of the squared radius of gyration is determined for knotted random walks and the associated equilibrium length is defined, see Figure 22.

5.3.3. *Influence of knotting on physical properties: average crossing number*

The average crossing number, ACN, is one of the most frequently used measures of *entanglement*, an intuitive but informal concept that plays a fundamental role in understanding the physical properties of many systems. For random polygons, Diao et al[51] show the scaling of the ACN is $\frac{3}{16}\, n\, ln(n) + O(n)$. Here, again, the increasing presence of more and more

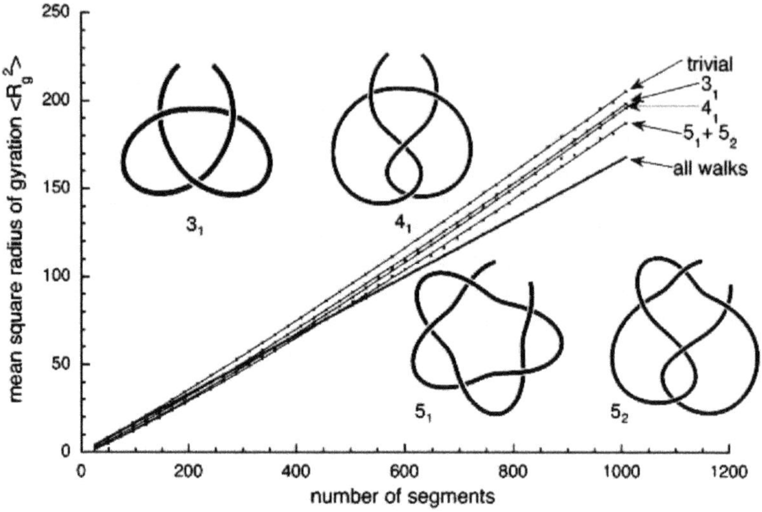

Fig. 22. Open polygons: average squared radius of gyration for all types and specified knot types in random walks, showing the difference in scaling.[7]

complex knots causes the existence of equilibrium lengths for individual knot types, see Figure 23.

5.3.4. *Influence of knotting on geometric properties: total curvature and total torsion*

The Fary-Milnor Theorem 3.1 states that the total curvature of a polygonal knot is bounded below by $2\pi Br(K)$, where $Br(K)$ is the bridge number of the knot type, a measure of complexity of the knot type. When one considers the family of equilateral polygons realizing the knot type, what is the average total curvature and, for that matter, the average total torsion? These give geometric measures of the complexity of the knotted polymer models.[56] To better understand the geometry and topology of ring polymer chains under θ conditions or in the melt phase, we consider the statistics of the total curvature and total torsion as the number of edges increases. A little reflection leads to the realization that the dominant term is $\frac{\pi}{2}n$, where n is the number of edges. Subtracting this dominant term allows one to analyze the second order terms, shown in Figure 24. There is visibly sufficient sensitivity to the presence of knotting to show again the existence of equilibrium lengths for the total curvature.

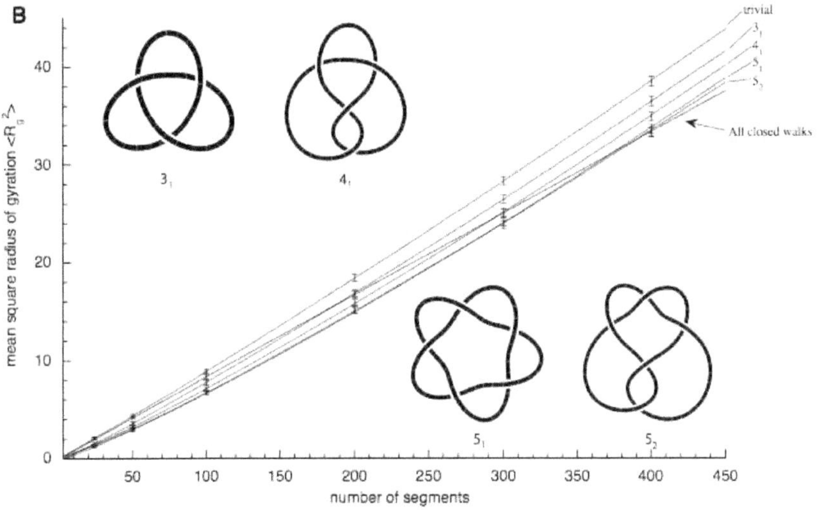

Fig. 23. The ACN and associated equilibrium lengths[51]

We noted that the average second order effect for total curvature and total torsion over all configurations appeared to be constant, only differing in sign. Indeed, this is the case as was proved by Grosberg[57] who showed that this value was exactly $\frac{3\pi}{8}$. Specifically, he showed that the average total curvature of phantom polygons is asymptotically equal to $\frac{n\pi}{2} + \frac{3\pi}{8} + O(\frac{1}{n})$. If we let et $\{e_i\}_{i=1,\dots,n}$ denote the set of edge vectors of a random polygon, then $\sum e_i = 0$. Taking the average of the inner product give $\langle e_i \cdot e_j \rangle = \frac{-1}{n-1}$ which leads to $\langle curv \rangle = \frac{n\pi}{2} + \frac{3\pi}{8}$.

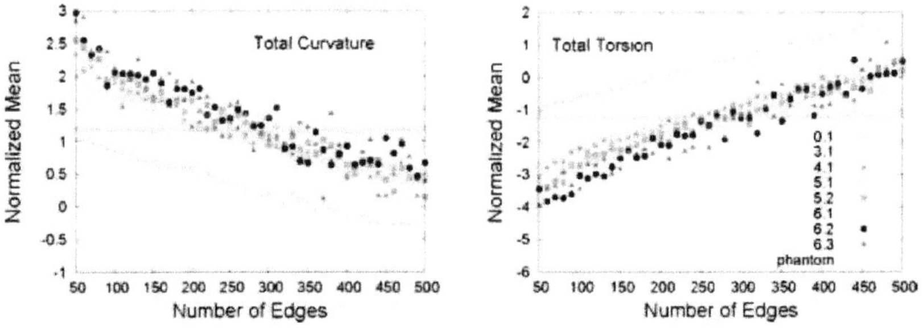

Fig. 24. The normalized total torsion and total curvature of random poygons.[56]

Figure 25 shows a comparision of these equilibrium lengths to experimentally observed gel separation of knotted DNA, a traditional means to compare the degree to which these quantities might correlate to physically observed behavior.

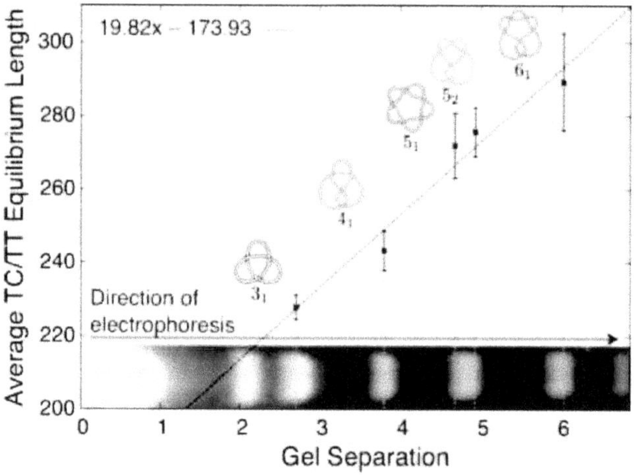

Fig. 25. Total curvature equilibrium lengths versus gel separation.[56]

Problem 5.7. *How does the shape of macromolecules depend upon knotting?*

6. Concluding remarks

In this report, I have described just a few of the more interesting features of the mathematical theory of knotting and linking that are deeply connected with natural phenomena. Some of these are truly geometric in nature, others are topological, and many of the concepts and problems arise due to the interest in biological and physical polymers, especially the structure of DNA and proteins. As rigorous mathematical analysis is often beyond present reach, much of the effort to forge and understand these relationships and their implications has been numerical in nature whereby one wishes to randomly sample the population of configurations associated to an appropriate mathematical model. As a consequence, we have employed two key procedures, the crankshaft rotations and the polygonal folds, each providing a method taking any equilateral polygon in 3-space to any other in a finite sequence of steps. These procedures can then be used to randomly sample the space of polygons and generate sets of data which are then used to illuminate a number of these questions. Finally, we have looked at several applications to questions concerning the influence of knotting on the spatial structure of macromolecules. In the course of discussing these results, we have cited outstanding problems and conjectures. Some are quite accesible while others, I believe, are very challenging and may well require the development of new mathematical methods.

I hope these lecture notes have provided an introduction to a few of the more interesting facets of physical knot theory and their connections with a topological and geometric knot theory, as well as some newer research areas which one might better describe as statistical and computational knot theory. The interactions with the students and my fellow particiants in the ICTP Advanced School and Conference on Knot Theory and its Applications to Physics and Biology was a truly memorable experience. I wish to express my appreciation to Le Dung Trang and his staff and especially to the organizers, Slavik Jablan, Louis Kauffman, Sofia Lambropoulou and Jozef Przytycki for their outstanding leadership in organizing this opportunity to share our passion for research with the most diverse group of students and researchers that I have even encountered.

References

1. J. Cantarella and H. Johnston, *J. Knot Theory Ramifications* **7**, 1027 (1998).
2. T. Biedl, E. Demaine, M. Demaine and et al., *Discrete Comput. Geom.* **26**, 269 (2001).
3. G. Toussaint, *Beiträge Algebra Geom.* **42**, 301 (2001).
4. E. D. Demaine, S. Langerman, J. O'Rourke and J. Snoeyink, *Comput. Geom.* **26**, 37 (2003), Thirteenth Canadian Conference on Computational Geometry—CCCG'01 (Waterloo, ON).
5. G. Aloupis, G. Ewald and G. Toussaint, *Beiträge Algebra Geom.* **45**, 429 (2004).
6. T. J. Clark and G. A. Venema, *J. Knot Theory Ramifications* **13**, 479 (2004).
7. K. Millett, A. Dobay and A. Stasiak, *Macromolecules* **38**, 601 (2005).
8. K. C. Millett and B. M. Sheldon, *Tying down open knots: a statistical method for identifying open knots with applications to proteins*, in *Physical and numerical models in knot theory*, , Ser. Knots Everything Vol. 36 (World Sci. Publ., Singapore, 2005), pp. 203–217.
9. K. C. Millett, *J. Knot Theory Ramifications* (2010).
10. D. W. Sumners and S. G. Whittington, *J. Phys. A* **21**, 1689 (1988).
11. E. J. Janse van Rensburg, D. A. W. Sumners, E. Wasserman and S. G. Whittington, *J. Phys. A* **25**, 6557 (1992).
12. M. Mansfield, *Structural Biology* **1**, 213 (1994).

13. Y. Diao, J. C. Nardo and Y. Sun, *Journal of Knot Theory and its Ramifications* **10**, 597 (2001).
14. W. Taylor, *Nature* **406**, 916 (2000).
15. B. Marcone, E. Orlandini, A. L. Stella and F. Zonta, *J. Phys. A* **38**, L15 (2005).
16. P. Virnau, L. A. Mirny and M. Kardar, *PLoS Computational Biology* **2**, p. e122(September 2006).
17. G. Kolesov, P. Virnau, M. Kardar and L. A. Mirny, *Nucleic Acids Res.* **35**, w425 (2007).
18. B. Marcone, E. Orlandini, A. L. Stella and F. Zonta, *Physical Review E (Statistical, Nonlinear, and Soft Matter Physics)* **75**, p. 041105 (2007).
19. M. Laso, N. C. Karayiannis, K. Foteinopoulou, M. L. Mansfield and M. Kroger, *Soft Matter* **5**, 1762 (2009).
20. P. Freyd, D. Yetter, J. Hoste, W. B. R. Lickorish, K. Millett and A. Ocneanu, *Bull. Amer. Math. Soc. (N.S.)* **12**, 239 (1985).
21. M. P. do Carmo, *Differential geometry of curves and surfaces* (Prentice-Hall Inc., Englewood Cliffs, N.J., 1976). Translated from the Portuguese.
22. R. S. Millman and G. D. Parker, *Elements of differential geometry* (Prentice-Hall Inc., Englewood Cliffs, N. J., 1977).
23. J. W. Milnor, *Ann. of Math.* **52**, 248 (1950).
24. O. Krötenheerdt and S. Veit, On the theory of solid knots, in *Physical and Numerical Models in Knot Theory*, eds. J. A. Calvo, K. C. Millett, E. J. Rawdon and A. Stasiak, Ser. Knots Everything, Vol. 36 (World Sci. Publishing, Singapore, 2005) pp. 1–18. Translated by Ted Ashton.
25. O. Gonzalez and J. H. Maddocks, *Proc. Natl. Acad. Sci. USA* **96**, 4769 (1999).
26. E. J. Rawdon, The thickness of polygonal knots, PhD thesis, University of Iowa1997.
27. K. C. Millett and E. J. Rawdon, *J. Comput. Phys.* **186**, 426 (2003).
28. J. Baranska, P. Pieranski and E. J. Rawdon, Ropelength of tight polygonal knots, in *Physical and Numerical Models in Knot Theory*, eds. J. A. Calvo, K. C. Millett, E. J. Rawdon and A. Stasiak, Ser. Knots Everything, Vol. 36 (World Sci. Publishing, Singapore, 2005) pp. 293–321.
29. R. Randell, A molecular conformation space, in *Studies in Physical and Theoretical Chemistry*, ed. R. Lacher, MATH/CHEM/COMP 1987, Vol. 54 (Elsevier Science, Amsterdam, 1988) pp. 125–140.
30. R. Randell, Conformation spaces of molecular rings, in *Studies in Physical and Theoretical Chemistry*, ed. R. Lacher, MATH/CHEM/COMP 1987, Vol. 54 (Elsevier Science, Amsterdam, 1988) pp. 141–156.
31. J. A. Calvo, Geometric knot theory: the classification of spatial polygons with a small number of edges, PhD thesis, University of California, Santa Barbara1998.
32. J. A. Calvo, *J. Knot Theory Ramifications* **10**, 245 (2001), Knots in Hellas '98, Vol. 2 (Delphi).
33. J. A. Calvo, Characterizing polygons in $\mathbb{R}^3$, in *Physical knots: knotting, linking, and folding geometric objects in* $\mathbb{R}^3$ *(Las Vegas, NV, 2001)*, , Contemp. Math. Vol. 304 (Amer. Math. Soc., Providence, RI, 2002) pp. 37–53.

34. J. A. Calvo and K. C. Millett, Minimal edge piecewise linear knots, in *Ideal knots*, (World Sci. Publishing, Singapore, 1998) pp. 107–128.
35. J. A. Calvo, Characterizing polygons in $\mathbb{R}^3$, in *Physical knots: knotting, linking, and folding geometric objects in BbbR^3 (Las Vegas, NV, 2001)*, , Contemp. Math. Vol. 304 (Amer. Math. Soc., 2001).
36. S. Alvarado, J. Calvo and K. Millett, *Journal of Statistical Physics* **143**, 102 (2011), 10.1007/s10955-011-0164-4.
37. K. C. Millett, *J. Knot Theory Ramifications* **3**, 263 (1994), Random knotting and linking (Vancouver, BC, 1993).
38. M. Kapovich and J. Millson, *J. Differential Geom.* **44**, 479 (1996).
39. G. Toussaint, *Computational Geometry* **31**, 219 (2005).
40. R. Larsen and M. Marx, *An Introduction to Mathematical Statistics and Its Applications*, 3 edn. (Prentice Hall, Upper Saddle River, New Jersey, 2001).
41. B. Ewing and K. C. Millett, Computational algorithms and the complexity of link polynomials, in *Progress in knot theory and related topics*, (Hermann, Paris, 1997) pp. 51–68.
42. B. Ewing and K. C. Millett, A load balanced algorithm for the calculation of the homfly polynomial, in *The Mathematical Heritage of C. F. Gauss*, (World Scientific Publishing Co., 1991) pp. 225–266.
43. W. B. R. Lickorish and K. C. Millett, *Topology* **26**, 107 (1987).
44. J. H. Przytycki and P. Traczyk, *Proc. Amer. Math. Soc.* **100**, 744 (1987).
45. K. C. Millett and D. W. Sumners (eds.), *Random knotting and linking* (World Sci. Publishing, Singapore, 1994). Papers from the AMS Special Session held in Vancouver, BC, August 15–19, 1993.
46. K. C. Millett, Monte carlo explorations of polygonal knot spaces, in *Proceedings of the International Conference on Knot Theory and its Ramifications held in Delphi, August 7–15, 1998*, eds. C. M. Gordon, V. F. R. Jones, L. H. Kauffman, S. Lambropoulou and J. H. Przytycki (World Sci. Publishing, Singapore, 2001).
47. K. C. Millett and E. J. Rawdon, Universal characteristics of polygonal knot probabilities, in *Physical and Numerical Models in Knot Theory*, eds. J. A. Calvo, K. C. Millett, E. J. Rawdon and A. Stasiak, Ser. Knots Everything, Vol. 36 (World Sci. Publishing, Singapore, 2005) pp. 247–274.
48. N. Pippenger, *Discrete Appl. Math.* **25**, 273 (1989).
49. Y. Diao, N. Pippenger and D. W. Sumners, *J. Knot Theory Ramifications* **3**, 419 (1994), Random knotting and linking (Vancouver, BC, 1993).
50. E. Orlandini, A. L. Stella, C. Vanderzande and F. Zonta, *J. Phys. A* **41**, 122002, 7 (2008).
51. Y. Diao, A. Dobay, R. B. Kusner, K. C. Millett and A. Stasiak, *J. Phys. A* **36**, 11561 (2003).
52. A. Dobay, J. Dubochet, K. Millett, P.-E. Sottas and A. Stasiak, *Proc. Natl. Acad. Sci. USA* **100**, 5611 (2003).
53. E. J. Rawdon, A. Dobay, J. C. Kern, K. C. Millett, M. Piatek, P. Plunkett and A. Stasiak, *Macromolecules* **41** (2008).
54. K. C. Millett, Plunkett, M. Patrick, Piatek, E. J. Rawdon and A. Stasiak, *The Journal of Chemical Physics* **130**, 165104 (2009).

55. A. Y. Grosberg, *Phys. Rev. Lett.* **85**, 3858 (2000).
56. P. Plunkett, M. Piatek, A. Dobay, J. Kern, K. Millett, A. Stasiak and E. Rawdon, *Macromolecules* **40**, 3860 (2007).
57. A. Y. Grosberg, *Macromolecules* **41**, 4524 (2008).

KNOTS, SATELLITES AND QUANTUM GROUPS

H. R. Morton
Department of Mathematical Sciences, University of Liverpool
Peach Street, Liverpool L69 7ZL

Keywords: Satellite; Homfly polynomial; skein; annulus; quantum invariant.

1. Introduction

These notes are an expanded version of a series of lectures given in ICTP Trieste and at Universidad Complutense, Madrid in April and May 2009.

They are primarily an account of knot and link invariants derived from the Homfly polynomial by the use of satellites. They are based mainly on work by myself and former students at Liverpool University. Much of this can be found in greater detail through the Liverpool University knot theory publications list, including the doctoral theses of Aiston, Lukac and Hadji.

I refer readers also to an extended expository article [11] where a similar approach, focused more on satellite invariants for the Jones polynomial, is adopted. This article contains earlier work on the Homfly invariants, which culminated in two papers with Aiston [2, 13].

More recent papers are those on the Murphy operators [9, 10] and work with Lukac and Hadji [3, 7, 14], where the technique of using the meridian maps has been developed and refined. The latest paper with Manchon [15] gathers together results about the Homfly skein of the annulus, which is the prime tool for organising satellite invariants.

1.1. Setting the scene

We are all able to tie a knot in a piece of rope. The pictures in figure 1 show some examples.

What exactly do we mean though when we say that a piece of rope is knotted?

Let us first stop the knots escaping by joining up the ends of the rope, as in figure 2. Compare what happens in the three cases.

Fig. 1.

Fig. 2.

In the first case we get a simple, or 'unknotted' circle, while in the second case we have a circle with what appears to be a knot in it.

Let us say that the rope is *knotted* if no possible manipulation of it will result in the unknotted circle. We do not allow cutting and rejoining.

The third example can clearly be undone by a little manipulation to form the simple circle, so again the rope is unknotted.

We model this notion of a knot mathematically by referring to a closed curve in $\mathbf{R}^3$ as a knot, with the special case of the simple circle, lying say as the unit circle in a plane, known as the *trivial knot* or *unknot.* Knot theory in the mathematical sense is then the study of closed curves in space.

We call two knots *equivalent* if one can be manipulated, without passing one strand through another, to become the other knot. I give a more formal technical description of this below, but essentially anything is allowed which could be done with a rather stretchable piece of rope. The one manoeuvre which must be excluded is the analogue of the bachelor's technique for ignoring knots on a piece of cotton – pull it so tight that you can hardly see it! Using this technique on a curve with no physical thickness would get rid of any knot.

Fig. 3. Bachelor's unknotting

We would like to know for a start if there are any knots which are not equivalent to the trivial knot. If so, are there lots of different knots, and how might we distinguish between them?

It is easy to imagine that you have been given two knots and by a little patient work you manage to manipulate one to look like the other, e.g. the first and third knots in figure 2. What happens though if you find that even after a lot of trying you can't make them look the same – does it follow that the knots are inequivalent, or have you just not been dextrous enough? There is clearly a problem here, and something else will be needed, as there is no way that failure to manipulate can show that it is actually impossible to do so.

It should be realised that the question of how the rope is knotted isn't an intrinsic question about the rope alone, but rather a matter of how the rope is placed in space. Every closed loop of rope looks the same to an ant inside the rope. Some of the techniques developed for the study of knots have proved fruitful in other 'placement problems', i.e. in studying the different ways in which one particular geometric object, here a closed curve, may lie inside a larger one.

1.2. *Background*

The idea of looking at knotted and unknotted closed curves goes back to Gauss and beyond. Kelvin had some idea of trying to relate different types of atoms to knotted curves in the ether; this was taken up by a Scottish physicist Tait, who set out to enumerate all possible different knots in the hope of tallying them against different atoms. His lists of knots soon showed that the task of systematically enumerating all knots was hopelessly complicated; among other problems there are infinitely many. It is still true today that no practical framework exists for producing a comprehensive list, although Thistlethwaite has devised a fairly good means of handling the simpler knots. Various mathematicians in the 1920s and 1930s developed methods to show up a number of general properties shared by all knots, using some very elegant geometrical techniques and exploiting the growing interplay between algebra and this style of geometry. From this period has come the Alexander polynomial, and interpretations of it, as well as group theoretic invariants. Much more recently knot theory and theoretical physics have again had close contacts.

Definition 1.1. A *knot* is a simple closed curve $K \subset \mathbf{R}^3$ or in S^3.

Definition 1.2. The *complement* of K is $S^3 - K$.

We shall only deal with *tame* knots, e.g. smooth or polygonal curves, and we assume that K has a solid torus neighbourhood V with

$$(V, K) \cong (S^1 \times D^2, S^1 \times \{0\}).$$

This is like insisting on using a piece of rope, although one whose exact thickness will not matter.

It is often convenient to deal with $S^3 - \text{int}V = \text{ext}K$, the *exterior* of K, which is a *compact* 3-manifold with boundary $\partial(\text{ext}K) = \partial V \cong$ torus $S^1 \times S^1$.

From the point of view of topological invariants there is not much difference between $S^3 - K$, $\text{ext}K$ and $S^3 - V$.

Definition 1.3. Knots K_0 and K_1 are *homeomorphic* if there exists a homeomorphism $h : \mathbf{R}^3 \to \mathbf{R}^3$ such that $h(K_0) = K_1$.

If h is orientation preserving we can deform K_0 to K_1 through a family of knots $K_t = h_t(K_0)$. We shall call K_0 and K_1 *equivalent* when they are related in this way. (The term *ambient isotopic* is also used.)

Conversely a 1-parameter sliding of a neighbourhood V of K_0 to one of K_1 through $\mathbf{R}^3$ can be extended to such a family h_t of homeomorphisms, and models quite well the physical notion of equivalence by manipulation of a closed loop of rope.

We then have the result, by composing with a reflection if necessary, that two knots K_0 and K_1 are homeomorphic if and only if K_0 is equivalent to K_1 or its mirror-image.

Remark 1.1. Some knots, for example the trefoil, are not equivalent to their mirror image, while others such as the figure-eight knot are.

1.3. *Knot diagrams and moves*

For our subsequent analysis we concentrate on tame knots, i.e. knots equivalent to finite polygonal curves or equally to regular smooth curves.

Diagrams connected by a sequence of Reidemeister's moves, seen in figure 4, represent equivalent knots.

The converse is also true.

Theorem 1.1 (Reidemeister). *If two diagrams represent equivalent knots then one diagram can be converted to the other by a finite sequence of Reidemeister moves, along with isotopy (deformation) of the image within the projection plane.*

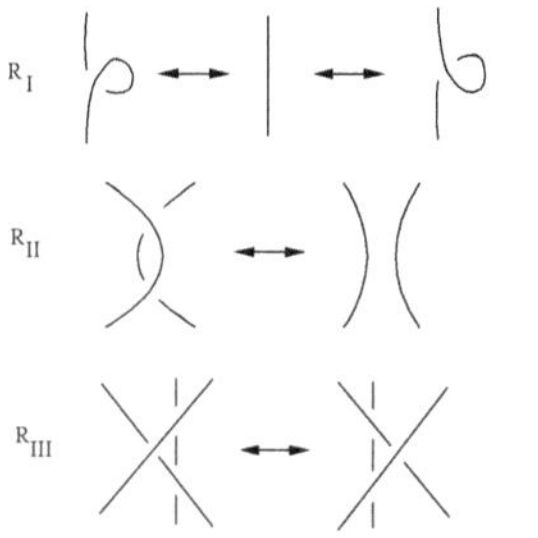

Fig. 4. Reidemeister's moves

1.4. *Links and linking number*

We may enlarge our scope slightly and look, as Gauss did, not just at a single closed curve but at several at once.

Definition 1.4. A *link* of r components is a collection $L = L_1 \cup L_2 \cup \ldots \cup L_r$ of r closed non-intersecting curves.

When $r = 1$ we have a knot. In the case $r = 2$ we can very simply associate an integer with a link, which is the same for every equivalent link. This is called the *linking number* of the two components.

To define the linking number $\mathrm{lk}(L_1, L_2)$ we must first choose an orientation of each of the components, which we note on a diagram of the link by drawing arrows on the curves. Now look at one diagram of the link and consider only the crossings where L_1 crosses over L_2. Each of these crossings c_i can be given a sign $\varepsilon_i = \pm 1$, according to a conventional choice. The sum of these signs $\sum \varepsilon_i$ is unaltered when the diagram is changed by Reidemeister moves. For crossings of L_1 over L_2 are not affected by moves I and III, while if there are any involved in a move of type II they occur as a pair with opposite sign, so that the sum is unchanged.

Reidemeister's theorem holds also for links. We may then set $\mathrm{lk}(L_1, L_2) = \sum \varepsilon_i$ for any choice of diagram.

Proposition 1.1. *$lk(L_2, L_1) = lk(L_1, L_2)$.*

Proof. To calculate $\mathrm{lk}(L_2, L_1)$ we must count the crossings of L_2 over L_1 in some diagram. Start with a diagram in which we count the crossings c_i of L_1 over L_2. If we turn this diagram over and view it from the other side we get a new diagram of the link in which the crossings c_i become the crossings of L_2 over L_1. Each crossing, viewed from the other side has the same sign as it had initially, so the sum needed to calculate $\mathrm{lk}(L_2, L_1)$ from this diagram is identical to the sum calculating $\mathrm{lk}(L_1, L_2)$ in the original diagram. □

1.5. *Framed links*

Framed links are made from pieces of ribbon rather than rope, so that each component has a preferred annulus neighbourhood. Combinatorially they can be modelled by diagrams in S^2 up to R_{II} and R_{III}, excluding R_I, by use of the 'blackboard framing' convention. The ribbons are determined by taking parallel curves on the diagram.

Reidemeister moves II and III on a diagram give rise to isotopic ribbons. Any apparent twists in a ribbon can be flattened out using Reidemeister I.

Oriented link diagrams D have a *writhe* $w(D)$ which is the sum of the signs of all crossings. This is unchanged by moves II and III.

For a framed knot the writhe is sometimes called its 'self-linking number', which is independent of the orientation of the diagram. Generally a framing of a link is determined by a choice of writhe for each component.

1.6. *Satellites*

A *satellite* of a framed knot K is determined by choosing a diagram Q in the standard annulus, and then drawing Q on the annular neighbourhood of K determined by the framing, to give the satellite knot $K * Q$. We refer to this construction as *decorating* K *with the pattern* Q (see figure 5).

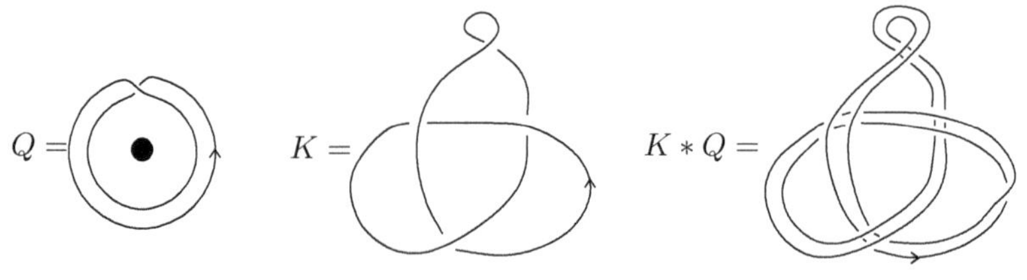

Fig. 5. Satellite construction

It is often possible to use satellites with some fixed choice of pattern Q in comparing two framed knots K and K'. When K and K' are equivalent then $K * Q$ and $K' * Q$ are equivalent. If we can find some knot invariant I for which $I(K * Q) \neq I(K' * Q)$ we can conclude that K' is not equivalent to K.

The Conway polynomial $\nabla(K)$ is not useful in this context, since

$$\nabla(K * Q) = \nabla(K' * Q)$$

for every choice of Q, if $\nabla(K) = \nabla(K')$.

This limitation does not hold in general. In particular the extension of the Conway polynomial known as the Homfly polynomial will often give useful extra information when applied to satellites.

Remark 1.2. The use of satellites is sometimes known as *cabling*. I prefer to reserve the term 'cable' for satellites where the pattern Q is based on some (p, q) torus knot.

2. Homfly invariants

In 1984 V.F.R.Jones constructed a new invariant of oriented links $V_L(t) \in \mathbf{Z}[t^{\pm\frac{1}{2}}]$, which turned out to have the property that

$$t^{-1}V_{L_+} - tV_{L_-} = (\sqrt{t} - 1/\sqrt{t})V_{L_0}$$

where the links

$L_+ =$ , $L_- =$ , $L_0 =$ 

differ only as shown. This was quickly extended to a 2-variable invariant $P_L(v, z) \in \mathbf{Z}[v^{\pm 1}, z^{\pm 1}]$, with the property that

$$v^{-1}P_{L_+} - vP_{L_-} = zP_{L_0}.$$

The name 'Homfly polynomial' has come to be attached to P, being the initial letters of six of the eight people involved in this further development. The polynomial P contains both the Conway/Alexander polynomial, and Jones' invariant, and can be shown to contain more information in general than both of these taken together. We have

$$\begin{aligned} P(1, z) &= \nabla(z) \\ P(1, s - s^{-1}) &= \Delta(s^2) \\ P(s^2, s - s^{-1}) &= V(s^2) \\ P(s, s - s^{-1}) &= \pm 1 \end{aligned}$$

The skein relation (1) can readily be shown to determine P and V once its value on the trivial knot is given. It has been usual to take $P = 1$ on the trivial knot, although in some recent applications a different normalisation can be more appropriate.

Given the existence of V and P we can then make some calculations. For example, the unlink with two components has

$$P = \frac{v^{-1} - v}{z},$$
$$V(s^2) = -(s + s^{-1}),$$

while the Hopf link with linking number $+1$ has

$$P = vz + (v^{-1} - v)v^2 z^{-1},$$
$$V(s^2) = s^3 - s - (s + s^{-1})s^4 = -s(1 + s^4).$$

The Hopf link with linking number -1 has

$$P = -v^{-1}z + (v^{-1} - v)v^{-2}z^{-1},$$
$$V(s^2) = -s^{-1}(1 + s^{-4}).$$

This illustrates the general feature that for the mirror image $\overline{L}$ of a link L, (where the signs of all crossings are changed), we have $P_{\overline{L}}(v, z) = P_L(v^{-1}, -z)$ and so $V_{\overline{L}}(s^2) = V_L(s^{-2})$. It is thus quite possible to use V in many cases to distinguish a knot from its mirror-image, while there will be no difference in their Conway polynomials. It is worth noting that although there are still knots which cannot be distinguished from each other by P in spite of being inequivalent, no non-trivial knot has so far been found for which $P = 1$, or even $V = 1$.

The original Homfly polynomial is invariant under *all* Reidemeister moves, but there is a convenient version which is an invariant of a framed oriented link

In its most adaptable form, $P_L(v, s)$, it lies in the ring

$$\Lambda = \mathbf{Z}[v^{\pm 1}, s^{\pm 1}, (s^r - s^{-r})^{-1}], r > 0.$$

Its defining characteristics are the local skein relations.

(1)

(2)

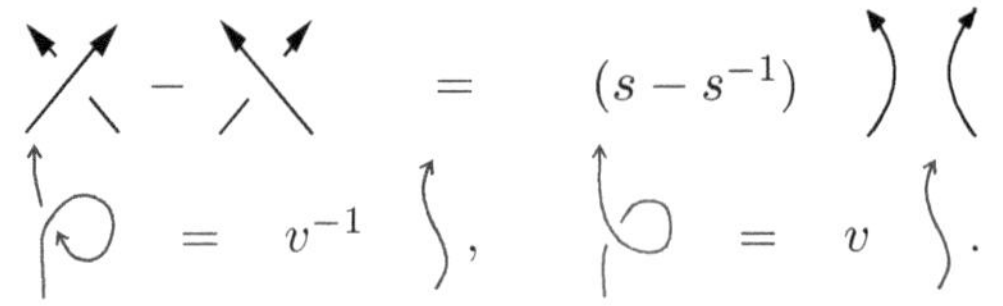

These relate the invariants of links whose diagrams differ only locally as shown.

They are enough to allow its recursive calculation from simpler diagrams in terms of the value for the unknot.

The unframed version for L, invariant under all Reidemeister moves, is given by $v^{w(D)}P_L(v, s)$ where D is a diagram for the framed link.

The local nature of the skein relations between invariants allows us to make a useful simplification in studying them.

Compare for example three patterns $Q_\pm$ and Q_0.

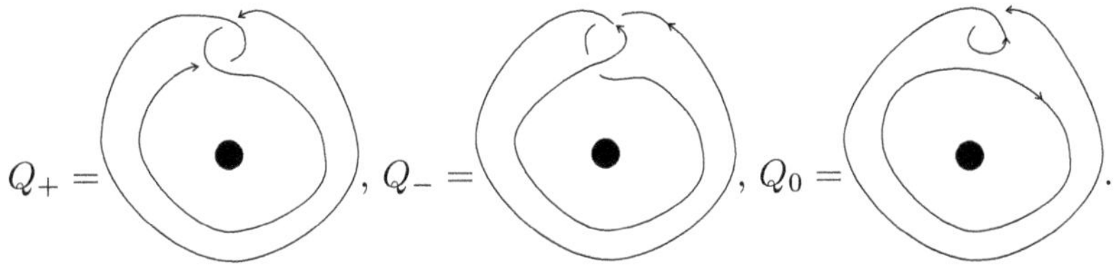

The framed Homfly invariants of $K * Q_\pm$ and $K * Q_0$ then satisfy

$$P(K * Q_+) - P(K * Q_-) = (s - s^{-1})P(K * Q_0).$$

Since $K * Q_-$ is the unknot for any K, this relates the invariants of the Whitehead double $K * Q_+$ of K and those of its reverse parallel.

More generally, consider the linear space $\mathcal{C}$ of Λ-linear combinations of diagrams in the annulus (up to R_{II}, R_{III}) and impose the local relations

(1)

(2)

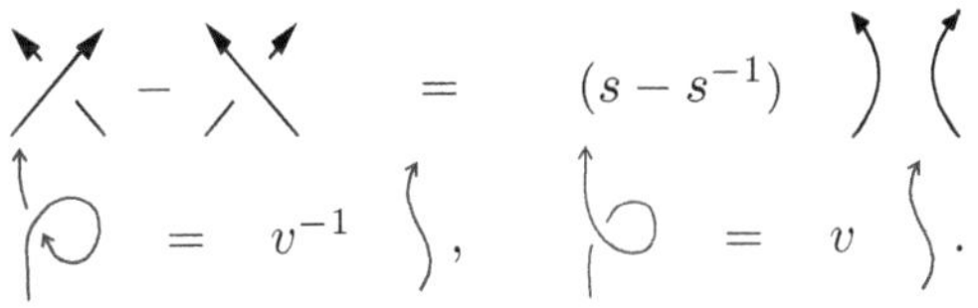

Decorating K by an element $\sum a_i Q_i$ of $\mathcal{C}$ gives a well-defined Homfly invariant $\sum a_i P(K * Q_i)$ since the skein relations are respected when the Homfly polynomials of the satellites are compared.

We could summarise our calculation above by saying that in the skein $\mathcal{C}$ we have

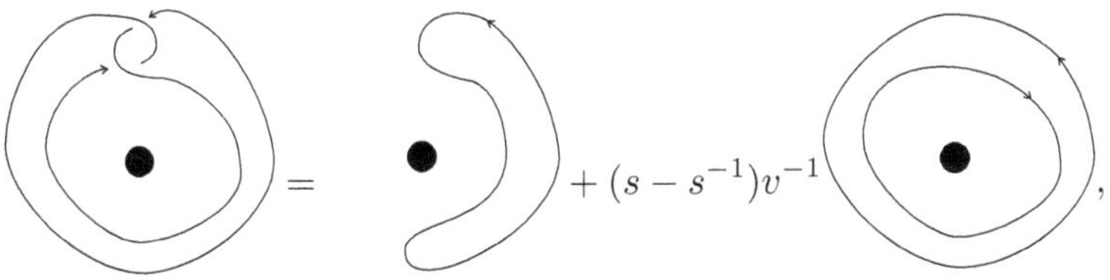

and hence

$$P(K * Q_+) = P(\text{unknot}) + (s - s^{-1})vP(\text{reverse parallel}).$$

The space $\mathcal{C}$, called the *Homfly skein of the annulus*, then gives a more effective parameter space for satellite invariants, as we only need to know the pattern as an element of $\mathcal{C}$.

For example, any of the twist patterns

is a linear combination of the reverse parallel and the trivial pattern, so the Homfly polynomial of any twisted double can be found from the reverse parallel.

We will return to look at more details of $\mathcal{C}$ later. For now, I will look at a further skein formulation which results in interesting models of certain algebras.

3. General Homfly skein theory

For a surface F with some designated input and output boundary points the (linear) Homfly skein of F is defined as linear combinations of framed oriented diagrams in F, up to Reidemeister moves II and III, modulo the skein relations

(1)
(2)

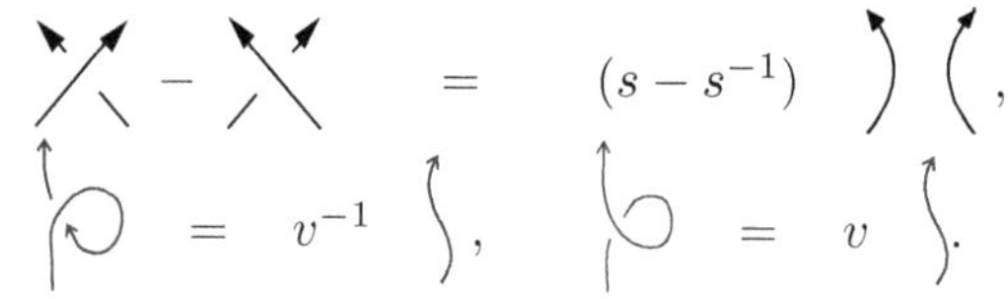

It is an immediate consequence that

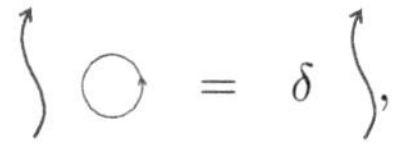

where $\delta = \dfrac{v^{-1}-v}{s-s^{-1}} \in \Lambda$. The coefficient ring Λ is taken as $Z[v^{\pm 1}, s^{\pm 1}]$, with denominators $\{r\} = s^r - s^{-r}, r \geq 1$.

We have already met the skein of the annulus, $\mathcal{C}$.

In the skein of $\mathbf{R^2}$ or S^2 every diagram D is equivalent to a multiple of the trivial diagram. Explicitly,

$$D = P(D)$$

where $P(D)$ is the framed Homfly polynomial of D.

As a general rule, geometric operations induce linear maps on the corresponding skeins. For example, given a framed knot K there is a linear map $K* : \mathcal{C} \to \mathcal{S}(\mathbf{R}^2)$ induced by $Q \mapsto K * Q$.

The skein of the rectangle with m inputs at the top and m outputs at the bottom is denoted by H_m. Elements are represented by combinations of diagrams in the rectangle made up of m arcs joining the input and output points, and possibly some further closed curves. Such diagrams are known as m-tangles.

A simple example of an m-tangle is an m-braid, while another important m-tangle is the tangle

$T^{(m)} =$ 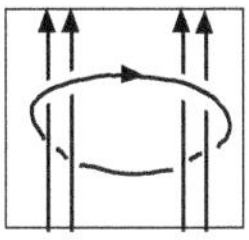 .

3.1. *Composition*

Putting one m-tangle above another defines an associative product with identity.

Theorem 3.1. *The set of invertible tangles consist of the m-braids, which form the braid group B_m.*

Artin's braid group B_m has a presentation in terms of elementary braids, $\{\sigma_i\}, i = 1, \ldots, m-1$ satisfying the braid relations

$$\sigma_i \sigma_j = \sigma_j \sigma_i, \quad |i-j| > 1,$$
$$\sigma_i \sigma_{i+1} \sigma_i = \sigma_{i+1} \sigma_i \sigma_{i+1},$$

where

$\sigma_i =$  .

Tangle composition induces a product in the skein H_m. This skein is spanned by a finite set of m-braids. One such spanning set consists of the $m!$ 'totally descending' braids in which the m arcs of the tangle are numbered from the bottom left, and each crossing is met first as an overcrossing on going along the arcs in order. These braids are sometimes termed 'positive permutation braids', and they each realise one of the permutations of the endpoints.

Then H_m forms a finite-dimensional algebra, with a presentation on generators $\{\sigma_i\}, i = 1, \ldots, m-1$ satisfying the braid relations

$$\sigma_i \sigma_j = \sigma_j \sigma_i, \quad |i-j| > 1,$$
$$\sigma_i \sigma_{i+1} \sigma_i = \sigma_{i+1} \sigma_i \sigma_{i+1},$$

and the quadratic relations $\sigma_i^2 = (s - s^{-1})\sigma_i + 1$, which result from the skein relation

$$\sigma_i - \sigma_i^{-1} = (s - s^{-1})\mathrm{Id}.$$

The resulting algebra is also known as the Hecke algebra $H_m(z)$, when $z = s - s^{-1} = \{1\}$ and the coefficients are extended to Λ. The Hecke algebra H_m can be also seen as the group algebra of Artin's braid group B_m generated by the elementary braids σ_i, $i = 1, \ldots, m-1$, modulo the further quadratic relation $\sigma_i^2 = z\sigma_i + 1$.

In the special case $z = 0$ the Hecke algebra reduces to the group algebra of the symmetric group, $\mathbf{C}[S_m]$, with σ_i becoming the transposition $(i\ i{+}1)$.

3.2. *Closure*

The closure map from H_m to $\mathcal{C}$ is the Λ-linear map induced by considering the closure $\hat{T}$ of a tangle T in the annulus (see figure 6). The image of this map is denoted by $\mathcal{C}_m$.

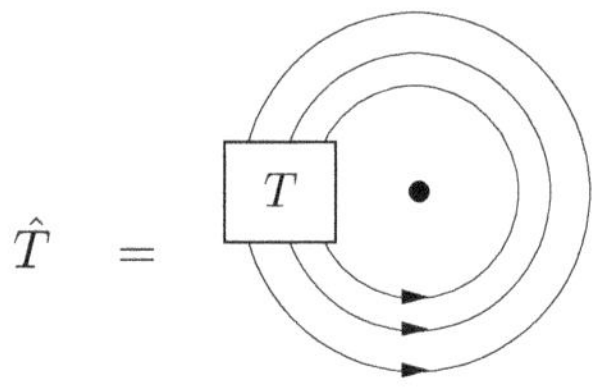

Fig. 6. The closure map

4. The skein of the annulus

The skein $\mathcal{C}$ of the annulus has been used formally for some time as a parameter space for the Homfly satellite invariants of a knot.

It has a product structure induced at the level of diagrams by placing one annulus outside another. This defines a bilinear product under which $\mathcal{C}$ becomes an algebra. This algebra is clearly commutative (lift the inner annulus up and stretch it so that the outer one will fit inside it).

Remark 4.1. The Kauffman bracket skein of the annulus gives a quotient of this algebra which is combinatorially convenient to use when working with the specialisation to the Jones polynomial.

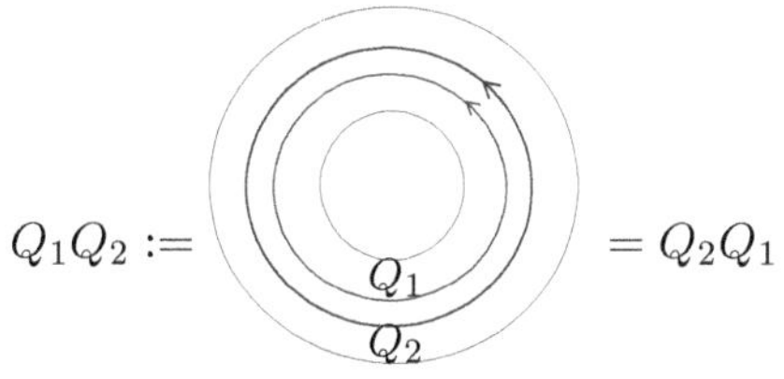

Fig. 7. The product Q_1Q_2

Turaev [17] showed that $\mathcal{C}$ is freely generated as an algebra by the elements A_m, which are the closures of the m-braids $\sigma_{m-1}\cdots\sigma_2\sigma_1 \in H_m$, and the elements A_m^*, with the reverse orientation. The identity element in the algebra is represented by the empty diagram in the annulus.

So the pattern Q_+ for the Whitehead double is a linear combination of 1 and $A_1A_1^*$ in this notation.

The linear subspace $\mathcal{C}_m$ spanned by the closures of m-tangles has a linear basis of monomials in $\{A_i\}$ with total weight m, where A_i has weight i. There are $p(m)$ of these, where $p(m)$ is the number of partitions of m. Thus $\mathcal{C}_3$ is spanned by A_1^3, A_1A_2 and A_3.

Although the satellite invariants of a knot K behave additively under addition of patterns, there is no relation between the invariants with patterns Q_1, Q_2 and Q_1Q_2. It may then happen that there are $p(m)$ independent invariants of a knot arising from decorations in $\mathcal{C}_m$.

In the interests of relating these to other invariants it is good to work with a rather different basis for $\mathcal{C}_m$, and indeed for the whole skein $\mathcal{C}$, which has the advantage of behaving well when the framing of the knot is changed.

For example when an extra twist is added to the framing of a knot K to form K' the satellite $K' * A_1^2$ becomes $K * Q$ with $Q = v^{-2}A_1^2 + zv^{-2}A_2$ in the skein $\mathcal{C}_2$ so that $P(K' * A_1^2) = v^{-2}P(K * A_1^2) + v^{-2}zP(K * A_2)$.

The two basis elements $Q_1 = A_1^2 + sA_2$ and $Q_2 = A_1^2 - s^{-1}A_2$ are much better for framing changes, in the sense that $P(K' * Q_1) = v^{-2}s^2P(K * Q_1)$ while$P(K' * Q_2) = v^{-2}s^{-2}P(K * Q_2)$

The framing change map is illustrated in figure 8 by its effect on the 2-parallel element $(A_1)^2$.

The framing change map is connected to the element in the Hecke algebra H_m represented by the full twist Δ_m^2 on m strings. This braid commutes with all m-braids, and hence represents an element in the centre of the Hecke algebra H_m.

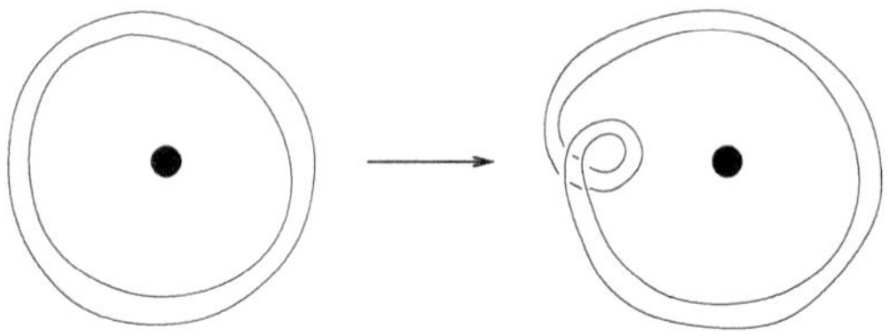

Fig. 8. The framing change map on a 2-parallel

A further important central element of H_m is represented by the tangle

$T^{(m)} =$ 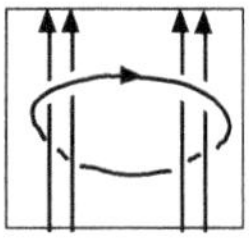

consisting of an oriented meridian curve around m parallel strings.

Closely related to the elements $T^{(m)}$ are the meridian maps $\varphi, \overline{\varphi} : \mathcal{C} \to \mathcal{C}$ in the skein of the annulus.

4.1. *Meridian maps*

The *meridian map* $\varphi : \mathcal{C} \to \mathcal{C}$ is induced by including a single meridian curve around a diagram Q in the thickened annulus to give the diagram shown in figure 9.

$\varphi(Q) =$ 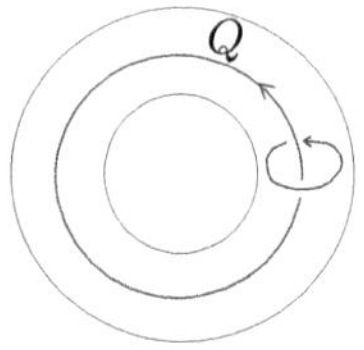

Fig. 9. The meridian map

The map $\overline{\varphi}$ is given similarly, using the opposite orientation on the meridian curve.

When Q is the closure $Q = \hat{T}$ of an m-tangle T then $\varphi(Q)$ is the closure of $T^{(m)}T$.

The subspace $\mathcal{C}_m$ spanned by closed m-tangles is invariant under the meridian map φ. The map $\varphi|\mathcal{C}_m$ has $p(m)$ distinct eigenvalues, one for each partition of m, with a 1-dimensional space of eigenvectors for each eigenvalue.

While $\mathcal{C}_m$ is also invariant under the framing change map, this map has fewer distinct eigenvalues than φ for $m \geq 6$. The eigenvectors for φ are also eigenvectors for the framing change map, and indeed the basis given above for $\mathcal{C}_2$ consisted of eigenvectors for φ.

4.2. *Partitions*

Partitions are widely used in descriptions of irreducible representations of the symmetric groups.

A partition λ of m into k parts $\lambda_1 \geq \lambda_2 \geq \ldots \geq \lambda_k > 0$ can be represented combinatorially by a *Young diagram* with m cells arranged in k rows. Successive rows have $\lambda_1, \lambda_2, \ldots, \lambda_k$ cells starting from a fixed left-hand end.

Theorem 4.1 (Hadji, Morton). *There is a basis for the skein $\mathcal{C}$ consisting of eigenvectors $Q_{\lambda,\mu}$ of the meridian map φ. Here λ and μ run through the set of all partitions, and the corresponding eigenvalues $s_{\lambda,\mu}$ are all* distinct.

The basis $Q_{\lambda,\mu}$ is thus very natural, and it shows up in many different ways.

For example the basis vectors are then also eigenvectors for any other linear endomorphism of $\mathcal{C}$ which commutes with φ. These include $\overline{\varphi}$ and the framing change map.

A further example is given by drawing a given knot K as the closure of a 1-tangle in the annulus.

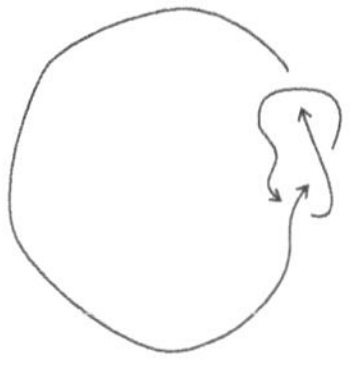

Decorate this with a pattern Q to get a diagram for $K * Q$ in the annulus,

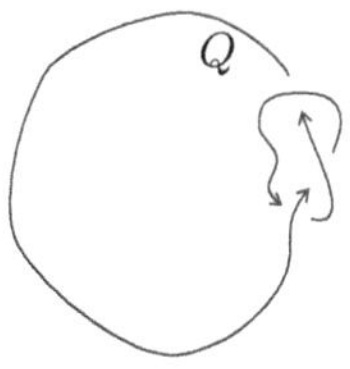

inducing a map $T_K : \mathcal{C} \to \mathcal{C}$.

Now T_K commutes with φ,

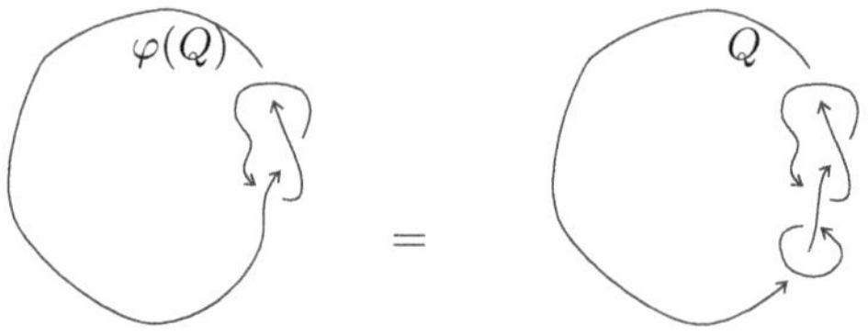

so $T_K(Q_{\lambda,\mu}) = a(K, \lambda, \mu)Q_{\lambda,\mu}$.

Theorem 4.2 (Morton). *The eigenvalues* $a(K, \lambda, \mu) \in \Lambda$ *are* integral *in* Λ, *and are the ratio of the Homfly invariants*

$$\frac{P(K * Q_{\lambda,\mu})}{P(\bigcirc * Q_{\lambda,\mu})}.$$

4.3. Branching rules in $\mathcal{C}$

The basis $Q_{\lambda,\mu}$ for $\mathcal{C}$ also behaves well under the product operation, namely the product of two basis elements is always a non-negative integer combination of basis elements. These can be found explicitly by combinatorial formulae from classical work with partitions.

Besides the identity element in $\mathcal{C}$, which is represented by the empty diagram, and forms the basis element $Q_{\lambda,\mu}$ with $|\lambda| = |\mu| = 0$, the simplest basis elements are the single oriented core curves A_1 and A_1^*. These represent $Q_{1,\phi}$ with $|\lambda| = 1, |\mu| = 0$ and $Q_{\phi,1}$ respectively.

The branching rules for these can be summarised as

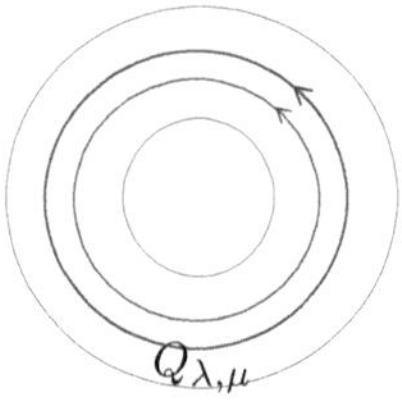

$$= A_1 Q_{\lambda,\mu} = \sum_{\rho \in \lambda^+} Q_{\rho,\mu} + \sum_{\nu \in \mu^-} Q_{\lambda,\nu}.$$

Here λ^+ is the set of partitions given from the Young diagram of λ by adding one further cell, and λ^- is the set of partitions given by removing a single cell.

5. Symmetric functions and the skein of the annulus

This section is concerned primarily with the subspace $\mathcal{C}^+$ of the skein of the annulus which is spanned by the elements $Q_{\lambda,\mu}$ with μ empty, or equally by the subspaces $\mathcal{C}_m$ spanned by the closures of directly oriented m-tangles. We write $Q_\lambda := Q_{\lambda,\phi}$ for the spanning basis elements.

An algebraic model of $\mathcal{C}^+$ that fits particularly well with these basis elements $\{Q_\lambda\}$ and also connects with the ideas of quantum group representations is that of symmetric functions.

5.1. Symmetric functions

We consider polynomials in N commuting variables $x_1, \ldots, x_N$ which are unchanged by permutation of the variables. The most familiar are the *elementary symmetric functions*

$$e_m = \sum_{i_1<i_2<\ldots<i_m} x_{i_1} x_{i_2} \ldots x_{i_m}.$$

These appear as the coefficients of the polynomial

$$E(t) = \prod_{i=1}^{N} (1 + x_i t) = 1 + e_1 t + \cdots + e_m t^m + \cdots$$

The *complete symmetric functions* are the coefficients of

$$H(t) = \prod_{i=1}^{N} \frac{1}{1 - x_i t} = 1 + h_1 t + \cdots + h_m t^m + \cdots$$

The generating series for these two sets of functions satisfy the relation $E(t)H(-t) = 1$.

Other familiar symmetric functions are the *power sums* $P_m = x_1^m + \cdots + x_N^m$.

A classical result says that every symmetric integer polynomial in $x_1, \ldots, x_N$ is an integer polynomial in $\{e_m\}$ and also in $\{h_m\}$. Indeed the polynomial is independent of the number of variables N for large enough N. For example $p_2 = e_1^2 - 2e_2$ for $N > 1$.

There is an extensive body of literature about symmetric functions. They occur in the representation theory of symmetric groups and the related representation theory of unitary groups. One substantial reference is the book of Macdonald [8].

The *character* of a representation of the unitary group $U(N)$ is the trace of the representing matrices considered as a function on the representation of the diagonal matrices (which form a maximal torus in the group). The trace gives a function on diagonal matrices which is symmetric in the entries. Characters add under sum of representations, and multiply under tensor product.

The irreducible representations correspond to certain symmetric functions called the *Schur* functions. The Schur functions $s_\lambda(\mathbf{x})$ of degree m form a basis for all degree m symmetric polynomials in $\mathbf{x} = (x_1, \ldots, x_N)$, and they correspond directly with the partitions λ of m. By the general result above each Schur function can be expressed as a polynomial in the elementary symmetric functions $\{e_m\}$, or the complete symmetric functions $\{h_m\}$. The functions e_m and h_m themselves are Schur functions, corresponding to the partitions of m into a single column or row respectively.

In the skein of the annulus a choice of elements to represent the complete symmetric functions $\{h_m\}$ can be made in such a way that the resulting Schur polynomial s_λ represents the basis element Q_λ [7]. The interpretation of $\mathcal{C}^+$ as symmetric functions based on this choice of representatives for $\{h_m\}$ then leads to a natural role for $\{Q_\lambda\}$ as the Schur functions. It allows the known formulae for products of Schur functions to tell us how to write a product of basis elements $Q_\lambda Q_\rho$ as a sum of basis elements. It also suggests a relation to the irreducible representations of the unitary groups. It is striking that the elements representing the power sums also play a significant role in satellite constructions and have satisfying geometric representatives [10].

5.2. Construction of the basis elements

The elements h_m and e_m can be constructed readily in terms of the simplest idempotents of the Hecke algebra H_m.

The element $h_m \in \mathcal{C}_m$, which is taken to represent the complete symmetric function of degree m, is the closure of the element $\frac{1}{\alpha_m} a_m \in H_m$ where

$$a_m = \sum_{\pi \in S_m} s^{l(\pi)} \omega_\pi$$

is one of the two basic quasi-idempotent elements of H_m. Here ω_π is the positive permutation braid associated to the permutation $\pi \in S_m$ with length $l(\pi)$, which is the writhe of the braid w_π. The scalar α_m is given by the equation $a_m a_m = \alpha_m a_m$ [2, 7, 9]. Using the other quasi-idempotent

$$b_m = \sum_{\pi \in S_m} (-s)^{-l(\pi)} \omega_\pi$$

in a similar way determines the element e_m which represents the elementary symmetric function. These elements are related by the power series equation $H(t)E(-t) = 1$. These two idempotents arise from the 1-dimensional representations of H_m in which $\sigma_i \to s$ or $\sigma_i \to -s^{-1}$ respectively.

Aiston's view of the elements Q_λ is in a more 3-dimensional context of combinations of diagrams in a solid torus, rather than an annulus. We show below a diagrammatic view of a linear combination e_λ of 3-dimensional braids, whose endpoints lie on the cells of a Young diagram λ, rather than in the conventional straight line. In this illustration $m = 9$ and λ is the partition $4, 3, 2$.

$e_\lambda \quad =$ 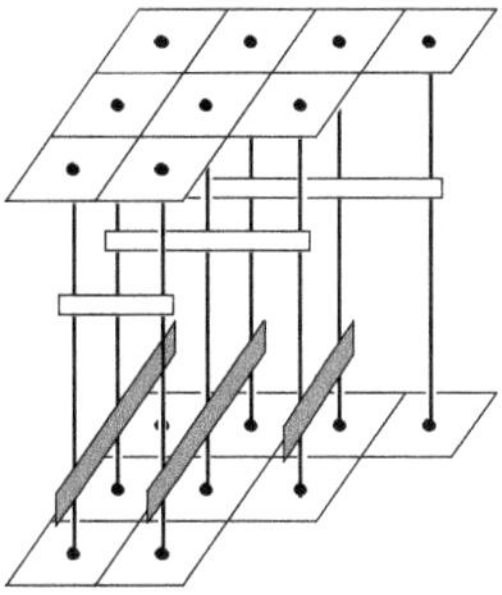

Here e_λ should be regarded as an element of the Homfly skein of $D^2 \times I$, with endpoints at the top and bottom on the template λ, and with some implicit choice of parallel for each strand to determine a framing. The white boxes, following the rows of λ, contain the braid combination a_j when the box has length j, while the grey boxes, following the columns, similarly contain combinations b_j. The whole combination will be denoted by e_λ. (The notation e_λ is used in [1] for a closely related element of the Hecke algebra $H_{|\lambda|}$ given by making a specific arrangement of the $|\lambda|$ endpoints in a straight line.) In either context the element e_λ can naturally be composed with itself, and satisfies the relation $e_\lambda^2 = \alpha_\lambda e_\lambda$ for some scalar $\alpha_\lambda \in \Lambda_1$. Aiston defines the element Q_λ by

$$Q_\lambda = \frac{1}{\alpha_\lambda} \hat{e}_\lambda,$$

where the closure of e_λ is an element of the skein $\mathcal{C}$.

In defining Q_λ in this way we have to ensure that the coefficient ring includes denominators α_λ. There is an explicit formula

$$\alpha_\lambda = \prod_{x \in \lambda} s^{c(x)} \frac{s^{h(x)} - s^{-h(x)}}{s - s^{-1}}$$

for α_λ as a product over the cells x of λ.

In this formula the *content*, $c(x)$, of a cell x in position (i, j) in a Young diagram for λ is $c(x) = j - i$. The *hook length*, $h(x)$, of the cell x is the total number of cells immediately to its right and immediately below it in the Young diagram.

Thus the denominators in Q_λ are indeed of the form $s^k - s^{-k}$, where the largest value of k is the largest hook length of any cell. This occurs for the cell in position $(1,1)$, at the top left of λ.

One striking feature [2] of the elements e_λ is their 'internal stability', namely that if any tangle T is inserted in $D^2 \times I$ between the white and the grey boxes, as shown schematically here, the resulting element of the skein is just some scalar multiple $t_\lambda e_\lambda$ of e_λ.

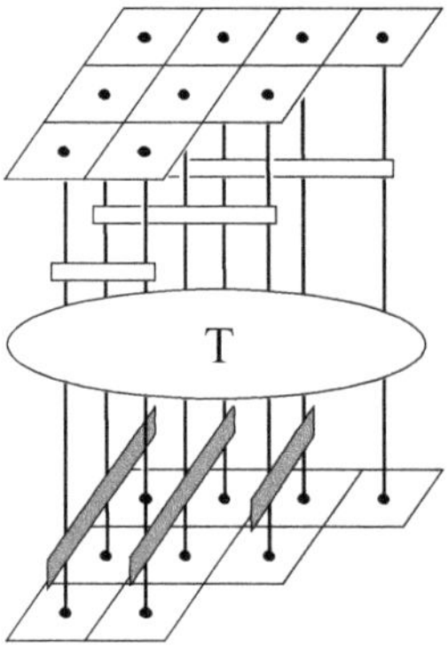

$= \quad t_\lambda e_\lambda$

The fact that $e_\lambda^2 = \alpha_\lambda e_\lambda$ for some $\alpha_\lambda \in \Lambda$ is an immediate consequence of this, although we need to know also that $\alpha_\lambda \neq 0$ in order to construct Q_λ.

An important case is when T is the complete right-hand curl on $|\lambda|$ strings. The resulting scalar $f_\lambda \in \Lambda$ is known as the *framing factor* for λ. When the invariant $P(L;\ldots,Q_\lambda,\ldots)$ is calculated with one component of the link L decorated by Q_λ, and the framing on that component is increased by 1, keeping the decorations of all other components unchanged, then the value of the invariant is multiplied by f_λ. This can be readily seen because the two invariants to be compared can be calculated from diagrams which differ only in having e_λ with or without the full curl inside it as one part of the complete diagram. A direct skein theory calculation [2] gives a cell-based formula

$$f_\lambda = v^{-|\lambda|} s^{n_\lambda}, \text{ where } n_\lambda = 2\sum_{x\in\lambda} c(x),$$

twice the sum of the content of the cells.

Besides the full curl, another important central element of the Hecke algebra H_m is the element $T^{(m)}$. The internal stability ensures that placing $T^{(m)}$ inside e_λ results in a multiple $s_\lambda e_\lambda$. Then the closure of e_λ is an eigenvector of the meridian map with eigenvalue s_λ, and can therefore be identified with one of the basis elements $Q_{\lambda,\mu}$, up to a scalar. This is the argument adopted by Lukac [7] to identify his element Q_λ, originally constructed in terms of Schur functions as a determinant of a matrix with entries drawn from the elements $\{h_k\}$, with Aiston's element constructed from the idempotent e_λ.

6. Unitary quantum invariants

Quantum groups give rise to 1-parameter invariants $J(K;W)$ of an oriented framed knot K depending on a choice of finite dimensional module W over the quantum group, following constructions of Turaev and others [2, 17, 19]. This choice is referred to as *colouring* K by W, and can be extended for a link to allow a choice of colour for each component.

Fix a natural number N. When we colour K by a finite dimensional module W over the quantum group $sl(N)_q$, its invariant $J(K;W)$ depends on one variable s. The invariant J is linear under direct sums of modules and all the modules over $sl(N)_q$ are semi-simple, so we can restrict our attention to the irreducible modules $V_\lambda^{(N)}$. For $sl(N)_q$ these are indexed by partitions λ with at most N parts, without distinguishing two partitions which differ in some initial columns with N cells.

To help in our comparison between Homfly satellite invariants and quantum invariants of K we write $P(K;Q)$ for $P(K*Q)$ and more generally $P(L;Q_1,Q_2,\ldots,Q_k)$ for the Homfly polynomial of a link L when its components are decorated by $Q_1,\ldots,Q_k$ respectively.

Theorem 6.1 (Comparison theorem). *(1) The $sl(N)_q$ invariant for the irreducible module $V_\lambda^{(N)}$ is the Homfly invariant for the knot decorated by Q_λ with $v=s^{-N}$, suitably normalised as in [6]. Explicitly,*

$$P(K;Q_\lambda)|_{v=s^{-N}} = x^{k|\lambda|^2} J(K;V_\lambda^{(N)})$$

where k is the writhe of K, and $x=s^{1/N}$.

(2) Each invariant $P(K;Q)|_{v=s^{-N}}$ is a linear combination of quantum invariants $\sum c_\alpha J(K;W_\alpha)$.

(3) Each $J(K;W)$ is a linear combination of Homfly invariants

$$\sum d_j P(K;Q_j)|_{v=s^{-N}}.$$

Remark 6.1.

- In the special case when $N = 2$ we can interpret quantum invariants of K in terms of Kauffman bracket satellite invariants, using the skein of the annulus based on the Kauffman bracket relations. This simpler skein is a quotient of the algebra $\mathcal{C}$. More generally the $sl(N)_q$ invariants depend only on a quotient of the algebra $\mathcal{C}$ for each N.
- The 2-variable invariant $P(K;Q)$ can be recovered from the specialisations $P(K;Q)|_{v=s^{-N}}$ for sufficiently many N.
- If the pattern Q is a closed braid on m strings then we only need use partitions $\lambda \vdash m$, since $\mathcal{C}_m$ is spanned by $\{Q_\lambda\}_{\lambda \vdash m}$. Conversely, to realise $J(K; V_\lambda^{(N)})$ with $\lambda \vdash m$ we can use closed m-braid patterns.

6.1. *Basic constructions of quantum invariants*

A quantum group $\mathcal{G}$ is an algebra over a formal power series ring $\mathbf{Q}[[h]]$, typically a deformed version of a classical Lie algebra. We write $q = e^h, s = e^{h/2}$ when working in $sl(N)_q$. A finite dimensional module over $\mathcal{G}$ is a linear space on which $\mathcal{G}$ acts.

Crucially, $\mathcal{G}$ has a coproduct Δ which ensures that the tensor product $V \otimes W$ of two modules is also a module. It also has a *universal R-matrix* (in a completion of $\mathcal{G} \otimes \mathcal{G}$) which determines a well-behaved module isomorphism

$$R_{VW} : V \otimes W \to W \otimes V.$$

This has a diagrammatic view indicating its use in converting coloured tangles to module homomorphisms.

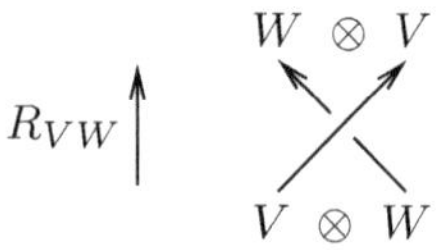

A braid β on m strings with permutation $\pi \in S_m$ and a colouring of the strings by modules $V_1, \ldots, V_m$ leads to a module homomorphism

$$J_\beta : V_1 \otimes \cdots \otimes V_m \to V_{\pi(1)} \otimes \cdots \otimes V_{\pi(m)}$$

using $R^{\pm 1}_{V_i,V_j}$ at each elementary braid crossing. The homomorphism J_β depends *only on the braid* β itself, not its decomposition into crossings, by the Yang-Baxter relation for the universal R-matrix.

When $V_i = V$ for all i we get a module homomorphism $J_\beta : W \to W$, where $W = V^{\otimes m}$. Now any module W decomposes as a direct sum

$\bigoplus(W_\mu \otimes V_\mu^{(N)})$, where $W_\mu \subset W$ is a linear subspace consisting of the *highest weight vectors* of type μ associated to the module $V_\mu^{(N)}$. Highest weight subspaces of each type are preserved by module homomorphisms, and so J_β determines (and is determined by) the restrictions $J_\beta(\mu) : W_\mu \to W_\mu$ for each μ, where μ runs over partitions with at most N parts.

If a knot (or one component of a link) K is decorated by a pattern T which is the closure of an m-braid β, then its quantum invariant $J(K*T;V)$ can be found from the endomorphism J_β of $W = V^{\otimes m}$ in terms of the quantum invariants of K and the restriction maps $J_\beta(\mu) : W_\mu \to W_\mu$ by the formula

$$J(K*T;V) = \sum c_\mu J(K;V_\mu^{(N)}) \tag{1}$$

with $c_\mu = \operatorname{tr} J_\beta(\mu)$. This formula follows from lemma II.4.4 in [18]. We set $c_\mu = 0$ when W has no highest weight vectors of type μ.

More generally the methods of Reshetikhin and Turaev allow the quantum groups $\mathcal{G} = SU(N)_q$ to be used to represent oriented tangles whose components are coloured by $\mathcal{G}$-modules as $\mathcal{G}$-module homomorphisms. One additional feature is needed, namely the use of the dual module V^* defined by means of the antipode in $\mathcal{G}$, (an antiautomorphism of $\mathcal{G}$ which is part of its structure as a Hopf algebra). When the components of the tangle are coloured by modules the tangle itself is represented by a homomorphism from the tensor product of the modules which colour the strings at the bottom to the tensor product of the modules which colour the strings at the top, provided that the string orientations are inwards at the bottom and outwards at the top. The dual module V^* comes into play in place of V when an arc of the tangle coloured by V has an output at the bottom or an input at the top.

For example, the $(4,2)$-tangle below, when coloured as shown, is represented by a homomorphism $U \otimes W^* \to U \otimes X^* \otimes X \otimes W^*$.

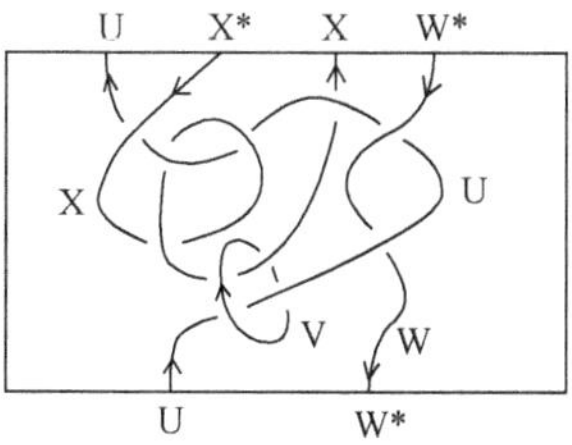

It is possible to build up the definition so that consistently coloured tangles are represented by the appropriate composite homomorphisms, starting from a definition of the homomorphisms for the elementary oriented tangles. Two cases, depending on the orientation, must be considered for both the local maximum and the local minimum, and a little care is needed here to ensure consistency. The final result is a definition of a homomorphism which is invariant when the coloured tangle is altered by R_{II} and R_{III}. When applied to an oriented k-component link diagram L regarded as an oriented $(0,0)$-tangle it gives an element $J(L; V_1, \ldots, V_k) \in \Lambda = \mathbf{Q}[[h]]$ for each colouring of the components of L by $\mathcal{G}$-modules, which is an invariant of the framed oriented link L.

The construction is simplified in the case of $sl(2)_q$ by the fact that all modules are isomorphic to their dual, and so orientation of the strings plays no role.

The quantum group invariants based on $sl(3)_q$ also admit a combinatorial simplification due to Kuperberg to allow an easier diagrammatic calculation of them. At the same time the quantum group itself is straightforward enough to make it possible to work directly with some of the smaller dimensional modules [12, 16].

7. Manifold invariants

Following work of Reshetikhin and Turaev, in response to ideas of Witten, there are increasingly sophisticated ways to construct invariants of oriented 3-dimensional manifolds based on quantum groups, and correspondingly on knot invariants such as the Homfly satellite invariants. The basic principles come from the original paper of Reshetkhin and Turaev, adapted at various times to give easier details in special cases, notably the case of the quantum $SU(2)$ invariants, as for example in [11].

7.1. Surgery presentation

The strategy is to present the manifold M by *surgery on a framed link* L with k components. This means that M is given by removing a neighbourhood of M from S^3 to give a manifold with k torus boundary components and then reattaching a solid torus to each of the k boundary tori in a way determined by the framing. The resulting manifold $M = M(L)$ depends on the choice of L. Any other link L' which also determines the same manifold M is related to L by a sequence of *Kirby moves* and their inverses.

These can be summarised as operations on framed link diagrams regarded in some way as a satellite of the unknot U_0 with framing 0. Then we can replace a link $L = U_0 * Q$ by $U_{\pm 1} * A_1 Q$, where $U_{\pm 1}$ is the unknot with framing ± 1 and $A_1 Q$ is the decoration Q with one extra parallel strand.

The strategy for finding invariants of M is then to identify knot invariants of framed links which are unchanged by the two basic Kirby moves.

7.2. *Manifolds with boundary*

The whole setting of manifold invariants is extended to include manifolds with boundary, regarded as cobordisms between two subsets of their boundary components. The wider setting envisages a standard vector space for each boundary component, associating a vector space to the incoming and outgoing boundary, with a linear map between them determined by the manifold itself, in such a way that pasting together manifolds corresponds to composition of linear maps. This is sometimes termed a 'modular functor' or 'topological quantum field theory' (TQFT).

Associated to the empty boundary component is the 1-dimensional space of scalars. A closed manifold then produces a linear map from scalars to scalars, in other words a scalar.

The exterior of a link L, with k torus boundary components thought of as the incoming boundary, and empty outgoing boundary, fits in to this general scheme. We could take the skein $\mathcal{C}$ (or $\mathcal{C}^+$) as the linear space associated to a torus and use the Homfly satellite invariants of L to provide a linear map from the k-fold tensor product of $\mathcal{C}$ to the scalars. One difficulty in trying to extend this to give a TQFT is that there is no immediate candidate for handling outgoing torus boundaries and hence no scope for gluing manifolds together along torus boundary components.

All the same, it suggests that what might be needed when attaching a solid torus to a boundary component would be to find an invariant for a solid torus, regarded as having empty incoming boundary and a torus as outgoing boundary. According to the proposed scheme we would need a linear map from the scalars to the linear space $\mathcal{C}$, which simply means the choice of one preferred element Ω, say, of $\mathcal{C}$. The resulting scalar for the manifold given by attaching a solid torus to each boundary component of the exterior of L would then be the evaluation of the satellite invariant of L where each component is decorated by Ω, giving $P(L:\Omega,\ldots,\Omega)$ as the invariant of the manifold $M(L)$.

Although it is *not* possible to find such a universal element Ω to carry through this plan it turns out that *a restricted version of this idea, sketched below, can be made to work.*

7.3. *Evaluation of knot invariants*

The comparison between $sl(N)_q$ invariants and Homfly satellite invariants uses the evaluation of the Homfly invariants by putting $v = s^{-N}$. This is the ring homomorphism $s_N : \Lambda \to \mathbf{Z}[s^{\pm 1}, (s^r - s^{-r})^{-1}]$ with $s_N(v) = s^{-N}, s_N(s) = s$.

For the trivial knot U write

$$\delta(Q) = P(U; Q) \in \Lambda.$$

Now $\delta(Q_1Q_2) = \delta(Q_1)\delta(Q_2)$. There is a nice formula

$$\delta(Q_\lambda) = \prod_{x\in\lambda} \frac{v^{-1}s^{c(x)} - vs^{-c(x)}}{s^{h(x)} - s^{-h(x)}}. \tag{2}$$

Since $s_N(\delta(Q_\lambda)) = J(U; V_\lambda^{(N)})$ is the 'quantum dimension' of the module $V_\lambda^{(N)}$ it is common to call $\delta(Q)$ the *quantum dimension* of $Q \in \mathcal{C}$.

It follows from (2) that $s_N(\delta(Q_\lambda)) = 0$ if λ has more than N rows. It is then also true that $s_N(P(L; \ldots, Q_\lambda, \ldots)) = 0$ if λ has more than N rows.

To find the s_N evaluation of a Homfly satellite invariant we then only need to know its value for decorations with at most N rows. This can be simplified further, as decorations by λ and λ' give the same s_N evaluation when λ and λ' differ by a number of columns with exactly N cells in each.

For example in calculating the s_2 evaluation (to get the Jones polynomial) we only need to use decorations with one row.

7.4. Level invariants for manifolds

Following Witten, Reshetikhin and Turaev we can use quantum group invariants to get a sequence of manifold invariants, along the general lines proposed above.

Choose $N \geq 2$ and a further positive integer l, termed the *level*.

Write

$$\Omega_{N,l} = \sum_{\lambda\in(N-1,l)} \delta(Q_\lambda)Q_\lambda,$$

where $(N-1, l)$ is the finite set of partitions with at most $N-1$ rows and at most l columns.

Take $s \in \mathbf{C}$ to satisfy $s^{2(l+N)} = 1$.

Theorem 7.1. *The evaluation $s_N(P(L; \Omega_{N,l}, \ldots, \Omega_{N,l})) \in \mathbf{C}$ is an invariant of the manifold $M(L)$, up to a normalising factor depending on the linking numbers of L.*

In this form the result is shown by Aiston [1] with some refinements by Lukac [7]; a brief account can also be found in [13], where the more natural use of dual modules in places makes for a slightly easier argument. In this context, having chosen N, the Young diagram of the partition λ^* dual to λ is the complement of the diagram of λ in an $N \times \lambda_1$ rectangle.

The amended definition for $\Omega_{N,l}$ is

$$\Omega_{N,l} = \sum_{\lambda \in (N-1,l)} \delta(Q_{\lambda^*}) Q_\lambda.$$

Since $s_N(\delta(Q_{\lambda^*})) = s_N(\delta(Q_\lambda))$, the result above is unaltered.

The main technical fact needed about $\Omega_{N,l}$ is that the product

$$S\Omega_{N,l} = \delta(S)\Omega_{N,l}$$

for any $S \in \mathcal{C}^+$, modulo elements of an ideal which contribute 0 to the s_N evaluation when $s^{2(N+l)} = 1$.

Proof.

To show Kirby move invariance, when we change $L = U_0 * Q$ to $U_{\pm 1} * A_1 Q$, decorate the components of the diagram Q in the annulus by $\Omega_{N,l}$ to determine an element $S \in \mathcal{C}^+$.

We need to compare $P(U_0; S)$ and $P(U_{\pm 1}; S\Omega_{N,l})$. Now $P(U_0; S) = \delta(S)$ and

$$P(U_{\pm 1}; S\Omega_{N,l}) = P(U_{\pm 1}; \delta(S)\Omega_{N,l}) = \delta(S) P(U_{\pm 1}; \Omega_{N,l}),$$

after evaluation.

The factors $c_\pm = P(U_{\pm 1}; \Omega_{N,l})$ are dealt with by the normalisation. □

Remark 7.1. When evaluating invariants under s_N with the additional restriction that $s^{2(l+N)} = 1$ it is possible to replace $\mathcal{C}^+$ as the decorating space by the finite dimensional space spanned by $\{Q_\lambda\}, \lambda \in (N-1, l)$ in a straightforward way, since the space can be interpreted as a Verlinde algebra, given by factoring out a suitable ideal from $\mathcal{C}^+$.

In fact this space can be interpreted as the ring of polynomials in $e_1, \ldots, e_{N-1}$ modulo the ideal generated by the polynomials $h_{l+1}, \ldots, h_{l+N-1}$ written as polynomials in the elementary symmetric function, setting $e_N = 1, e_m = 0, m > N$.

In the case $N = 2$ we only have to use polynomials in $e_1 = A_1$ as decoration when evaluating satellite invariants, in other words linear combinations of parallels of our given link will provide all the satellite invariants needed.

References

1. Aiston, A. K.: Skein theoretic idempotents of Hecke algebras and quantum group invariants. PhD. thesis, University of Liverpool, 1996.
2. Aiston, A. K. and Morton, H. R.: Idempotents of Hecke algebras of type A, *J. Knot Theory Ramifications* **7** (1998), 463–487.

3. Hadji, R. J. and Morton, H. R.: A basis for the full Homfly skein of the annulus. *Math. Proc. Camb. Phil. Soc.* **141** (2006), 81–100.
4. Kawagoe, K.: On the skeins in the annulus and applications to invariants of 3-manifolds. *J. Knot Theory Ramifications* **7** (1998), 187–203.
5. Labastida, J. M. F. and Mariño, M.: A new point of view in the theory of knot and link invariants. *J. Knot Theory Ramifications* **11** (2002), 173 –197.
6. Lukac, S. G.: Homfly skeins and the Hopf link. PhD. thesis, University of Liverpool, 2001.
7. Lukac, S. G.: Idempotents of the Hecke algebra become Schur functions in the skein of the annulus. *Math. Proc. Camb. Phil. Soc.* **138** (2005), 79–96.
8. Macdonald, I. G.: Symmetric functions and Hall polynomials. Clarendon Press, Oxford. Second edition, 1995.
9. Morton, H. R.: Skein theory and the Murphy operators. *J. Knot Theory Ramifications* **11** (2002), 475–492.
10. Morton, H. R.: Power sums and Homfly skein theory. In *Geometry and Topology Monographs, Volume 4: Invariants of knots and* 3*-manifolds (Kyoto 2001)*, (2002), 235–244.
11. Morton, H.R. : Invariants of links and 3-manifolds from skein theory and from quantum groups. In *Topics in Knot Theory; Proceedings of the NATO Summer Institute in Erzurum 1992*, NATO ASI Series C 399, ed. M.Bozhüyük. Kluwer 1993, 107–156.
12. Morton, H. R.: Mutant knots with symmetry. *Math. Proc. Camb. Phil. Soc.* **146** (2009), 95–107.
13. Morton, H.R. and Aiston, A. K.: Young diagrams, the Homfly skein of the annulus and unitary invariants. In *Proceedings of Knots96*, ed.Shin'ichi Suzuki, World Scientific, Singapore. (1997), 31–45.
14. Morton, H. R. and Lukac, S. G.: The Homfly polynomial of the decorated Hopf link. *J. Knot Theory Ramifications* **12** (2003), 395–416.
15. Morton, H. R. and Manchon, P. M. G.: Geometrical relations and plethysms in the Homfly skein of the annulus. *J. London Math. Soc.* **78** (2008), 305–328.
16. Morton, H.R. and Ryder, H. J.: Mutants and $SU(3)_q$ invariants. In *Geometry and Topology Monographs, Vol.1: The Epstein Birthday Schrift*, ed. C. Rourke, I. Rivin, C. Series, University of Warwick (1998), 365–381.
17. Turaev, V. G.: The Conway and Kauffman modules of a solid torus. Zap. Nauchn. Sem. Leningrad. Otdel. Mat. Inst. Steklov. (LOMI) 167, 1988, Issled. Topol. 6, 79–89 (Russian), English translation: *J. Soviet Math.* **52** (1990), 2799–2805.
18. Turaev, V. G.: Quantum invariants of knots and 3-manifolds. De Gruyter Studies in Mathematics, 18. Walter de Gruyter and Co., Berlin, 1994.
19. Wenzl H.: Representations of braid groups and the quantum Yang-Baxter equation. *Pacific J. Math.* **145** (1990), 153–180.

THE TRIESTE LOOK AT KNOT THEORY

Józef H. Przytycki (Washington)

Abstract This paper is base on talks which I gave in May, 2010 at Workshop in Trieste (ICTP). In the first part we present an introduction to knots and knot theory from an historical perspective, starting from Summerian knots and ending on Fox 3-coloring. We show also a relation between 3-colorings and the Jones polynomial. In the second part we develop the general theory of Fox colorings and show how to associate a symplectic structure to a tangle boundary so that tangles becomes Lagrangians (a proof of this result has not been published before). We also discuss rational moves on links and their relation to Fox colorings.

1 Classical Roots of Knot Theory

Knots have fascinated people from the dawn of human history. One of the oldest examples of knots in art or religion is the cylinder seal impression (c. 2600-2500 B.C.) from Ur, Mesopotamia (see Figure 1.1). It is described in the book *Innana* by Diane Wolkstein and Samuel Noah Kramer [Wo-Kr] (page 7), illustrating the text:

"Then a serpent who could not be charmed made its nest in the roots of the tree."

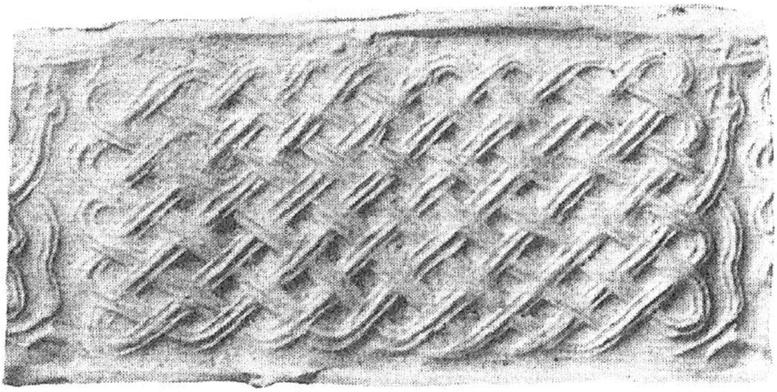

Fig. 1.1. *Snake with Interlacing Coil.*
Cylinder seal. Ur, Mesopotamia. The Royal Cemetery, Early Dynastic period, c. 2600-2500 B.C. Lapis lazuli. Iraq Museum. Photograph courtesy of the British Museum, UI 9080, [Wo-Kr].

In today's audience we have students from all over the world, including Iraq, in which Ur lies today. I encourage you to send me an example of an ancient knot from your culture or country.

In the 19'th century knot theory was an experimental science. Topology (or geometria situs) had not developed enough to offer tools that allow precise definitions and proofs. Johann Benedict Listing (1808-1882), a student of Gauss and pioneer of knot theory, writes in [Lis]: *In order to reach the level of exact science, topology will have to translate facts of spatial contemplation into easier notion which, using corresponding symbols analogous to mathematical ones, we will be able to do corresponding operations following some simple rules.* A combinatorial definition of the Gaussian linking number (initially defined by Gauss in 1833 as an integral [Gaus]) was the first step in realizing Listing's program, [Brunn-1892].

Much of early knot theory was motivated by physics and chemistry. In the 1860s, it was believed that a substance called the ether pervaded all of space. In an attempt to explain the different types of matter, William Thomson later known as Lord Kelvin (1824-1907) hypothesized that atoms were merely knots in the fabric of this ether. Different knots would then correspond to different elements.

Thus, in the second half of the 19'th century knot theory was developed primarily by physicists (Thomson, James Clerk Maxwell (1831-1879), Peter Guthrie Tait(1831-1901)), and one can argue that a high level of precision was not always appreciated[1].

1.1 Felix Klein observation on knotting in dimension four

Tait write in his paper of 1877 [Tait-2]: *Klein himself made the very singular discovery that in space of four dimensions there cannot be knots. Klein observation was noticed in non-mathematical circles and it became part of popular culture. For example, the American magician and medium Henry Slade was performing "magic tricks" claiming that he solves knots in fourth dimension. He was taken seriously by a German astrophysicist J.K.F.Zoellner who had with him a number of seances in 1877 and 1878..* Tait is referring to "Mathematische Annalen, ix. 478" and later authors often cite that paper ([Klein], 1875), and even give the page 476 ([As]). Most likely it is a misunderstanding as Klein discusses there intrinsic and ambient topological properties (of curves and surfaces) but never in context of knots in dimension four. More likely explanation is that Klein described to Tait the observation in a private correspondence. For many of you German is a native language. I would challenge you to go through Klein papers and correspondence to find a root of Klein's "singular discovery".

[1] For an outline of the global history of knot theory see [P-4] or the second chapter of my book on knot theory [P-Book].

1.2 Precision comes to knot theory

Throughout the 19'th century knots were understood as closed curves in space up to a natural deformation, described as a movement in space without cutting and pasting. This understanding allowed scientists (Tait, Thomas Penyngton Kirkman, Charles Newton Little, Mary Gertrude Haseman) to build tables of knots, but did not lead to precise methods to distinguish knots that cannot be be practically deformed into each other. In a letter to O. Veblen, written in 1919, young J. Alexander expressed his disappointment[2]: *"When looking over Tait* On Knots *among other things, He really doesn't get very far. He merely writes down all the plane projections of knots with a limited number of crossings, tries out a few transformations that he happens to think of and assumes without proof that if he is unable to reduce one knot to another with a reasonable number of tries, the two are distinct. His invariant, the generalization of the Gaussian invariant ...for links is an invariant merely of the particular projection of the knot that you are dealing with, - the very thing I kept running up against in trying to get an integral that would apply."*

In 1907, in the famous *Mathematical Encyclopedia*, Max Dehn and Poul Heegaard outlined a systematic approach to topology. In particular, they precisely formulated the subject of knot theory [D-H]. To bypass the notion of deformation of a curve in a space (which was not well defined at the time) they introduced lattice knots and a precise definition of (lattice) equivalence. Later on, Reidemeister and Alexander considered more general polygonal knots in a space, with equivalent knots related by a sequence of Δ-moves; they also explained Δ-moves via elementary moves on link diagrams – Reidemeister moves (see Figure 1.6). The approach of Dehn and Heegaard was long ignored, however recently there has been interest in the study of lattice knots[3] (e. g. [B-L]).

---

[2]We should remember that this was written by a young revolutionary mathematician forgetting that he was "standing on the shoulders of giants.". In fact, the knot invariant Alexander outlined in the letter is closely related to the Kirchhoff matrix, and the numerical invariant he also obtained is equivalent to complexity of the signed graph corresponding to the link via Tait translation; see Subsection 1.3.

[3]I am aware of two exceptions: in 1954, a popular article of Alan Turing (1912-1954) considers elementary moves on knots that lie on the unit lattice in R^3. He concludes: "A similar decision problem which might well be unsolvable is the one concerning knots which has already been mentioned." [Turing],[Gor-2]. In 1962, the biophysicist Max Delbrück (1906 – 1981) winner of the Nobel Prize in Physiology or Medicine in 1969, proposed that long molecules discovered in living organisms can be knotted, and asked about the shortest length of such a knot [Del]. In his model, lattice knots are restricted to those having straight segments of length 1. Delbrück found a realization of the trefoil knot of length 36. Delbrück's problem was popularized by Martin Gardner (1914-2010) in the November 1970 issue of Scientific American, where Gardner had a popular "Mathematical Games" column. Gardner comments that it is still unknown whether 36 is the minimal number of segments for Delbrück's (molecule) lattice nontrivial knot and

1.3 Early invariants of links

The fundamental problem in knot theory was, until recently,[4] to distinguish non-equivalent knots. Even in the case of the unknot and the trefoil knot, this was not achieved until the fundamental work of Jules Henri Poincaré (1854-1912) was applied. In his seminal paper "Analysis Situs" ([Po-1] 1895) he laid the foundations for algebraic topology. According to W. Magnus [Mag]: *Today, it appears to be a hopeless task to assign priorities for the definition and the use of fundamental groups in the study of knots, particularly since Dehn had announced [De-0] one of the important results of his 1910 paper (the construction of Poincaré spaces with the help of knots) already in 1907.* Wilhelm Wirtinger (1865-1945), in his lecture delivered at a meeting of the German Mathematical Society in 1905 outlined a method of finding a knot group presentation (now called the Wirtinger presentation of a knot group, but examples using his method were given only after the work of Dehn.

1.4 Tait's relation between knots and graphs

Tait was the first to notice the relation between knots and planar graphs. He colored the regions of a knot diagram alternately white and black so that the infinite region is black. He then constructed a graph by placing a vertex inside each white region, and connecting vertices by edges going through the crossing points of the diagram (see Figure 1.2).

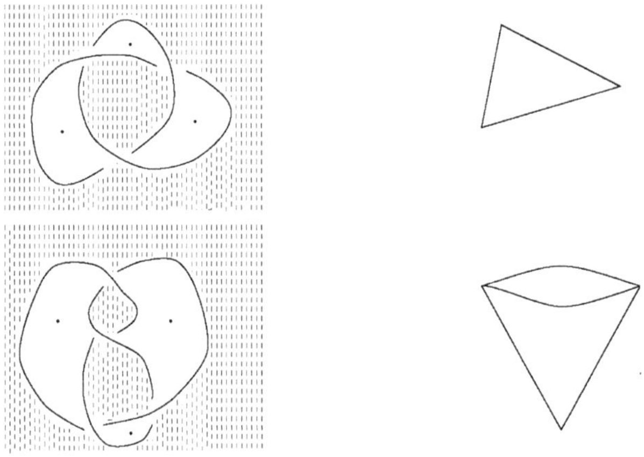

Fig. 1.2. Tait's construction of graphs from link diagrams as described in [D-H].

---

he comments that if a segment can be of any length, then 24 is possible. We know now that this is the smallest number [Dia].

[4]There are now algorithms that allow recognition of any knot, but they are very slow [Mat]. Modern knot theory, on the other hand, looks for structures on a space of knots or for mathematical or physical meanings of knot invariants.

It is useful to mention the Tait construction in the opposite direction, going from a signed planar graph G to a link diagram $D(G)$. We replace every signed edge of a graph by a crossing according to the convention of Figure 1.3, and connect endpoints along edges as in Figures 1.4 and 1.5.

Fig. 1.3. convention for crossings of signed edges (edges without markers are assumed to be positive).

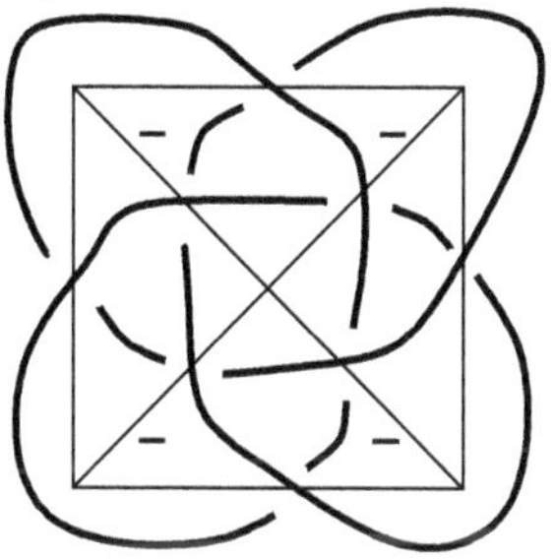

Fig. 1.4. The knot 8_{19} and its Tait graph (8_{19} is the first non-alternating knot in tables).

We should mention here one important observation already known to Tait (and in explicit form to Listing):

Proposition 1.1 *The diagram $D(G)$ of a connected graph G is alternating if and only if G is positive (i. e. all edges of G are positive) or G is negative.*

A proof is illustrated in Figure 1.5.

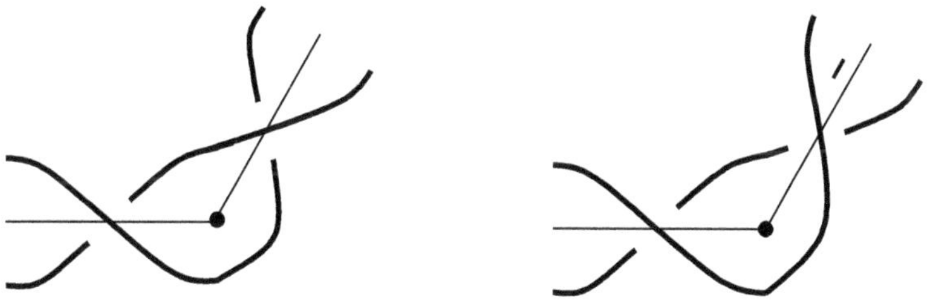

Fig. 1.5. Alternating and non-alternating parts of a diagram.

Exercise 1.2 *Draw all connected plane graphs of up to 7 edges without loops and isthmuses (edges whose removal disconnects a graph). Identify related Tait diagrams with knots and links in tables of knots [Rol].*

Maxwell was the first person to consider the question of two projections representing equivalent knots. He considered some elementary moves (reminiscent of the future Reidemeister moves), but never published his findings.

The formal interpretation of equivalence of knots in terms of diagrams was described by Reidemeister [Rei-1], 1927, and Alexander and Briggs [A-B], 1927.

Theorem 1.3 (Reidemeister theorem) *Two link diagrams are equivalent*[5] *if and only if they are connected by a finite sequence of Reidemeister moves* $R_i^{\pm 1}, i = 1, 2, 3$ *(see Fig. 1.6) and isotopy of the diagram inside the plane. The theorem also holds for oriented links and diagrams. One then has to take into account all possible coherent orientations of the diagrams involved in the moves.*

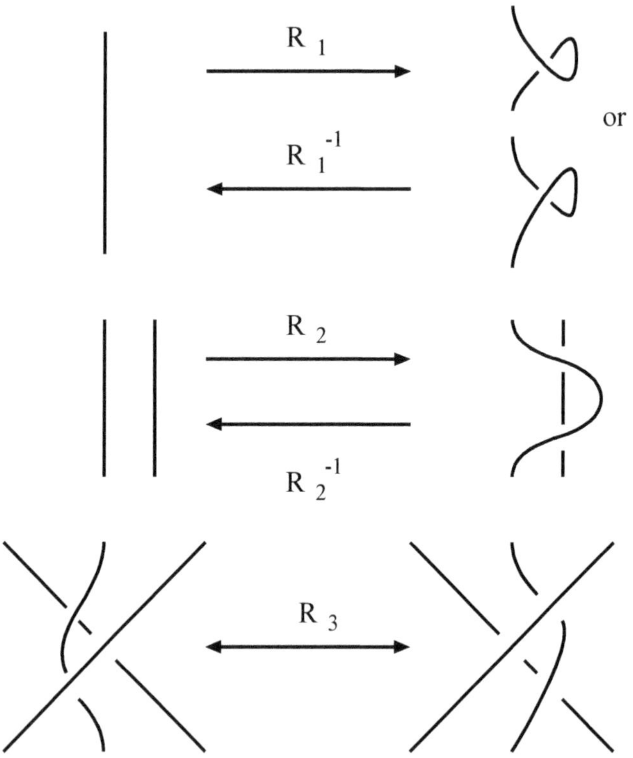

Fig. 1.6. Three Reidemeister moves: R_1, R_2 and R_3.

[5]In modern knot theory, especially after the work of R. Fox, we use usually the equivalent notion of ambient isotopy in R^3 or S^3.

1.5 Fox 3-colorings of link diagrams

The simplest invariant of links which distinguishes the trefoil knot and the trivial knot ( $\neq$ 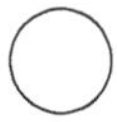) is the Fox tricoloring invariant (denoted $tri(L)$). It is an invariant which does not require much more than counting. The idea of tricoloring was introduced by Ralph Hartzler Fox (1913 -1973) around 1956 when he was explaining knot theory to undergraduate students at Haverford College ("in an attempt to make the subject accessible to everyone" [C-F]); [C-F,Chapter VI,Exercises 6-7], [Fo-2]. It was also popularized in articles directed toward middle and high school teachers and students [Cr, Vi, P-6].

Definition 1.4 ([P-1]) *We say a link diagram D is Fox tricolored if every arc is colored r (red), b (blue) or y (yellow) (we consider arcs of the diagram literally, so that in the undercrossing one arc ends and the second starts; compare Fig.1.7, 1.9), and at any given crossing either all three colors appear or only one color appears. The number of different Fox tricolorings is denoted by $tri(D)$. If a tricoloring uses only one color we say that it is a trivial Fox tricoloring.*

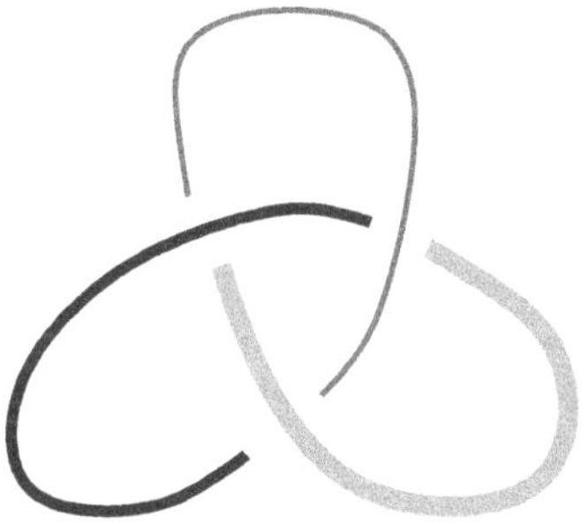

Fig. 1.7. Different colors are marked by lines of different thickness.

Proposition 1.5 *The number of Fox tricolorings of D, $tri(D)$ is an (ambient isotopy) link invariant. In particular, the tricolorability, that is the existence of a non-trivial Fox tricoloring, is a link invariant.*

Proof:
We have to check that $tri(D)$ is preserved under the Reidemeister moves. The invariance under R_1 and R_2 is illustrated in Fig. 1.8, and the invariance under R_3 is illustrated in Fig. 1.9. □

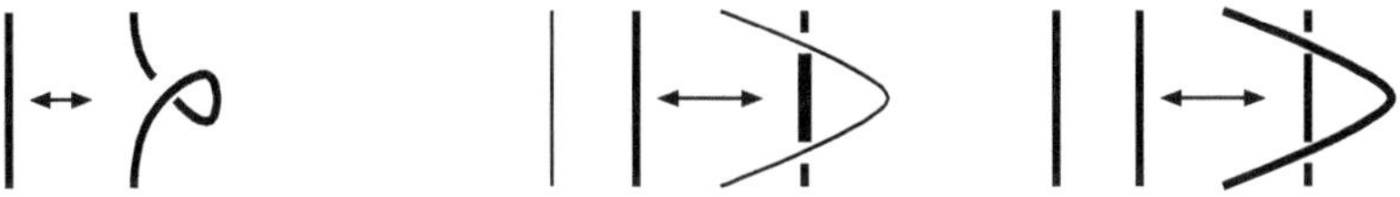

Fig. 1.8

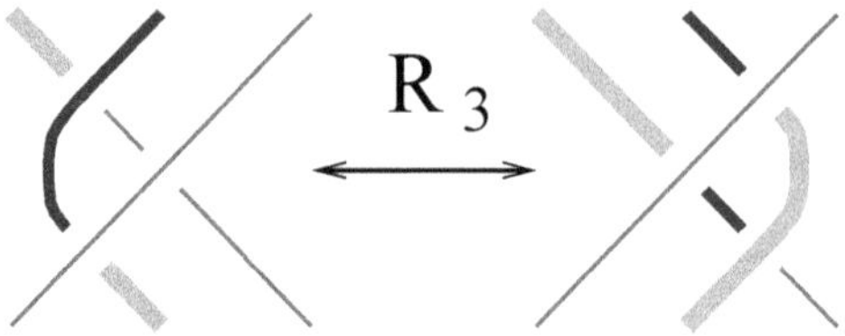

Fig. 1.9

Because the trivial knot has only trivial tricolorings, $tri(T_1) = 3$, and the trefoil knot allows a nontrivial tricoloring (Fig.1.7), it follows that the trefoil knot is a nontrivial knot.

Exercise 1.6 *Find the number of tricolorings for the trefoil knot* (3_1)*, the figure eight knot*[6] (4_1)*, and the square knot* ($3_1\#\bar{3}_1$*, see Fig.1.10). Then deduce that these knots are pairwise different.*

It is very difficult to prove any nontrivial result using our previous definition of tricoloring. For example how would you prove the following statement?

Proposition 1.7 $tri(L)$ *is always a power of* 3.

We can see immediately that if we tricolor arcs of a diagram D without Fox coloring conditions we get 3^λ possibilities, where λ is the number of arcs of D. Thus for a diagram without a crossing proposition 1.7 holds but if D has a crossing we only can say that $tri(D) \leq 3^\lambda$

Proposition 1.7 becomes easy to prove if we introduce some basic language of linear algebra or abstract algebra. Namely:

Proof: Denote the colors of the Fox tricoloring by $0, 1$ and 2 and treat them modulo 3, that is, as elements of the group (or field) $\mathbb{Z}_3$. All colorings of the arcs of a diagram using colors $0, 1$,and 2 (not necessarily permissible Fox tricolorings) can be identified with the group $\mathbb{Z}_3^\lambda$ (or the linear space over $\mathbb{Z}_3$). The (permissible) Fox tricolorings can be characterized by the property that at each crossing, the sum of the colors is equal to zero modulo 3. Thus Fox tricolorings form a subgroup (linear subspace) of Z_3^λ. We denote this group $Tri(D)$.

I encourage you to play around with this concept. Notice that trivial colorings form a one dimensional subspace, so one can should consider the quotient space of all Fox 3-colorings by the subspace of trivial tricolorings $\mathbb{Z}_3^{tr}$. We call this quotient space the space of reduced Fox tricolorings; $Tri^{rd}(D) = Tri(D)/\mathbb{Z}_3^{tr}$. $\square$

[6]The figure eight knot is often called the Listing knot, as Listing noticed in 1849 that it is equivalent to its mirror image. The notation 4_1 refers to the fact that it is the first knot of 4 crossings in knot tables.

Given our an easy success with the proof of Proposition 1.7, let us try our skills on the following fact and its useful corollary. Recall that an n-tangle is a part of a link diagram placed in a 2-disk with $2n$ points on the disk boundary: n inputs and n outputs (however only if a tangle is oriented we have unique notion of inputs and outputs); see examples in Figures 1.10 – 1.12.

Proposition 1.8 *(i) For any Fox 3-coloring of a 1-tangle; see Fig. 1.12(a), boundary arcs share a color .*

(ii) $tri(L_1)tri(L_2) = 3tri(L_1\#L_2)$, *where # denotes the connected sum of links*[7]*; see Fig. 1.10.*

Proof: (i) Let T be our Fox tricolored tangle and let the 1-tangle T' be obtained from T by adding a trivial component C below T, close enough to the boundary of the tangle, so that it cuts T only near the boundary points; Fig.1.11(b). Obviously the tricoloring of T can be extended to a tricoloring of T' (in three different ways) because the tangle T' is ambient isotopic to a tangle obtained from T by adding a small trivial component disjoint from T. However, if we try to color C, we see immediately that it is possible if and only if the input and the output arcs of T have the same color. Namely, if x is the color of a point on C and a and b colors of the input and the output then following C and using Fox tricoloring rules at two crossings of C with T we get $x = a - b + x$, so $a = b$; see Figure 1.11.
(ii) If we consider the connected sum $L_1\#L_2$, we see from the part (i) that the arcs joining L_1 and L_2 have the same color. Therefore the formula $tri(L_1\#L_2) = \frac{1}{3}tri(L_1)tri(L_2)$ follows. □

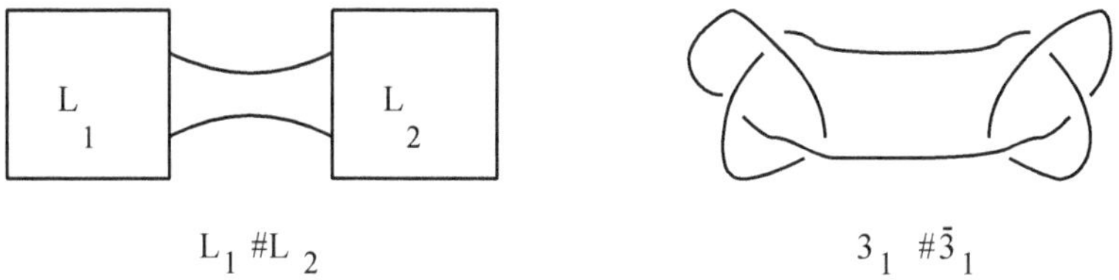

Fig. 1.10. Connected sum of link diagrams.

[7]A diagram $D_1\#D_2$ is a connected sum of diagrams D_1 and D_2 if there is a simple closed curve cutting $D_1\#D_2$ in exactly two points and 1-tangles obtained by cutting $D_1\#D_2$ by the curve have D_1 and D_2 as their closures. A link $L_1\#L_2$ is a connected sum of links L_1 and L_2 if there is a diagram of $L_1\#L_2$ which is a connected sum of diagrams of L_1 an L_2. Connected some maybe not unique, and may depend on components of links connected in the connected sum and on orientation of links.

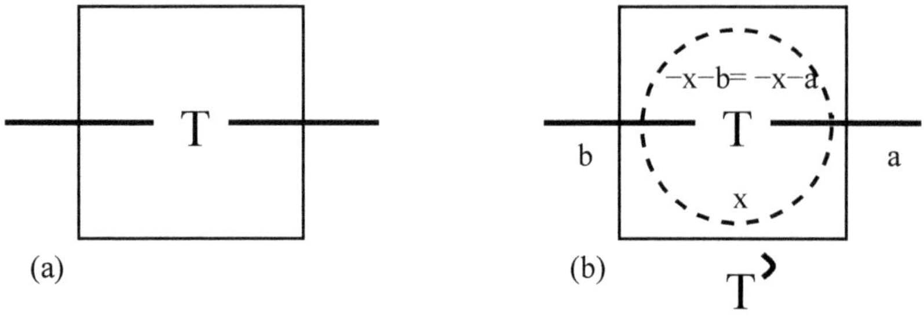

Fig. 1.11. 1-tangle and tricoloring of its boundary points; $a = b$.

The next proposition gives a very interesting property relating the number of Fox tricolorings of four unoriented links which differ in a small neighborhood as in Figure 1.12. Using basic algebra we can only partially prove Proposition 1.9. Tomorrow I will show you how to place more structure on colorings (symplectic structure) to fully prove the proposition and its generalizations.

Proposition 1.9 *[P-3] Let* D_+, D_-, D_0 *and* D_∞ *denote four unoriented link diagrams (of links* L_+, L_-, L_0 *and* L_∞*) as in Fig. 1.12. Then among the four numbers* $tri(L_+), tri(L_-), tri(L_0)$ *and* $tri(L_\infty)$*, three are equal and the fourth is* 3 *times bigger then the rest.*

Fig. 1.12

We first prove here the weaker fact that among these four numbers either all 4 are equal or 3 of them are equal and the 4'th is 3 times bigger then the rest (the rest of the proof will wait till tomorrow).

Proof: Consider a crossing p of the diagram D. If we cut a neighborhood of p out of D, we are left with the 2-tangle T_D (see Fig.1.13(a)). The set of Fox tricolorings of T_D, $Tri(T_D)$, forms a linear space over $\mathbb{Z}_3$ with subspaces $Tri(D_+), Tri(D_-), Tri(D_0)$ and $Tri(D_\infty)$. Let x_1, x_2, x_3, x_4 be elements of $Tri(T_D)$ corresponding to arcs cutting the boundary of the tangle; see Fig.1.13(b). Then any element of $Tri(T_D)$ satisfies the equality $x_1 - x_2 + x_3 - x_4 = 0$. To show this, we proceed as in part (i) of Proposition 1.8, see Figure 1.13(b). Any element of $Tri(D_+)$ (resp. $Tri(D_-)$, $Tri(D_0)$ and $Tri(D_\infty)$) satisfies additionally the equation $x_1 = x_3$ (resp. $x_2 = x_4$, $x_2 = x_3$, and $x_2 = x_1$). Thus $Tri(D_+)$ (resp. $Tri(D_-)$, $Tri(D_0)$ and

$Tri(D_\infty)$) is a subspace of $Tri(T_D)$ of codimension at most one. Let F be the subspace of $Tri(T_D)$ given by the equations $x_1 = x_2 = x_3 = x_4$, that is, the space of 3-colorings monochromatic on the boundary of the tangle. F is a subspace of codimension at most one in any of the spaces $Tri(D_+)$, $Tri(D_-)$, $Tri(D_0)$, $Tri(D_\infty)$. Furthermore the common part of any two of $Tri(D_+)$, $Tri(D_-), Tri(D_0), Tri(D_\infty)$ is equal to F. To see this, we just compare the defining relations for these spaces. Finally, notice that $Tri(D_+) \cup Tri(D_-) \cup Tri(D_0) \cup Tri(D_\infty) = Tri(T_D)$.

We have the following possibilities:

(1) F has codimension 1 in $Tri(T_D)$.
Then by the above considerations:
One of $Tri(D_+), Tri(D_-), Tri(D_0), Tri(D_\infty)$ is equal to $Tri(T_D)$.
The remaining three spaces are equal to F and Proposition 1.9 holds.

(2) $F = Tri(D_+) = Tri(D_-) = Tri(D_0) = Tri(D_\infty) = Tri(T_D)$,

(3) F has codimension 2 in $Tri(T_D)$. Then $3|F| = tri(D_+) = tri(D_-) = tri(D_0) = tri(D_\infty) = \frac{1}{3}tri(T_D)$

This completes the weaker statement of Proposition 1.9. To prove Proposition 1.9 fully, one must exclude cases (2) and (3). To exclude (2) and (3) one can use the Goeritz matrix of the link diagram; see [P-Book]. In the second part we show how to use the concept of Lagrangian tangles to show essential generalization of Proposition 1.9 (the concept was introduced in [DJP]). □

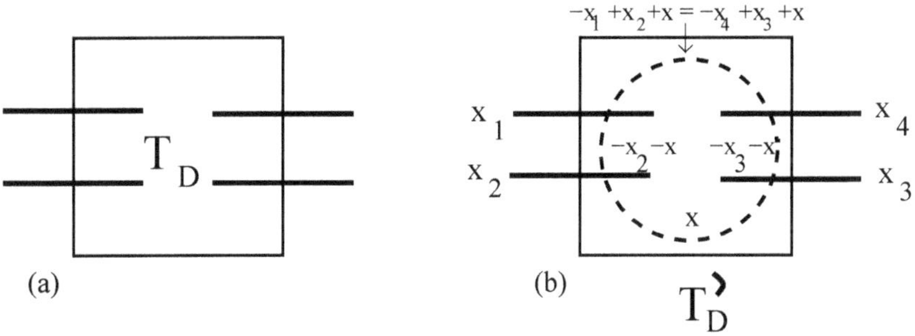

Fig. 1.13. 1-tangle and tricoloring of its boundary points; $x_1 - x_2 + x_3 - x_4 = 0$.

We will show below that the dimension of the space of Fox 3-colorings of a link is bounded from above by the bridge index of the link. For this we need few basic definitions:
Let L be a link embedded in R^3 which meets a plane $E \subset R^3$ in $2k$ points such that the arcs of L contained in each half-space relative to E possess orthogonal projections onto E which are simple and disjoint. (L, E) is

called a k-bridge presentation of L; [B-Z]. The bridge index of a link L, denoted $bridge(L)$, is a minimal number k such that L has a k-bridge presentation. Notice that the k-bridge presentation of L can be interpreted as an embedding of L with exactly k minima and k maxima (in the z direction).

Proposition 1.10 *For any link L we have*

$$tri(L) \leq 3^{bridge(L)}.$$

Proof: If we color the bridges of a diagram, then the 3-coloring of the other arcs is uniquely determined. It may happen however, that we get "contradictions" at some minima; which leads to the inequality in Proposition 1.10. □

Remark 1.11 *We can look at links or tangles with n bridges from a different perspective, by organizing diagrams along the y axis that, is we deal with maxima (and minima) along the y axis. In the case of a 2-tangle we also have 4 minimal (boundary) points, in addition to n maxima ($\cap$) and $n-2$ minima ($\cup$); compare Figure 1.14.*

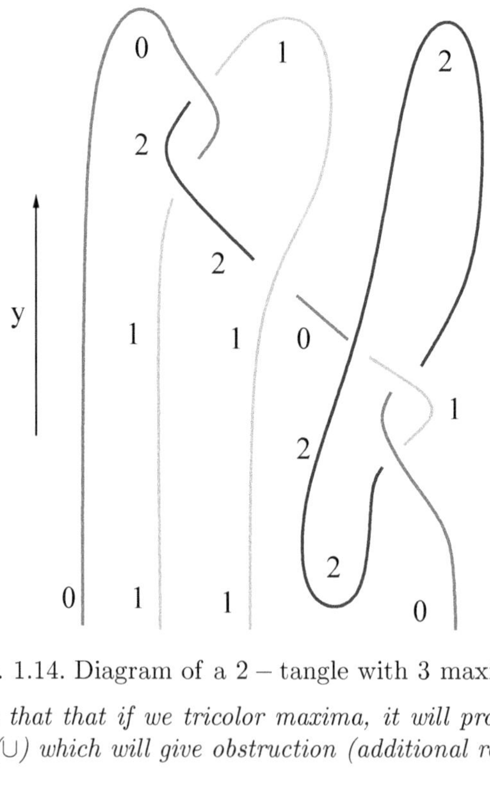

Fig. 1.14. Diagram of a $2-$tangle with 3 maxima.

We observe that that if we tricolor maxima, it will propagate until we reach minima ($\cup$) which will give obstruction (additional relations) to pos-

sible 3-colorings. If we start with n-maxima, we also start with $\mathbb{Z}_3^n$ as the space of colorings. When we move along our diagram down, with respect to the y axis (like a braid) we uniquely color the arcs of the diagram and, at any level, keeping n-dimensional space of boundary colorings until we reach minima. Assume we deal with a 4-tangle T_D. Then we have $n-2$ minima leading to $n-2$ relations on $\mathbb{Z}_3^n$. Thus we are left with at least a 2-dimensional space. In Section 2 we show more: the colorings of boundary points span exactly a 2-dimensional space. To express this algebraically, we consider a linear map $\psi : Tri(T_D) \to \mathbb{Z}_3^4$, in which the coloring of the 2-tangle T_D yields a coloring of the four boundary points. If we start from an n-tangle with n-maxima, we have an isomorphism $Tri(T_D) \to \mathbb{Z}_3^n$ and after adding the $n-2$ relations, the image of ψ is 2-dimensional.

In conclusion, this shows that any 2-tangle has a 3-coloring that is not monochromatic on the boundary. This will be discussed, given additional tools, more generally, tomorrow[8].

1.6 Fox 3-colorings and the Jones polynomial

In many talks we heard about the Jones polynomial – the great breakthrough in knot theory, in 1984.

I noticed a connection between Fox tricolorings and the Jones polynomial when I analyzed the influence of 3-moves on 3-colorings and the Jones polynomial [P-1].

Definition 1.12
The local change in a link diagram which replaces parallel lines by n positive half-twists is called an n-move; see Fig.1.14.

---

[8]Tomorrow's talk will introduce a symplectic structure on the space of colorings of a tangle boundary which does not apply to virtual links and tangles (which we heard about today). Therefore one should mention that the considerations in the observation above apply partially to virtual tangles as well. On the other hand, part of the proof of Proposition 1.9 does not work for virtual links: the equality $x_1 - x_2 + x_3 - x_4 = 0$ does not always hold for diagrams with virtual crossings, and is related with the fact that a virtual crossing alone does not satisfies the property for any nontrivial coloring. For the virtual crossing 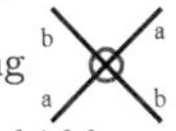

we have $a - b + a - b = 2(a-b)$. In the virtual knot theory we have two forbidden moves as an arc cannot be moved under or over a virtual crossing (the first forbidden move: 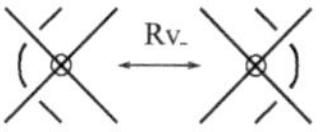

) and the second forbidden move: 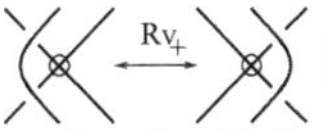

). The theory of Fox colorings works for virtual links or tangles. It works also if the second forbidden move is allowed (so can be used for welded knots described in Kauffman's talk). Fox colorings and more generally quandle colorings (see Section 3) are preserved by this move.

Fig. 1.15. 3-move and n-move.

Lemma 1.13 *Let the diagram D_{+++} be obtained from D by a 3-move (Fig.1.14(a)). Then:*

(a) $tri(D_{+++}) = tri(D)$,

(b) $V_{D_{+++}}(e^{2\pi i/6}) = \pm i^{(com(D_{+++})-com(D))}V_D(e^{2\pi i/6})$, *where V is the Jones polynomial, and $com(D)$ denotes the number of link components of D,*

(c) $F_{D_{+++}}(1,-1) = F_D(1,-1)$, *where F is the Kauffman polynomial.*

Before we prove Lemma 1.13 let us recall definition of the Jones polynomial (1984) and the specialization of the Kauffman polynomial first introduced by Brandt-Lickorish-Millett and Ho [BLM, Ho] (1985).

Definition 1.14 *(J) The Jones polynomial $V_L(t)$ of an oriented link L is a link invariant ($V_L(t) \in \mathbb{Z}[t^{\pm 1/2}]$) normalized to be one for the trivial knot and satisfies the skein relation*

$$t^{-1}V_{L_+}(t) - tV_{L_-}(t) = (t^{\frac{1}{2}} - t^{-\frac{1}{2}})V_{L_0}(t).$$

(K) The Brandt-Lickorish-Millett-Ho polynomial $Q_L(x)$ is normalized to be one for the trivial knot and satisfies the skein relation

$$Q_{L_+}(x) + Q_{L_-}(x) = xQ_{L_0}(x) + xQ_{L_\infty}(x).$$

The Kauffman 2-variable polynomial $F(a,x)$ satisfies $F(1,x) = Q_L(x)$.

Exercise 1.15 *(i) Show that $V_{T_n} = (-t^{1/2} - t^{-1/2})^{n-1}$, for T_n being the trivial link of n components.*

(ii) Show that $V_L(t) \in \mathbb{Z}[t^{\pm 1}]$ if L has odd number of components and $t^{1/2}V_L(t) \in \mathbb{Z}[t^{\pm 1}]$ if L has even number of components.

(iii) Show that $V_K(t) - 1$ is divisible by $(t-1)(t^3-1)$ for any knot K.

(iv) Show that $V_L(t) - V_{T_{com(L)}}$ is divisible by (t^3-1) for any link L. Here $com(L)$ denotes the number of components of L and T_k is the trivial link of k components

Proof of Lemma 1.13.
We prove (a) and (c) and partially (b) (one of two possible orientation choices).

(a) The bijection between 3-colorings of D and D_{+++} is illustrated in Fig. 1.15.

Fig. 1.16

(c) $F_{D_{+++}}(1,-1) = -F_{D_+}(1,-1) - F_{D_{++}}(1,-1) - F_{D_\infty}(1,-1) = -F_{D_+}(1,-1) + F_D(1,-1) + F_{D_+}(1,-1) + F_{D_\infty}(1,-1) - F_{D_\infty}(1,-1) = F_D(1,-1)$.

(b) Assume that arcs in Figure 1.15(a) have parallel orientation. Then for $t = e^{2\pi i/6}$ ($t^{1/2} = e^{\pi i/6}$) we have:
$V_{D_{+++}} = t^2 V_{D_+} + t(t^{1/2} - t^{-1/2})V_{D_{++}} = t^2 V_{D_+} + t^3(t^{1/2} - t^{-1/2})V_D + t^2(t^{1/2} - t^{-1/2})^2 V_{D_+} = t^2(t - 1 + \frac{1}{t})V_{D_+} + t^{1/2}(t^3 - t^2)V_D = \frac{t^3+1}{t+1}V_{D_+} + t^{3/2}\frac{t^3+1}{t+1}V_D - t^{3/2}V_D = -e^{\pi i/2}V_D = -iV_D$, as needed.
In the case when a 3-move is not preserving orientation, we would have to consider several involved cases, but we can make shortcut using so called Jones reversing result[9], that is if one changes an orientation of some components of a link, then its Jones polynomial is changed in a precisely described way, in particular by multiplying by a number being the power of t^3 (in our case the power of -1).

One can easily check that for a trivial n-component link, T_n, $tri(T_n) = 3^n = 3V^2_{T_n}(e^{2\pi i/6}) = 3(-1)^{n-1}F_{T_n}(1,-1)$. Furthermore it follows from Lemma 1.9 that as long as a link L can be obtained from a trivial link by 3-moves we have: $tri(L) = 3|V^2_L(e^{2\pi i/6})| = 3|F_L(1,-1)|$.

These immediately lead to three questions:

(1) (**Montesinos-Nakanishi 3-move conjecture**). Any link can be reduced to a trivial link by a finite sequence of 3-moves.

---

[9] It was initially proven in a series of involved papers but now it has an easy proof using the Kauffman bracket polynomial which do not depend on a link orientation; compare [P-Book]. Precisely, we have: Suppose that L_i is a component of an oriented link L and $\lambda = \text{lk}\,(L_i, L - L_i)$. If L' is a link obtained from L by reversing the orientation of the component L_i then $V_{L'}(t) = t^{-3\lambda}V_L(t)$.

(2) Is it true that $tri(L) = 3|F_L(1,-1)|$?

(3) Is it true that $tri(L) = 3|V_L^2(e^{2\pi i/6})|$?

Formulas (2) and (3) follow immediately from (1) as equalities from (2) and (3) hold for trivial links and it is propagated by 3-moves. Thus we proved (2) and (3) for any link which can be reduced by 3-moves to a trivial link.

Y.Nakanishi first considered the conjecture (i) in 1981. J.Montesinos analyzed 3-moves before, in connection with 3-fold dihedral branch coverings, and asked a related but different question. The conjecture was proved in many special cases (e.g. [Che]) but it was an open problem for over 20 years. In 2002 it was showed by M.K.Dąbkowski and the author that the conjecture does not hold. The smallest counterexample we found, suggested first by Q. Chen, has 20 crossings, see Figure 1.16, [D-P-1].. We conjecture that it is in fact the smallest counterexample, that is every link up to 19 crossings can be reduced to a trivial link by 3-moves, furthermore we predict that every link of 20 crossing is reduced by a 3-move either to the trivial link or to the Chen link (up to the mirror image). With todays computers it should be laborious but doable exercise – please try it!

Fig. 1.17. The Chen link, the closure of the 5-string braid $(\sigma_2\sigma_1^{-1}\sigma_2\sigma_3\sigma_4^{-1})^4$.

The Montesinos-Nakanishi conjecture does not hold but the formulas (2) and (3) linking tricoloring with the Jones and Kauffman polynomials holds for any link.

The proof of (a) in [P-1] uses Fox's interpretation of 3-coloring and the connection with the first homology group of the branched 2-fold cover of S^3 branched over the link. However, a simple, totally elementary proof follows from Proposition 1.9.

Proof: Because $tri(L)$ is a power of 3, we can consider the signed version of the tricoloring defined by: $tri'(L) = (-1)^{log_3(tri(L))}tri(L)$. It follows from Proposition 1.9 that

$$tri'(L_+) + tri'(L_-) = -tri'(L_0) - tri'(L_\infty).$$

This is however exactly the recursive formula for the Kauffman polynomial $F_L(a,x)$ at $(a,x) = (1,-1)$. Comparing the initial data (for the unknot) of tri' and $F(1,-1)$ we get generally that: $-3F_L(1,-1) = tri'(L) = (-1)^{log_3(tri(L))}tri(L)$, which proves part (b) of Theorem 1.13. Part (a) follows from Lickorish's observation [Li], that $F_L(1,-1) = (-1)^{com(L)}V_L^2(e^{2\pi i/6})$. This observation can be directly proven from the Kauffman bracket polynomial version of the Jones polynomial. For people who attended Lou Kauffman talk it should be a pleasure exercise: just consider the difference of squares of the Kauffman bracket relation for $L+$ and L_- (that is ⟨  ⟩^{2} − ⟨  ⟩^{2}). You will get the relation of the, so called, Dubrovnik version of the Kauffman polynomial which can be converted to the standard one. □

I would challenge you to find completely elementary proof of Proposition 1.9 or directly formulas (b) and (c) (as we noted all three facts are related by elementary consideration). As a prize I offer a copy of my book [P-Book].

Tomorrow I will define general Fox k-colorings and Fox coloring group, and I will place the theory of Fox coloring in more general (sophisticated) context, and apply it to the analysis of k-moves (and rational and braid moves) of n-tangles. Interpretation of tangle colorings as Lagrangians in symplectic spaces is our main (and new) tool. In the second lecture tomorrow, I will also mention another motivation for studying 3-moves: to understand skein modules based on their deformation.

2 Fox Colorings, Rational Moves, and Lagrangian Tangles

Many of you, likely, wondered yesterday why we consider only 3-colorings not, say generally n-colorings. Some of you probably tried to replace the relation $a+b+c \equiv 0$ mod 3 by the relation $a+b+c \equiv 0$ mod k, and noticed that it does not work well with Reidemeister moves. In fact, as observed by Fox, the proper relation to generalize is $2b - a - c \equiv 0$ mod 3. This leads to Fox k-colorings:

Definition 2.1 *(i) We say that a link (or a tangle) diagram is Fox k-colored if every arc is colored by one of the numbers* $0, 1, ..., k-1$ *(forming a group* Z_k*) in such a way that at each crossing the sum of the colors of the undercrossings is equal to twice the color of the overcrossing modulo k;algebraically* $c \equiv 2b - a$ *mod* k *as illustrated in Fig.2.1.*

(ii) The set of Fox k*-colorings forms an abelian group (or* $\mathbb{Z}_k$*-module), denoted by* $Col_k(D)$*. The cardinality of the group will be denoted by* $col_k(D)$*. For an* n*-tangle* T *each Fox* k*-coloring of* T *yields a*

coloring of boundary points of T and we have the homomorphism $\psi : Col_k(T) \to Z_k^{2n}$

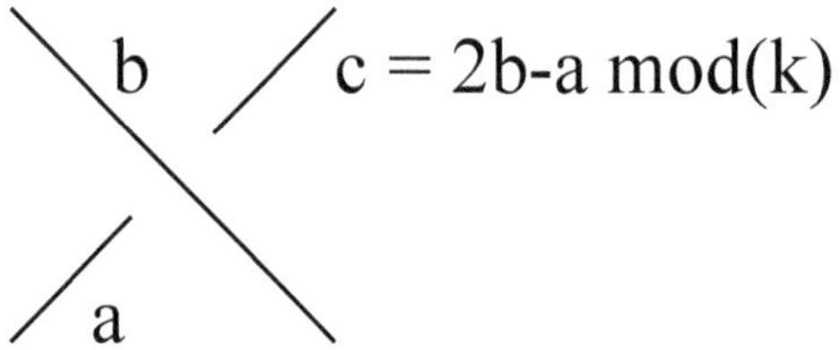

Fig. 2.1

It is a pleasant exercise to show that $Col_k(D)$ is unchanged by Reidemeister moves (see Figure 2.2),

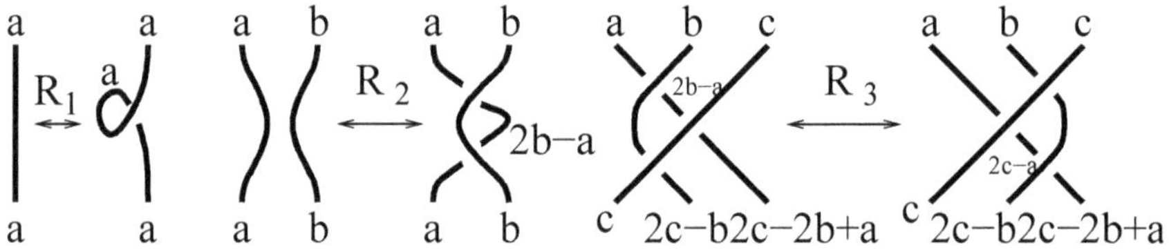

Fig. 2.2

I will start this part from the basic observations on Fox k-colorings analogous to those proven yesterday for 3-colorings. The talk will culminate by the introduction of the symplectic structure on the boundary of a tangle in such a way that tangles yields Lagrangians in the symplectic space. We end with some corollaries, in particular the method to recognize often that a virtual tangle is not a classical tangle (by boundary k-coloring comparison).

We follow here [P-3] and [DJP] (see [P-5] for historical introduction).

Proposition 2.2 ([P-3]) *The space of Fox k-colorings is preserved by k-moves.*

Proof: Figure 2.3 illustrates the bijection between $col_k(D)$ and $col(m_k(D))$ where $m_k(D)$ is obtained from D by a k-move. This bijection is an isomorphism of groups $Col_k(D)$ and $Col_k(m_k(D))$ □

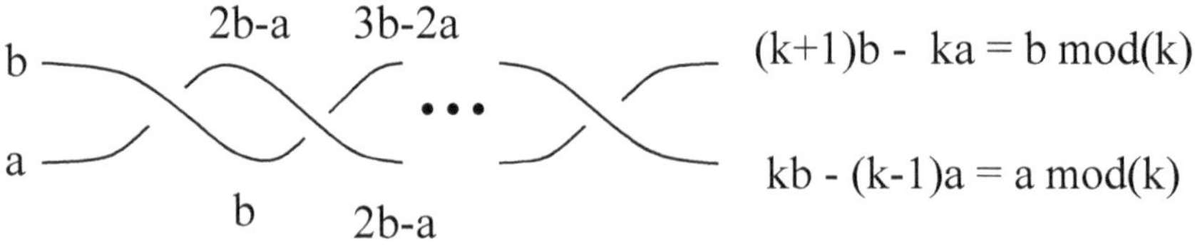

Fig. 2.3. from (b, a) to $(k(b - a) + b, k(b - a) + a)$.

The following properties of k-colorings, are a straightforward generalizations from 3-colorings and can be proved in a similar way. However, an elementary proof of the part (c) is, as before, more involved and the simplest proof (not involving double branched covers), I am aware of, requires an interpretation of k-colorings using the Goeritz matrix [Goe, Gor-1, P-5] or use of Lagrangian tangles (see below).

Lemma 2.3 *(a) $col_k(L)$ is a divisor of a power of k and for a link with b bridges, $col_k(D)$ divides k^b. More precisely. $Col_k(L)$ is a subgroup of Z_k^b.*

(b) $col_k(L_1)col_k(L_2) = k(col_k(L_1\#L_2))$ (notice that our yesterday's proof works only for odd k as we use the fact that 2 is invertible in Z_k),

(c) Consider $k+1$ diagrams $L_0, L_1, ..., L_{k-1}, L_\infty$; see Fig. 2.4 . If k is a prime number then among the $k+1$ numbers $col_k(L_0), col_k(L_1), ...,$ $col_k(L_{k-1})$ and $col_k(L_\infty)$ k are equal one to another and the $(k+1)$'th is k times bigger.

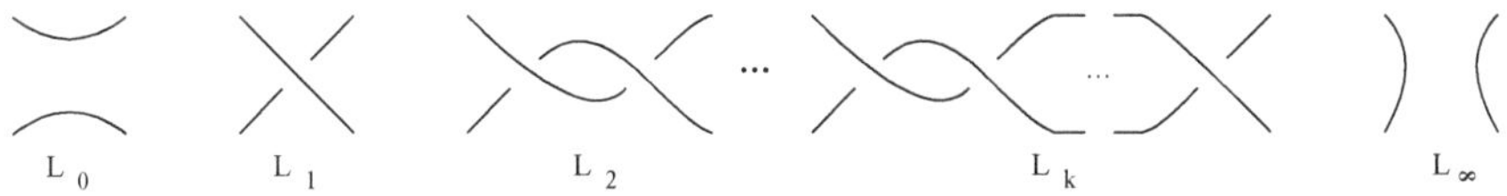

Fig. 2.4

Notice, that (c) can be interpreted as follows:
Let k be a prime number and $col'_k(L) = (-1)^{col_k(L)}col_k(L)$ then

$$col'_k(L_0) + col'_k(L_1) + ... + col'_k(L_{k-1}) + col'_k(L_\infty) = 0,$$

a skein relation of $k+1$ terms often called (k,∞) skein relation.

Example 2.4 *(i) For the figure eight knot, 4_1, one has $col_5(4_1) = 25$, so the figure eight knot is a nontrivial knot; compare Figure 2.5.*

(ii) For the knot 5_2 we have $col_7(5_2) = 49$ (more precisely $Col_7(5_2) = \mathbb{Z}_7^2$); compare Figure 2.5.

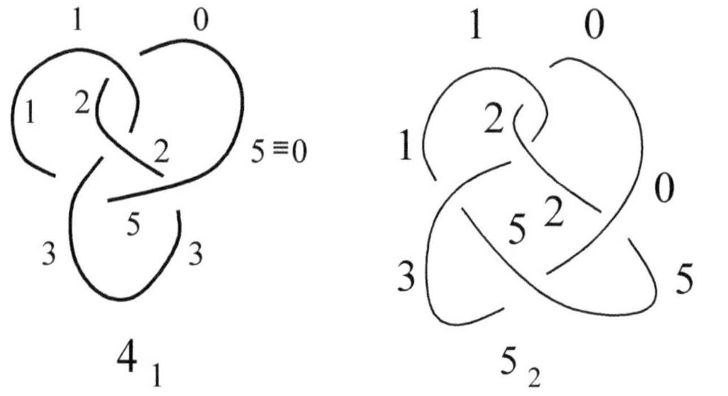

Fig. 2.5

Let us look closer at the observation that a k-move preserves the space of Fox k-colorings. One should consider a general *rational moves*, that is, a rational $\frac{p}{q}$-tangle of Conway is substituted in place of the identity tangle[10]. The important observation for us is that $Col_p(D)$ is preserved by $\frac{p}{q}$-moves. Fig.2.6 illustrates the fact that $Col_{13}(D)$ is unchanged by a $\frac{13}{5}$-move.

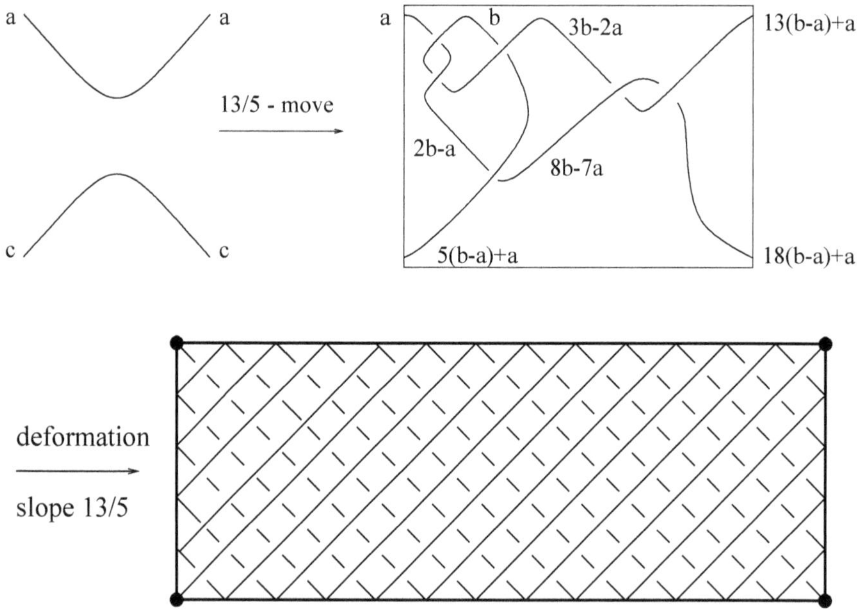

Fig. 2.6. $\frac{13}{5}$-move and $\frac{13}{5}$ tangle in Conway's and pillow case forms.

[10]The move was first considered by J.M.Montesinos[Mo-2]; compare also Y.Uchida [Uch].

We just have heard about the Conway's classification of rational tangles at the Lou's and Sofia's talks, so I only briefly sketch definitions and notation. The 2-tangles shown in Figure 2.7 are called rational tangles with Conway's notation $T(a_1, a_2, ..., a_n)$. A rational tangle is the $\frac{p}{q}$-tangle if $\frac{p}{q} = a_n + \frac{1}{a_{n-1}+...+\frac{1}{a_1}}$.[11] Conway proved that two rational tangles are ambient isotopic (with boundary fixed) if and only if their slopes are equal (compare [Kaw]).

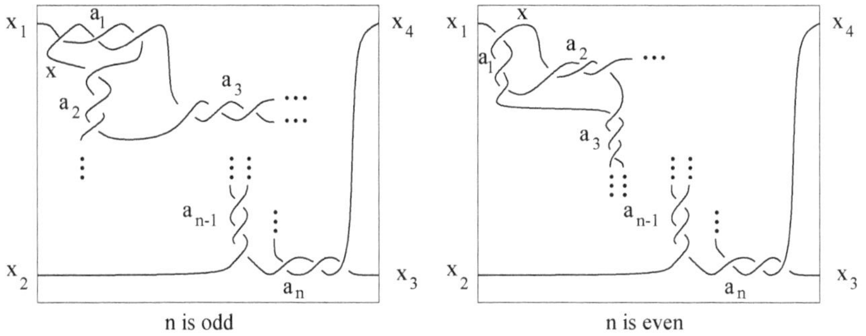

Fig. 2.7

For a Fox coloring of a rational $\frac{p}{q}$-tangle with boundary colors x_1, x_2, x_3, x_4 (Fig.2.5), one has $x_4 - x_1 = p(x - x_1)$, $x_2 - x_1 = q(x - x_1)$ and $x_3 = x_2 + x_4 - x_1$. If a coloring is nontrivial ($x_1 \neq x$) then $\frac{x_4 - x_1}{x_2 - x_1} = \frac{p}{q}$ as has been explained in the talk by Lou Kauffman.

Corollary 2.5 *$\frac{p}{q}$-move on a link or a tangle is preserving the group of p-colorings.*

2.1 Symplectic structure on Fox colorings, Lagrangian tangles

The usefulness of the symplectic structure in the knot theory, was probably first observed by R. Fox in his review of the A. Plans paper of 1953 [Pla]. In this part we follow [DJP] showing how to define a symplectic form on the space of Fox colorings of the boundary of n-tangles so that every tangle corresponds to a Lagrangian (in the case of a field of colors) or a virtual Lagrangian (for PID) of the symplectic structure (that is, a subspace of a maximal dimension on which the form vanishes). Inversely, for a field $R = Z_p$, $p > 2$, every Lagrangian can be realized by a tangle. It does not hold for Z_2 and $n > 3$.[12]

---

[11] $\frac{p}{q}$ is called the slope of the tangle and can be easily identified with the slope of the meridian disk of the solid torus being the branched double cover of the rational tangle.

[12] In [DJP] we draws from the construction several far fetching conclusions: first, it allows us to understand the space of colored tangles as a Tits building. Second, it provides

2.2 Alternating form on colorings of a tangle boundary

We work with modules over a commutative ring with identity, R. We concentrate our attention on the finite field $R = Z_p$.

Consider $2n$ points on a circle (or a square, Fig. 2.8). Let the ring R be treated as a set of colors (e.g. a field $\mathbf{Z}_p$).

For $R = \mathbf{Z}_p$ the colorings of $2n$ points form a linear space $V = \mathbf{Z}_p^{2n}$. Let $e_1, \ldots, e_{2n}$ be its basis, $e_i = (0, \ldots, 1, \ldots, 0)$, where 1 occurs in

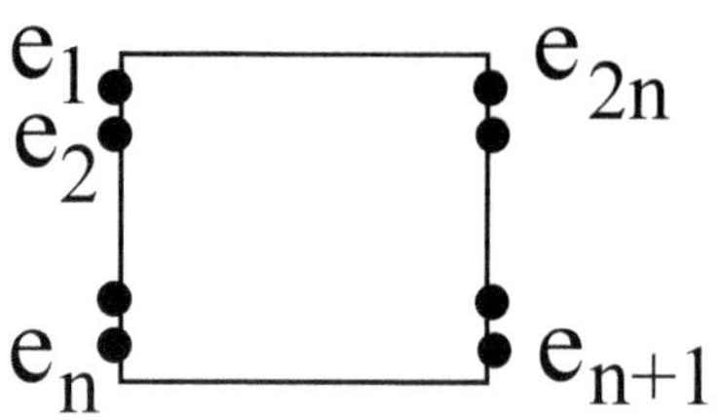

Fig. 2.8

the i-th position. Let $V' = \mathbf{Z}_p^{2n-1} \subset \mathbf{Z}_p^{2n}$ be the subspace of vectors $\sum a_i e_i$ satisfying $\sum(-1)^i a_i = 0$ (the alternating condition). Consider the basis $f_1, \ldots, f_{2n-1}$ of $\mathbf{Z}_p^{2n-1}$ where $f_k = e_k + e_{k+1}$. We can also introduce the vector $f_{2n} = e_{2n} + e_1$ and then $f_{2n} = f_1 - f_2 + f_3 \pm \ldots - f_{2n-2} + f_{2n-1}$. Consider an alternating form[13] ϕ on $\mathbf{Z}_p^{2n-1}$ of nullity 1 given by the matrix

$$\phi = \begin{pmatrix} 0 & 1 & 0 & 0 & \ldots & 0 & 0 & 0 \\ -1 & 0 & 1 & 0 & \ldots & 0 & 0 & 0 \\ 0 & -1 & 0 & 1 & \ldots & 0 & 0 & 0 \\ \ldots & \ldots & \ldots & \ldots & \ldots & \ldots & \ldots & \ldots \\ \ldots & \ldots & \ldots & \ldots & \ldots & \ldots & \ldots & \ldots \\ 0 & 0 & 0 & 0 & \ldots & -1 & 0 & 1 \\ 0 & 0 & 0 & 0 & \ldots & 0 & -1 & 0 \end{pmatrix}$$

that is,

applications to 3-manifold topology. In particular, we show that our symplectic space is related (via double branched cover) to the symplectic structure on homology on a surface (with the symplectic form given by the intersection number). It relates our results with a known fact that 3-manifolds yield Lagrangians in $H_1(\partial M; Q)$. One application is to use Lagrangians to find obstructions for embedding n-tangles into links. Rotation of a tangle yields an isometry of our symplectic space, and we analyze invariant subspaces of the map, in particular we look for invariant Lagrangians of the rotation by $2\pi/n$ (along z-axis). We use our analysis to answer, partially the question whether rotation of a link (as described in [APR]) preserves the homology of the double branch cover of S^3 with the link as branching set.

[13]That is for any $a \in V$ one has $\phi(a, a) = 0$. From this anti-symmetry follows ($\phi(a, b) = -\phi(b, a)$).

$$\phi(f_i, f_j) = \begin{cases} 0 & \text{if } |j-i| \neq 1 \\ 1 & \text{if } j = i+1 \\ -1 & \text{if } j = i-1. \end{cases}$$

An alternating form of nullity one is called a pre-symplectic form. A pre-symplectic form on $\mathbf{Z}_p^{2n-1}$ leads to a symplectic (i.e. alternating, non-degenerated) form on $\mathbf{Z}_p^{2n-2}$ as follows:
The vector $e_1 + e_2 + \ldots + e_{2n}$ $(= f_1 + f_3 + \ldots + f_{2n-1} = f_2 + f_4 + \ldots + f_{2n})$ is ϕ-orthogonal to any other vector. If we consider $\mathbf{Z}_p^{2n-2} = \mathbf{Z}_p^{2n-1}/\mathbf{Z}_p^{tr}$, where the subspace $\mathbf{Z}_p^{tr}$ is generated by $e_1 + \ldots + e_{2n}$, that is, $\mathbf{Z}_p^{tr}$ consists of monochromatic (i.e. trivial) colorings, then ϕ descends to a symplectic form $\hat{\phi}$ on $\mathbf{Z}_p^{2n-2}$. Now we can analyze the isotropic subspaces of $(\mathbf{Z}_p^{2n-2}, \hat{\phi})$, that is, subspaces on which $\hat{\phi}$ is 0 ($W \subset \mathbf{Z}_p^{2n-2}, \phi(w_1, w_2) = 0$ for $w_1, w_2 \in W$). The maximal isotropic ($(n-1)$-dimensional) subspaces of $\mathbf{Z}_p^{2n-2}$ are called Lagrangian subspaces (or maximal totally degenerated subspaces) and there are $\prod_{i=1}^{n-1}(p^i + 1)$ of them. We use the term pre-Lagrangian for a maximal totally degenerated subspace of $\mathbf{Z}_p^{2n-1}$. Of course, $\mathbf{Z}_p^{tr}$ lies in every pre-Lagrangian. Lagrangians in $\mathbf{Z}_p^{2n-2}$ are $(n-1)$-dimensional and pre-Lagrangians in $\mathbf{Z}_p^{2n-1}$ are n-dimensional [O'M].

Let $\psi = \psi_T : Col_p(T) \to \mathbf{Z}_p^{2n}$ be the homomorphism which sends colorings of T into colorings of boundary points of the tangle. Our local condition on Fox colorings (Fig.2.1) guarantees that for any n-tangle T, $\psi(Col_p(T)) \subset \mathbf{Z}_p^{2n-1}$.[14] Furthermore, the space of trivial colorings, $\mathbf{Z}_p^{tr}$, always lies in $Col_p(T)$. The quotient space $Col_p(T)/\mathbf{Z}_p^{tr}$ is called the reduced space of Fox colorings and denoted by $Col_p^{rd}(T)$. Thus ψ descends to $\hat{\psi} : Col_p^{rd}(T) \to \mathbf{Z}_p^{2n-2} = \mathbf{Z}_p^{2n-1}/\mathbf{Z}_p$. Now we have a fundamental question: which subspaces of $\mathbf{Z}_p^{2n-2}$ are yielded by n-tangles? We answer this question below.

Theorem 2.6 *$\hat{\psi}(Col_p^{rd}(T))$ is a Lagrangian subspace of $\mathbf{Z}_p^{2n-2}$ with the symplectic form $\hat{\phi}$. In particular, $dim(\hat{\psi}(Col_p^{rd}(T))) = n-1$. Equivalently, $\psi(Col_p(T))$ is a pre-Lagrangian subspace of $\mathbf{Z}_p^{2n-1}$ with the alternating form ϕ. In particular, $dim(\psi(Col_p(T))) = n$.*

A natural question would be whether every Lagrangian subspace can be realized by a tangle. The answer is negative for $p = 2$, but for $p > 2$ we have

[14]We checked it before for a ring R in which 2 is not a zero divisor; the general case follows from considerations given later.

Theorem 2.7 *For p an odd prime number every Lagrangian subspace of $\mathbf{Z}_p^{2n-2}$ can be realized by a tangle, in fact, by an n-rational tangle*[15].

Theorem 2.7 follows from the work of J. Assion [As] (for $p = 3$), and B. Wajnryb [Wa-1, Wa-2] (for $p > 2$). Wajnryb constructs the natural epimorphism from the odd braid group B_{2n+1} to the symplectic group $\mathrm{Sp}(n, p)$, that is, the group of isometries of the symplectic space $\mathbf{Z}_p^{2n}$.

As a corollary to Theorems 2.6 and 2.7 we obtain a fact which was considered difficult before, even for 2-tangles.

Corollary 2.8 *For any p-coloring of a tangle boundary satisfying the alternating property (i.e., is an element of $\mathbf{Z}_p^{2n-1}$) there is an n-tangle and its p-coloring yielding the given coloring on the boundary. In other words: $\mathbf{Z}_p^{2n-1} = \bigcup_T \psi_T(Col_p(T))$. Furthermore, the space $\psi_T(Col_p(T))$ is n-dimensional.*

In [DJP] we give short, high-tech proof of Theorems 2.6 and 2.7. Here we provide a longer but elementary proof based on the presentation of an n-tangle as a tangle with N maxima and $N - n$ minima as in Figure 1.13. The proof of Theorem 2.6 is straightforward: we check that the theorem holds for a trivial n-tangle, that it holds when we add crossings, and finally that it is still valid after applying minima. On the way we show that Theorem 2.7 follows from the second part of the proof (the slogan will be that braid-like transvections generate a symplectic group over $\mathbf{Z}_p$, $p > 2$; as follows from Wajnryb [Wa-2]).

Step 0: Consider the trivial tangle, T_0, in which the point v_{2i-1} is connected to v_{2i}; see Figure 2.9. Clearly $\psi(Col_p(T_0))$ is the n-dimensional subspace of $\mathbf{Z}_p^{2n}$ generated by $e_{2i-1} + e_{2i}$ $(1 \leq i \leq n)$, and thus it is a pre-Lagrangian in $\mathbf{Z}_p^{2n-1}$ generated by $f_1, f_3, \ldots, f_{2n-1}$. In effect, $\hat{\psi}(Col_p^{rd}(T_0))$ is the Lagrangian subspace of $\mathbf{Z}_p^{2n-2}$ generated by $f_1, f_3, \ldots, f_{2n-3}$.

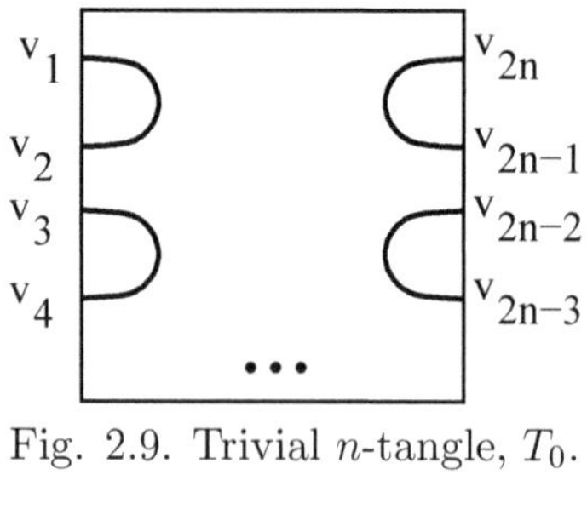

Fig. 2.9. Trivial n-tangle, T_0.

[15] An n-rational tangle is an n-tangle having presentation so it is often called an n-bridge tangle.

Step 1: We show here that if $\hat{\psi}_T(Col_p^{rd}(T))$ is a Lagrangian in $\mathbf{Z}_p^{2n-2}$ then $\hat{\psi}(Col_p^{rd}(T'))$ is also a Lagrangian where T' is a tangle obtained from T by adding one crossing to it (without loss of generality we can assume that the crossing is between arcs from v_{2n} and v_{2n-1}, see Figure 2.10; the relevant observation is that the rotation of the tangle by $\frac{2\pi}{2n}$ is an isometry, that is $\phi(f_i, f_j) = \phi(f_{i+1}, f_{j+1})$, where indices are taken modulo $2n$).

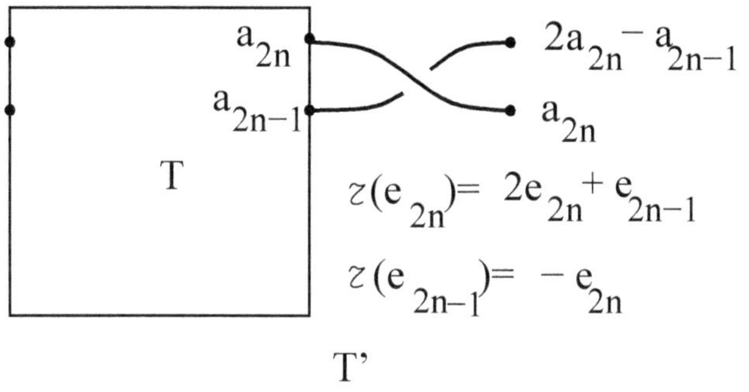

Fig. 2.10. Adding a crossing to T induces isometry on $(\mathbf{Z}_p^{2n-1}, \phi)$.

Moving from a coloring of the boundary of T to the boundary of T' induces the linear map $\tau : \mathbf{Z}_p^{2n} \to \mathbf{Z}_p^{2n}$ given by:
$\tau(e_{2n}) = 2e_{2n} + e_{2n-1}$, $\tau(e_{2n-1}) = -e_{2n}$ and $\tau(e_i)) = e_i$ otherwise. τ sends an alternating sum to an alternating sum so it preserves the subspace $\mathbf{Z}_p^{2n-1}$ on which the alternating form ϕ is defined. In the basis $f_1, f_2, ..., f_{2n-1}$ it is defined by:

$$\tau(f_{2n-2}) = \tau(e_{2n-2} + e_{2n-1}) = e_{2n-2} - e_{2n} = f_{2n-2} - f_{2n-1}$$

$$\tau(f_{2n-1}) = \tau(e_{2n-1} + e_{2n}) = e_{2n} + e_{2n-1} = f_{2n-1}$$

$$\tau(f_{2n}) = \tau(e_{2n} + e_1) = 2e_{2n} + e_{2n-1} + e_1 = f_{2n} + f_{2n-1}$$

$$\tau(f_i) = \tau(f_i) \ \ otherwise.$$

In summary, we get $\tau(f_i) = f_i - \phi(f_i, f_{2n-1})f_{2n-1}$. Such a linear map is called a symplectic transvection with respect to vector f_{2n-1} and denoted by $\tau_{f_{2n-1}}$. Generally the transvection $\tau_b(a) = a - \phi(a, b)b$ is an isometry with respect to the form ϕ (for completeness here is the check:

$$\phi(\tau_b(a_1), \tau_b(a_2)) = \phi(a_1 - \phi(a_1, b)b, a_1 - \phi(a_2, b)b) =$$

$$\phi(a_1, a_2) - \phi(a_1, \phi(a_2, b)b) - \phi(\phi(a_1, b)b, a_2) = \phi(a_1, a_2).)$$

Notice that if we change the crossing in Figure 2.10 to its mirror image then τ is replaced by τ^{-1} with $\tau^{-1}(f_i) = f_i + \phi(f_i, f_{2n-1})f_{2n-1}$.

For us it is important that transvection, as an isometry, is sending pre-Lagrangians to pre-Lagrangians and Lagrangians to Lagrangians.

Step 2:[16] Consider a minimum (here right cup, see Figure 2.11).

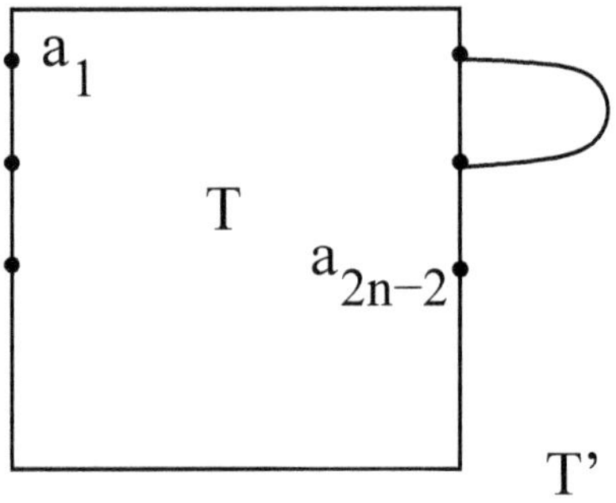

Fig. 2.11. Adding a minimum (right cup).

Initially let us consider adding a right cup in general, without assuming that $\psi_T(Col_p(T))$ is a pre-Lagrangian in $\mathbf{Z}_p^{2n-1}$, (this may be useful in a less restricted setting of virtual or welded tangles).
Consider the linear space V with a basis $\{e_1, ..., e_{2n}\}$ (corresponding to $\mathbf{Z}_p^{2n}$ and $\mathbf{Z}_p$ colorings of the boundary of a n-tangle) and consider the right cup of Figure 2.11. We analyze induced $\mathbf{Z}_p$-colorings of the boundary of a $(n-1)$-tangle in two steps:
(I) Let F be a subspace of V:

(1) We consider the subspace F_1 of F defined by $F_1 = \{a = \sum_{i=1}^{2n} a_i \in F \mid a_{2n-1} = a_{2n}\}$. We have two cases for the dimension of F_1.

   (i) $dim(F_1) = dim(F)$. This is the case iff $F_1 \subset span\{e_1, ..., e_{2n-2}, f_{2n-1}\}$, where $f_{2n-1} = e_{2n-1} + e_{2n}$.

   (ii) $dim(F_1) = dim(F) - 1$. This is the case iff there is $a \in F$ such that $a_{2n-1} \neq a_{2n}$.

(2) We consider the projection $p : V \to W = span\{e_1, ..., e_{2n-2}\}$ (here $p(e_{2n-1}) = p(e_{2n}) = 0$ and $p(e_i) = e_i$ for $1 \leq i \leq 2n-2$.). Let $F_2 = p(F_1) \subset W$. We have two cases for the dimension of F_2:

   (i) $dim(F_2) = dim(F_1)$. This holds iff f_{2n-1} is not in F_1.

   (ii) $dim(F_2) = dim(F_1) - 1$ iff $f_{2n-1} \in F_1$.

[16]It is related to the "contraction lemma" described by Turaev [Tur], p.180.

We show that both (1)(i) and (2)(i) cannot hold if F is a pre-Lagrangian. Similarly (1)(ii) and 2(ii) cannot hold for such an F.

If elements of F satisfy the alternating condition ($a \in F \Rightarrow \sum_{i=1}^{2n}(-1)^i a_i = 0$), then (1) can be reformulated as:

(1)(i') $dim(F_1) = dim(F)$ iff $F_1 \subset span\{f_1, ..., f_{2n-3}, f_{2n-1}\}$,

(1)(ii') $dim(F_1) = dim(F) - 1$ iff there is $a \in F$ such that $a = f_{2n-2} + v$, $v \in span\{f_1, ..., f_{2n-3}, f_{2n-1}\}$.

Let V' be the subspace of V of elements satisfying the alternating condition, thus V' is generated by $f_1, ..., f_{2n-1}$. Assume that F is a pre-Lagrangian in V'. Then 1(i') and 2(i) cannot hold as in that case F is not a pre-Lagrangian – it is not maximal: adding f_{2n-1} still gives a totally degenerated space. . Similarly, if 1(ii') and 2(ii) hold then $a \in F$ and $f_{2n-1} \in F$ but $\phi(a, f_{2n-1}) = \phi(f_{2n-2}, f_{2n-1}) = 1$, so F could not be a totally degenerate space.

Thus we proved that F_2 is $n-1$ dimensional in W. It is also a totally degenerated space (as an embedding of W' in V' is an isometry: W' is a subspace of W satisfying the alternating condition, so it has a basis $f_1,...f_{2n-2}$). Thus F_2 is a pre-Lagrangian in W'. The proof of Theorem 2.6 is completed.

As we mentioned before, Theorem 2.7 follows from the result of Wajnryb that the symplectic group is generated by braid-like generators in which braid generators act as transvections [Wa-2]. I our situation it means that adding crossings to a tangle allows us to realize any symplectic map on the symplectic space $\mathbf{Z}_p^{2n-2}$ of boundary coloring. In particular any Lagrangian is an image of the Lagrangian $span\{f_1, f_3, ..., f_{2n-2}\}$ associated to the trivial n-tangle T_0.

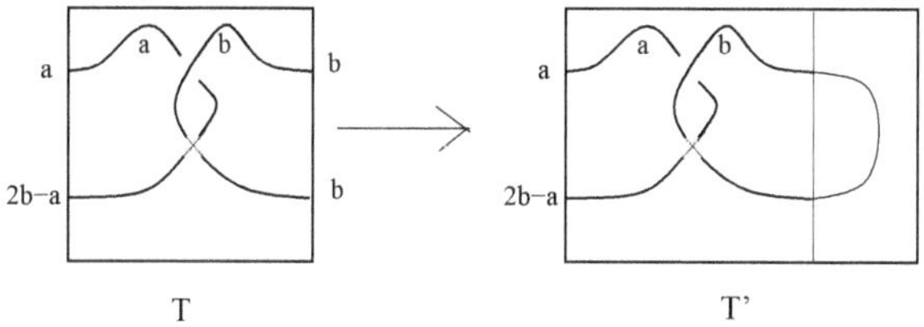

Fig. 2.12. non-classical colorings of virtual tangles.

On Figure 2.12 we have an example of a virtual 1-tangle T' such that $\psi(Col_p(T')) = \mathbf{Z}_p^2$. Combining n tangles T' together we get a virtual n-tangle $T^{(n)}$ with n virtual crossings and $\psi(Col_p(T^{(n)}) = \mathbf{Z}_p^{2n}$. Combining T' tangles and trivial tangles we can get a virtual tangle T with

$dim(\psi(Col_p(T)))$ any number between n and $2n$. Can we get dimension smaller from n? In particular, is there a virtual 2-tangle such that boundary coloring is always monochromatic?

Let me complete this presentation by mentioning two generalizations of the Fox k-colorings.

In the first generalization we consider any commutative ring with the identity in place of $\mathbf{Z}_k$. We construct $Col_R T$ in the same way as before with the relation at each crossing, Fig.2.1, having the form $c = 2a - b$ in R. The skew-symmetric form ϕ on R^{2n-1}, the symplectic form $\hat{\phi}$ on R^{2n-2} and the homomorphisms ψ and $\hat{\psi}$ are defined in the same manner as before. Theorem 2.4 generalizes as follows ([DJP]):

Theorem 2.9 *Let R be a Principal Ideal Domain (PID) then, $\hat{\psi}(Col_R T/R)$ is a virtual Lagrangian submodule of R^{2n-2} with the symplectic form $\hat{\phi}$. That is $\hat{\psi}(Col_R T/R)$ is a finite index submodule of a Lagrangian in R^{2n-2}.*

The second generalization leads to racks and quandles [Joy, F-R] but we restrict our setting to the abelian case – Alexander-Burau-Fox colorings[17]. An ABF-coloring uses colors from a ring, R, with an invertible element t (e.g. $R = \mathbb{Z}[t^{\pm 1}]$). The relation in Fig.2.1 is modified to the relation $c = (1-t)a + tb$ in R at each crossing of an oriented link diagram; see Fig. 2.13.

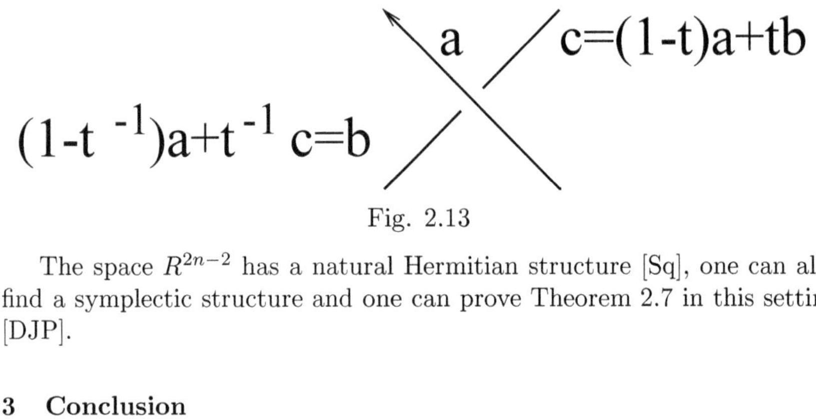

Fig. 2.13

The space R^{2n-2} has a natural Hermitian structure [Sq], one can also find a symplectic structure and one can prove Theorem 2.7 in this setting [DJP].

3 Conclusion

I hope that our snapshot of knot theory will inspire you to consider ideas described in the last two days. I am sure you are already asking: what

[17]The related approach was first outlined in the letter of J.W.Alexander to O.Veblen, 1919 [A-V]. Alexander was probably influenced by P.Heegaard dissertation, 1898, which he reviewed for the French translation [Heeg]. Burau was considering a braid representation but locally his relation was the same as that of Fox. According to J.Birman, Burau learned of the representation from Reidemeister or Artin [Ep], p.330.

about other n-move conjectures? Why should we use only abelian groups? Can we use more general structures following the Fox approach? I wish you fruitful thoughts, and you can compare your ideas with that in my book, that has been in preparation for over 20 years, and whose few chapters are available in arXiv [P-Book]. But here concisely:

(1) The oldest n-move conjecture is the Nakanishi 4-move conjecture: every knots can be reduced by 4-moves to the trivial knot. Formulated in 1979, it is still an open problem [Kir].

(2) One can look for an universal algebra (magma), $(X, *)$ where $* : X \times X \to X$ such that coloring of arcs of the diagram by elements of X is consistent (Fig. 2.14) and is preserved by Reidemeister moves. For example the third Reidemeister move leads to right self-distributivity $(a * b) * c = (a * c) * (b * c)$, Fig. 2.14. This leads to keis, racks, quandles, and shelfs as was explained in Scott Carter talk [Ca]. The simplest example is $\mathbf{Z}_p$ with $a * b = 2b - a$.

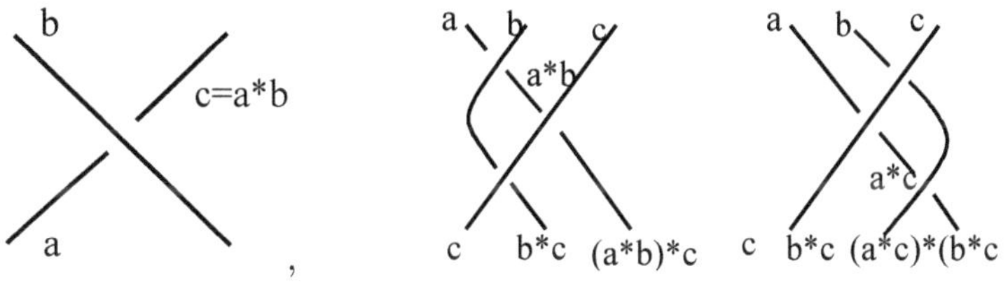

Fig. 2.14. coloring a crossing by elements of X and the third Reidemeister move.

Acknowledgement

The author was partially supported by the NSA grant (# H98230-08-1-0033), by the Polish Scientific Grant: Nr. N-N201387034, by the GWU REF grant, and by the CCAS/UFF award.

References

[A-B] J. W. Alexander, G. B. Briggs, On types of knotted curves, *Ann. of Math.*, 28(2), 1927/1928, 563-586.

[A-V] J. W. Alexander, Letter to Oswald Veblen, 1919, Papers of Oswald Veblen, 1881-1960 (bulk 1920-1960), Archival Manuscript Material (Collection), Library of Congress. (I would like to thank Jim Hoste for providing me with a copy of the letter).

[APR] R. P. Anstee, J. H. Przytycki, D. Rolfsen. Knot polynomials and generalized mutation. *Topology and Applications* 32, 1989, 237-249;
e-print: `http://xxx.lanl.gov/abs/math.GT/0405382`

[As] C. W. Ashley, *The Ashley book of knots*, Doubleday, 1944.

[As] J. Assion, Einige endliche Faktorgruppen der Zopfgruppen, *Math. Z.*, 163, 1978, 291-302.

[B-L] S. Baldridge, A. Lowrance, Cube diagrams and a homology theory for knots;
e-print: `arXiv:0811.0225v1 [math.GT]`

[BLM] R. D. Brandt, W. B. R.Lickorish, K. C. Millett, A polynomial invariant for unoriented knots and links, *Invent. Math.* 84, 1986, 563-573.

[Brunn-1892] H. Brunn, Topologische Betrachtungen, *Zeitschrift für Mathematik und Physik*, 37, 1892, 106-116.

[B-Z] G. Burde, H. Zieschang, *Knots*, De Gruyter, 1985 (second edition 2003).

[Ca] S. Carter, A Survey of Quandle Ideas, This Book.

[Che] Q.Chen, The 3-move conjecture for 5-braids, Knots in Hellas' 98; The Proceedings of the International Conference on Knot Theory and its Ramifications; Volume 1. In the Series on Knots and Everything, Vol. 24, September 2000, pp. 36-47.

[Cr] R.H.Crowell, Knots and Wheels, National Council of Teachers of Mathematics (N.C.T.M.) Yearbook, 1961.

[C-F] R.H.Crowell, R.H.Fox, *An introduction to knot theory*, Ginn and Co., 1963.

[D-P-1] M. K. Dąbkowski, J. H. Przytycki, Burnside obstructions to the Montesinos-Nakanishi 3-move conjecture, *Geometry and Topology (G&T)*, 6, June, 2002, 335-360;
e-print: `http://front.math.ucdavis.edu/math.GT/0205040`

[De-0] M. Dehn, *Jahresbericht der deutschen Mathematiker–Vereinigung*, Vol.16, 1907, p.573.

[D-H] M. Dehn, P.Heegaard, Analysis situs, *Encykl. Math. Wiss.*, vol. III AB3 Leipzig, 1907, 153-220.

[Del] M. Delbrück, Knotting problems in biology, in *Mathematical Problems in the Biological Sciences:* Proceedings of Symposia in Applied Mathematics, 14, 1962, 55-68.

[Dia] Y. Diao, Minimal knotted polygons on the cubic lattice, *J. Knot Th. Ramifications*, **2**, 1993, 413-425.

[DJP] J. Dymara, T. Januszkiewicz, J. H. Przytycki, Symplectic structure on Colorings, Lagrangian tangles and Tits buildings, preprint, May 2001.

[Ep] M. Epple, Geometric aspects in the development of knot theory, History of topology (ed. I.M.James) , 301–357, North-Holland, Amsterdam, 1999.

[F-R] R. Fenn, C. Rourke, Racks and links in codimension two, *Journal of Knot Theory and its Ramifications*, 1(4) 1992, 343-406.

[Fo-1] R. H. Fox, A quick trip through knot theory, *In: Top. 3-manifolds*, Proc. 1961 Top.Inst.Univ. Georgia (ed. M.K.Fort, jr), 120-167. Englewood Cliffs. N.J.: Princeton-Hall, 1962.

[Fo-2] R. H. Fox, Metacyclic invariants of knots and links, Canadian J. Math., XXII(2) 1970, 193-201.

[Gaus] K. F. Gauss, Zur mathematischen Theorie der electrodynamischen Wirkungen, 1833, *Werke*, Königlichen Gesellschaft der Wissinchaften zu Gottingen, 5, 1877, 602-629.

[Goe] L. Goeritz, Knoten und quadratische Formen, *Math. Z.*, 36,1933, 647-654.

[Gor-1] C. McA. Gordon, Some aspects of classical knot theory, *In: Knot theory*, L.N.M. 685, 1978, 1-60.

[Gor-2] C. McA. Gordon, 3-Dimensional Topology up to 1960, *History of topology* (ed. I.M.James), 449–489, North-Holland, Amsterdam, 1999.

[Heeg] P. Heegaard, Forstudier til en Topologisk Teori for de algebraiske Fladers Sammenhæng, København, 1898, Filosofiske Doktorgrad; French translation: Sur l'Analysis situs, *Soc. Math. France Bull.*, 44 (1916), 161-242. English translation, by A.Przybyszewska, of the topological part of Poul Heegaard Dissertation is in: J.H.Przytycki, Knot theory from Vandermonde to Jones, Preprint 43, Odense University 1993.

[Ho] C. F. Ho, A new polynomial for knots and links; preliminary report, *Abstracts AMS* 6(4), 1985, p. 300.

[Joy] D. Joyce, A classifying invariant of knots: the knot quandle, *Jour. Pure Appl. Alg.*, 23, 1982, 37-65.

[Kaw] A. Kawauchi, A survey of Knot Theory, Birkhäusen Verlag, Basel-Boston-Berlin, 1996.

[Kir] R. Kirby, Problems in low-dimensional topology; Geometric Topology (Proceedings of the Georgia International Topology Conference, 1993), Studies in Advanced Mathematics, Volume 2 part 2., Ed. W.Kazez, AMS/IP, 1997, 35-473.

[Klein] Felix Klein, Úeber den zusammenhang der flächen. Mathematische Annalen, 9:476482,
(some authors refer to this paper as containing a remark about no knotting in dimension four, on pages 476 or 478; however most likely the remark is not there.)

[Li] W.B.R.Lickorish, A relationship between link polynomials, *Math. Proc. Cambridge Phil. Soc.*, 100 (1986), 109-112.

[Lis] J. B. Listing, *Vorstudien zur Topologie*, Göttinger Studien (Abtheilung 1) 1, 1847, 811-875.

[Mag] W. Magnus, **Max Dehn**, *Math. Intelligencer* 1 (1978), 132–143. (Also in: Wilhelm Magnus Collected Paper, Edited by G.Baumslag and B.Chandler, Springer-Verlag, 1984).

[Mat] S. V. Matveev, Algorithmic Topology and Classification of 3-Manifolds *Algorithms and Computation in Mathematics*, Vol 9, Second edition, Springer 2010 (First edition 2007).

[Mo-1] J. M. Montesinos, Variedades de Seifert que son cubiertas ciclicas ramificadas de dos hojas, *Bol. Soc. Mat. Mexicana (2)*, 18, 1973, 1–32.

[Mo-2] J. M. Montesinos, Lectures on 3-fold simple coverings and 3-manifolds, Combinatorial methods in topology and algebraic geometry (Rochester, N.Y., 1982), Contemp. Math., 44, Amer. Math. Soc., Providence, RI, 1985, 157–177.

[O'M] O. T. O'Meara, Symplectic groups, Mathematics Surveys, Number 16, Published by AMS, 1978.

[Pla] A. Plans, Contribution to the study of the homology groups of the cyclic ramified coverings corresponding to a knot (Spanish), *Revista Acad. Ci. Madrid* 47, 1953, 161–193. Math.Rev. 15,147a 56.0X (R.H.Fox).

[Po-1] H. Poincaré, Analysis Situs (&12), *Journal d'Ecole Polytechnique Normale*, 1, 1895, 1-121.

[P-1] J. H. Przytycki, t_k-moves on links, In *Braids*, ed. J.S.Birman and A.Libgober, Contemporary Math. Vol. 78, 1988, 615-656.

[P-2] J. H. Przytycki, Elementary conjectures in classical knot theory, in *Quantum Topology*, Ed. L.J.Kauffman, R.A.Baadhio, Series on Knots and Everything - Vol.3, World Scientific, 1993, 292-320.

[P-3] J. H. Przytycki, 3-coloring and other elementary invariants of knots, Banach Center Publications, Vol. 42, *Knot Theory*, 1998, 275-295;
e-print: `http://arxiv.org/abs/math.GT/0608172`

[P-4] J. H. Przytycki, Classical roots of Knot Theory, *Chaos, Solitons and Fractals*, Vol. 9 (No. 4-5), 1998, 531-545.

[P-5] J. H. Przytycki, From 3-moves to Lagrangian tangles and cubic skein modules, *Advances in Topological Quantum Field Theory*, Proceedings of the NATO ARW on New Techniques in Topological Quantum Field Theory, Kananaskis Village, Canada from 22 to 26 August 2001; John M. Bryden (ed), October 2004, 71-125;
e-print: `http://front.math.ucdavis.edu/math.GT/0405248`

[P-6] J. H. Przytycki, Kolorowanie splotów (Coloring of knots), *Delta* 7, July 2003, 8-10.

[P-Book] J. H. Przytycki, **KNOTS:** From combinatorics of knot diagrams to the combinatorial topology based on knots,

Cambridge University Press, accepted for publication, to appear 2013, pp. 600.
Chapter II, e-print: `http://arxiv.org/abs/math/0703096`
Chapter V, e-print: `http://arxiv.org/abs/math.GT/0601227`
Chapter IX, e-print: `http://arxiv.org/abs/math.GT/0602264`
Chapter X, e-print: `http://arxiv.org/abs/math.GT/0512630`

[Rei-1] K. Reidemeister, Elementare Begrundung der Knotentheorie, *Abh. Math. Sem. Univ. Hamburg*, 5 (1927), 24-32.

[Rol] D. Rolfsen, *Knots and links*, Publish or Perish, 1976 (second edition, 1990).

[Sq] C. Squier, The Burau representation is unitary, *Proc. Amer. Math. Soc.*, 90(2), 1984, 199-202.

[Uch] Y. Uchida, Proc. of Knots 96 ed. S. Suzuki 1997, 109-113 *Knots 96*, Proceedings of the Fifth International Research Institute of MSJ, edited by Shin'ichi Suzuki, 1997 World Scientific Publishing Co., 109-113.

[Tait-1] P. G. Tait, Preliminary note on a new method of investigating the properties of knots Proc. Royal Soc. Edinburgh, Vol. 9, 98 (1876-7), 403 (Tait writes here: "As we cannot have knots in two dimensions, and as Prof. Klein has proved that they cannot exist in space of four dimensions;...")

[Tait-2] P. G. Tait, On knots I, II, III, Scientific Papers, Cambridge University Press, 1898-1900. Including: *Trans. R. Soc. Edin.*, 28, 1877, 35-79. Reprinted by Amphion Press, Washington D.C., 1993, (On Knots I, pages 273-317; remark about Klein is on the page 316).

[Tur] V. G. Turaev, Quantum invariants of knots and 3-manifolds, de Gruyter Studies in Mathematics, 18, Walter de Gruyter & Co., Berlin, 1994. x+588 pp.

[Turing] A. M. Turing, Solvable and unsolvable problems, Science News, 31, 1954, 7-23.

[Vi] O. Ya. Viro, Raskrasiennyje uzly, in Russian (Colored knots), Kvant 3, 1981, 8-14 (Russian); English translation: Tied into Knot Theory: unraveling the basics of mathematical knots, Quantum 8(5), 1998, 16-20.

[Wa-1] B. Wajnryb, Markov classes in certain finite symplectic representations of braid groups, in: Braids (Santa Cruz, CA, 1986), 687–695, Contemp. Math., 78, Amer. Math. Soc., Providence, RI, 1988.

[Wa-2] B. Wajnryb, A braidlike presentation of $\mathrm{Sp}(n,p)$, *Israel J. Math.* 76(3), 1991, 265–288.

[Wir] W. Wirtinger, Über die Verzweigungen bei Funktionen von zwei Veränderlichen, *Jahresbericht d. Deutschen Mathematiker Vereinigung*, 14 (1905), 517. (The title of the talk supposedly given at September 26 1905 at the annual meeting of the German Mathematical Society in Meran).

[Wo-Kr] D. Wolkstein, S. N. Kramer, Inanna: Queen of heaven and earth, Harper Collins Publisher, 1983.

Department of Mathematics
George Washington University
Washington, DC 20052
USA
e-mail: przytyck@gwu.edu

DETECTION OF CHIRALITY AND MUTATIONS OF KNOTS AND LINKS

Ramadevi Pichai

Department of Physics, Indian Institute of Technology Bombay,
Mumbai, Maharashtra-400 076, INDIA
**E-mail: ramadevi@phy.iitb.ac.in*

In this brief presentation, we would like to present our attempts of detecting chirality and mutations from Chern-Simons gauge theory. The results show that the generalised knot invariants, obtained from Chern-Simons gauge theory, are more powerful than Jones, HOMFLYPT and Kauffman polynomials. However the classification problem of knots and links is still an open challenging problem.

Keywords: Chirality, mutation, Chern-Simons field theory invariants.

1. Introduction

The classification of three and four manifolds is one of the open questions which has been addressed by both mathematicians and physicists. In particular, physicists have shown that a class of quantum field theories called topological field theories provides an elegant approach to solve these problems.

The main idea in any quantum field theory is to represent the theory by an action S which gives information about the particle content and their interactions. The interaction strengths are given by coupling constants. For capturing the topological features of knots or links as shown in Fig. 1, we need a theory which does not change if we alter the shape or size of these knots or links. One such theory is the Chern-Simons gauge theory where the action S is explicitly metric independent. Hence, Chern-Simons field theory provides a natural framework to study knots,links and three manifolds. The action S defining the Chern-Simons theory on a three manifold M based on a gauge group G is

$$\begin{aligned} S &= \frac{k}{4\pi}\int_M Tr\left(A\wedge dA+\frac{2}{3}A\wedge A\wedge A\right) \\ &= \frac{k}{4\pi}\int_M \epsilon_{\mu\nu\lambda}\, d^3x\; Tr\left(A_\mu\partial_\nu A_\lambda+\frac{2}{3}A_\mu A_\nu A_\lambda\right)\,, \end{aligned} \tag{1}$$

Fig. 1. Both knots C (though different sizes) are trefoil knots

where k is the coupling constant and A_μ's are the gauge fields or connections matrix-valued in group G.

The knots or links - for example, the trefoil knot C carrying representation R of the gauge group G are described by the expectation value of Wilson loop operators $W_R(C) = Tr[Pexp \oint A_\mu dx^\mu]$:

$$V_R[C] = \langle W_R(C) \rangle \; = \; \frac{\int_M [\mathcal{D}A] \; W_R(C) \; exp(iS)}{\mathcal{Z}[M]} \,, \tag{2}$$

where

$$\mathcal{Z}[M] = \int_M [\mathcal{D}A] \; exp(iS)$$

is the partition function and $V_R[C]$ are the knot invariants.
Witten's pioneering work[1] established a three-dimensional definition for knots and links. In particular, Jones and HOMFLYPT polynomials and their recursion relations were obtained from Chern-Simons gauge theory based on $SU(2)$ and $SU(N)$ gauge groups. We can relate coupling constant k and the rank N to the polynomial variables of HOMFLYPT polynomials. Similarly, the Jones' polynomial variable q will be related as $q = \exp{(2\pi i/k + 2)}$.
The two main ingredients which go into the evaluation of the polynomial invariants $V_R[C]$ of knots and links are

(1) Connection between Chern-Simons theory on the three-dimensional ball to the two-dimensional Wess-Zumino conformal field theory on the boundary of the three-ball.
(2) Using Alexander's theorem, any knot or link can be obtained as a closure of braid.

In Fig. 2(a), we illustrate Alexander's theorem by re-drawing the trefoil knot as a closure of two-strand braid with three crossings. Also, we have diagrammatically shown in Fig. 2(b) that these knots or links in S^3 can be viewed as gluing two three-balls with oppositely oriented S^2 boundaries. In this particular trefoil knot example, the S^2 boundary has four-punctures. The connection between Chern-Simons theory and Wess-Zumino conformal field theory states that the Chern-Simons functional integral over a three-ball with a four-punctured S^2 boundary corresponds to state $|\Psi_3\rangle$ which represents four-point correlator conformal block in the Wess-Zumino conformal field theory. The suffix 3 on the state is to indicate that the middle two-strands are braided thrice. In fact, the punctures get exchanged whenever the middle two-strands get braided. We can denote the no-crossing four-punctured boundary state as $|\Psi_0\rangle$ and apply a braiding operator $\mathcal{B}$ thrice to get the state $|\Psi_3\rangle$:

$$|\Psi_3\rangle = \mathcal{B}^3|\Psi_0\rangle \ . \tag{3}$$

Similarly, the state for the oppositely oriented boundary will be in the dual space. For the above example in Fig. 2(b), the state is $\langle\Psi_0|$. The knot invariant is

$$V_R[C] = \langle\Psi_0|\Psi_3\rangle = \langle\Psi_0|\mathcal{B}^3|\Psi_0\rangle \ . \tag{4}$$

In order to see the polynomial form, we need to expand the state $|\Psi_0\rangle$ in an eigenbasis of the braiding operator $\mathcal{B}$. In general for the four-punctured S^2 boundary, the braiding can be either on the side two-strands or on the middle two strands. For clarity, we will take the gauge group $G = SU(2)$. For the four-punctured S^2 boundary, we can chose eigenbasis $|\phi_s\rangle$ if the braiding is in the side two-strands. That is,

$$\mathcal{B}_1|\phi_s\rangle = \lambda_s(R,R)|\phi_s\rangle \ , \tag{5}$$

where the suffix 1 on the braiding operator denotes the braiding between first and the second strands and the eigenvalue is $\lambda_s(R,R)$. The basis is shown diagrammatically in Fig. 3(a) where R denotes the $SU(2)$ representation placed on the strands. From the picture, the representation s will be an element in the tensor product $R\otimes R$. Similarly, for braiding middle two

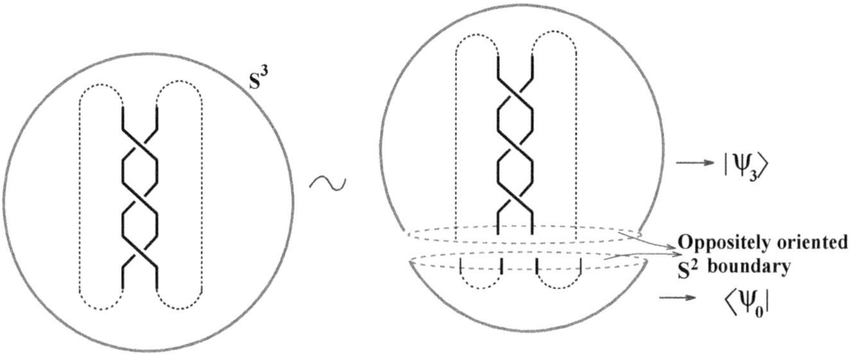

Fig. 2. (a) Trefoil in S^3 drawn as a closure of braid ≡ (b) gluing of two three-balls with oppositely oriented S^2 boundaries

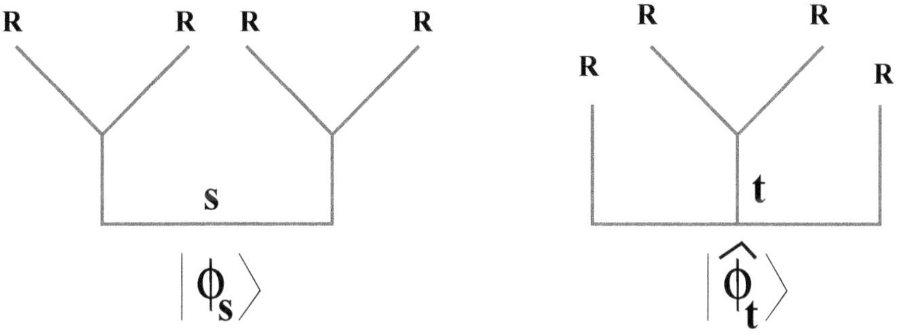

Fig. 3. eigenbasis when braiding (a) side two-strands (b) middle two-strands.

strands, we choose the basis $|\hat{\phi}_t\rangle$ as shown in Fig. 3(b) where $t \in R \otimes R$. Clearly, these two basis states must be related by a duality matrix:

$$|\hat{\phi}_t\rangle = a_{ts} \begin{bmatrix} R & R \\ R & R \end{bmatrix} |\phi_s\rangle \ . \tag{6}$$

When the four strands carry the same representation, we can write in short-hand notation the duality matrix as

$$a_{ts} \begin{bmatrix} R & R \\ R & R \end{bmatrix} \equiv a_{ts} \ . \tag{7}$$

These duality matrices turns out to be proportional to quantum Racah co-efficients. The explicit form of $SU(2)_q$ Racah coefficients and their identities satisfied are available in Ref. 2. See also papers[3–5] . In this example, the braiding involves middle two strands. Hence the state $|\Psi_0\rangle$ can be expanded in the middle-strand basis $|\hat{\Phi}_t\rangle$:

$$|\Psi_0\rangle = \sum_t \mu_t |\hat{\Phi}_t\rangle . \tag{8}$$

Interestingly, the coefficients μ_t has to satisfy

$$\mu_t = \sqrt{V_t[U]} = \sqrt{S_{0t}/S_{00}} \equiv \sqrt{dim_q t} , \tag{9}$$

so that two equivalent knots share the same polynomial invariant. Here, $V_t[U]$ denotes the polynomial invariant for unknot carrying representation t whose form can be written as the ratio of elements of the modular transformation matrix S in Wess-Zumino conformal field theory or in terms of quantum dimensions of the representation t of the quantum group as indicated in the above equation. The knot invariant (4) will be

$$V_R[C] = \langle\Psi_0|\mathcal{B}^3|\Psi_0\rangle = \sum_t dim_q t(\lambda_t(R,R))^3 . \tag{10}$$

So far, we have not introduced orientation on the strands. In general, the braiding eigenvalue depends on the **framing** and also on the relative orientation on the two braiding strands. Two conventional framing are standard framing and blackboard framing. Standard framing is one where the self-linking number of the knot with its frame is zero. This is useful to obtain ambient isotopy invariants. The self-linking number matches the crossing number in the blackboard framing. Hence the braiding eigenvalue in the blackboard framing is useful to obtain regular isotopy invariants. In the trefoil example, we could place parallel orientation in the middle two-strands. As the crossing sign due to braiding is positive, we called such a braiding as right-handed braiding. Similarly, an inverse braiding leading to mirror of trefoil (T^*) will be called left-handed braiding.

For parallely oriented strands, the right-handed braiding eigenvalue in standard framing is

$$\lambda_t^{(+)}(R,R) = (-1)^{\epsilon} q^{2C_R - C_t/2},\ q = e^{\frac{2\pi i}{k+C_v}} . \tag{11}$$

where C_R, C_t, C_v denotes the quadratic casimirs in the respective R, t and adjoint representation. ϵ will be ± 1 depending on the representation t appears symmetrically or antisymmetrically in the tensor product $R \otimes R$. Similarly, the left-handed braiding eigenvalue for antiparallely oriented strands is

$$\lambda_t^{(-)}(R,R) = (-1)^{\epsilon} q^{C_t/2},\ q = e^{\frac{2\pi i}{k+C_v}} . \tag{12}$$

Now using the appropriate braiding eigenvalues, the knot invariants can be written as polynomials in the variable q. The method for a four-punctured

S^2 boundary is generalisable for r such four-punctured S^2 boundaries as shown in Fig. 4. We will see in the next section that this building block will be useful to redraw knots like knot 9_{42} and knot 10_{71} as gluing of three-balls with one or more four-punctured S^2 boundaries. The basis state for such a

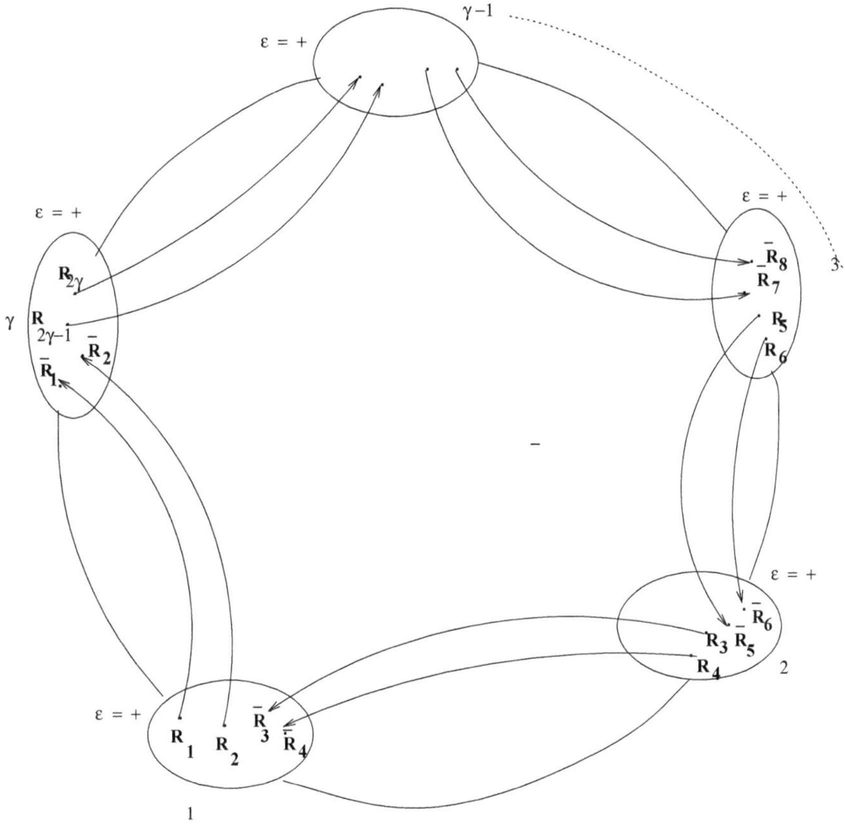

Fig. 4. Three-ball with r S^2 boundaries each with four-punctures.

r-S^2 boundaries is

$$\nu_r = \sum_{R_s} \frac{|\phi^{(1)side}_{R_s}\rangle |\phi^{(2)side}_{R_s}\rangle \cdots |\phi^{(r)side}_{R_s}\rangle}{(dim_q R_s)^{\frac{r-2}{2}}}, \tag{13}$$

where $R_s \in R_i \otimes R_{i+1}$ for any i. Sometimes, it is useful to keep S^2 boundaries with more than four-puctures. Then the basis state for a S^2 boundary with n puctures will be a n-point conformal block. For braiding $\mathcal{B}_{2i+1}$'s and $\mathcal{B}_{2i}$'s, we choose the basis shown in Fig. 5(a) and (b) respectively. The

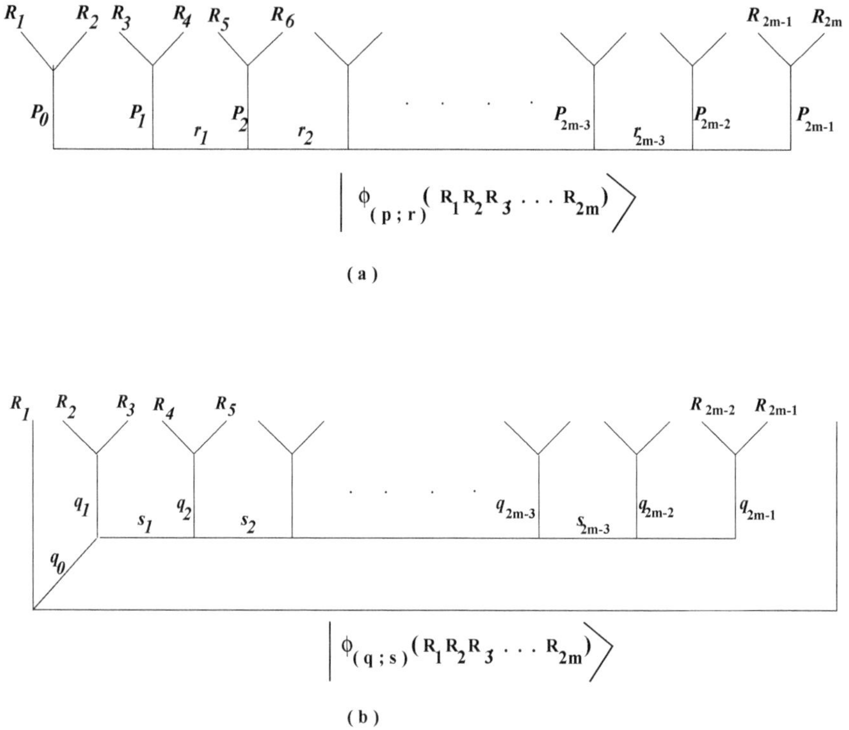

Fig. 5. Basis states for a n-punctured S^2 boundary.

procedure we elaborated for the trefoil obtained from gluing two three-balls each with a four-punctured S^2 boundary is generalisable for any knot. That is, using the building blocks in Fig. 4 or three-balls with n-punctured S^2 boundaries, it is not difficult to see that any knot can be obtained either from gluing two three-balls each with a n-punctured S^2 boundaries or from gluing many three-balls with one or more r four-puctured S^2 boundaries. The method presented here enables direct evaluation of any knot/link polynomial directly without going through the recursive procedure.

Though we have concentrated on the gauge group $SU(2)$, it is straightforward to generalise for any compact semi-simple gauge group.[6] The representation R must be replaced by conjugate representation $\bar{R}$ ($R \equiv \bar{R}$ for $SU(2)$) depending on the oriented strand is outgoing from or incoming to a S^2 boundary as shown in Fig. 4. Thus, we can place any representation R of any compact semi-simple gauge group on the knot and obtain **generalised knot invariants.** If we place the defining representation on the strands, we recover some of the well-known polynomials as tabulated below:

Gauge Group	Polynomial
$SU(2)$	Jones'
$SU(N)$	Two-variable HOMFLYPT
$SO(N)$	Two-variable Kauffman

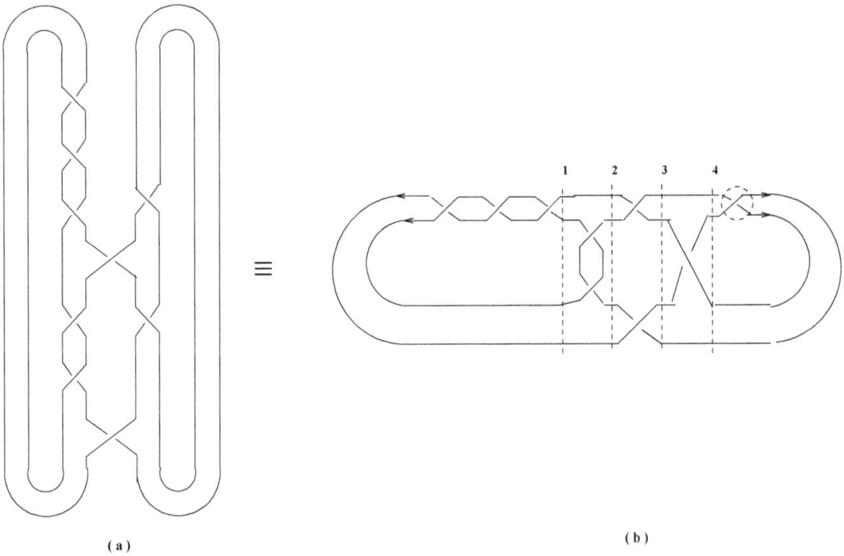

Fig. 6. Chiral Knot 9_{42}

We know that these well-known polynomials do not solve the classification problem. Apart from these special cases, Chern-Simons field theory gives a huge pool of generalised polynomials depending on placing representation R of any gauge group other than the defining representation on the strands. We believe that at least one of these generalised Chern-Simons invariant will be able to distinguish two inequivalent knots which are not distinguished by the well-known polynomials.

Knot theory literature gives a list of chiral knots and mutant knots which are not distinguished by Jones', HOMFLYPT and Kauffman. We tried to check the ability of generalised knot invariants, from Chern-Simons field theory, to detect **chirality and mutations**. We address the chirality detection in the following section.

2. Chirality Detection

Upto 10 crossings, there are two knots : knot 9_{42} and knot 10_{71} which are chiral but their chirality is not detected by the well-known polynomials. In Fig. 6, we have drawn **knot** 9_{42} in two equivalent ways. Clearly, the knot 9_{42} can be obtained as gluing of five building blocks as shown in Fig. 7.

For $G = SU(2)$, $R_n = \overbrace{\square\square\square\square\square\square}^{n}$ (spin $n/2$ representation) placed on knot, the states for the building blocks can be written down following the methods presented in the previous section and also using the properties of the duality matrix.[3–5] The states for these five building blocks are[7]

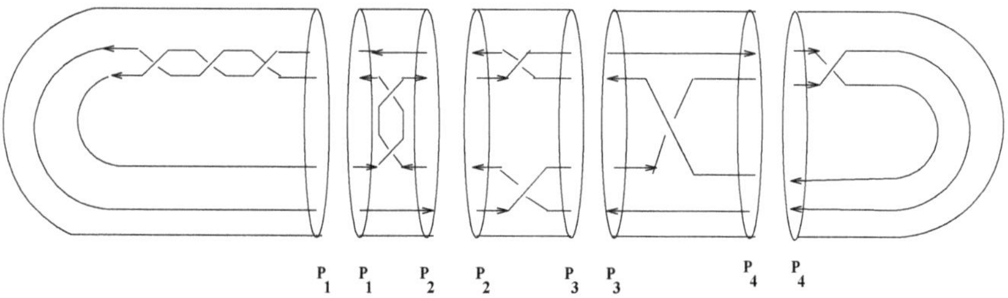

Fig. 7. Gluing of five building blocks

$$\nu_1(P_1) = \sum_{l_1=0}^{n} \sqrt{[2l+1]}(-1)^{3(n-l_1)} q^{-3/2[n(n+2)-l_1(l_1+1)]} |\phi^{(1)}_{l_1}\rangle$$

$$\nu_1(P_4) = \sum_{l_5=0}^{n} (-1)^{n-l_5} q^{-1/2[n(n+2)-l_5(l_5+1)]} |\phi^{(1)}_{l_5}\rangle$$

$$\nu_2(P_1; P_2) = \sum_{i_1,j_1,l_2,r=0}^{n} \frac{a_{l_1 r} a_{j_1 r} a_{l_2 r} \sqrt{[2l_2+1]}}{\sqrt{[2r+1]}} \times q^{n(n+2)-l_2(l_2+1)} |\phi^{(1)}_{i_1}\rangle |\phi^{(2)}_{j_1}\rangle$$

$$\nu_2(P_2; P_3) = \sum_{l_3=0}^{n} q^{l_3(l_3+1)} |\phi^{(1)}_{l_3}\rangle |\phi^{(2)}_{l_3}\rangle$$

$$\nu_2(P_3; P_4) = \sum_{i_2,j_2,l_4=0}^{n} (-1)^{l_4} q^{-l_4(l_4+1)/2} a_{l_4 i_2} a_{l_4 j_2} |\phi^{(1)}_{i_2}\rangle |\phi^{(2)}_{j_2}\rangle\ .$$

Here P_i's denote the S^2 boundaries as indicated in Fig. abg:fig7. Using the above states, the knot invariant is

$$V_n[9_{42}] = (-1)^n q^{\frac{-3}{2}[n(n+2)]} \sum_{r,l_1,l_2,j_1,j_2=0} \sqrt{[2l_1+1]} \times \sqrt{[2l_2+1]}\sqrt{[2j_2+1]} a_{l_1 r} a_{l_2 r} a_{j_1 r} a_{j_1 j_2} \times (-1)^{l_1} q^{\frac{3}{2}[l_1(l_1+1)]} q^{\frac{3}{2}[j_1(j_1+1)]} q^{-l_2(l_2+1)} q^{j_2(j_2+1)}$$

We checked the general result for the special cases. That is, $n = 1$ gives Jones' polynomial and $n = 2$ gives Akutsu-Wadati/Kauffman polynomial.[8] Interestingly, for $n = 3$, the polynomial is

$$\begin{aligned} V_3[9_{42}] = {} & q^{45/2} - q^{41/2} - q^{39/2} + q^{35/2} + q^{23/2} + q^{21/2} - q^{19/2} \\ & - q^{17/2} + q^{13/2} - q^{9/2} + q^{5/2} + q^{3/2} + q^{-3/2} + q^{-5/2} \\ & - q^{-13/2} - q^{-15/2} + q^{-21/2} + 2q^{-23/2} - q^{-27/2} + 2q^{-31/2} \\ & - 3q^{-35/2} - q^{-37/2} + q^{-39/2} + q^{-41/2} . \end{aligned}$$

Clearly, $V_3[9_{42}](q) \neq V_3[9_{42}](q^{-1})$ indicating that $SU(2)$ Chern-Simons spin $3/2$ ($n = 3$ in representation R_n) knot polynomial is powerful to detect chirality. Similar exercise was performed for knot 10_{71} by gluing the four building blocks as shown in Fig. 8. The knot invariant[7] is

$$V_n[10_{71}] = (-1)^n q^{\frac{n(n+2)}{2}} \sum_{i,r,s,u,m=0} \sqrt{\frac{[2r+1][2s+1][2u+1]}{[2m+1]}} a_{im} a_{ms} a_{rm} a_{iu} (-1)^s \, q^{-i(i+1)} q^{m(m+1)} q^{-r(r+1)} q^{u(u+1)} q^{\frac{3}{2}s(s+1)} .$$

For $n = 3$, we have checked that $V_3[10_{71}](q) \neq V_3[10_{71}](q^{-1})$ confirming the ability of generalised Chern-Simons invariant in detecting chirality. With such positive results for chiral knots, we attempted to check whether generalised Chern-Simons invariant is capable of detecting mutation operation which we shall present in the following section.

3. Mutation and mutant knots

Remove a two-tangle region from any knot and do a rotation by π about any of the three perpendicular axis. Replace the rotated two-tangle back with suitable reversal of orientation of the strands to give another knot. This operation is called mutation and the two knots are said to be mutants. In Fig. 9, we have indicated the three mutation operation γ_i's. An example for the mutant knots is the well-known eleven-crossing Kinoshita-Terasaka and Conway knots. We can formally represent mutant knots shown in Fig. 10 as

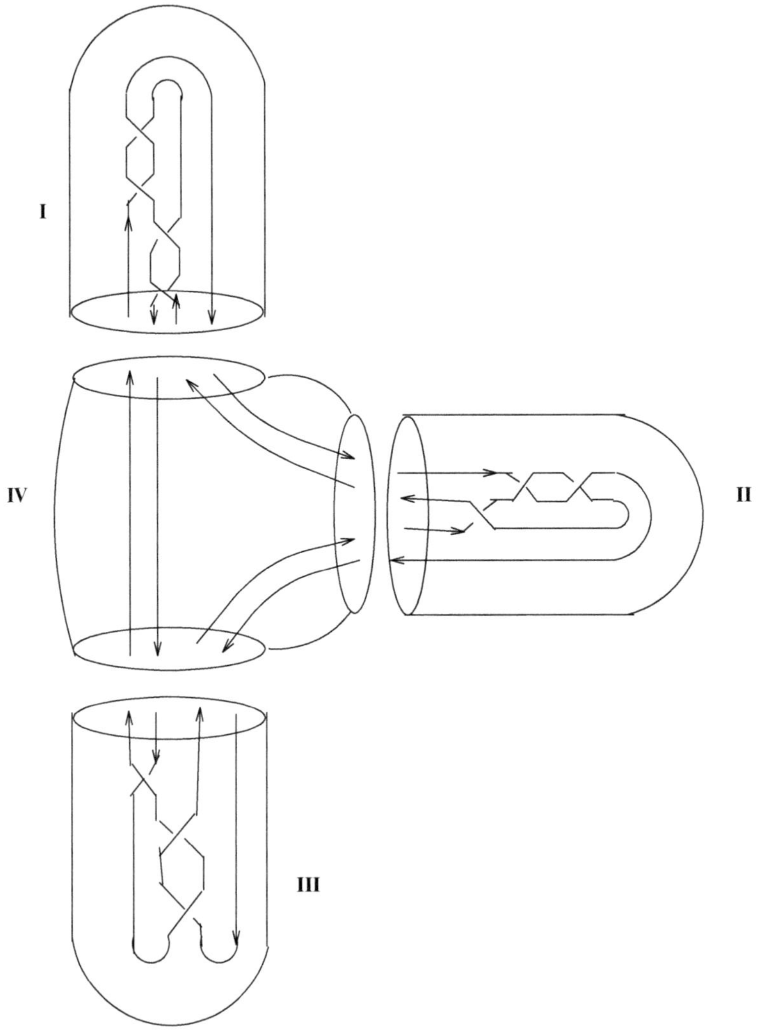

Fig. 8. Chiral Knot 10_{71} obtained from gluing the building blocks

gluing two three-balls with the two-tangle rooms as shown in Fig. 11. That is, gluing Fig. 11(d) with any of the Fig. 11(a),(b) and (c) gives mutant knots. Interestingly, the states (a), (b), (c) in Fig. 11 can be obtained by gluing Fig. 11(a) with the two-boundary states (a), (b) and (c) in Fig. 12 respectively. Clearly, ν_2 and ν_3 represent mutation γ_1 and γ_2 respectively in Fig. 9 respectively. As braid words, Fig. 12 (a), (b), (c) are different but

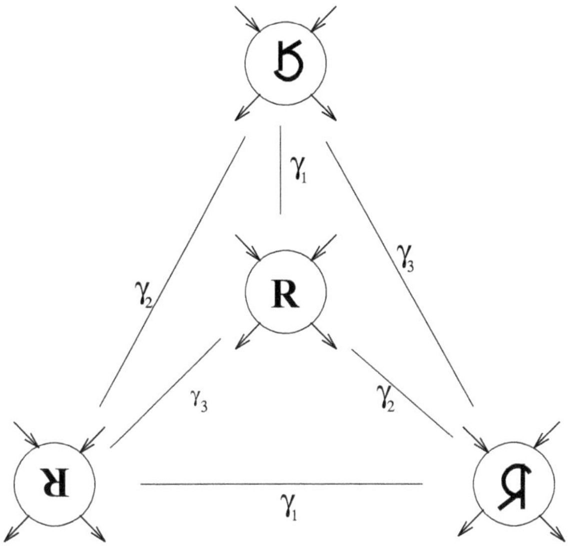

Fig. 9. Mutation Operation

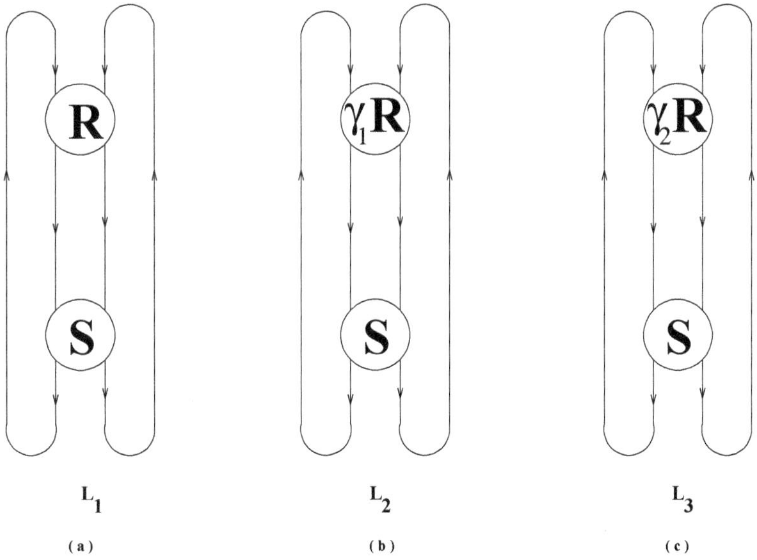

Fig. 10. Mutant Knots

the states ν_i's are same:

$$\nu_2 \ = \ \sum_l \ |\phi_l^{side(1)}\rangle b_1 b_3^{-1} |\phi_l^{side(2)}\rangle \ = \ \mathcal{C}\nu_1 \ . \tag{14}$$

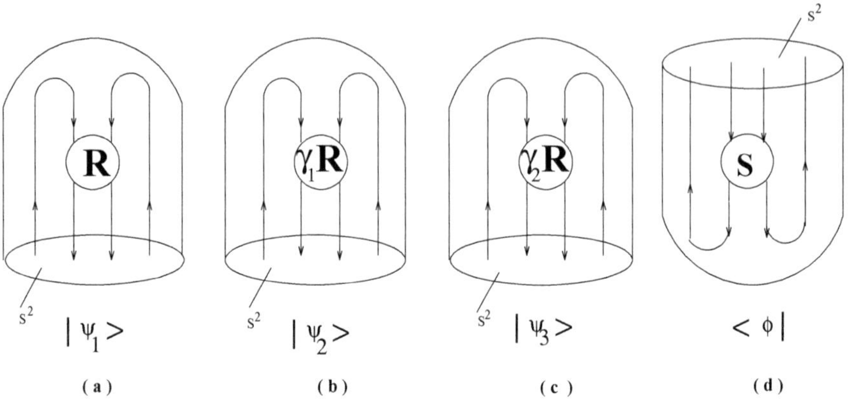

Fig. 11. Building blocks

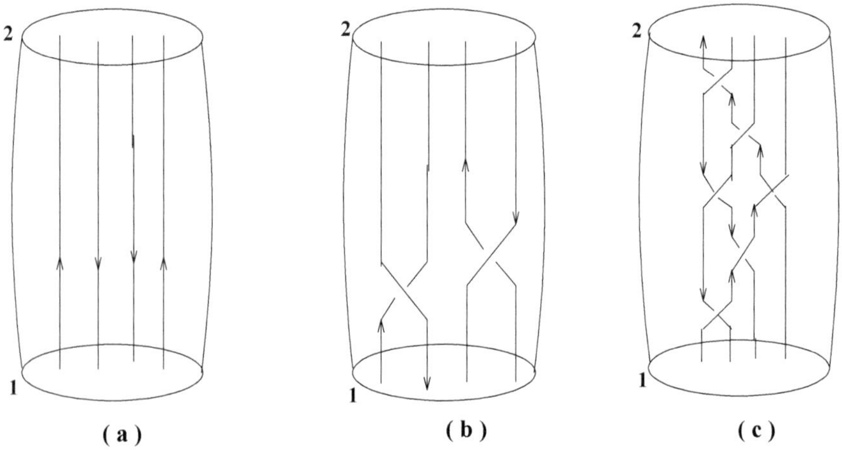

Fig. 12. state (a) ν_1 (b) ν_2 and (c) ν_3 .

where we have used

$$b_1|\phi_l^{side}\rangle = b_3|\phi_l^{side}\rangle = \lambda_l^{(-)}(R,\bar{R})|\phi_l^{side}\rangle \ ,$$

and the operator $\mathcal{C}$ interchanges the representations on the first and second, the third and fourth punctures in that basis. Similarly, we can show

$$\nu_3 \ = \ \sum_l |\phi_l^{side(1)}\rangle b_1 b_2 b_1 b_3 b_2 b_1 |\phi_l^{side(2)}\rangle = \mathcal{C}\nu_1 \ . \tag{15}$$

So, the **generalised invariants** of the mutants L_1, L_2, L_3 obtained by gluing Fig. 11(a),(b) and (c) with (d) are **same**.

It appears that the identities of the states for **four-punctured boundaries** play a crucial role in making the states representing the mutation operation (ν_2, ν_3) to be similar to identity braid ν_1. In order to go beyond four-punctured boundary state, we studied composite braiding[9] . Using the representation theory of composite braids, we showed that the composite invariant for knots are sum of the generalised knot invariants. This implies composite invariants cannot detect mutations in knots. However, some mutant links can be distinguished by composite invariants.

4. Summary and Discussion

In this article, we have briefly presented the direct evaluation of generalised invariants of knots and links from Chern-Simons field theory. For $SU(2)$ gauge group in Chern-Simons theory, we get the the colored Jones' polynomials for the knots and links carrying higher dimensional $SU(2)$ representations. There is a huge pool of generalised Chern-Simons invariants for other gauge groups like $SU(N)$, $SO(N)$ etc with knots and links carrying arbitrary representations. The hope is that at least one of the invariants will be able to distinguish two inequivalent knots.

We showed that the chiral knots upto 10 crossings are distinguished by the Chern-Simons field theory invariants. However, we gave a proof that the process of mutation cannot be detected within the Chern-Simons field theoretic framework. The proof is for knots and links carrying arbitrary representation of any compact semi-simple gauge group.

Discussions during the knot theory conference revealed that there are approaches from quantum groups[10] and Floer homology[11] whose invariants does distinguish 11-crossing Kinoshita-Terasaka and Conway mutant knot. It will be interesting to to check whether this method distinguishes other mutant knots.

References

1. E. Witten, Commun. Math. Phys. **121** (1989) 351.
2. A.N. Kirillov, N.Yu. Reshetikhin, *Representation algebra $U_q(sl_2)$, q-orthogonal polynomials and invariants of links*, LOMI preprint E-9-88; see also in: New Developments in the Theory of Knots. ed. T. Kohno, World Scientific (Singapore, 1989).
3. R.K. Kaul and T.R. Govindarajan, Nucl. Phys. **B380** (1992) 293.
4. R.K. Kaul and T.R. Govindarajan, Nucl. Phys. **B393** (1993) 392.

5. R.K. Kaul, Commun. Math. Phys.**162** (1994) 289.
6. P. Ramadevi, T.R. Govindarajan and R.K. Kaul, Nucl. Phys. **B402** (1993) 548
7. P. Ramadevi, T.R. Govindarajan and R.K. Kaul, Mod. Phys. Lett **A9** (1994) 3205.
8. M.Wadati, T.Deguchi and Y.Akutsu, Phys. Rep. **180** (1989) 247 and references therein.
9. P. Ramadevi, T.R. Govindarajan and R.K. Kaul, Mod. Phys. Lett **A10** (1995) 1635.
10. H.R.Morton and H.J.Ryder, Geom. Topol. Monogr.**1** (1998)365.
11. John A. Baldwin, W.D.Gilam, *Computations of Heegard-Floer knot homology*, arXiv:math/0610167.

PHYSICAL KNOT THEORY: THE STUDY OF SIZES AND SHAPES OF POLYMERS

Eric J. Rawdon

Department of Mathematics, University of St. Thomas,
Saint Paul, MN 55105, USA
E-mail: ejrawdon@stthomas.edu
http://george.math.stthomas.edu/rawdon/

We present some commonly-used models for polymers in different experimental conditions, including some discussion of implementation issues. We then analyze how the size and shape of polymers are affected by length and the existence of knotting.

Keywords: Physical knot theory; knots; polymers; asphericity; prolateness

1. Introduction

One of the goals of Physical Knot Theory is to study the configurations within a given knot type to glean information which can be used to model physical systems. The main goal of this paper is to summarize our thoughts and efforts in understanding how knotting affects the size and shape of modeled polymers.

Our work is motivated by curiosity stemming from the early 1990s. At the time, there was some hope that energy functionals for knots could be used to determine the knot type of a configuration. The process would be to enter the configuration into a computer, run some energy minimization technique on the configuration, and arrive at a unique global minimum energy configuration. Then one could compare that global minimum to a table of knot configurations and determine the knot type. As it turns out, complications (e.g. the existence of a host of local minima) derailed the plans, but researchers continue to be interested in energy functionals for the sake of energy functionals.

In 1988, Fukuhara[1] published the first "knot energy" paper (although others, for example Krotenheerdt and Veit[2,3] in 1976 had proposed similar ideas without the moniker). Soon thereafter, several other knot energies were defined, see for example Refs. 4–6. In the late 1990s, we published Ref. 7 on the thickness/ropelength of knots (the ideas of which were inspired and rediscovered from works such as Refs. 2,8–10). As a part of my dissertation,[11] I defined a version of thickness/ropelength for polygonal knots and used simulated annealing to find (nearly) ropelength-minimal configurations. At the same time, Pieranski had invented SONO[12–14] (a sort of pseudo gradient-descent algorithm) for minimizing ropelength. In discussions with Kenneth Millett, we became interested in understanding what these ropelength-minima look like, i.e. are they short and squat, long and skinny, spherical? This led to our analysis of the shape of ropelength-minima (often referred to as *ideal* or *tight knots*) in Ref. 15.

Additionally, inspired by several interdisciplinary conferences with colleagues in Biology, Chemistry, and Physics, we became interested in polymers modeled as random chains. In particular, we were drawn to the closed chains, which could then form knots. Much of the analysis of the size and shape of polymers was (and still is) concentrated on quantities familiar to physicists which can be measured experimentally (e.g. radius of gyration). However, new experimental techniques make it possible to get 3D coordinates for "frozen" configurations (i.e. snapshots of a polymer at a given time) which can be analyzed more deeply. Still, these experimental techniques are expensive (in time, money, and labor) so simulation provides a data set that simply cannot be reproduced in the lab. We decided to apply the spatial measurements from Ref. 15 to the polymer models to gain additional information about the configurations. Furthermore, we invented new measurements and adapted others to analyze polymer structure.

In this paper, we wish to give the reader a sense for what research looks like in studying entanglement in polymers. It is certainly impossible to complete this goal. Nevertheless, we present many topics, some of which we discuss in only a cursory way. However, we hope to provide enough references that an interested reader could find an insertion point into the literature. We start broadly and make choices in problems and models in each section to arrive at the results of a specific problem in the final section, namely analyzing the size and shape of knotted configurations within a certain polymer model. Specifically, in Section 2, we introduce some models for polymers in different experimental conditions and discuss how one generates numerical data sets to study these models. In Section 3, we discuss how topology affects the size and shape of polymer configurations. In Section 4, we define some spatial solids which can be used to analyze polymer structure. In Section 5, we discuss recent progress in understanding how knotting affects the size and shape of polymers.

2. Primer on polymer models

In this section, we give a short explanation of some polymer models. Kenneth Millett's chapter in this volume has additional information about the models and how to generate sets of random polygons. We attempt to minimize the overlap here while still providing a description of different methods and their implementations.

In our work, we concentrate on closed polymers in Θ-solvent. The Θ-conditions are an equilibrium state where there is no repulsion nor attraction within the polymer. These can be modeled by *freely jointed chains* or *Gaussian polygons.* For freely jointed chains, the polygons created are equilateral (up to computer precision) but for Gaussian polygons, the edges are allowed to have different lengths. Gaussian chains are called Gaussian because the coordinates of the edge vectors follow a normal distribution, e.g. the x-coordinates of the edge vectors. In our work, we use freely jointed chains.

There are two other types of solvent regimes: *good solvents* and *bad solvents.* In the case of good solvents, the polymer cannot get too close to itself, and the typical implementation method is to give a little thickness to the polygon. Polymers in a good solvent are usually studied on the standard unit lattice $\mathbb{Z}^3$ although one can also work out of the lattice with freely jointed chains by eliminating the configurations where non-consecutive edges are closer than some prescribed small distance. For bad solvents, the chain is self-attracting although one imposes a minimal thickness condition to keep the chains from self-intersecting. There is no "standard" way of generating chains in a bad solvent. However, the interested reader might see Ref. 16 for one method of generating self-attracting models using Hamiltonian cycles on the unit lattice.

Polymers in a good solvent are slightly thickened so they tend to be more spread out. Physicists refer to polymers in a good solvent as being *swollen* or to have *excluded volume.* Polymers in a bad solvent are compact and are referred to as being *collapsed.* It is not surprising that the size of the polymers differ depending on the solvent type. In fact, for open chains, the squared radius of gyration (the mean squared distance from the vertices to the center of mass) behaves like $An^{2\nu}$ where $\nu = 1/2$ in Θ-solvent, $\nu = 1/3$ in a bad solvent, $\nu \approx 0.588$ in a good solvent, and A is a constant specific to the model. This is often referred to as the *scaling,* and the growth powers tend to be the element of scientific interest. Note that above we say that chains in a good solvent can be modeled with polygons on the unit lattice or with the subset of freely jointed chains that are sufficiently "thick". The choice of model within a given regime (i.e. good solvent here) does not affect the scaling exponent, but merely the amplitude A.

For DNA, modeling is a bit more challenging due to the double helix and its twisting. The main model for DNA is that of the *wormlike chain*,[17] which can be generated using Monte Carlo methods. An energy is imposed on the chain, typically with some combination of

(1) An energy to keep the edges roughly the same length;
(2) Bending energy (i.e. $\sum(\text{bending angles})^2$);
(3) Twisting energy (based on the writhing of the chain);
(4) Excluded volume condition (i.e. the chain is given a small thickness so that non-consecutive pairs of cylinders of a small radius cannot intersect).

During the simulation, one modifies the chain and measures its energy. When the energy drops, one always accepts the new configuration and when the energy rises, one accepts the configuration with a probability $\exp(-\Delta E/T)$ where T is the absolute temperature of the system. Typically, the wormlike chains are free to change topology, the only requirement being that the new chain still satisfies the excluded volume condition. The goal is to sample uniformly from the space of wormlike chains, but this is a bit of an inexact science. As a surrogate for uniform sampling, researchers strive to generate an independent set of chains by measuring the autocorrelation time of some measurement of the chain (e.g. the radius of gyration) and then trimming the data set by only taking every k^{th} chain (where k is sufficiently large to eliminate the autocorrelation). See Ref. 18 for a thorough investigation of generating wormlike chains.

For what appears in the remainder of this paper, we will concentrate on the case of freely jointed closed chains, i.e. the polygons are equilateral and there is no repulsion nor attraction between edges. When one generates open chains (referred to as *random walks or random flights*), the technique is straightforward, just sample uniformly from the unit sphere and use these as the edge vectors. When generating circular chains (referred to as *random polygons*), the closure condition makes things more complicated.

We have two sets of freely jointed chains that we have studied. In the first, we use the Hedgehog method[19] to generate the chains and in the second, we used a modified version of the Hedgehog method. For the

Hedgehog method, one generates $n/2$ (assuming n is even) vectors uniformly on the sphere and then adds the opposite vectors to create a set of n vectors whose sum is the zero vector $\mathbf{0}$. If we take this list of vectors $v_1, -v_1, \cdots, v_{n/2}, -v_{n/2}$ and generate vertices by using these as edge vectors, the resulting polygon resembles a hedgehog. Clearly, we need to "mix up" the edge vectors to break the inter-correlations and generate realistic chains. So one randomly takes pairs of vectors from the list and rotates them by a random angle about the sum of the vectors (which keeps the sum of the two vectors fixed, and thus the sum of all of the vectors at $\mathbf{0}$). If we do this enough times, we get chains as in Fig. 1.

Fig. 1. Three examples of freely jointed chains with 200 edges.

Of course, the question is how many times do we need to do these rotations to adequately sample knot space. The answer is not so clear. In Ref. 20, we found that $3n$ rotations, where n is the number of edges, appeared to generate adjacent angle distributions that were consistent with distributions for much larger numbers of rotations, and thus, we concluded, must generate random polygons. However, Millett (in this volume) reports that the number may be higher. This remains an open question. Currently, we are using a slight variation of Hedgehog where instead of rotating two random edge vectors along their sum, we take some random number of edge vectors (between 2 and $n/2$) and rotate them randomly along the sum of the vectors. Again this keeps the sum of the angles at $\mathbf{0}$ so we generate a closed chain. See Ref. 21 for another Hedgehog-like method.

In the traditional Monte Carlo crankshaft method, one completes some number of crankshaft moves (rotate an arc of the polymer by a random angle about the chord connecting two randomly chosen vertices) between sampling the population. For this method, the concern is whether the next chain is independent from the current chain. In the case of the Hedgehog algorithm, each chain is clearly independent from the one before it because a new set of $n/2$ vectors and their opposites are generated at each step.

The concern with the Hedgehog algorithm is whether we have decorrelated from the subspace that contains many edge vector pairs that are opposite each other. With the exception of Calvo's analysis of equilateral six-edge polygons,[22–24] there has been no rigorous analysis of the space of freely jointed polygons (note that for lattice polygons there are many mathematical results). Instead researchers rely on autocorrelation measures and other numerical methods to test whether the algorithms appear to be working.

Now that we have introduced the freely jointed chain model, in addition to some models for other situations, we analyze the effects of knotting on the structure of polymers.

3. Topology and its effects in polymers

In the last section, we showed that repulsion or attraction within the physical system affects the size of the polymer. Of course, topology also affects the size (and shape) of the polymer. In Fig. 2, we see an extreme case with an unknot and trefoil made with six edges. For three to five edges, the unknot is the only knot type that can be realized.[25] At six edges, one can create equilateral trefoils (for a more full treatise on the subject of minimal edge knots, see e.g. Ref. 26) but the configurations are quite restricted in size. Observe that the unknot is more enlongated and the trefoil appears more spherical.

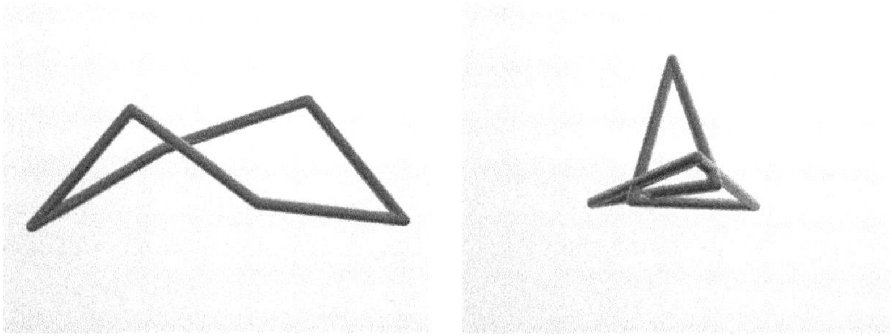

Fig. 2. For a six-edge unknot and trefoil, one can see clearly that the topology affects the size and shape of the polygon. Note that the trefoil (right) has been thickened so that the reader can see it.

As the length of the polygon (in other words, the number of edges) grows, does the difference in size and shape between the unknots and trefoils remain? Clearly, the act of being a trefoil will consume some amount of the length of the polygon, so one would expect the trefoils always to be slightly smaller, on average, than the set of unknots. This brings to mind the *knot localization conjecture*, namely that for knot configurations of a given type but with increasing lengths, the "knotted portion" of the curves grows at a rate slower than the length. For example, Orlandini et al.[27] found that the length of the knotted portion increases at a rate of $n^{3/4}$ where n is the length of the lattice polygon (i.e. for polymers in a good solvent). The "length of the knotted portion" computation is still up for debate (see Millett's chapter for a more in-depth discussion) but it appears that however one reasonably defines when an open chain is knotted, the length of the open portion of a closed polygon that contains the actual knotting grows at some fractional power of the closed polygon length.

For long chains, it has been shown that the probability of being knotted tends to one exponentially.[28–30] The proofs of these theorems rely on showing that there exist trefoil summands within long chains with probability tending to one as the length increases. Thus, one might expect that (typically) long chains resemble necklaces with many small knotted regions, and tend to be composite knots. As such, it becomes very difficult to be long and unknotted (or to be long and simply a trefoil, or any other simple knot type). In fact, one could think of a long trefoil as being a long unknot with a small defect, and thus it seems reasonable to assume that a long trefoil will look much like a long unknot, on average, and that asymptotically their average spatial structures will be indistinguishable. Using a similar argument, the chains forming any given knot type, for very long lengths, will eventually look like the unknotted chains. For complicated knots, it may take many more edges, but the result will be the same. This is the essence of knot localization: for closed chains of a given knot type, the size of the knotted region grows at a slower rate than the overall length and the structure of the configurations resemble that of unknotted chains asymptotically.

When we analyze polymers of a given knot type, the configurations are said to be *topologically constrained*. This is a bit of a misnomer. In particular, when we generate the chains, edges are allowed to pass through each other, thereby possibly changing the topology. The polymers are assumed to be randomly moving in space and our sample sets are simply snapshots of the polymer at different times. We extract configurations of a given knot type and these are our snapshots of a polymer within this topological state. When the set is not constrained with respect to topology (i.e. we look at all polymer states within a given model), the polygons are called *phantom chains*. Below we analyze the differences in size and shape between phantom chains and those topologically constrained to the most simple knot types 0_1, 3_1, 4_1, and 5_1.

Next we ask how the topological constraint, in general, affects the size of the molecules. In particular, does the average size of the trefoil knot population grow at the same rate as the size of the conglomerate population (i.e. the phantom polygons) with increasing length? The answer appears to be no. In Ref. 31, des Cloizeaux and Mehta argue that the topological constraint (polygons restricted to a fixed knot type) should act like excluded volume (the knot has a little thickness). In other words, for knots of a given knot type and increasing length n, asymptotically we should see, for example, the radius of gyration grow at a rate of $n^{0.588}$ for freely jointed chains. This has been verified numerically to some extent (see e.g. Refs. 32–36) although the difficulty of computing the knot type has kept researchers from verifying the result for long polymers.

One cannot compute the exact knot type for populations of knots. Researchers rely on knot polynomials (e.g. Alexander or HOMFLYPT) as a surrogate for the knot type (see also Refs. 37 and 38 for different strategies to determine knot types). For very small numbers of edges, these polynomials are perfect (e.g. with six-edge polygons where we can only see the unknot and trefoil), but contamination occurs for larger numbers of edges where we begin to see different knot types arise with the same knot polynomial. Still, at the lengths at which researchers currently work, contamination is not a concern because the probabilities fall off significantly as a function of the crossing number. For example, other knot types with the same HOMFLYPT polynomial as the trefoil (if they exist, which is yet unknown) are so much less probable, in this range, than the trefoils that the trefoil probability is not significantly affected. Strictly speaking, we divide the configurations by knot polynomial and then use the knot polynomial as a surrogate for knot type. Issues such as properly generating the knots appear to be more of a concern.

We have now discussed polymer models and the effect of topology on the size of these polymers. Recall that we wish to explore the size and shape of these polymers, with a particular interest in seeing how topology changes the behavior. The next section focuses on this topic.

4. Size and shape of polymers

In this section, we summarize some of our simulation results regarding the size and shape of polymers modeled as freely-jointed polygons. We want to know whether polymers are, on average, short and squat, long and skinny, spherical, or some other shape. In particular, we wish to explore how having a particular knot in the chain affects the spatial extent of the configurations. But how do we get this information quantitatively, especially since we are analyzing millions of configurations?

Our strategy has been to create spatial solids that describe the shape in some manner and analyze measurements of these solids. For example, we have analyzed smallest boxes, convex polyhedra, spheres, ellipsoids that envelope the polymers, and ellipsoids describing the inertial properties of the polymers. We next describe the solids we have used to measure the spatial extent of the polymer models.

4.1. *Solids measuring the size and shape of polymers*

In this section, we describe the spatial solids we have used to measure the size and shape of polymers. In particular, we define the convex hull, standard and skinny boxes, miniball, enveloping ellipsoid, and characteristic inertial ellipsoid. We have explored the structure of these solids in Refs. 39–42 for polymers in Θ-solution and in Ref. 15 for tight knots. In Figs. 3, 4, and 6, we show these solids for a trefoil knot from Ref. 43.

4.1.1. *Convex hull*

The convex hull is the smallest convex set containing the vertices of the given polygon. Intuitively, it is equivalent to shrink-wrapping the configuration. We use the algorithm from `Qhull`.[44]

4.1.2. *Minimal enclosing boxes*

We have two types of enclosing boxes: standard box and skinny box.

For the standard box, we compute the largest distance between vertices and this is the *box length*. When then project the knot onto a plane perpendicular to the length-chord. The largest distance between the vertices on the projection defines the *box width*. Again, we project the (already projected) knot onto a line perpendicular to the width-chord and the largest distance between vertices defines the *box height*.

For the skinny box, we compute the height, width, and then the length, just the opposite of the standard box. For the height, we search for two parallel planes that enclose the knot and are the minimal distance apart. These planes can be supported by three points on one plane and one on the other (the more likely case), or two points on each plane (which happens less frequently, but is observed for example on the vertices of a tetrahedron). The distance between the planes is the *skinny box height.* We then project the knot onto one of these planes and search for the pair of parallel lines that enclose the projected knot and are minimal distance apart. The distance between these lines defines the *skinny box width.* We then project the (already projected) knot onto one of these lines and determine the maximum distance between the points, which defines the *skinny box length.*

Given the length, width, and height, we can measure the surface area, volume, and several other shape descriptors of the boxes.

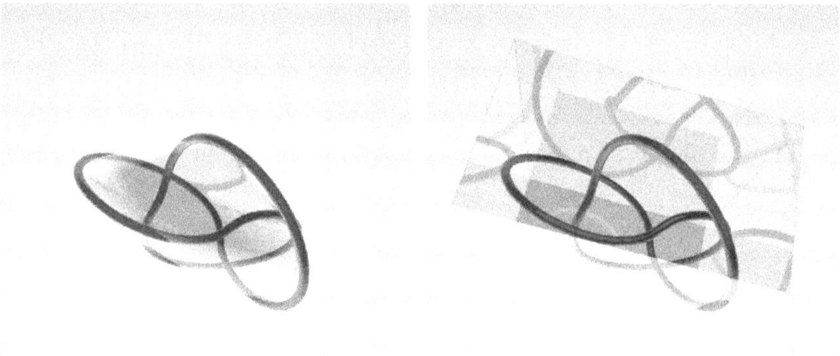

(a) Convex hull (b) Standard box

Fig. 3. The convex hull and standard box for a trefoil configuration.

4.1.3. *Miniball*

The *miniball* is the minimal enclosing sphere. It is computed using freely available (and very easy to use) code from the CGAL libraries.[45]

4.1.4. *Enveloping ellipsoid*

The *enveloping ellipsoid* is the minimal volume enclosing ellipsoid. We use the algorithm due to Schönherr.[46]

4.1.5. *Inertial ellipsoid*

The *ellipsoid of inertia* can be determined using the gyration tensor. This is a matrix whose eigenvectors yield the principal axes and corresponding eigenvalues yield the squared semi-axis lengths of the ellipsoid of inertia.

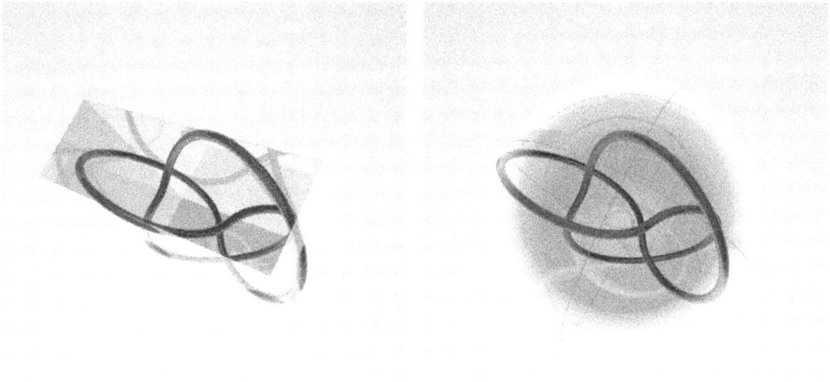

(a) Skinny box (b) Miniball

Fig. 4. The skinny box and miniball for a trefoil configuration.

The gyration tensor, which is an 3×3 symmetric matrix, is defined as follows:

$$a_{ij} = \frac{1}{2n^2} \sum_{k=1}^{n} \sum_{l=1}^{n} (v_k^i - v_l^i) \cdot (v_k^j - v_l^j)$$

where v_k^i denotes the ith coordinate of the kth vertex.

Traditionally, researchers study the ellipsoid of inertia, whose axes are the principal axes and whose semi-axis lengths are the square roots of the eigenvalues. However, this definition is a little non-intuitive in that the inertial ellipsoid of an ellipsoid is not the ellipsoid itself. Instead, the inertial ellipsoid has the same axes, but the semi-axis lengths are scaled by a factor of $1/\sqrt{3}$. See Fig. 5. As a result, we define the *characteristic inertial ellipsoid* as the ellipsoid whose axes are the principal axes and whose semi-axis lengths are given by $\alpha_i = \sqrt{3\lambda_i}$, $i = 1, 2, 3$, where λ_i are the eigenvalues of the gyration tensor. The characteristic inertial ellipsoid of an ellipsoid is the ellipsoid itself. Note that this is simply a rescaling, so the "shape" of the ellipsoid is unchanged.

5. Size and shape

Now that we have defined solids that capture spatial information about the configurations, we wish to analyze their shapes. We will concentrate on the characteristic inertial ellipsoid and enveloping ellipsoid (the other shapes were analyzed in Ref. 39). We want to separate the calculation of the size of the ellipsoids from the calculation of shape. A standard way to

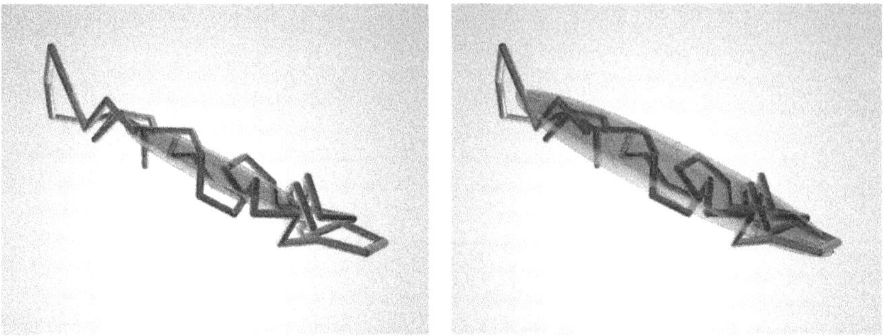

(a) Ellipsoid of inertia (b) Characteristic inertial ellipsoid

Fig. 5. The traditional ellipsoid of inertia for a 50 edge trefoil knot and its characteristic inertial ellipsoid.

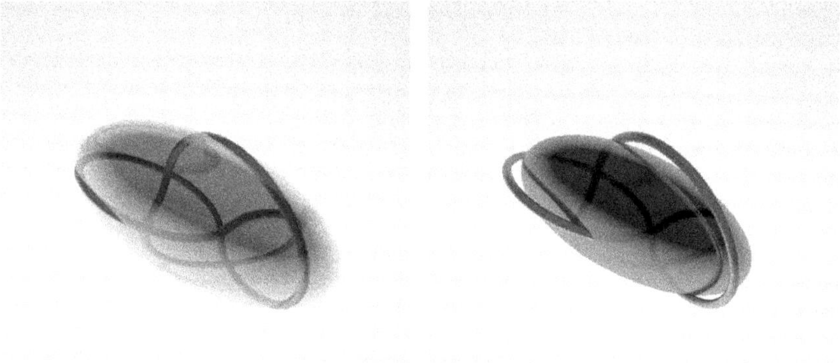

(a) Enveloping ellipsoid (b) Characteristic inertial ellipsoid

Fig. 6. The enveloping ellipsoid and characteristic inertial ellipsoid for a trefoil configuration.

measure the size is via the squared radius of gyration (i.e. the mean squared distance from the vertices of the polygon to the center of mass). Note that we assume that the mass is uniformly distributed over the vertices of the polygons.

5.1. *Squared radius of gyration*

The squared radius of gyration of the vertex set can be computed from the eigenvalues λ_1, λ_2, λ_3 of the gyration tensor via the formula

$$R' = \lambda_1 + \lambda_2 + \lambda_3 \,. \tag{1}$$

If we think in terms of the semi-axis lengths a, b, c of the characteristic inertial ellipsoid then we can write

$$R = \frac{a^2 + b^2 + c^2}{3} \,. \tag{2}$$

For the enveloping ellipsoid, we use the same formula to measure size although we no longer have the same physical connection (i.e. we cannot compute the radius of gyration of the vertex set from the semi-axis lengths of the enveloping ellipsoid).

To analyze the shape of the ellipsoids, we use the *asphericity*[47,48] and *prolateness* (originally called the *nature of asphericity*).[49]

5.2. *Asphericity*

The *asphericity* measures the extent to which an ellipsoid is non-spherical. It is a number $0 \le A \le 1$, where $A = 0$ means that the ellipsoid is perfectly spherical and $A = 1$ means that the ellipsoid is a line segment. The asphericity is symmetric in the arguments and scale invariant. See Fig. 7 for example ellipsoids.

Traditionally, asphericity is defined by

$$A' = \frac{(\lambda_1 - \lambda_2)^2 + (\lambda_1 - \lambda_3)^2 + (\lambda_2 - \lambda_3)^2}{2(\lambda_1 + \lambda_2 + \lambda_3)^2} \tag{3}$$

where λ_i are the eigenvalues of the gyration tensor. However, in Ref. 40, we defined the asphericity in terms of the semi-axis lengths of the characteristic inertial ellipsoid via

$$A = \frac{(a - b)^2 + (a - c)^2 + (b - c)^2}{2(a + b + c)^2} \,. \tag{4}$$

Note that this is equivalent to using the square roots of the λ_i. Ideally, we would like an orthogonal set of measures, i.e. a set of size and shape measures that are uncorrelated. If we use the measurements in terms of semi-axis lengths, we get that

$$\nabla R \cdot \nabla A = 0 \,. \tag{5}$$

However, if we use the measurements in terms of the eigenvalues of the gyration tensor, we have

$$\nabla R' \cdot \nabla A' \neq 0\,. \tag{6}$$

Thus, using the semi-axis lengths, the asphericity and squared radius of gyration are independent measures, but this is not the case when we use eigenvalues for the arguments.

5.3. *Prolateness*

We typically think of asphericity as the primary measure of shape, although (as is demonstrated in Fig. 7) quite different shapes can be made with the same asphericity. A secondary shape measure, the *prolateness*, is a modifier which tells us how a given shape achieves a given asphericity value. It was originally named the *nature of asphericity*.[49] A prolate ellipsoid has one long axis and two shorter axes and resembles a rugby ball. An oblate ellipsoid (the opposite of prolate) has two long axes and one shorter axis and resembles M&M candy. The prolateness is a number $-1 \leq P \leq 1$ where $P = 1$ means that the ellipsoid is perfect prolate and $P = -1$ means the ellipsoid is perfectly oblate. Again, see Fig. 7 for prolate and oblate ellipsoids with the same asphericity.

The prolateness is defined by

$$P = \frac{(2a-b-c)(2b-a-c)(2c-a-b)}{2((a-b)^2+(a-c)^2+(b-c)^2)^{3/2}}\,. \tag{7}$$

The transition point from oblate to prolate ellipsoids (i.e. $P = 0$) occurs when one of the semi-axis lengths is the average of the other two. For a perfect sphere, the prolateness is undefined. It is interesting to note that regardless of whether we use eigenvalues or semi-axis lengths as arguments for P, we still get independence:

$$\nabla P' \cdot \nabla A' = \nabla P' \cdot \nabla R' = 0; \tag{8}$$

$$\nabla P \cdot \nabla A = \nabla P \cdot \nabla R = 0\,. \tag{9}$$

5.4. *Independence of size and shape measures*

As discussed above, using semi-axis lengths for the arguments of the squared radius of gyration, asphericity, and prolateness generates an independent set of orthogonal measures. See Fig. 8 for images of the level surfaces with respect to these quantities. Note that the level surfaces intersect orthogonally.

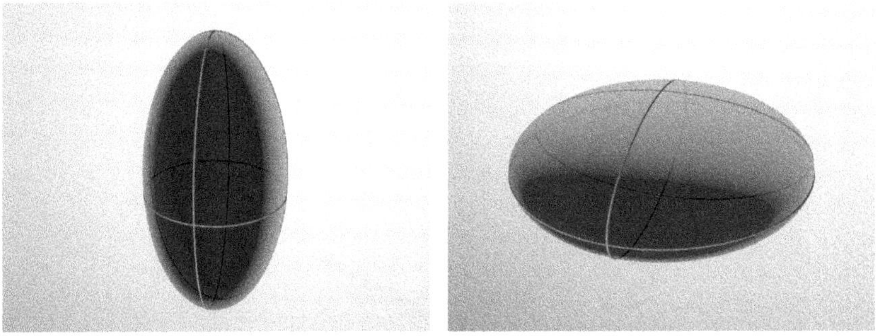

(a) Prolate ellipsoid with semi-axis lengths $(1, 0.5, 0.5)$, $A = 0.0625$, and $P = 1$.

(b) Oblate ellipsoid with semi-axis lengths $(1, 1, 0.4)$, $A = 0.0625$, and $P = -1$.

Fig. 7. Examples of a prolate and oblate ellipsoid with matching asphericity.

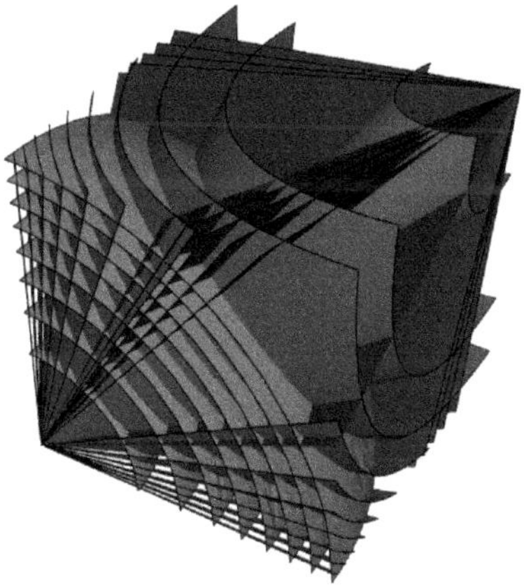

Fig. 8. (Color online) Level surfaces for the squared radius of gyration (green), asphericity (red), and prolateness (blue) intersect orthogonally and, therefore, form an independent system of size and shape measures. The axes for this plot are the semi-axis lengths for different ellipsoids.

5.5. *Size and shape as a function of length and topology*

We now present the results of our analysis of the size and shape of polymers. These chains were generated using the Hedgehog algorithm with 400,000 samples per number of edges, where the number of edges ranges from 6 to 50 by steps of 2 and 50 to 500 by steps of 10. The number of samples for a given knot type changes for each number of edges.

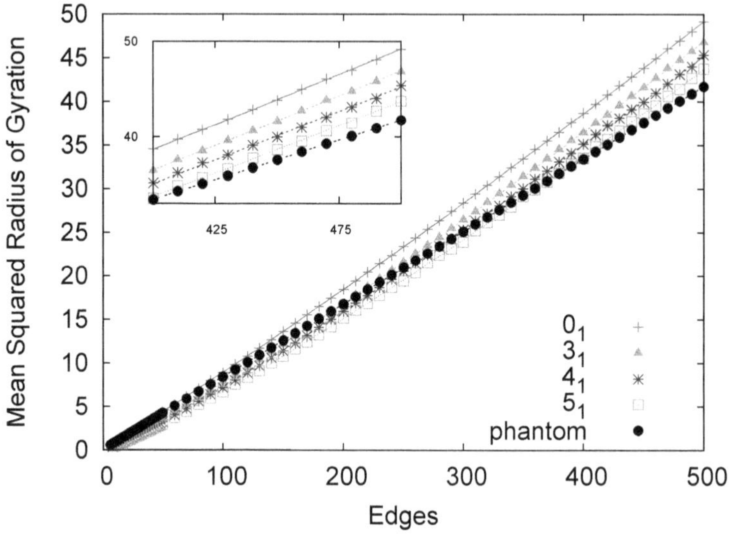

(a) Characteristic inertial ellipsoid

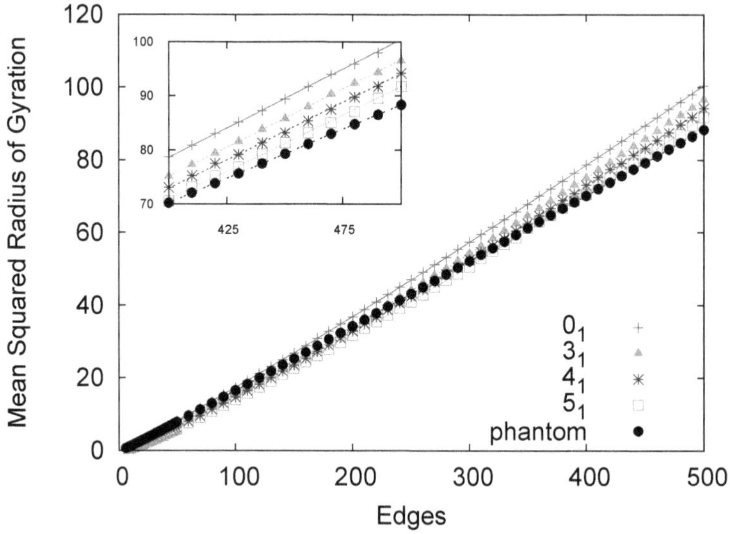

(b) Enveloping ellipsoid

Fig. 9. The figures present the mean squared radius of gyration as a function of polymer length and topological type for the (a) characteristic inertial ellipsoid and (b) enveloping ellipsoid. In the insets, we zoom in on the span of edges from 400 to 500. We see good agreement in the fitting.

For the size measurement, we plot the mean squared radius of gyration (the R value defined above) versus edges in Fig. 9. Recall that asymptotically, the squared radius of gyration for phantom chains (i.e. all 400,000 configurations per number of edges) should scale as n^1 (the radius of gyration scales as $n^{0.5}$, and the squared radius of gyration with twice the power). For individual knot types, we see that the scaling has a form of $n^{1.176}$ (where $1.176 = 2(0.588)$, the scaling exponent for polymers with excluded volume), which is consistent with the conjecture that the topological constraint behaves like excluded volume. This scaling holds for both for the characteristic inertial ellipsoid and the enveloping ellipsoid. We fit the graph using the formula $n^{2\nu}(A + B/\sqrt{n} + C/n)$ where A, B, and C are fitting parameters. The formula takes into account small edge effects, i.e. when there are few edges the closure condition restricts the spatial extent of the polymers. Note that for a fixed number of edges, more complicated knots have smaller values for the mean squared radius of gyration due to the fact that more of the polymer length is consumed in creating the knot.

For the asphericity (Fig. 10), the phantom chains appear to approach asymptotic values of around 0.074 and 0.044. The knot types all begin more spherical on average than the unknotted and phantom chains. If one has very few edges but wants to create maximal knotting, the optimal strategy is to send a lot of edges through the center region (near the center of mass) in different directions. This results in spherical configurations (i.e. where A is close to 0). Thus, for a fixed number of edges, the configurations of more complicated knot types are more spherical than those of less complex knot types. Furthermore, we see the mean asphericity for the trefoil knot approach the mean asphericity for the unknotted configurations due to the effects of knot localization. The unknotted configurations (and those of other knot types, asymptotically) are more enlongated for larger numbers of edges since the more spherical configurations contain the types of interactions that would be likely to create more highly entangled configurations. Ref. 41 contains an argument for why the mean asphericity of the enveloping ellipsoids all decrease while the mean asphericity of the characteristic ellipsoids increase towards an asymptotic value based on the fact that the enveloping ellipsoids must take into account extremal values whereas those values are suppressed in the inertial case (which is more of an averaging procedure).

For the prolateness (Fig. 11), the situation is more complicated than for asphericity. The phantom chains quickly reach asymptotic values of around 0.33 and 0.26, confirming the arguments of Kuhn[50] that polymers

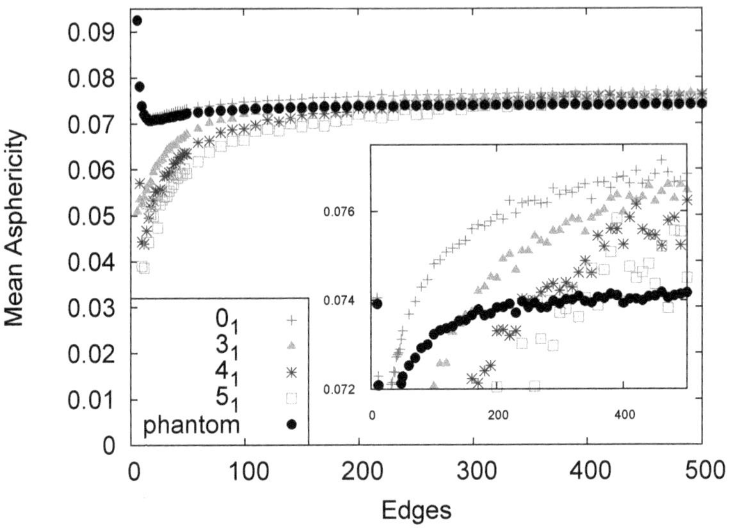

(a) Characteristic inertial ellipsoid

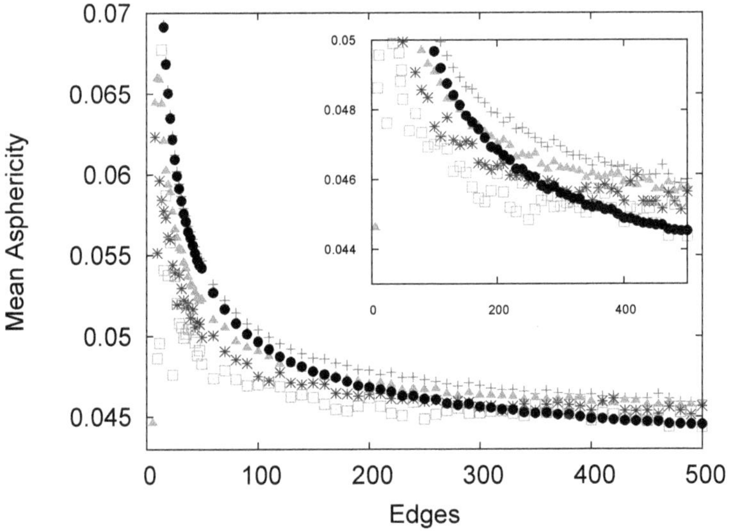

(b) Enveloping ellipsoid

Fig. 10. The figures present the mean asphericity as a function of polymer length and topological type for the (a) characteristic inertial ellipsoid and (b) enveloping ellipsoid. In the insets, we zoom in vertically to see the asymptotic behavior. In particular, we see the trefoil data approach the unknot data, consistent with knot localization.

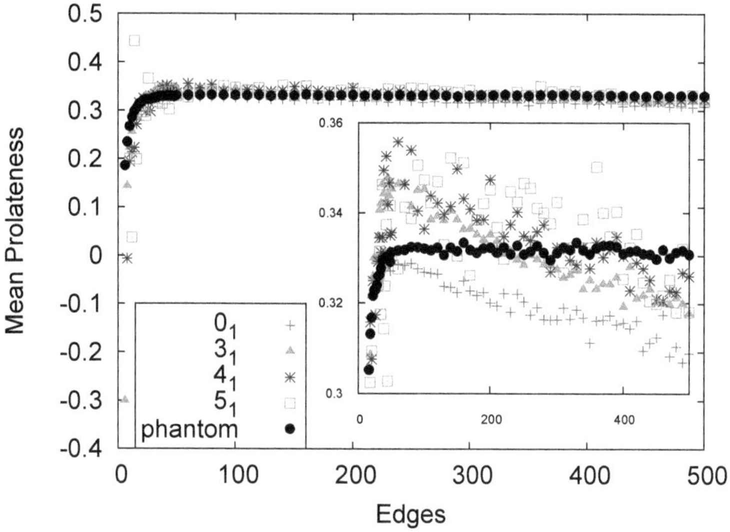

(a) Characteristic inertial ellipsoid

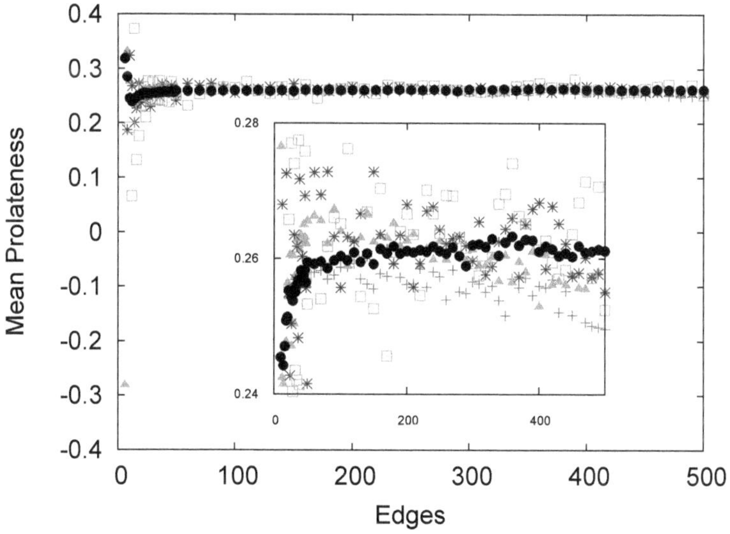

(b) Enveloping ellipsoid

Fig. 11. The figures present the mean prolateness as a function of polymer length and topological type for the (a) characteristic inertial ellipsoid and (b) enveloping ellipsoid. In the insets, we zoom in vertically to see the complex interaction between the prolateness of the topologically constrained polymers versus the phantom polymers.

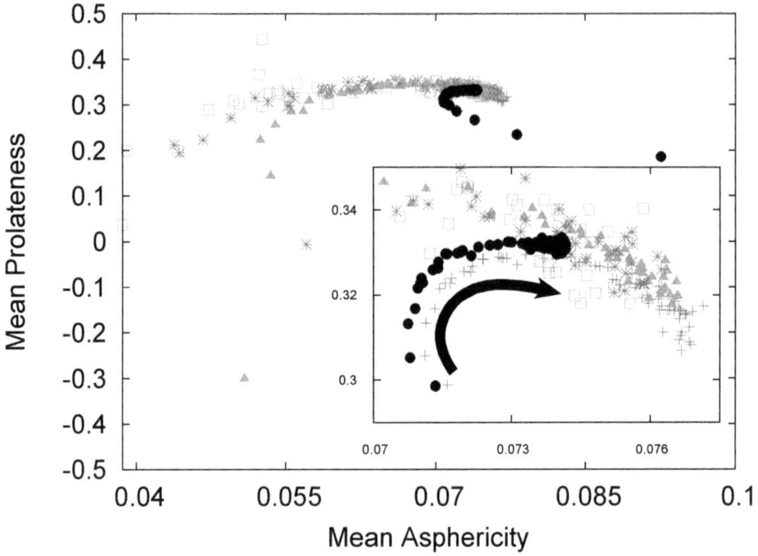

(a) Characteristic inertial ellipsoid

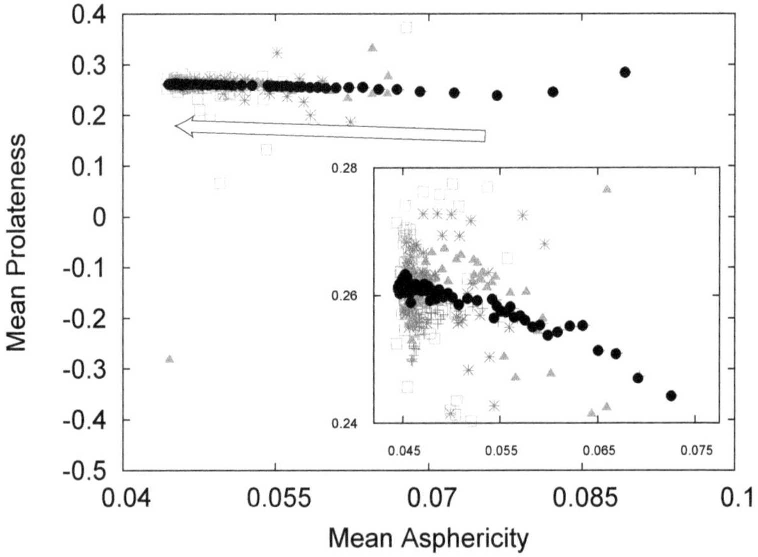

(b) Enveloping ellipsoid

Fig. 12. The figures present the mean prolateness versus mean asphericity for the different topological types in the (a) characteristic inertial ellipsoid and (b) enveloping ellipsoid. In the insets, we zoom in near the asymptotic values of asphericity and prolateness for the phantom chains. Each point represents the (mean asphericity, mean prolateness) at a given number of edges. The arrow shows the direction with increasing numbers of edges.

on average take on a prolate shape. However, the data within the knot types has more noise than the data for asphericity, likely due to the fact that prolateness has cubic terms in the numerator and denominator whereas asphericity has quadratic terms. One would have to push the data set to a higher number of edges to analyze the asymptotics in the topologically constrained regime. It is interesting to note that the configurations within a knot type start out less prolate than the phantom configurations, become more prolate, and then again become less prolate. More analysis needs to be completed to understand this behavior.

We can view the asphericity and prolateness data on one graph by plotting the mean prolateness versus mean asphericity (Fig. 12). For the characteristic inertial ellipsoids, the phantom chains congregate at an asymptotic value, implying a characteristic shape to the "average" polymer. For the enveloping ellipsoids, the difference is more difficult to see. However, note that the mean values for the unknotted chains congregate to the lower right of the mean values for the phantom chains. We expect that the mean values for the knot types are moving to similar region but are not as clearly resolved.

From these graphs, we see that the phantom chains have a different size and shape from configurations with a fixed knot type, i.e. the topological constraint changes the size and shape properties of polymers. Furthermore, we see evidence that the size and shape of the non-trivial configurations converge to the corresponding values for the unknotted configurations.

In our final images (Figs. 13 and 14) we show the "average ellipsoids" for different topological types so that we can see the differences in both size and shape simultaneously. The ellipsoids have semi-axis lengths corresponding to the mean semi-axis lengths at 500 edges and the scale of both sets of pictures is identical.

6. Conclusion

The previous section demonstrates how knotting within circular polymers affects their size and shape. We show that the characteristic inertial ellipsoid and enveloping ellipsoid sometimes tell us different stories (e.g. different asymptotic values) and sometimes tell us similar stories (e.g. more complex knots tend to be more spherical). The utility of these observations depends on what one is trying to understand about the polymers.

More generally, we have explored some different models for polymers and some of the subtleties in generating sets of random closed chains. We have shown how topology and interactions between polymer edges can affect

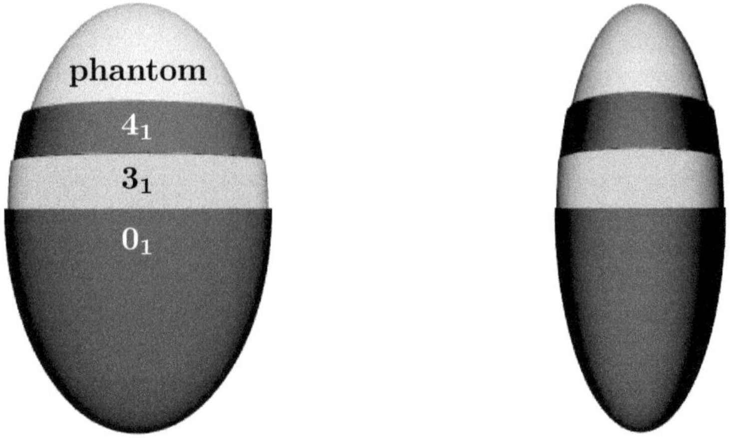

Fig. 13. (Color online) Average characteristic inertial ellipsoids for 500 edge 0_1 (blue), 3_1 (green), 4_1 (red), and phantom polygons (yellow) as seen along the two shortest axes of inertia.

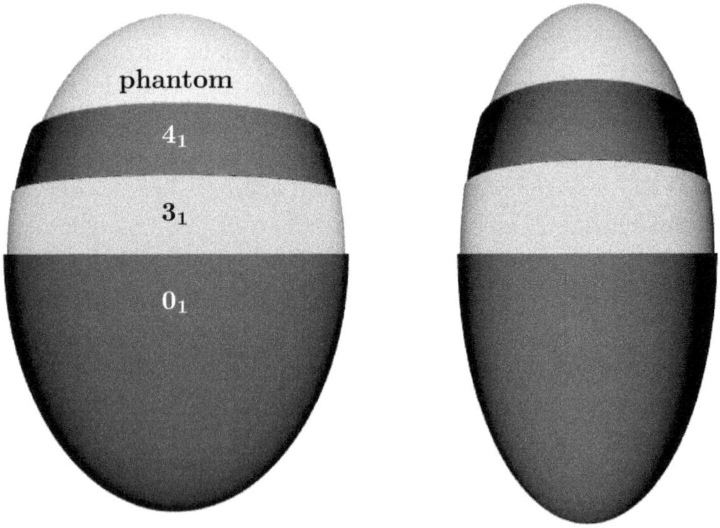

Fig. 14. (Color online) Average enveloping ellipsoids for 500 edge 0_1 (blue), 3_1 (green), 4_1 (red), and phantom polygons (yellow) as seen along the two shortest axes of inertia. Note that the scale of these ellipsoids is same as in Fig. 13.

the spatial extent of configurations. In particular, we concentrated on the size and shape of polymers measured via the squared radius of gyration, asphericity, and prolateness. These three quantities, when using semi-axis lengths as input, describe an orthogonal system that separates the size calculation from the measurement of shape. Still, there is certainly more to be learned about these polymer models in the years to come.

Acknowledgments

I would like to thank my collaborators on the projects that inspired this manuscript, namely Akos Dobay, John Kern, Ken Millett, Michael Piatek, Patrick Plunkett, and Andrzej Stasiak. I would also like to thank the referee for the careful review. This material is based upon work supported by the National Science Foundation under Grant No. 0810415.

References

1. S. Fukuhara, Energy of a knot, in *A fête of topology*, (Academic Press, Boston, MA, 1988) pp. 443–451.
2. O. Krötenheerdt and S. Veit, *Wiss. Beitr. Martin-Luther-Univ. Halle-Wittenberg Reihe M Math.* **7**, 61 (1976).
3. O. Krötenheerdt and S. Veit, On the theory of solid knots, in *Physical and Numerical Models in Knot Theory*, , Ser. Knots Everything Vol. 36 (World Sci. Publishing, Singapore, 2005) pp. 1–18. Translated by Ted Ashton.
4. J. O'Hara, *Topology* **30**, 241 (1991).
5. M. H. Freedman, Z.-X. He and Z. Wang, *Ann. of Math. (2)* **139**, 1 (1994).
6. G. Buck and J. Orloff, *Topology Appl.* **51**, 247 (1993).
7. R. A. Litherland, J. Simon, O. Durumeric and E. Rawdon, *Topology Appl.* **91**, 233 (1999).
8. H. Federer, *Trans. Amer. Math. Soc.* **93**, 418 (1959).
9. A. Nabutovsky, *Comm. Pure Appl. Math.* **48**, 381 (1995).
10. G. Buck and J. Orloff, *Topology Appl.* **61**, 205 (1995).
11. E. J. Rawdon, The thickness of polygonal knots, PhD thesis, University of Iowa1997.
12. P. Pierański, *Pro Dialog* **5**, 111 (1997), (in Polish).
13. V. Katritch, J. Bednar, D. Michoud, R. G. Scharein, J. Dubochet and A. Stasiak, *Nature* **384**, 142 (1996).
14. A. Stasiak, J. Dubochet, V. Katritch and P. Pieranski, Ideal knots and their relation to the physics of real knots, in *Ideal knots*, (World Sci. Publishing, Singapore, 1998) pp. 1–19.
15. K. C. Millett and E. J. Rawdon, *J. Comput. Phys.* **186**, 426 (2003).
16. R. C. Lua, N. T. Moore and A. Y. Grosberg, Under-knotted and over-knotted polymers: 2. compact self-avoiding loops, in *Physical and Numerical Models in Knot Theory*, , Ser. Knots Everything Vol. 36 (World Sci. Publishing, Singapore, 2005) pp. 385–398.
17. A. V. Vologodskii, S. D. Levene, K. V. Klenin, M. D. Frank-Kamenetskii and N. R. Cozzarelli, *J. Mol. Biol.* **227**, 1224 (1992).
18. Z. Liu and H. S. Chan, *J. Chem. Phys.* **128**, p. 145104 (2008).

19. K. V. Klenin, A. V. Vologodskii, V. V. Anshelevich, A. M. Dykhne and M. D. Frank-Kamenetskii, *J. Biomol. Struct. Dyn.* **5**, 1173 (1988).
20. P. Plunkett, M. Piatek, A. Dobay, J. C. Kern, K. C. Millett, A. Stasiak and E. J. Rawdon, *Macromolecules* **40**, 3860 (2007).
21. R. Varela, K. Hinson, J. Arsuaga and Y. Diao, *J. Phys. A* **42**, 1 (2009).
22. J. A. Calvo, *Topology Appl.* **112**, 137 (2001).
23. J. A. Calvo, *J. Knot Theory Ramifications* **10**, 245 (2001), Knots in Hellas '98, Vol. 2 (Delphi).
24. J. A. Calvo, Characterizing polygons in $\mathbb{R}^3$, in *Physical knots: knotting, linking, and folding geometric objects in* $\mathbb{R}^3$ *(Las Vegas, NV, 2001)*, , Contemp. Math. Vol. 304 (Amer. Math. Soc., Providence, RI, 2002) pp. 37–53.
25. R. Randell, *J. Knot Theory Ramifications* **3**, 279 (1994), Random knotting and linking (Vancouver, BC, 1993).
26. E. J. Rawdon and R. G. Scharein, Upper bounds for equilateral stick numbers, in *Physical knots: knotting, linking, and folding geometric objects in* $\mathbb{R}^3$ *(Las Vegas, NV, 2001)*, , Contemp. Math. Vol. 304 (Amer. Math. Soc., Providence, RI, 2002) pp. 55–75.
27. O. E., S. A. L. and V. C., *Phys Biol.* **6**, p. 25012 (2009).
28. N. Pippenger, *Discrete Appl. Math.* **25**, 273 (1989).
29. D. W. Sumners and S. G. Whittington, *J. Phys. A* **21**, 1689 (1988).
30. Y. Diao, *J. Knot Theory Ramifications* **4**, 189 (1995).
31. J. des Cloizeaux and M. Mehta, *Journal de Physique* **40**, 665 (1979).
32. J. des Cloizeaux, *J. Phys. Lett. (France)* **42**, L433 (1981).
33. J. M. Deutsch, *Phys. Rev. E* **59**, R2539 (1999).
34. A. Y. Grosberg, *Phys. Rev. Lett.* **85**, 3858 (2000).
35. M. K. Shimamura and T. Deguchi, *J. Phys. A* **35**, 241 (2002).
36. T. Deguchi and M. K. Shimamura, Topological effects on the average size of random knots, in *Physical Knots: Knotting, Linking, and Folding Geometric Objects in* $\mathbb{R}^3$, , Contemp. Math. Vol. 304 (Amer. Math. Soc., Providence, RI, 2002) pp. 93–114.
37. M. L. Mansfield, *J. Chem. Phys.* **127**, p. 244901 (2007).
38. C. Micheletti, D. Marenduzzo, E. Orlandini and D. W. Summers, *J. Chem. Phys.* **124**, p. 064903 (2006).
39. E. Rawdon, A. Dobay, J. C. Kern, K. C. Millett, M. Piatek, P. Plunkett and A. Stasiak, *Macromolecules* **41**, 4444 (2008).
40. E. J. Rawdon, J. C. Kern, M. Piatek, P. Plunkett, A. Stasiak and K. C. Millett, *Macromolecules* **41**, 8281 (2008).
41. K. C. Millett, P. Plunkett, M. Piatek, E. J. Rawdon and A. Stasiak, *J. Chem. Phys.* **130**, p. 165104 (2009).
42. E. J. Rawdon, J. C. Kern, M. Piatek, P. Plunkett, A. Stasiak and K. C. Millett, *Bussei Kenkyu* **92**, 32 (2009).
43. E. J. Rawdon, *Experiment. Math.* **12**, 287 (2003).
44. Qhull http://www.qhull.org, Program for computing the convex hull.
45. Miniball http://miniball.sourceforge.net/, Program for computing smallest enclosing ball.

46. S. Schönherr, Berechnung kleinster Ellipsoide um Punktemengen, PhD thesis, Freie Universität Berlin1994.
47. J. A. Aronovitz and D. R. Nelson, *J. Phys. I France* **47**, 1445 (1986).
48. J. Rudnick and G. Gaspari, *J. Phys. A: Math. Gen.* **19**, L191 (1986).
49. J. W. Cannon, J. A. Aronovitz and P. Goldbart, *J. Phys. I France* **1**, 629 (1991).
50. W. Kuhn, *Kolloid Zeitschrift* **68**, 2 (1934).

DERIVATION AND INTERPRETATION OF THE GAUSS LINKING NUMBER

Renzo L. Ricca
Department of Mathematics and Applications, U. Milano-Bicocca
Via Cozzi 53, 20125 Milano, Italy
E-mail: renzo.ricca@unimib.it
URL: http://www.matapp.unimib.it/~ricca

Bernardo Nipoti
Department of Mathematics "F. Casorati", U. Pavia
Via Ferrata 1, 27100 Pavia, Italy
E-mail: bernardo.nipoti@unipv.it

In these lectures we present for the first time a mathematical reconstruction of what might have been Gauss' *own* derivation of the linking number of 1833, and we provide also an alternative, explicit proof of its modern interpretation in terms of degree, signed crossings and intersection number. The reconstruction offered here is entirely based on an accurate study of Gauss' own work on terrestrial magnetism and it is complemented by the independent analysis and discussion made by Maxwell in 1867. Since the linking number interpretations in terms of degree, signed crossings and intersection index play such an important rôle in modern mathematical physics, we offer a direct proof of their equivalence, and we provide some examples of linking number computation based on oriented area information. The material presented in these lectures forms an integral part of a paper that will appear in the *Journal of Knot Theory and Its Ramifications.*

Keywords: Linking number; magnetic potential; solid angle; degree; signed crossings; intersection number; oriented area.

1. Introduction

The concept of linking number was introduced by Gauss in a brief note on his diary in 1833 (see §2 below), but no proof was given, neither of its derivation, nor of its topological meaning. Its derivation remained indeed a mystery. Nevertheless this concept was seminal, and proved to be fundamental in the subsequent development of knot theory, general topology and modern topological field theory (see, for example, the survey by Spera[1]).

In this paper we provide a plausible mathematical proof of what Gauss himself might have done, by bridging his possible derivation with a modern proof of the alternative, equivalent definitions of linking number in terms of degree, crossing signs and intersection number.

Reconstructions of possible derivations made by Gauss have been offered by historians, in particular Epple,[2–4] and physicists.[5] These reconstructions rely on ideas that can be certainly traced back to Gauss, even though direct derivation of the linking number from Gauss's own work is left to speculation. Here we offer a plausible reconstruction entirely based on Gauss' own work on terrestrial magnetism; this is done in §3 by providing a mathematical reconstruction that relies step by step on specific, published results of Gauss. This reconstruction is complemented by the independent analysis and discussion of the linking number put forward by Maxwell as early as 1867, and briefly re-examined in §4.

It is well known that the Gauss linking number formula admits several alternative interpretations. Rolfsen[6] gives eight of them, proving their equivalence, even though at times by very concise statements. This is the case, for instance, for the interpretation of the linking number in terms of degree, where his brief reference to Spivak's *Calculus on Manifolds* (p. 135) seems all too demanding for non-experts. Since this equivalence is still source of inspiration for modern mathematics,[7–10] we believe that a proof of the interpretation in terms of degree deserves a discussion somewhat more accessible to non-specialists. This is done in §5. Similar considerations apply to its interpretation in terms of signed crossings and intersection number, two different ways of computing the linking number, that have proven to be very useful in many physical contexts.[11–15] This discussion is also presented in §5. Finally in §6 direct inspection of the linking number by the Gauss map is presented for three different links. The material presented in these lectures forms an integral part of a paper[16] that will be published in the *Journal of Knot Theory and Its Ramifications.*

2. Gauss' note on the linking number

It is well known that Gauss had a great interest in topology — then known as "geometry of position" (*geometria situs*) — even though he never published anything on it. From his personal correspondence with Olbers it is known that he became deeply interested in the subject from the early years of 1800 (Ref. 17, p. 221): a few notes on knots are from that period. One of the oldest is dated 1794, bearing the heading *A Collection of Knots*, and it contains thirteen sketches of knots with English names written beside them.

[4.]

Von der *Geometria Situs*, die LEIBNITZ ahnte und in die nur einem Paar Geometern (EULER und VANDERMONDE) einen schwachen Blick zu thun vergönnt war, wissen und haben wir nach anderthalbhundert Jahren noch nicht viel mehr wie nichts.

Eine Hauptaufgabe aus dem *Grenzgebiet* der *Geometria Situs* und der *Geometria Magnitudinis* wird die sein, die Umschlingungen zweier geschlossener oder unendlicher Linien zu zählen.

Es seien die Coordinaten eines unbestimmten Punkts der ersten Linie x, y, z; der zweiten x', y', z' und

$$\iint \frac{(x'-x)(dy\,dz'-dz\,dy')+(y'-y)(dz\,dx'-dx\,dz')+(z-z')(dx\,dy'-dy\,dx')}{[(x'-x)^2+(y'-y)^2+(z'-z)^2]^{\frac{3}{2}}} = V$$

dann ist dies Integral durch beide Linien ausgedehnt

$$= 4m\pi$$

und m die Anzahl der Umschlingungen.

Der Werth ist gegenseitig, d. i. er bleibt derselbe, wenn beide Linien gegen einander umgetauscht werden. 1833. Jan. 22.

[4.]

Of the *geometria situs*, which was imagined by LEIBNITZ, and into which only a pair of geometers (EULER and VANDERMONDE) were granted a bare glimpse, we know and have, after a century and a half, little more than nothing.

A principal problem at the *interface* of *geometria situs* and *geometria magnitudinis* will be to count the intertwinings of two closed or endless curves.

Let x, y, z be the coordinates of an undetermined point on the first curve; x', y', z' those of a point on the second and let

$$\iint \frac{(x'-x)(dydz'-dzdy')+(y'-y)(dzdx'-dxdz')+(z'-z)(dxdy'-dydx')}{\left[(x'-x)^2+(y'-y)^2+(z'-z)^2\right]^{\frac{3}{2}}} = V$$

then this integral taken along both curves is

$$= 4m\pi$$

m being the number of intertwinings.

The value is reciprocal, i.e. it remains the same if the curves are interchanged. 1833. Jan. 22.

Fig. 1. Top: excerpt from the *Nachlass zur Electrodynamik*, published in the V volume of Gauss' *Werke* (1867). In this note Carl Friedrich Gauss introduces the concept of linking number. There is a misprint in the notation $(z - z')$ at the numerator. Bottom: English translation.

Two other sheets of paper with sketches of knots are one dated 1819, and the other bearing the notation "Riedl, Beiträge zur Theorie des Sehnenwinkels, Wien, 1827" (Ref. 17, p. 222). Few other remarks referring to the knotting of closed curves are reproduced in his *Werke* (VIII, pp. 271–285).

The concept of linking number was introduced by Carl Friedrich Gauss in a brief note on a page of his personal diary — a kind of logbook of his most important discoveries (Ref. 18, p. 19) — dated January 22, 1833 (see Fig. 1). This note remained unknown for 34 years, until it was discovered and published in the *Nachlass zur Elektrodynamik*, as part of Gauss' posthumous *Werke*.[19]

In this note the notion of linking number is merely introduced to count the number of times a closed (or endless) curve encircles a second closed (or endless) curve in space. There is no explicit reference neither to a physical system, electrical or magnetic, nor to an application. Thus, inclusion of this note in the *Nachlass zur Electrodynamik* (supported by Schäfer, one of Gauss scientific biographers[20]) generated some controversy. The mathematical reconstruction presented in the next section, however, advocates in favor of Schäfer's choice.

3. Derivation of the linking number based on Gauss' work

Gauss developed an interest in the Earth magnetism and its mysterious origin as early as 1806, but any publication on the subject was postponed until the '30s, when firm observational data were readily available. The first paper on the subject, entitled *Intensitas vis magneticae terrestris ad mensuram absoluta revocata*,[21] was read before the Royal Society of Göttingen on December 15, 1832. In this paper Gauss presents the first systematic use of absolute units to measure magnetic quantities. Stimulated by Faraday's discovery of induced currents, made in 1831, Gauss and Weber started to collect electromagnetic measurements from October 22, 1832. During this work, in 1833, they were led to anticipate the discovery of Kirchoff's laws of branched circuits.[22] Apparently, all the fundamental concepts for a mathematical theory of terrestrial magnetism were conceived as early as 1806, and were readily formulated in 1822 (Ref. 22, p. 306). Gauss' theory on terrestrial magnetism was therefore ready for a test in 1832 (Ref. 17, p. 158), but work on its publication was postponed till the winter of 1838, when experimental observations were finally confirmed. Gauss' paper on this, entitled *Allgemeine Theorie des Erdmagnetismus*, finally appeared in 1839.[23]

In this paper the magnetic potential induced at any point on the Earth's surface is worked out in terms of an infinite series of spherical functions. In considering the magnetic effects produced by some "magnetized fluid", supposedly present in the Earth interior, Gauss examines the potential $V = V(P')$ induced at an exterior point $P' = P'(x', y', z')$, placed at a distance $r = |P - P'|$ from a magnetic source at $P = P(x, y, z)$. The theory of magnetic potential, derived there from first principles, is applied then to calculate the value of the magnetic effects at various geographical

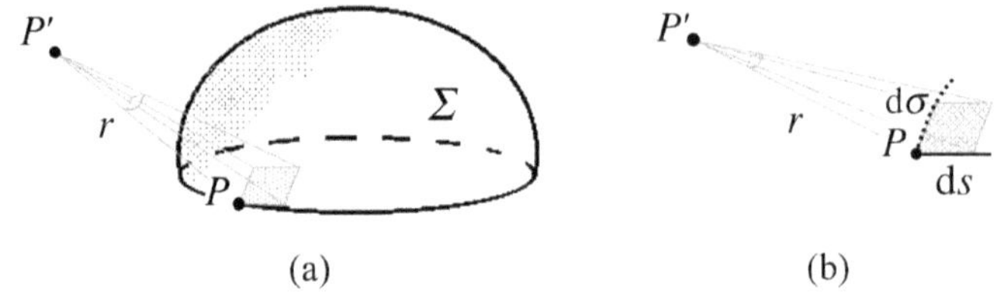

Fig. 2. (a) The potential $V = V(P')$ induced by a magnetic shell of unit strength and surface Σ at an exterior point P', placed at a distance r from a source point P, can be calculated by integrating the elementary contribution $\mathrm{d}\Sigma$ over Σ, via the solid angle subtended by $\mathrm{d}\Sigma$. (b) The volume of the elementary solid angle subtended by $\mathrm{d}\Sigma$ can be worked out in terms of the volume of the pyramid of base $\mathrm{d}\Sigma$ and side r.

locations. By acknowledging Ampère's result on the interpretation of the magnetic effects in terms of electric currents (Ref. 23, p. 188, §2; pp. 229–230, §36–37), Gauss states the following theorem (p. 230, §38):

Theorem 3.1 (Gauss, 1839). *The potential of a magnetic shell Σ at any point is equal to the solid angle which it subtends at that point multiplied by its magnetic strength.*

"Magnetic shell" means here an infinitely thin magnetized layer given by an orientable surface, whose opposite sides have opposite polarity. Theorem 3.1 is derived from geometric considerations of certain elementary expressions for the magnetic potential. Of central importance here is indeed the result of Ampère, mentioned by Gauss. This states (cf., for example, Thomson[24]) that:

Theorem 3.2 (Ampère, 1831). *A closed galvanic circuit C produces the same effect as a magnetic shell Σ of any form having the circuit C for its edge.*

A "galvanic circuit" is indeed an electrically charged wire carrying electric current. By using this theorem, the value of the potential at P', induced by a magnetic surface Σ, bounded by $C = \partial\Sigma$, is equivalent to the potential at P' induced by an electric current flowing in the circuit identified by C. According to Ampère's theorem, the magnetic effects are evidently independent of the shape of Σ, so that Σ (given, for example, by a spherical cap, as in Fig. 3a) can be replaced by any other (topologically equivalent) surface bounded by C: the laminar disk on the r.h.s of Fig. 3a, for instance, is an example. Hence, we have (Thomson,[24] p. 272, Corollary 2) :

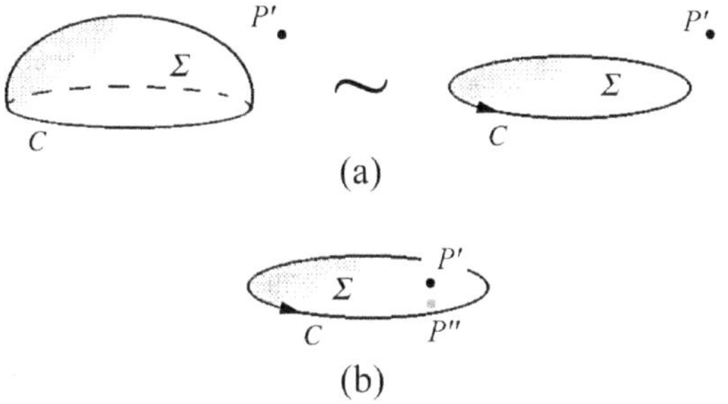

Fig. 3. (a) By Ampère's theorem the potential induced by a magnetized shell Σ bounded by C at an exterior point P' is equivalent to that induced at the same point by an electric current flowing in the circuit identified by C. (b) The potential of a point P', infinitely close to the magnetized shell Σ, and that of its antipodal P'', placed on the opposite side of Σ, differ by 4π.

Corollary 3.1 (Thomson, 1850). *The potential of a magnetic shell at any point is independent of the form of the shell itself, and depends solely on the bounding line or edge.*

Thinking of the Earth surface as a magnetized sphere threaded by a south-north oriented magnetization, Gauss gives the following example (Ref. 24, §38, p. 231): consider two magnetic shells (representing the Earth hemispheres) Σ_1 and Σ_2, having common boundary (that is the equator) and unit magnetic strength. The value of the potential at an external point P' depends on the position of this point with respect to the two shells. By considerations on the solid angle and the respective orientation of Σ_1, whose normal vector is positively oriented towards the exterior, and Σ_2, negatively oriented towards the exterior, Gauss states that the potential due to Σ_1, at any point in the region interior to the two shells, exceeds the potential due to Σ_2 by 4π (the solid angle of the unit sphere). When P' moves from a position very near to the positive side of the shell to its antipodal P'', infinitely close to the start point on the negative side of the shell, through a path around the edge, (see Fig. 3b and Fig. 4a), the solid angle thus measured must continuously increase by 4π. We know that indeed the potential is a multi-valued function of the position. Following Gauss, Lord Kelvin (then W. Thomson) states this result as follows (Ref. 24, p. 272, Corollary 3):

Corollary 3.2 (Thomson, 1850). *Of two points infinitely near one another on the two sides of a magnetic shell, but not infinitely near its edge, the potential at that one which is on the north polar side exceeds the potential at the other by the constant 4π.*

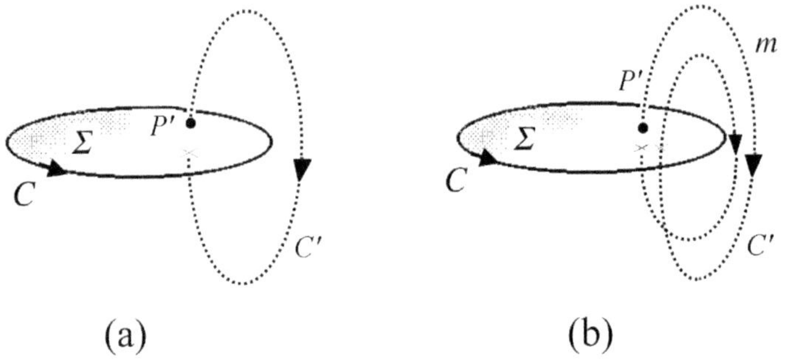

Fig. 4. (a) The potential $V(P')$ jumps by 4π as P' passes through the surface Σ. (b) If the point P' passes through Σ m times, then $V(P')$ increases by $m\,4\pi$.

Gauss' note on the linking number can now be proven in full. Let $V = V(P')$ the potential induced by Σ (with $\partial\Sigma = C$) at P'.

Proposition 3.1 (Gauss, 1832). *Let C and C' be two smooth, simple, closed curves in $\mathbb{R}^3$, with C' encircling C m times. Let $P = P(x, y, z) \in C$, $P' = P'(x', y', z') \in C'$. We have:*

(i)

$$\iint \frac{(x'-x)(\mathrm{dydz'}-\mathrm{dzdy'})+(\mathrm{y'}-\mathrm{y})(\mathrm{dzdx'}-\mathrm{dxdz'})+(\mathrm{z'}-\mathrm{z})(\mathrm{dxdy'}-\mathrm{dydx'})}{[(x'-x)^2+(y'-y)^2+(z'-z)^2]^{3/2}} = V \ ;$$

(ii)

$$V = 4m\pi \ ,$$

where $m = m(C, C')$ is the linking number of C and C';
(iii)

$$m(C, C') = m(C', C) \ .$$

Proof. The potential induced by Σ at the point P' is given by Gauss' Theorem 3.1. The volume of the elementary solid angle $\mathrm{d}\omega$ at P' (see Fig. 2a) can be worked out in terms of the volume of the pyramid of solid angle $\mathrm{d}\omega$ and side r. The base of the pyramid has elementary area $\mathrm{d}\Sigma$. We now use Ampère's theorem: since the induction due to Σ is equivalent to that due to the electric current flowing in C (see Fig. 3), the elementary solid angle can be worked out in terms of direction cosines (cf. the analysis done by Maxwell in Ref. 25, §419–420 eqs. (1) and (6), and reproduced in §4 below) by

$$\frac{1}{3}r^3\mathrm{d}\omega = \frac{1}{3}\mathrm{r}^3\Pi\,\mathrm{ds}\,\mathrm{d}\sigma \ , \tag{1}$$

where

$$r = \left[(x'-x)^2+(y'-y)^2+(z'-z)^2\right]^{1/2} \tag{2}$$

is the distance between P' and P, and Π is given by

$$\Pi = \frac{1}{r^3} \begin{vmatrix} x' - x & y' - y & z' - z \\ \frac{\mathrm{dx}}{\mathrm{ds}} & \frac{\mathrm{dy}}{\mathrm{ds}} & \frac{\mathrm{dz}}{\mathrm{ds}} \\ \frac{\mathrm{dx}'}{\mathrm{d}\sigma} & \frac{\mathrm{dy}'}{\mathrm{d}\sigma} & \frac{\mathrm{dz}'}{\mathrm{d}\sigma} \end{vmatrix} . \tag{3}$$

The potential is thus given by

$$V = \int_{\Omega} \mathrm{d}\omega = \int_{\mathrm{C}} \int_{\mathrm{C}'} \Pi \, \mathrm{ds} \, \mathrm{d}\sigma$$

$$= \int_C \int_{C'} \frac{(x' - x)(\mathrm{dydz}' - \mathrm{dzdy}') + (\mathrm{y}' - \mathrm{y})(\mathrm{dzdx}' - \mathrm{dxdz}') + (\mathrm{z}' - \mathrm{z})(\mathrm{dxdy}' - \mathrm{dydx}')}{[(x' - x)^2 + (y' - y)^2 + (z' - z)^2]^{3/2}} ,$$

where the double integral is extended to C and C'. Statement (i) is thus proven. Since C' encircles C m times (m integer; see Fig. 4(b)), at each turn around C the potential $V = V(p')$ increases by 4π (cf. Corollary 3.2; see also Ref. 25, §421); hence, after m times, we have $V = m\,4\pi$; thus (ii) is proven. Since m is an algebraic number, it does not depend on the shape of the two curves in space: it is therefore an invariant of the topology of the embedded curves. By direct inspection of the integrand of (i) we can see that $m(C, C') = m(C', C)$. Thus (iii) is proven, together with Proposition 3.1.□

4. Maxwell's rediscovery of the linking number

Maxwell developed an interest in knots as early as November 1867, prompted by Thomson's theory of vortex atoms and Tait's engagement in classification work. Indeed, several concepts introduced and discussed by Tait in his first paper on knots[26] were actually suggested by Maxwell, as their private correspondence demonstrate. Maxwell's primary interests lay then squarely in foundational aspects of electricity and magnetism, an interest that kept him well acquainted with Gauss' work on terrestrial magnetism and his solid angle interpretation of the magnetic potential. As we shall see from the material presented below, it seems almost certain that he could be unaware of Gauss' note by December, 1867. In any case, the introductory material presented in Chapter III, Vol. II of his *Treatise*[25] is entirely devoted to Gauss' work on magnetic potential. After introducing Gauss' theorem for magnetic shells (§409), Maxwell proceeds to apply the solid angle interpretation to calculate the potential. Starting from the very definition of solid angle (§417), Maxwell puts forward three alternative methods (§§417–420) to compute this quantity, by relying on geometric and physical arguments.

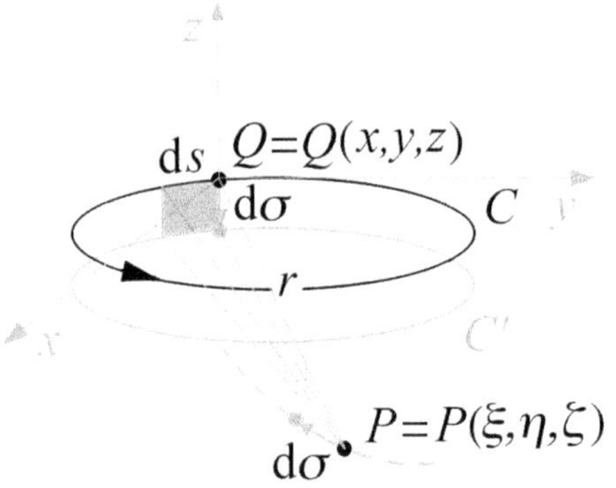

Fig. 5. Maxwell's interpretation of the solid angle in terms of apparent displacement of the boundary curve C.

The physical reasoning of §419, for instance, is quite illuminating: by interpreting the potential at a point P as the work done by a unit magnetic pole at P against the force exerted by a magnetic shell, while it moves from an infinite distance to the point P, he notes that the potential (i.e. the solid angle) must be the result of a double line integration: one performed along the boundary curve C of the shell (by making use of Theorem 3.2), and the other one performed along the path followed by the pole at P, as it approaches C. If we denote by s and σ the respective arc-lengths of these two paths, the elementary contribution $\mathrm{d}\omega$ to this solid angle can thus be expressed as

$$\mathrm{d}\omega = \Pi \, \mathrm{ds} \, \mathrm{d}\sigma \ , \tag{4}$$

where Π remains to be determined.

Now, consider a small displacement $\mathrm{d}\sigma$ of P towards C (see Figure 5). As the pole approaches C from infinity, the circuit C will be seen (from P) to move apparently to C' of the same distance $\mathrm{d}\sigma$, in the opposite direction. The elementary arc ds of C will thus be seen to span an elementary surface of area $\mathrm{ds}\,\mathrm{d}\sigma$. The volume of the pyramid with base this surface, vertex in P and solid angle $\mathrm{d}\omega$ is indeed given by eq. (1), where, with reference to Figure 5, $r = \overline{PQ}$ denotes the distance between the point $P = P(\xi, \eta, \zeta)$ and a point $Q = Q(x, y, z)$ on C. This volume may equally be expressed in terms of the direction cosines of ds, $\mathrm{d}\sigma$ and r with respect to the triad

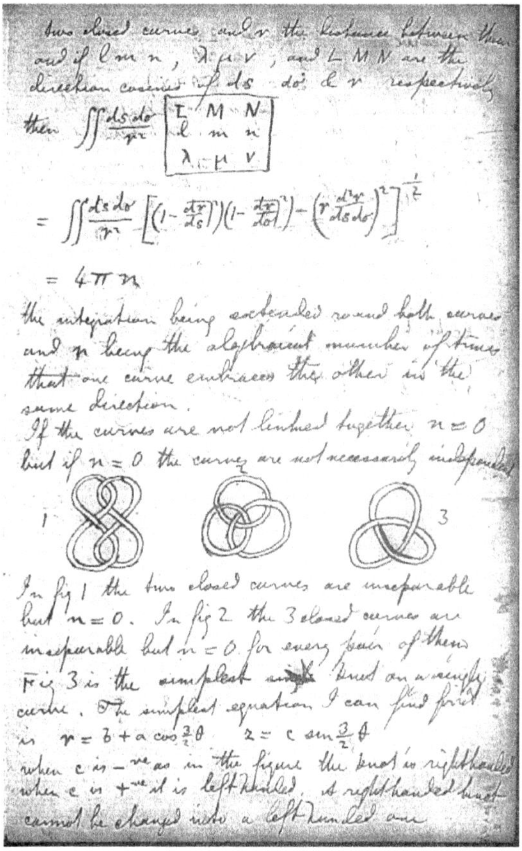
two closed curves, and r the distance between them
and if l m n, λ μ ν, and L M N are the
direction cosines of ds dσ & r respectively
then $\iint \frac{ds\,d\sigma}{r^2} \begin{vmatrix} L & M & N \\ l & m & n \\ \lambda & \mu & \nu \end{vmatrix}$

$= \iint \frac{ds\,d\sigma}{r^2}\left[\left(1-\frac{dr^2}{ds^2}\right)\left(1-\frac{dr^2}{d\sigma^2}\right)-\left(r\frac{d^2r}{ds\,d\sigma}\right)^2\right]^{\frac{1}{2}}$

$= 4\pi n$

the integration being extended round both curves
and n being the algebraical number of times
that one curve embraces the other in the
same direction.
If the curves are not linked together n = 0
but if n = 0 the curves are not necessarily independent

In fig 1 the two closed curves are inseparable
but n = 0. In fig 2 the 3 closed curves are
inseparable but n = 0 for every pair of them
Fig 3 is the simplest knot on a single
curve. The simplest equation I can find for it
is $r = b + a\cos\frac{3}{2}\theta$ $z = c\sin\frac{3}{2}\theta$
when c is −ve as in the figure the knot is righthanded
when c is +ve it is lefthanded. A righthanded knot
cannot be changed into a left handed one

Fig. 6. Third sheet of a letter by J.C. Maxwell to P.G. Tait, dated December 4, 1867. Calculation of the linking number "[...] n being the algebraical number of turns that one curve embraces the other in the same direction." The first two diagrams show examples of essential links with $n = 0$. The third diagram shows a trefoil knot, described by the parametric equation given at the bottom of the page. Cambridge University Library, papers of James Clerk Maxwell, Ms. Add. 7655, Box **1**(Ib), 7 (unpublished). Courtesy of the Syndics of Cambridge University Library.

(x, y, z), so that

$$\mathrm{d}\omega = \frac{\mathrm{ds}\,\mathrm{d}\sigma}{\mathrm{r}^2}\begin{vmatrix} L & M & N \\ l & m & n \\ \lambda & \mu & \nu \end{vmatrix} = \frac{\mathrm{ds}\,\mathrm{d}\sigma}{\mathrm{r}^2}\begin{vmatrix} \frac{\mathrm{dx}}{\mathrm{ds}} & \frac{\mathrm{dy}}{\mathrm{ds}} & \frac{\mathrm{dz}}{\mathrm{ds}} \\ \frac{\mathrm{d}\xi}{\mathrm{d}\sigma} & \frac{\mathrm{d}\eta}{\mathrm{d}\sigma} & \frac{\mathrm{d}\zeta}{\mathrm{d}\sigma} \\ \frac{\xi - x}{r} & \frac{\eta - y}{r} & \frac{\zeta - z}{r} \end{vmatrix}$$

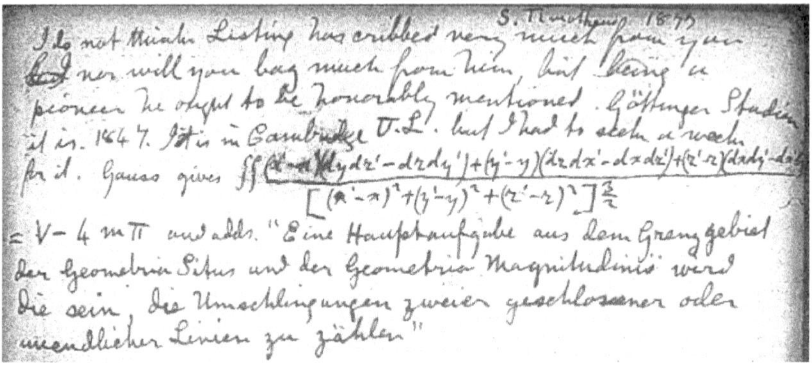

I do not think Listing has cribbed very much from you nor will you bag much from him, but being a pioneer he ought to be honorably mentioned. Göttinger Studien it is. 1847. It is in Cambridge U.L. but I had to seek a week for it. Gauss gives $\iint \frac{(x'-x)(dy\,dz'-dz\,dy')+(y'-y)(dz\,dx'-dx\,dz')+(z'-z)(dx\,dy'-dy\,dx')}{[(x'-x)^2+(y'-y)^2+(z'-z)^2]^{\frac{3}{2}}}$ $= V - 4m\pi$ and adds. "Eine Hauptaufgabe aus dem Grenzgebiet der Geometria Situs und der Geometria Magnitudinis wird die sein, die Umschlingungen zweier geschlossener oder unendlicher Linien zu zählen"

Fig. 7. Extract from a postcard to P.G. Tait dated January 24, 1877. After a few comments on Listing's work, Maxwell communicates the Gauss linking number formula, quoting Gauss' annotation (see Figure 1). The term $V - 4m\pi$ should be read $V = 4m\pi$. Cambridge University Library, papers of James Clerk Maxwell, Ms. Add. 7655, Box **1**(Ib), 86: # 497 (unpublished). Courtesy of the Syndics of Cambridge University Library.

$$= \frac{\mathrm{d}s\,\mathrm{d}\sigma}{r^2}\left[\left(1-\left(\frac{\mathrm{d}r}{\mathrm{d}s}\right)^2\right)\left(1-\left(\frac{\mathrm{d}r}{\mathrm{d}\sigma}\right)^2\right)-\left(r\frac{\mathrm{d}^2 r}{\mathrm{d}s\mathrm{d}\sigma}\right)^2\right]^{1/2} . \qquad (5)$$

Note that the determinant on the r.h.s. of the equation above can be reduced to that of eq. (3). These expressions are essentially those given by eqs. (5) and (6) of §420.

These calculations can be dated precisely, since they are communicated in a letter to Tait of December 4, 1867 (see Fig. 6). This letter is quite remarkable: not only Maxwell derives here the linking number, but he quickly proceeds to point out its shortcomings, by considering two cases (given by the first two diagrams shown) of zero linking number for two essential links. These considerations will be reproduced by Tait in his first paper on knots.[26]

A careful reading of this (and other) correspondence indicates that Maxwell must have derived this result independently, presumably giving Gauss due credit only afterwards, on the occasion of the publication of the *Treatise*. Further confirmation of this can be found in a postcard to Tait, dated January 24, 1877 (see Fig. 7), where explicit reference to Gauss' original note is made.

5. Interpretation of the linking number in terms of degree, signed crossings and intersection number

In this section we concentrate on the interpretation of the linking number in terms of degree, signed crossings and intersection number, by providing a more direct, explicit and, to some extent, elementary proof of their equivalence.

It is useful to recall some basic definitions. Let γ_1, γ_2 be two smooth, disjoint, closed, oriented curves in S^3, and $\mathbf{r}_1(t_1), \mathbf{r}_2(t_2)$ their parametrizations, with $\{t_1, t_2\} \in [0, 2\pi]$. To each pair $(Q_1, Q_2) \in \gamma_1 \times \gamma_2$ there corresponds a point (t_1, t_2) on the torus $\mathbb{T}$. The *Gauss map* $\psi : \mathbb{T} \to S^2$ associates to each point (t_1, t_2) the unit vector

$$\mathbf{n}(t_1, t_2) = \frac{\mathbf{r}_1(t_1) - \mathbf{r}_2(t_2)}{|\mathbf{r}_1(t_1) - \mathbf{r}_2(t_2)|} \ . \tag{6}$$

Now, let M and N be two compact, unbounded, oriented manifolds of same dimensions, and $\mathbf{f} : M \to N$ a continuous function. We have:

Definition 5.1. The *degree* of $\mathbf{f}$ is defined by

$$\deg(\mathbf{f}) := \sum_{x \in Y} \operatorname{sign}\left[\det\left(\frac{\partial \mathbf{f}}{\partial x}\right)_x\right] , \tag{7}$$

where Y denotes the set of regular points for which $\det(\partial \mathbf{f}/\partial x) \neq 0$.

The original formula of Gauss gives:

Definition 5.2. The *linking number* $Lk^{(1)} = Lk^{(1)}(\gamma_1, \gamma_2)$ of γ_1 and γ_2 is defined by

$$Lk^{(1)}(\gamma_1, \gamma_2) := \frac{1}{4\pi} \int_0^{2\pi} \int_0^{2\pi} \frac{(\mathbf{r}_2 - \mathbf{r}_1, \mathbf{r}_1', \mathbf{r}_2')}{|\mathbf{r}_2 - \mathbf{r}_1|^3} \, \mathrm{d}t_1 \mathrm{d}t_2 \ . \tag{8}$$

Here primes denote derivatives with respect to the argument. Gauss' original notation is recovered by taking $\mathbf{r}_1 = \mathbf{r}_1(s) = (x(s), y(s), z(s))$ for P and $\mathbf{r}_2 = \mathbf{r}_2(\sigma) = (\xi(\sigma), \eta(\sigma), \zeta(\sigma))$ for P'; hence (cf. eq. 3 above) $r = |\mathbf{r}_2(\sigma) - \mathbf{r}_1(s)|$ and

$$\frac{(\mathbf{r}_2 - \mathbf{r}_1, \mathbf{r}_1', \mathbf{r}_2')}{|\mathbf{r}_2 - \mathbf{r}_1|^3} = \Pi \ . \tag{9}$$

An alternative definition can be given in terms of degree:

Definition 5.3. The *linking number* $Lk^{(2)} = Lk^{(2)}(\gamma_1, \gamma_2)$ of γ_1 and γ_2 is defined by

$$Lk^{(2)}(\gamma_1, \gamma_2) := \deg(\psi) \ . \tag{10}$$

Consider now the indented, oriented diagram $D_{\boldsymbol{\nu}}(\mathcal{L})$ of the (tame) link $\mathcal{L} = \gamma_1 \sqcup \gamma_2$, obtained by projecting $\mathcal{L}$ along $\boldsymbol{\nu}$ onto the plane, allowing under- and over-crossings. Let $D_{\boldsymbol{\nu}}(\mathcal{L})$ be a good projection of $\mathcal{L}$, that is one for which the standard projection has nodal points of multiplicity at most two. We assign to each apparent crossing k of $\gamma_1 \sqcap \gamma_2$ the number $\epsilon(k) = \pm 1$ according to standard convention. We have

Definition 5.4. The *linking number* $Lk^{(3)} = Lk^{(3)}(\gamma_1, \gamma_2)$ of γ_1 and γ_2 is defined by

$$Lk^{(3)}(\gamma_1, \gamma_2) := \frac{1}{2} \sum_{k \in \gamma_1 \sqcap \gamma_2} \epsilon(k) \ . \tag{11}$$

Finally, let us choose a Seifert surface M of γ_1, such that $M \cap \gamma_2$ is made of a finite number of transversal intersections. According to their relative sign, the algebraic sum of these is given by $I(M, \gamma_2)$. We have:

Definition 5.5. The *linking number* $Lk^{(4)} = Lk^{(4)}(\gamma_1, \gamma_2)$ of γ_1 and γ_2 is defined by

$$Lk^{(4)}(\gamma_1, \gamma_2) := I(M, \gamma_2) \ . \tag{12}$$

In all these cases, we have $Lk^i(\gamma_1, \gamma_2) = Lk^i(\gamma_2, \gamma_1)$ $(i = 1, \ldots, 4)$.

5.1. *Equivalence of definitions*

Proposition 5.1. *The following holds true:*

$$Lk^{(1)} = Lk^{(2)} = Lk^{(3)} = Lk^{(4)} \ . \tag{13}$$

Proof. $(Lk^{(2)} = Lk^{(1)})$. Consider the set of pre-images

$$Y = \{\mathbf{T}_1 = (t_{1_1}, t_{1_2}), \ldots, \mathbf{T}_n = (t_{n_1}, t_{n_2})\}$$

of any regular point of ψ on S^2. By eq. (7) any point of Y contributes ± 1 to the value of $\deg(\psi)$, depending on the orientation of the surface $\psi(\mathbb{T})$. At any point $\psi(\mathbf{T}_i)$ $(i = 1, \ldots, n)$ this orientation is given by $\boldsymbol{\nu}(\mathbf{T}_i) = \left(\frac{\partial \mathbf{n}}{\partial t_1} \times \frac{\partial \mathbf{n}}{\partial t_2}\right)_{\mathbf{T}_i}$ normal to the surface at $\mathbf{T}_i$, thus contributing $+1$ or -1 depending on the orientation of $\boldsymbol{\nu}(\mathbf{T}_i)$ (outwards or inwards) with respect to the sphere. The sign is given by

$$\mathbf{n}(\mathbf{T}_i) \cdot \boldsymbol{\nu}(\mathbf{T}_i) = \left(\mathbf{n}, \frac{\partial \mathbf{n}}{\partial t_1}, \frac{\partial \mathbf{n}}{\partial t_2}\right)_{\mathbf{T}_i} \ ; \tag{14}$$

hence

$$\deg(\boldsymbol{\psi}) = \sum_{\mathbf{T}_i \in Y} \operatorname{sgn}\left(\mathbf{n}, \frac{\partial \mathbf{n}}{\partial t_1}, \frac{\partial \mathbf{n}}{\partial t_2}\right)_{\mathbf{T}_i} . \tag{15}$$

Since the latter does not depend on any particular regular value $\mathbf{y}$ of $\boldsymbol{\psi}$, and by Sard's theorem the critical values of $\boldsymbol{\psi}$ form a null set, we have

$$\deg(\boldsymbol{\psi}) = \frac{1}{4\pi}\int_{S^2} \sum_{\mathbf{T}_i \in Y} \operatorname{sgn}\left(\mathbf{n}, \frac{\partial \mathbf{n}}{\partial t_1}, \frac{\partial \mathbf{n}}{\partial t_2}\right)_{\mathbf{T}_i} \mathrm{d}^2\mathbf{y} , \tag{16}$$

where the r.h.s. is an average over S^2. If $\mathbf{y} \notin \boldsymbol{\psi}(\mathbb{T})$, then the set $\boldsymbol{\psi}^{-1}(\mathbf{y})$ is empty; thus, the integration domain can be reduced to $\boldsymbol{\psi}(\mathbb{T})$. By exploiting the continuity of $\boldsymbol{\psi}$ we can sub-divide $\boldsymbol{\psi}(\mathbb{T})$ into m regions $\mathcal{R}_k$, each having n_k number of pre-images (constant). Since each $\mathcal{R}_k$ is the image of the n_k regions $\mathcal{S}_{k,i}$ $(i = 1, \ldots, n_k)$, the orientation of each $\mathcal{R}_{k,i} = \boldsymbol{\psi}(\mathcal{S}_{k,i})$ is constant for any $i = 1, \ldots, n_k$. Moreover, since $\mathbb{T} = \bigcup_{k=1}^{m} (\bigcup_{i=1}^{n_k} \mathcal{S}_{k,i})$, then eq. (16) can be written as

$$\deg(\boldsymbol{\psi}) = \frac{1}{4\pi}\sum_{k=1}^{m}\sum_{i=1}^{n_k}\int_{\mathcal{R}_k} \operatorname{sgn}\left(\mathbf{n}, \frac{\partial \mathbf{n}}{\partial t_1}, \frac{\partial \mathbf{n}}{\partial t_2}\right)_{\boldsymbol{\psi}^{-1}(\mathbf{y})\cap\mathcal{S}_{k,i}} \mathrm{d}^2\mathbf{y} . \tag{17}$$

We now make a change of variable by taking $\boldsymbol{\tau} = \boldsymbol{\psi}^{-1}(\mathbf{y})$, so that

$$\deg(\boldsymbol{\psi}) = \frac{1}{4\pi}\sum_{k=1}^{m}\sum_{i=1}^{n_k}\int_{\mathcal{S}_{k,i}} \operatorname{sgn}\left(\mathbf{n}, \frac{\partial \mathbf{n}}{\partial t_1}, \frac{\partial \mathbf{n}}{\partial t_2}\right)_{\boldsymbol{\tau}} \left|\frac{\partial \mathbf{n}}{\partial t_1} \times \frac{\partial \mathbf{n}}{\partial t_2}\right|_{\boldsymbol{\tau}} \mathrm{d}^2\boldsymbol{\tau} . \tag{18}$$

Thus

$$\deg(\boldsymbol{\psi}) = \frac{1}{4\pi}\int_{\mathbb{T}} \left(\mathbf{n}, \frac{\partial \mathbf{n}}{\partial t_1}, \frac{\partial \mathbf{n}}{\partial t_2}\right)_{\boldsymbol{\tau}} \mathrm{d}^2\boldsymbol{\tau} . \tag{19}$$

By some tedious, but straightforward algebra we can see that

$$\left(\mathbf{n}, \frac{\partial \mathbf{n}}{\partial t_1}, \frac{\partial \mathbf{n}}{\partial t_2}\right) = \frac{(\mathbf{r}_2 - \mathbf{r}_1, \mathbf{r}_1', \mathbf{r}_2')}{|\mathbf{r}_2 - \mathbf{r}_1|^3} . \tag{20}$$

Eq. (20) is thus proven and eq. (19) becomes

$$\deg(\boldsymbol{\psi}) = \frac{1}{4\pi}\int_0^{2\pi}\int_0^{2\pi} \frac{(\mathbf{r}_2 - \mathbf{r}_1, \mathbf{r}_1', \mathbf{r}_2')}{|\mathbf{r}_2 - \mathbf{r}_1|^3}\, \mathrm{d}t_1 \mathrm{d}t_2 . \tag{21}$$

Definitions 5.2 and 5.3 are thus equivalent. □

Proof. $(Lk^{(2)} = Lk^{(3)})$. We have $\mathbf{T}_0 = (t_{1_0}, t_{2_0}) \in \boldsymbol{\psi}^{-1}(\mathbf{v})$ iff

$$\frac{\mathbf{r}_1(t_{1_0}) - \mathbf{r}_2(t_{2_0})}{|\mathbf{r}_1(t_{1_0}) - \mathbf{r}_2(t_{2_0})|} = \mathbf{v} , \tag{22}$$

i.e. iff $\pi_{\mathbf{v}}(\mathbf{r}_1(t_{1_0})) = \pi_{\mathbf{v}}(\mathbf{r}_2(t_{2_0}))$, i.e. iff $\pi_{\mathbf{v}}(\mathbf{T}_0)$ is a crossing in $D_{\boldsymbol{\nu}}(\mathcal{L})$, where γ_1 goes over γ_2. Let $\mathbf{P}_k(\gamma_i)$ be the unit vector tangent to $\pi_{\mathbf{v}}(\gamma_i)$ at k. Without loss of generality we assume $\pi_{\mathbf{v}}(\mathbf{r}_i'(t_{i_0})) \neq \mathbf{0}$. By standard sign convention, we have

$$\epsilon(k) = -(\mathbf{v}, \mathbf{P}_k(\gamma_1), \mathbf{P}_k(\gamma_2)) \ . \tag{23}$$

Now let $\mathbf{N}$ denote a vector normal to Π; then, we have

$$\operatorname{sgn}(\mathbf{N}, \mathbf{r}_1', \mathbf{r}_2')_{\mathbf{T}_0} = \operatorname{sgn}(\mathbf{N}, \pi_{\mathbf{v}}(\mathbf{r}_1'), \pi_{\mathbf{v}}(\mathbf{r}_2'))_{\mathbf{T}_0} \ . \tag{24}$$

Moreover

$$\frac{(\pi_{\mathbf{v}}(\mathbf{r}_1') \times \pi_{\mathbf{v}}(\mathbf{r}_2'))_{\mathbf{T}_0}}{|\pi_{\mathbf{v}}(\mathbf{r}_1') \times \pi_{\mathbf{v}}(\mathbf{r}_2')|_{\mathbf{T}_0}} = (\mathbf{P}_k(\gamma_1) \times \mathbf{P}_k(\gamma_2))_{\mathbf{T}_0} \ . \tag{25}$$

Hence, by eq. (23) and by using eqs. (22) and (25), we have

$$\epsilon(k) = \operatorname{sgn}(\mathbf{r}_2 - \mathbf{r}_1, \pi_{\mathbf{v}}(\mathbf{r}_1'), \pi_{\mathbf{v}}(\mathbf{r}_2'))_{\mathbf{T}_0} \ ; \tag{26}$$

since $\mathbf{r}_2(t_{2_0}) - \mathbf{r}_1(t_{1_0})$ is orthogonal to Π, by (24) we have

$$\epsilon(k) = \operatorname{sgn}(\mathbf{r}_2 - \mathbf{r}_1, \mathbf{r}_1', \mathbf{r}_2')_{\mathbf{T}_0} \ . \tag{27}$$

By assuming that $\mathbf{v} \in S^2$ is a regular value of $\boldsymbol{\psi}$, by eqs. (7) and (20), we have

$$\deg(\boldsymbol{\psi}) = \sum_{\mathbf{T}_i \in \boldsymbol{\psi}^{-1}(\mathbf{y})} \operatorname{sgn}\left(\mathbf{n}, \frac{\partial \mathbf{n}}{\partial t_1}, \frac{\partial \mathbf{n}}{\partial t_2}\right)_{\mathbf{T}_i} = \sum_{k \in K} \epsilon(k) \ . \tag{28}$$

where K is the number of over-crossings (γ_1 over γ_2). By considering the total number of crossings (over *and* under) we have that indeed $Lk^{(2)} = Lk^{(3)}$. Hence, Definitions 5.3 and 5.4 are equivalent. □

Proof. ($Lk^{(2)} = Lk^{(4)}$). We recall that the degree of a (continuous) function restricted to the boundary of the manifold where it is defined is zero. Let us construct a continuous function $\tilde{\boldsymbol{\psi}}$ from a manifold N to S^2 such that $\gamma_1 \times \gamma_2$ is a component of ∂N and $\tilde{\boldsymbol{\psi}}|_{\gamma_1 \times \gamma_2} = \boldsymbol{\psi}$. Let $\tilde{\boldsymbol{\psi}}(x, y) = (y - x)/\,|y - x|$ and $N = (M \times \gamma_2) \setminus \bigcup_{m \in M \cap \gamma_2} \mathcal{B}(m, \epsilon)$, where $\mathcal{B}_\epsilon(m)$ is a ball of small radius ϵ, centred on m: $\tilde{\boldsymbol{\psi}}$ is well-defined. One component of ∂N is $\gamma_1 \times \gamma_2$; the union of the other components of ∂N is given by $\bigcup_{m \in M \cap \gamma_2} S^2_\epsilon(m)$, the orientation of each sphere being opposite to that inherited from M. Now, from $\deg(\tilde{\boldsymbol{\psi}}|_{\partial N}) = 0$, we have

$$\deg(\tilde{\boldsymbol{\psi}}|_{\gamma_1 \times \gamma_2}) + \deg(\tilde{\boldsymbol{\psi}}|_{\bigcup_{m \in M \cap \gamma_2} S^2_\epsilon(m)}) = 0 \ . \tag{29}$$

Hence,

$$Lk^{(2)}(\gamma_1, \gamma_2) = \deg(\psi) = -\deg(\tilde{\psi}|_{\bigcup_{m\in M\cap\gamma_2} S^2_\epsilon(m)}) \; . \tag{30}$$

Moreover, we have

$$\deg(\tilde{\psi}|_{\bigcup_{m\in M\cap\gamma_2} S^2_\epsilon(m)}) = -I(M, \gamma_2) \; , \tag{31}$$

and therefore $Lk^{(2)} = Lk^{(4)}$. This completes the demonstration of Proposition 5.1.

□

6. Gauss' integral in terms of oriented area

As we saw from the discussion in §3, Gauss' integral can be interpreted in terms of oriented area contributions. Its evaluation by the Gauss map is indeed useful. We provide three elementary examples of direct inspection of $\psi(\mathbb{T})$ through the *oriented area* $\vec{A}$. Let us sub-divide $\mathbb{T}$ into regions $\mathbb{T}_j$, in which $\frac{\partial \mathbf{n}}{\partial t_1} \times \frac{\partial \mathbf{n}}{\partial t_2}$ has constant sign. The oriented area of each $\mathbb{T}_j$ is given by

$$\pm \iint \left| \frac{\partial \mathbf{n}}{\partial t_1} \times \frac{\partial \mathbf{n}}{\partial t_2} \right| \mathrm{d}t_1 \mathrm{d}t_2 \; ,$$

where the sign depends on the orientation of the surface. $\vec{A}$ is given by summing up (algebraically) the contributions from the positive and negative regions $\mathbb{T}_j$. Since $\left(\mathbf{n}, \frac{\partial \mathbf{n}}{\partial t_1}, \frac{\partial \mathbf{n}}{\partial t_2}\right) = \pm \left| \frac{\partial \mathbf{n}}{\partial t_1} \times \frac{\partial \mathbf{n}}{\partial t_2} \right|$ has the sign of the oriented surface $\psi(\mathbb{T})$, the oriented area is given by

$$\vec{A} = \int_0^{2\pi} \int_0^{2\pi} \left(\mathbf{n}, \frac{\partial \mathbf{n}}{\partial t_1}, \frac{\partial \mathbf{n}}{\partial t_2}\right) \mathrm{d}t_1 \mathrm{d}t_2 \; . \tag{32}$$

Thus, by dividing by the area of the unit sphere S^2, we have $\deg(\psi) = \vec{A}/4\pi$. Since $\psi(\mathbb{T})$ is a continuous mapping of a closed surface and $\psi(\mathbb{T}) \subseteq S^2$, then the oriented area of $\psi(\mathbb{T})$ will be a multiple integer of 4π and $\deg(\psi)$ an integer. Since positive and negative contributions cancel out, the degree is simply given by counting the number of times $\psi(\mathbb{T})$ covers S^2 *essentially* (see also, for example, Ref. 27).

6.1. *Un-link, Hopf link and link* 4_1^2

The linking number can be estimated by direct inspection of the oriented area of $\psi(\mathbb{T})$. To illustrate this, let us consider the following three examples.

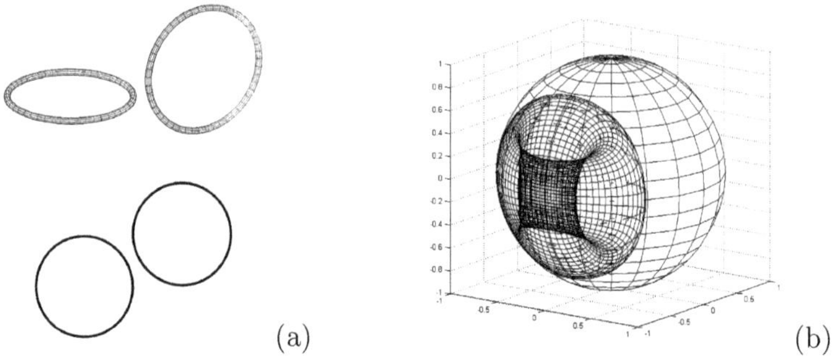

Fig. 8. (a) The un-link 0_1^2 with $Lk(0_1^2) = 0$ viewed (top) in space and (bottom) in its minimal diagram representation. (b) $\psi(\mathbb{T})$ resulting from the Gauss map of the un-link 0_1^2.

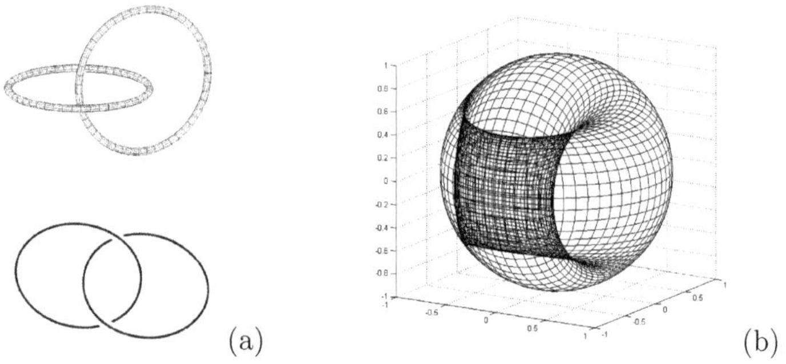

Fig. 9. (a) The Hopf link 2_1^2 with $Lk(2_1^2) = \pm 1$ viewed (top) in space and (bottom) in its minimal diagram representation. (b) $\psi(\mathbb{T})$ resulting from the Gauss map of the Hopf link 2_1^2.

First, let us consider the Gauss map associated with the trivial link 0_1^2 (the "un-link") (see Fig. 8c). Any point on the unit sphere has same degree, that is $\deg(\psi) = 0$. Different grid levels identify regions that contribute differently to the oriented area. The lighter region is not covered by $\psi(\mathbb{T})$. The intermediate region is doubly covered by $\psi(\mathbb{T})$, with opposite orientations and contributions to the degree (hence $\deg(\psi) = 0$). The darker region is covered four times: two inherited from the double covering of the intermediate region and two, of opposite orientations, of its own. Thus, the overall contribution to the oriented area is zero, thus $Lk(0_1^2) = 0$.

A second example is provided by the Hopf link 2_1^2 (see Fig. 9). Any point on the sphere has $\deg(\psi) = \pm 1$. Two distinct regions can be identified: the light region is covered once, depending on the surface orientation. The

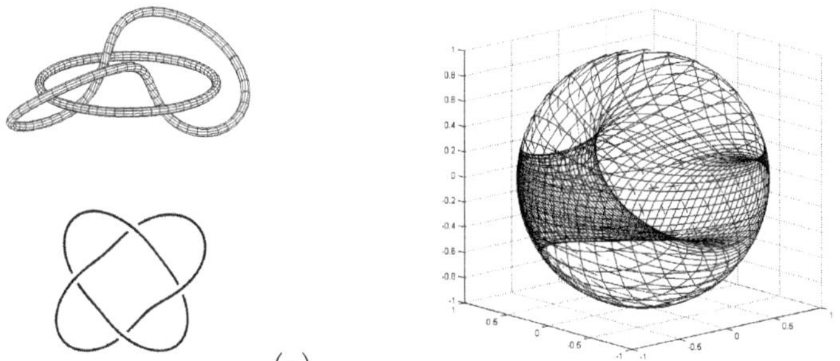

(a) (b)

Fig. 10. (a) The link 4_1^2 with $Lk(4_1^2) = \pm 2$ viewed (top) in space and (bottom) in its minimal diagram representation. (b) $\boldsymbol{\psi}(\mathbb{T})$ resulting from the Gauss map of the link 4_1^2.

dark region is covered twice, by two opposite orientations; one covering is inherited from the light region. The oriented area is thus given by $\pm 4\pi$, hence $Lk(2_1^2) = \pm 1$.

Finally we consider the link 4_1^2 of Fig. 10. From Fig. 10(c) we see that $\deg(\boldsymbol{\psi}) = \pm 2$, with two distinct regions: the light region is covered twice, with $\deg(\boldsymbol{\psi}) = \pm 2$, whereas the dark region is covered four times. The oriented area is thus given by $\pm 8\pi$, hence $Lk(4_1^2) = \pm 2$.

7. Conclusions

A mathematical reconstruction of a possible derivation made by Gauss of the linking number has been presented, together with explicit proofs of modern, equivalent interpretations of the linking number in terms of degree, signed crossings and intersection number. The mathematical reconstruction offered here is based entirely on first principles of potential theory due to Gauss. This proves two important points: i) that its derivation is entirely consistent with the contemporary work of Gauss on magnetic potential, and that indeed it can be worked out by using explicit results of Gauss on terrestrial magnetism; ii) that the concept of linking number is intimately related to the foundations of potential theory and modern mathematical physics, and that indeed contributes to the topological foundations of physical theory. Further support to this comes from the subsequent derivation made by Maxwell, that, as demonstrated by the documents provided here, is most probably authentically independent.

Since the interpretations of the linking number in terms of degree, signed crossings and intersection number find numerous applications in modern mathematical physics, the proofs of their equivalence presented here bridge an existing gap between the original definition and the alternative meanings present in literature, that prove sometimes hard to reconcile for non-experts. Moreover, direct estimate of the linking integral by the Gauss map of oriented areas is applied to three examples, the un-link, the Hopf link and the 4_1^2 link, to show how Gauss' original concepts and ideas in potential theory are still so fruitful in modern science. The material presented here forms an integral part of a paper[16] that will be published in the *Journal of Knot Theory and Its Ramifications.*

Acknowledgments

R.L. Ricca wishes to thank Claude Weber for many interesting discussions held during his visit to the U. Geneva, where part of this work originated. R.L.R. wishes to acknowledge financial support from *The Royal Society of London* and the *Fonds National Suisse (FNS).*

References

1. M. Spera, A survey on the differential and symplectic geometry of linking numbers, *Milan J. Math.* **74** (2006) 139–197.
2. M. Epple, Orbits of asteroids, a braid, and the first link invariant, *Math. Intelligencer* **20** (1998) 45–52.
3. M. Epple, Topology, matter, and space, I: topological notions in 19th-century natural philosophy, *Arch. Hist. Exact Sci.* **52** (1998) 297–392.
4. M. Epple, Geometric aspects in the development of knot theory, in *History of Topology*, ed. I.M. James (Elsevier Science B.V., 1999), pp. 301–357.
5. A. C. Hirshfeld, Knots and physics: old wine in new bottles. *Am. J. Phys.* **66** (1998) 1060–1066.
6. D. Rolfsen, *Knots and Links* (AMS Chelsea Publishing, Providence, Rhode Island, 1990).
7. V. I. Arnold, The asymptotic Hopf invariant and its applications, in *Proc Summer School in Diff. Eqs. at Dilizhan*, Armenian Acad. Sci. Erevan, (1974), pp. 229–256 (in Russian). [English translation: *Sel. Math. Sov.* **5** (1986) 327–345.]
8. M. H. Freedman and Z.-X. He, Divergence-free fields: energy and asymptotic crossing number, *Ann. Math.* **134** (1991) 189–229.
9. Y. Aminov, *Differential Geometry and Topology of Curves* (CRC Press, Boca Raton, 2000).
10. J. O'Hara, *Energy of Knots and Conformal Geometry* (Series of Knots and Everything **33**, World Scientific, Singapore, 2003).
11. A. Widom and T. D. Clark, Quantum electrodynamic uncertainty relations, *Phys. Lett.* **90**A (1982) 280.
12. A. Ashtekar and A. Corichi, Gauss linking number and the electromagnetic uncertainty principle, *Phys. Rev.* D **56** (1997) 2073–2079.

13. C. F. Barenghi, R. L. Ricca and D. C. Samuels, How tangled is a tangle?, *Physica* D **157** (2001) 197–206.
14. R. L. Ricca, Topology bounds energy of knots and links, *Proc. R. Soc.* A **464** (2008) 293–300.
15. R. L. Ricca, Structural complexity and dynamical systems, in *Lectures on Topological Fluid Mechanics*, ed. R.L. Ricca (Springer-CIME Lecture Notes in Mathematics **1973**, Springer-Verlag, Heidelberg, 2009), pp. 169–188.
16. R. L. Ricca and B. Nipoti, Gauss' linking number revisited. *J. Knot Theory & Its Ramifications*. In press.
17. G. W. Dunnington, *Carl Friedrich Gauss Titan of Science* (The Mathematical Association of America, Washington, 2004).
18. W. K. Bühler, *Gauss – A Biographical Study* (Springer-Verlag, New York, 1981).
19. C. F. Gauss, Note dated 22 Jan. 1833, in *Werke* V, ed. C. Schäfer (Königliche Gesellschaft der Wissenschaften zu Göttingen, Leipzig-Berlin, 1867), p. 605.
20. C. Schäfer, Gauss' investigations on electrodynamics, *Nature* **128** (1931) 339–341.
21. C. F. Gauss, Intensitas vis magneticae terrestris ad mensuram absoluta revocata. *Comment.* VIII (Göttingen, 1831). [English translation: *Proc. R. Soc. London* **11** (1833) 166-167.]
22. K. O. May, Gauss, Carl Friedrich, in *Dictionary of Scientific Biography* **5**, ed. C.C. Gillispie (Charles Scribner's Sons, New York, 1970), pp. 298–315.
23. C. F. Gauss, Allgemeine Theorie des Erdmagnetismus. ed. C.F. Gauss, (1839) pp. 1–57, pp. 146–148. [English translation: *Scientific Memoirs, Selected from the Transactions of Foreign Academies and Learned Societies and from Foreign Journals*, ed R. Taylor (London, 1841), pp. 184–251, pp. 313–316.]
24. W. Thomson, A mathematical theory of magnetism – Continuation of Part I. *Phil. Trans. R. Soc. London*, June (1850) 269–285.
25. J. C. Maxwell, *A Treatise on Electricity and Magnetism* (Clarendon Press, Oxford, 1873).
26. P. G. Tait, On Knots. in *Scientific Papers of P.G. Tait*, ed. C.G. Knott (Cambridge University Press, 2 Volumes, 1898), pp. 273–317.
27. M. R. Dennis and J. H. Hannay, Geometry of Călugăreanu's theorem. *Proc. R. Soc.* A **461** (2005) 3245–3254.

Introduction to Virtual Knot Theory

Louis H. Kauffman
Department of Mathematics, Statistics
and Computer Science (m/c 249)
851 South Morgan Street
University of Illinois at Chicago
Chicago, Illinois 60607-7045
<kauffman@uic.edu>

Abstract

This paper is an introduction to virtual knot theory and an exposition of new ideas and constructions, including the parity bracket polynomial, the arrow polynomial, the parity arrow polynomial and categorif cations of the arrow polynomial.

1. Introduction

Virtual knot theory is an extension of classical knot theory to stabilized embeddings of circles into thickened orientable surfaces of genus possibly greater than zero. Classical knot theory is the case of genus zero. There is a diagrammatic theory for studying virtual knots and links, and this diagrammatic theory lends itself to the construction of numerous new invariants of virtual knots as well as extensions of known invariants.

This paper gives an introduction to virtual knot theory and it def nes a new invariant of virtual links and f at virtual links that we call the *arrow polynomial* [4, 29].

One of the applications of this new invariant is to the category of f at virtual knots and links. In Section 5 we review the def nitions of f at virtuals and we prove that *isotopy classes of long f at virtual knots embed in the isotopy classes of all long virtual knots* via the ascending map $A : LVF \longrightarrow LVK$. See Section 5 for this result. This embedding of the long f at virtual knots means that there are many invariants of them obtained by appling any invariant Inv of virtual knots via the composition with the ascending map A. This situation is in direct contrast to closed long virtual f ats, where it is not so easy to def ne invariants. The arrow polynonmial is an invariant of both long virtual f ats and closed virtual f ats.

This paper is relatively self-contained, with Section 2 reviewing the def nitions of virtual knots and links, Section 3 reviewing the surface interpretations of virtuals and Section 4 reviewing the def nitions and properties of f at virtuals.

Section 5 proves that long f at virtual knots (virtual strings in Turaev's terminology) embed into long virtual knots. This result is a very strong tool for discriminating long virtual knots from one another. We give examples and computations using this technique by applying invariants of virtual knots to corresponding ascending diagram images of f at virtual knots. A long virtual can be closed just as a long knot can be closed. It is a remarkable property of long virtuals (both f at and not f at) that there are non-trivial long examples whose closures are trivial. Thus one would like to understand the kernel of the closure mapping from long virtuals to virtuals both in the f at and non-f at categories. It is hoped that the invariants discussed in this paper will further this question. We also introduce in Section 5 the notion of the parity (odd or even) of a crossing in a virtual knot and recall the *odd writhe* of a virtual knot – an invariant that is a kind of self-linking number for virtual knots.

Section 6 is a review of the bracket polynomial and the Jones polynomial for virtual knots and links. We recall the virtualization construction that produces inf nitely many non-trivial virtual knots with unit Jones polynomial. These examples are of interest since there are no known examples of classical non-trivial knots with unit Jones polynomial. One may conjecture that all non-trivial examples produced by virtualization are non-classical. It may be that the virtualization of some non-trivial classical knot is isotopic to a classical knot, but we have no evidence that this can happen. The arrow polynomial invariant developed in this paper may help in deciding this question of non-classicality.

Section 7 introduces the Parity Bracket, a modif cation of the bracket polynomial due to Vassily Manturov that takes into account the parity of crossings in a knot diagram. The Parity Bracket takes values in sums of certain reduced graphs with polynomial coeff cients. We show the power of this invariant by given an example of a virtual knot with unit Jones polynomial that is not obtained from a classical knot by the virtualization construction of the previous section. This is the f rst time such an example has been pinpointed and we do not know any method to detect such examples at this time other than the Parity Bracket of Vassily Manturov.

In Section 8 we give the def nition of the arrow polynomial $\mathcal{A}[K]$ and a number of examples of its calculation. These examples include a verif cation of the non-classicality of the simplest example of virtualization, a verif cation that the Kishino diagram [32] and the f at Kishino diagram are non-trivial. The arrow polynomial is an invariant of f at diagrams by taking the specialization of its parameters so that $A = 1$ and $\delta = -2$. We give an example showing that the arrow polynomial can detect a long virtual knot whose closure is trivial. This is a capability that is beyond the reach of the Jones polynomial. As the reader will see from the def nitions in section 8, the arrow polynomial is a direct generalization of the bracket model of Jones polynomial, using the oriented structure of the link diagrams. Section 8 ends with a short discussion of how to create a parity version of the arrow polynomial. This study will be taken up in a separate paper.

The arrow polynomial has inf nitely many variables and integer coeff cients. It has been discovered independently by Miyazawa, via a different def nition in [41, 44]. Miyazawa does not consider the applications of this polynomial to f at virtual knots. In [29] we use the interaction of the arrow polynomial with the extended bracket polynomial to obtain results on the genus of virtual knots. In this paper we go beyond our previous work on the arrow polynomial by extending it to long virtual knots and to examples involving categorif cation and parity.

In Section 9 we give a quick review of Khovanov homology, and in Section 10 we describe the work of Heather Dye, the Author and Vassily Manturov in categorifying the arrow polyomial. We include an example (one of many found by Aaron Kaestner) of a pair of virtual knots that are not distinguished by the Jones polynomial, the bracket polynomial, the arrow polynomial or by Khovanov homology, while the pair is distinguished by our categorif cation of the arrow polynomial. We also show that this pair is distinguished by the Parity Bracket.

Section 10 ends the paper with a discussion about futher directions in virtual knot theory.

2. Virtual Knot Theory

Knot theory studies the embeddings of curves in three-dimensional space. Virtual knot theory studies the embeddings of curves in thickened surfaces of arbitrary genus, up to the addition and removal of empty handles from the surface. Virtual knots have a special diagrammatic theory, described below, that makes handling them very similar to the handling of classical knot diagrams. Many structures in classical knot theory generalize to the virtual domain.

In the diagrammatic theory of virtual knots one adds a *virtual crossing* (see Figure 1) that is neither an over-crossing nor an under-crossing. A virtual crossing is represented by two crossing segments with a small circle placed around the crossing point.

Moves on virtual diagrams generalize the Reidemeister moves for classical knot and link diagrams. See Figure 1. One can summarize the moves on virtual diagrams by saying that the classical crossings interact with one another according to the usual Reidemeister moves while virtual crossings are artifacts of the attempt to draw the virtual structure in the plane. A segment of diagram consisting of a sequence of consecutive virtual crossings can be excised and a new connection made between the resulting free ends. If the new connecting segment intersects the remaining diagram (transversally) then each new intersection is taken to be virtual. Such an excision and reconnection is called a *detour move*. Adding the global detour move to the Reidemeister moves completes the description of moves on virtual diagrams. In Figure 1 we illustrate a set of local moves involving virtual crossings. The global detour move is a consequence of moves (B) and (C) in Figure 1. The detour move is illustrated in Figure 2. Virtual knot and link diagrams that can be connected by a f nite sequence of these moves are said to be *equivalent* or *virtually isotopic*.

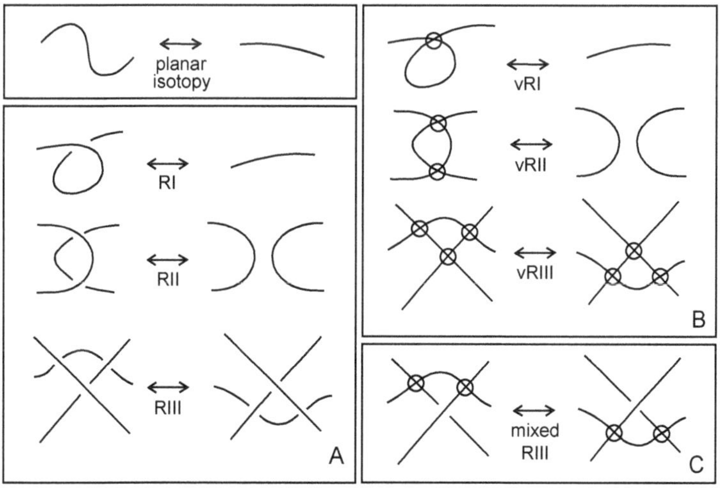

Fig. 1. **Moves**

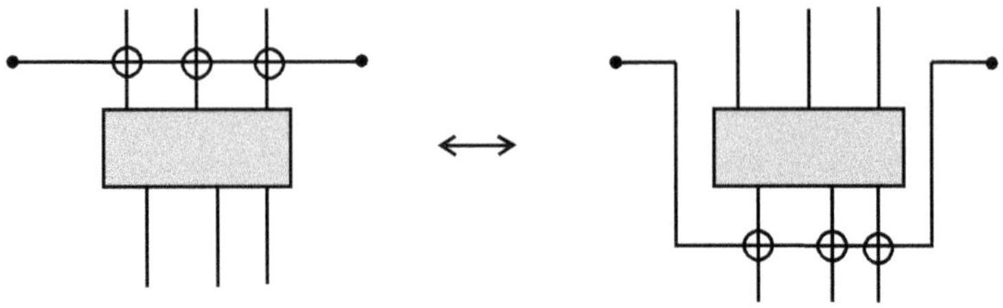

Fig. 2. **Detour Move**

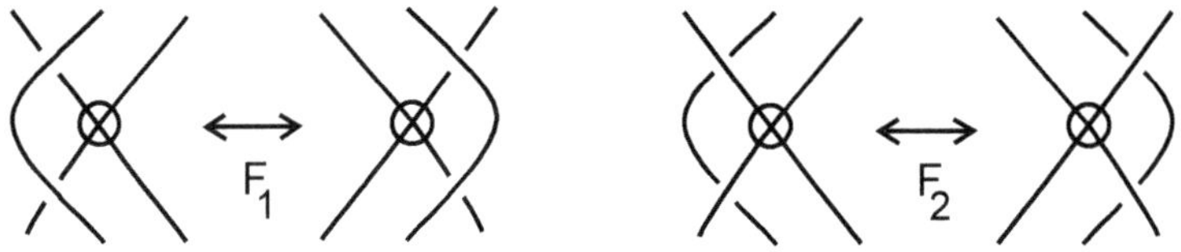

Fig. 3. **Forbidden Moves**

Another way to understand virtual diagrams is to regard them as representatives for oriented Gauss codes [8], [21, 22] (Gauss diagrams). Such codes do not always have planar realizations. An attempt to embed such a code in the plane leads to the production of the virtual crossings. The detour move makes the particular choice of virtual crossings irrelevant. *Virtual isotopy is the same as the equivalence relation generated on the collection of oriented Gauss codes by abstract Reidemeister moves on these codes.*

Figure 3 illustrates the two *forbidden moves*. Neither of these follows from Reidmeister moves plus detour move, and indeed it is not hard to construct examples of virtual knots that are non-trivial, but will become unknotted on the application of one or both of the forbidden moves. The forbidden moves change the structure of the Gauss code and, if desired, must be considered separately from the virtual knot theory proper.

3. Interpretation of Virtuals Links as Stable Classes of Links in Thickened Surfaces

There is a useful topological interpretation [21, 23] for this virtual theory in terms of embeddings of links in thickened surfaces. Regard each virtual crossing as a shorthand for a detour of one of the arcs in the crossing through a 1-handle that has been attached to the 2-sphere of the original diagram. By interpreting each virtual crossing in this way, we obtain an embedding of a collection of circles into a thickened surface $S_g \times R$ where g is the number of virtual crossings in the original diagram L, S_g is a compact oriented surface of genus g and R denotes the real line. We say that two such surface embeddings are *stably equivalent* if one can be obtained from another by isotopy in the thickened surfaces, homeomorphisms of the surfaces and the addition or subtraction of empty handles (i.e. the knot does not go through the handle).

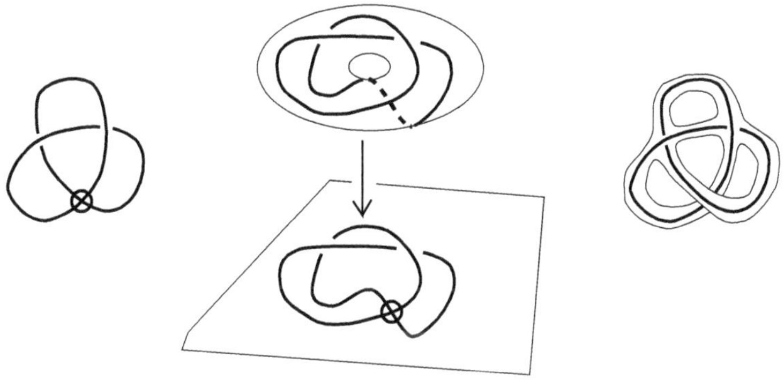

Fig. 4. **Surfaces and Virtuals**

We have the

Theorem 1 [3, 21, 23, 28]. *Two virtual link diagrams are isotopic if and only if their corresponding surface embeddings are stably equivalent.*

In Figure 4 we illustrate some points about this association of virtual diagrams and knot and link diagrams on surfaces. Note the projection of the knot diagram on the torus to a diagram in the plane (in the center of the f gure) has a virtual crossing in the planar diagram where two arcs that do not form a crossing in the thickened surface project to the same point in the plane. In this way, virtual crossings can be regarded as artifacts of projection. The same f gure shows a virtual diagram on the left and an “abstract knot diagram” [3, 45] on the right. The abstract knot diagram is a realization of the knot on the left in a thickened surface with boundary and it is obtained by making a neighborhood of the virtual diagram that resolves the virtual crossing into arcs that travel on separate bands. The virtual crossing appears as an artifact of the projection of this surface to the plane. The reader will f nd more

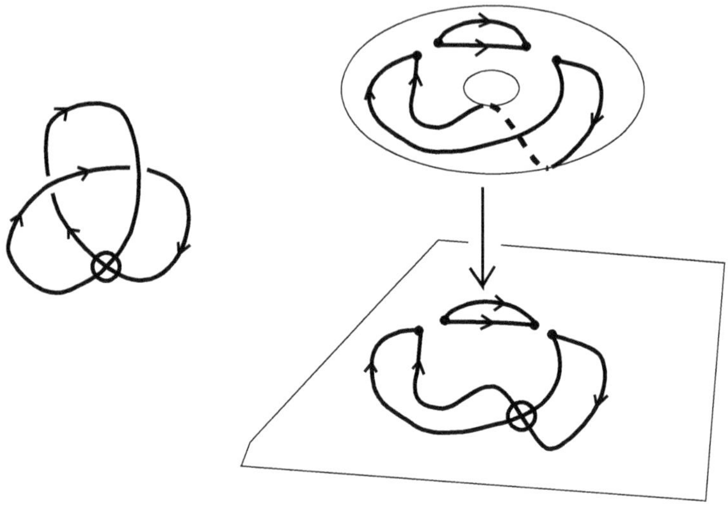

Fig. 5. **Surfaces and Virtual States**

information about this correspondence [21, 28] in other papers by the author and in the literature of virtual knot theory.

4. Flat Virtual Knots and Links

Every classical knot or link diagram can be regarded as a 4-regular plane graph with extra structure at the nodes. This extra structure is usually indicated by the over and under crossing conventions that give instructions for constructing an embedding of the link in three dimensional space from the diagram. If we take the f at diagram without this extra structure then the diagram is the shadow of some link in three dimensional space, but the weaving of that link is not specif ed. It is well known that if one is allowed to apply the Reidemeister moves to such a shadow (without regard to the types of crossing since they are not specif ed) then the shadow can be reduced to a disjoint union of circles. This reduction is no longer true for virtual links. More precisely, let a *f at virtual diagram* be a diagram with *virtual crossings* as we have described them and *f at crossings* consisting in undecorated nodes of the 4-regular plane graph. Two f at virtual diagrams are *equivalent* if there is a sequence of generalized f at Reidemeister moves (as illustrated in Figure 1) taking one to the other. A generalized f at Reidemeister move is any move as shown in Figure 1 where one ignores the over or under crossing structure. Note that in studying f at virtuals the rules for changing virtual crossings among themselves and the rules for changing f at crossings among themselves are identical. Detour moves as in Figure 1C are available for virtual crossings with respect to f at crossings and *not* the other way around. The analogs of the forbidden moves of Figure 3 remain forbidden when the classical crossings are replaced by f at crossings.

The theory of f at virtual knots and links is identical to the theory of all oriented Gauss codes (without over or under information) modulo the f at Reidemeister moves. Virtual crossings are an artifact of the realization of the f at diagram in the plane. In Turaev's work [49] f at virtual knots and links are called *virtual strings*. See also recent papers of Roger Fenn [6, 7] for other points of view about f at virtual knots and links.

We shall say that a virtual diagram *overlies* a f at diagram if the virtual diagram is obtained from the f at diagram by choosing a crossing type for each f at crossing in the virtual diagram. To each virtual diagram K there is an associated f at diagram $F(K)$, obtained by forgetting the extra structure at the classical crossings in K. Note that if K and K' are isotopic as virtual diagrams, then $F(K)$ and $F(K')$ are isotopic as f at virtual diagrams. Thus, if we can show that $F(K)$ is not reducible to a disjoint union of circles, then it will follow that K is a non-trivial virtual link. The f at virtual diagrams present a challenge for the construction of new invariants. They are fundamental to the study of virtual knots. A virtual knot is necessarily non-trivial if its f at projection is a non-trivial f at virtual diagram. We wish to be able to determine when a given virtual link is isotopic to a classcal link. The reducibility or irreducibility of the underlying f at diagram is the f rst obstruction to such an equivalence.

Def nition. A *virtual graph* is a f at virtual diagram where the classical f at crossings are not subjected to the f at Reidemeister moves. Thus a virtual graph is a $4-regular$ graph that is represented in the plane via a choice of cyclic orders at its nodes. The virtual crossings are artifacts of this choice of placement in the plane, and we allow detour moves for consecutive sequences of virtual crossings just as in the virtual knot theory. Two virtual graphs are *isotopic* if there is a combination of planar graph isotopies and detour moves that connect them. The theory of virtual graphs is equivalent to the theory of 4-regular graphs on oriented surfaces, taken up to empty handle stabilization, in direct analogy to our description of the theory of virtual links and f at virtual links.

5. Long Virtual Knots, Long Flats, Parity and the Odd Writhe

A long knot or link is a $1-1$ tangle. It is a tangle with one input end and one output end. In between one has, in the diagram, any knotting or linking, virtual or classical. Classical long knots (one component) carry essentially the same topological information as their closures. In particular, a classical long knot is knotted if and only if its closure is knotted. This statement is false for virtual knots. An example of the phenomenon is shown in Figure 5.

The long knots L and L' shown in Figure 5 are non-trivial in the virtual category. Their closures, obtained by attaching the ends together are unknotted virtuals. Concomittantly, there can be a multiplicity of long knots associated to a given virtual knot diagram, obtained by cutting an arc from the diagram and creating a $1-1$ tangle. *It is a fundamental problem to determine the kernel of the closure mapping from long virtual knots to virtual knots.* In Figure 5, the long knots L and L' are trivial as welded long knots (where the f rst forbidden move is allowed). The obstruction to untying them as virtual long knots comes from the f rst forbidden move. The matter of proving that L and L' are non-trivial distinct long knots is diff cult by direct attack. There is a fundamental relationship betweem long f at virtual knots and long virtual knots that can be used to see it.

Let LFK denote the set of long f at virtual knots and let LVK denote the set of long virtual knots. We def ne

$$A : LFK \longrightarrow LVK$$

by letting $A(S)$ be the ascending long virtual knot diagram associated with the long f at virtual diagram S. That is, $A(S)$ is obtained from S by traversing S from its left end to its right end and creating a crossing at each f at crossing so that *one passes under each crossing before passing over that crossing*. Virtual crossings are not changed by this construction. The idea of using the ascending diagram to def ne invariants of long f at virtuals is exploited in [47]. The following result is new.

Long Flat Embedding Theorem 2. The mapping $A : LFK \longrightarrow LVK$ is well-def ned on the corresponding isotopy classes of diagrams, and it is injective. Hence *long f at virtual knots embed in the class of long virtual knots.*

Proof. It is easy to see that if two f at long diagrams S, T are virtually isotopic then $A(S)$ and $A(T)$ are isotopic long virtual knots. We leave the verif cation to the reader. To see the injectivity, def ne

$$Flat : LVK \longrightarrow LFK$$

by letting $Flat(K)$ be the long f at diagram obtained from K by f attening all the classical crossings in K. By def nition $Flat(A(S)) = S$. Note that the map $Flat$ takes isotopic long virtual knots to isotopic f at virtual knots since any move using the under and over crossings is a legitimate move when this distinction is forgotten. This proves injectivity and completes the proof of the Theorem. //

The long f at theorem is easy to prove, and it has many good consequences. First of all, let $Inv(K)$ denote any invariant of long virtual knots. ($Inv(K)$ can denote a polynomial invariant, a number, a group, quandle, biquandle or other invariant structure.) Then we can def ne

$$Inv(S)$$

for any long f at knot S by the formula

$$Inv(S) = Inv(A(S)),$$

and this def nition yields an invariant of long f at knots.

Using the Odd Writhe J(K). View Figure 5. We show the long f at F and its image under the ascending map, $A(F)$ are non-trivial. In fact, $A(F)$ is non-trivial and non-classical. One computes that $J(A(F))$ is non-zero where $J(K)$ denotes the *odd writhe* of K. The odd writhe [27] is the sum of the signs of the odd crossings. A crossing is *odd* if it f anks an odd number of symbols in the Gauss code of the diagram. Classical diagrams have zero odd writhe. In this case, the f at Gauss code for F is 1212 with both crossings odd. Thus we see from the f gure that $J(A(F)) = 2$. Thus $A(F)$ is non-trivial, non-classical and inequivalent to its mirror image. Once we check that $A(F)$ is non-trivial, we know that the f at knot F is non-trivial, and from this we conclude that the long virtual L is also non-trivial. Note that $J(L) = 0$, so we cannot draw this last conclusion directly from $J(L)$. This same f gure illustrates a long f at G that is obtained by ref ecting F in a horizontal line. Then, as the reader can calculate from this f gure, $J(A(G)) = -2$. Thus F and G are distinct non-trivial long f ats. We conclude from these arguments that the long virtual knots L and L' in Figure 5 are both non-trivial, and that L is not virtually isotopic to L' (since such an isotopy would give an isotopy of F with G by the f attening map).

6. Review of the Bracket Polynomial for Virtual Knots

In this section we recall how the bracket state summation model [17] for the Jones polynomial [9, 51] is def ned for virtual knots and links. In the next section we give an extension of this model using orientation structures on the states of the bracket expansion. The extension is also an invariant of f at virtual links.

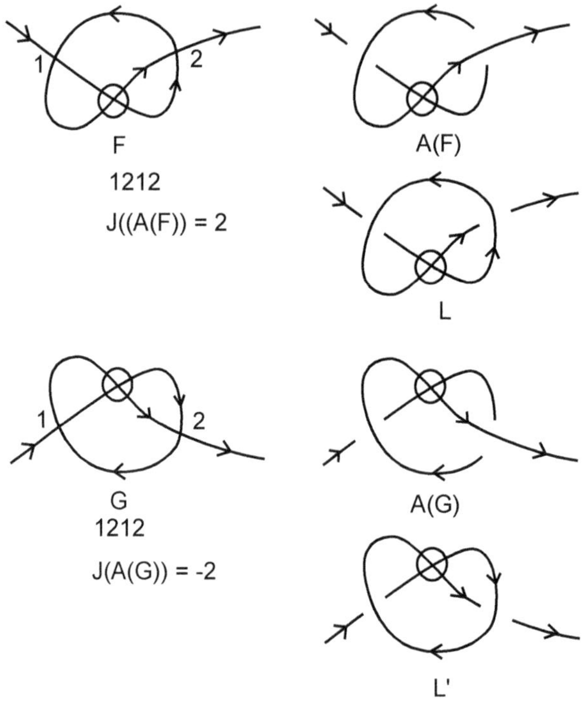

Fig. 6. **Ascending Map**

We call a diagram in the plane *purely virtual* if the only crossings in the diagram are virtual crossings. Each purely virtual diagram is equivalent by the virtual moves to a disjoint collection of circles in the plane.

A state S of a link diagram K is obtained by choosing a smoothing for each crossing in the diagram and labelling that smoothing with either A or A^{-1} according to the convention that a counterclockwise rotation of the overcrossing line sweeps two regions labelled A, and that a smoothing that connects the A regions is labelled by the letter A. Then, given a state S, one has the evaluation $< K|S >$ equal to the product of the labels at the smoothings, and one has the evaluation $||S||$ equal to the number of loops in the state (the smoothings produce purely virtual diagrams). One then has the formula

$$< K >= \Sigma_S < K|S > d^{||S||-1}$$

where the summation runs over the states S of the diagram K, and $d = -A^2 - A^{-2}$. This state summation is invariant under all classical and virtual moves except the first Reidemeister move. The bracket polynomial is normalized to an invariant $f_K(A)$ of all the moves by the formula $f_K(A) = (-A^3)^{-w(K)} < K >$ where $w(K)$ is the writhe of the (now) oriented diagram K. The writhe is the sum of the orientation signs (± 1) of the crossings of the diagram. The Jones polynomial, $V_K(t)$ is given in terms of this model by the formula

$$V_K(t) = f_K(t^{-1/4}).$$

This definition is a direct generalization to the virtual category of the state sum model for the original Jones polynomial. It is straightforward to verify the invariances stated above. In this way one has the Jones polynomial for virtual knots and links.

We have [23] the

Theorem 3. *To each non-trivial classical knot diagram of one component K there is a corresponding non-trivial virtual knot diagram $Virt(K)$ with unit Jones polynomial.*

Proof Sketch. This Theorem is a key ingredient in the problems involving virtual knots. Here is a sketch of its proof. The proof uses two invariants of classical knots and links that generalize to arbitrary virtual knots and links. These invariants are the *Jones polynomial* and the *involutory quandle* denoted by the notation $IQ(K)$ for a knot or link K.

Given a crossing i in a link diagram, we define $s(i)$ to be the result of *switching* that crossing so that the undercrossing arc becomes an overcrossing arc and vice versa. We define the *virtualization* $v(i)$ of the crossing by the local replacement indicated in Figure 7. In this figure we illustrate how, in virtualization, the original crossing is replaced by a crossing that is flanked by two virtual crossings. When we smooth the two virtual crossings in the virtualization we obtain the original knot or link diagram with the crossing switched.

Suppose that K is a (virtual or classical) diagram with a classical crossing labeled i. Let $K^{v(i)}$ be the diagram obtained from K by virtualizing the crossing i while leaving the rest of the diagram just as before. Let $K^{s(i)}$ be the diagram obtained from K by switching the crossing i while leaving the rest of the diagram just as before. Then it follows directly from the expansion formula for the bracket polynomial that

$$V_{K^{s(i)}}(t) = V_{K^{v(i)}}(t).$$

As far as the Jones polynomial is concerned, switching a crossing and virtualizing a crossing look the same. We can start with a classical knot diagram K and choose a subset S of crossings such that the diagram is unknotted when these crossings are switched. Letting $Virt(K)$ denote the virtual knot diagram obtained by virtualizing each crossing in S, it follows that the Jones polynomial of $Virt(K)$ is equal to unity, the Jones polynomial of the unknot. Nevertheless, if the original knot K is knotted, then the virtual knot $Virt(K)$ will be non-trivial. We outline the argument for this fact below.

The involutory quandle [18] is an algebraic invariant equivalent to the fundamental group of the double branched cover of a knot or link in the classical case. In this algebraic system one associates a generator of the algebra $IQ(K)$ to each arc of the diagram K and there is a relation of the form $c = ab$ at each crossing, where ab denotes the (non-associative) algebra product of a and b in $IQ(K)$. See Figure 8. In this f gure we have illustrated the fact that

$$IQ(K^{v(i)}) = IQ(K).$$

As far as the involutory quandle is concerned, the original crossing and the virtualized crossing look the same.

If a classical knot is actually knotted, then its involutory quandle is non-trivial [50]. Hence if we start with a non-trivial classical knot and virtualize any subset of its crossings we obtain a virtual knot that is still non-trivial. There is a subset A of the crossings of a classical knot K such that the knot SK obtained by switching these crossings is an unknot. Let $Virt(K)$ denote the virtual diagram obtained from A by virtualizing the crossings in the subset A. By the above discussion, the Jones polynomial of $Virt(K)$ is the same as the Jones polynomial of SK, and this is 1 since SK is unknotted. On the other hand, the IQ of $Virt(K)$ is the same as the IQ of K, and hence if K is knotted, then so is $Virt(K)$. We have shown that $Virt(K)$ is a non-trivial virtual knot with unit Jones polynomial. This completes the proof of the Theorem. //

See Figure 32 for an example of a virtualized trefoil, the simplest example of a non-trivial virtual knot with unit Jones polynomial. More work is needed to prove that the virtual knot T in Figure 32 is not classical. In the next section we will give a proof of this fact by using an extension of the bracket polynomial.

It is an open problem whether there are classical knots (actually knotted) having unit Jones polynomial. (There are linked links whose linkedness is unseen [5, 38] by the Jones polynomial.) If there exists a classical knot with unit Jones polynomial, then one of the knots $Virt(K)$ produced by this Theorem may

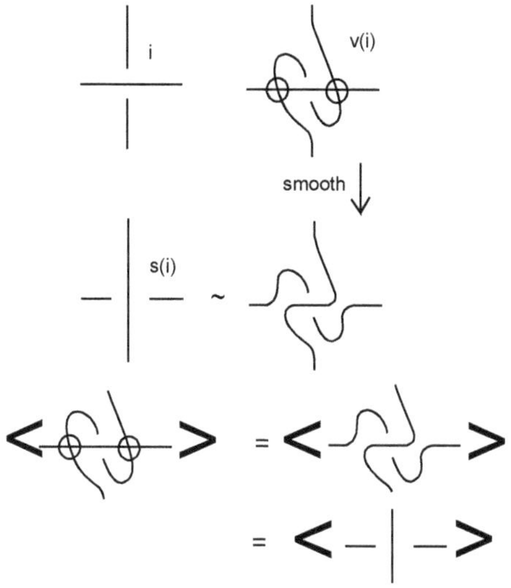

Fig. 7. **Switch and Virtualize**

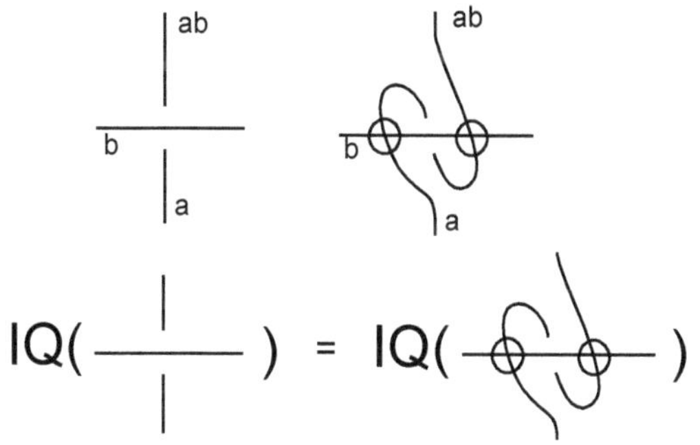

Fig. 8. **IQ(Virt)**

be isotopic to a classical knot. Such examples are guaranteed to be non-trivial, but they are usually also not classical. We do not know at this writing whether all such virtualizations of non-trivial classical knots, yielding virtual knots with unit Jones polynomial, are non classical. It is an intricate task to verify that specif c examples of $Virt(K)$ are not classical. This has led to an investigation of new invariants for virtual knots. In this way the search for classical knots with unit Jones polynomial expands to exploration of the structure of the inf nite collection of virtual knots with unit Jones polynomial.

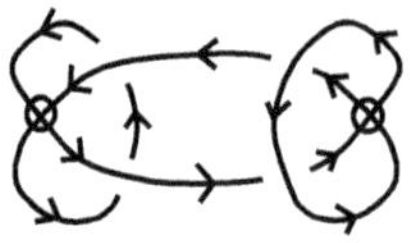

Fig. 9. **Kishino Diagram**

In Figure 9 we show the *Kishino diagram* K. This diagram has unit Jones polynomial and its fundamental group is infinite cyclic. The Kishino diagram was discovered by Kishino in [32]. Many other invariants of virtual knots fail to detect the Kishino knot. Thus it has been a test case for examining new invariants. Heather Dye and the author [15] have used the bracket polynomial defined for knots and links in a thickened surface (the state curves are taken as isotopy classes of curves in the surface) to prove the non-triviality and non-classicality of the Kishino diagram. In fact, we have used this technique to show that knots with unit Jones polynomial obtained by a single virtualization are non-classical. See the problem list by Fenn, Kauffman and Manturov [24] for other problems and proofs related to the Kishino diagram. In the next section we describe a new extension of the bracket polynomial that can be used to discriminate the Kishino diagram, and, in fact, shows that its corresponding flat virtual knot is non-trivial.

7. The Parity Bracket Polynomial

In this section we introduce the Parity Bracket Polynomial of Vassily Manturov [36]. This is a generalization of the bracket polynomial to virtual knots and links that uses the parity of the crossings. We define a *Parity State* of a virtual diagram K to be a labeled virtual graph obtained from K as follows: For each odd crossing in K replace the crossing by a graphical node. For each even crossing in K replace the crossing by one of its two possible smoothings, and label the smoothing site by A or A^{-1} in the usual way. Then we define the parity bracket by the state expansion formula

$$\langle K \rangle_P = \sum_S A^{n(S)}[S]$$

where $n(S)$ denotes the number of A-smoothings minus the number of A^{-1} smoothings and $[S]$ denotes a combinatorial evaluation of the state defined as follows: First reduce the state by Reidemeister two moves on nodes as shown in Figure 10. The graphs are taken up to virtual equivalence (planar isotopy plus detour moves on the virtual crossings). Then regard the reduced state as a disjoint union of standard state loops (without nodes) and graphs that irreducibly contain nodes. With this we write

$$[S] = (-A^2 - A^{-2})^{l(S)}[G(S)]$$

where $l(S)$ is the number of standard loops in the reduction of the state S and $[G(S)]$ is the disjoint union of reduced graphs that contain nodes. In this way, we obtain a sum of Laurent polynomials in A multiplying reduced graphs as the Manturov Parity Bracket. It is not hard to see that this bracket is invariant under regular isotopy and detour moves and that it behaves just like the usual bracket under the f rst Reidemeister move. However, the use of parity to make this bracket expand to graphical states gives it considerable extra power in some situations. For example, consider the Kishino diagram in Figure 9. We see that all the classical crossings in this knot are odd. Thus the parity bracket is just the graph obtained by putting nodes at each of these crossings. The resulting graph does not reduce under the graphical Reidemeister two moves, and so we conclude that the Kishino knot is non-trivial and non-classical. Since we can apply the parity bracket to a f at knot by taking $A = -1$, we see that this method shows that the Kishino f at is non-tirival.

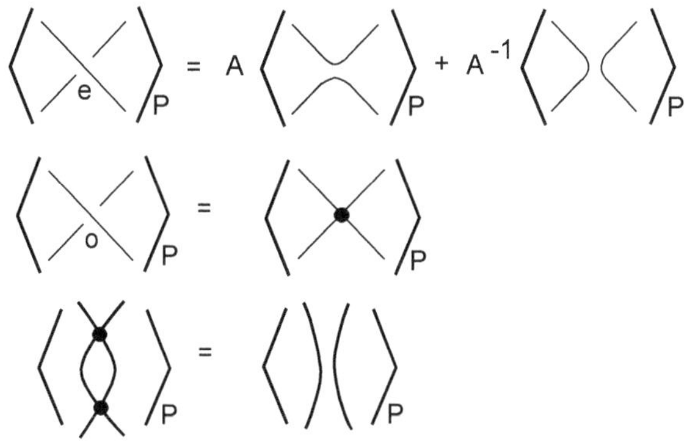

Fig. 10. **Parity Bracket Expansion**

In Figure 11 we illustrate the *Z-move* and the *graphical Z-move*. Two virtual knots or links that are related by a Z-move have the same standard bracket polynomial. This follows directly from our discussion in the previous section. We would like to analyze the structure of Z-moves using the parity bracket. In order to do this we need a version of the parity bracket that is invariant under the Z-move. In

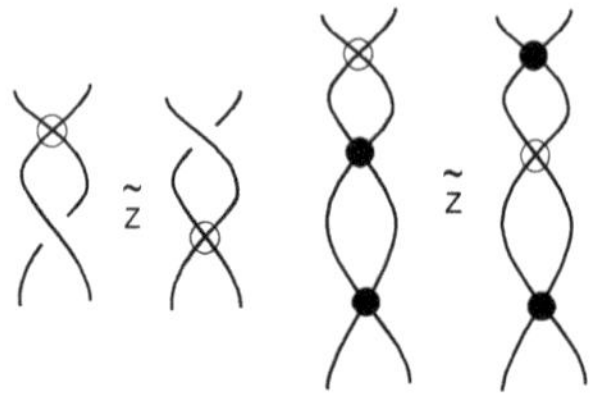

Fig. 11. **Z-Move and Graphical Z-Move**

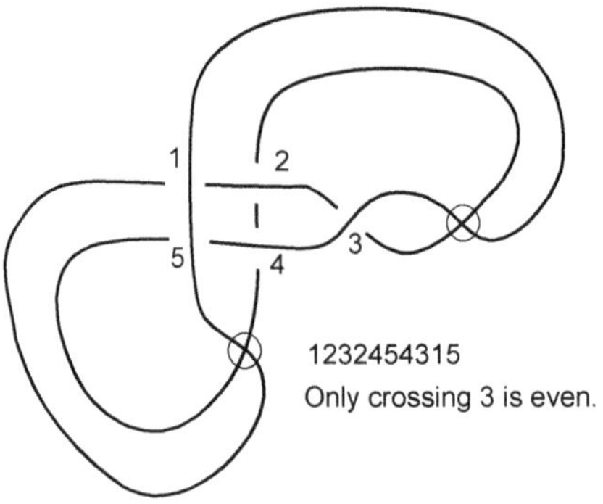

Fig. 12. **A Knot KS With Unit Jones Polynomial**

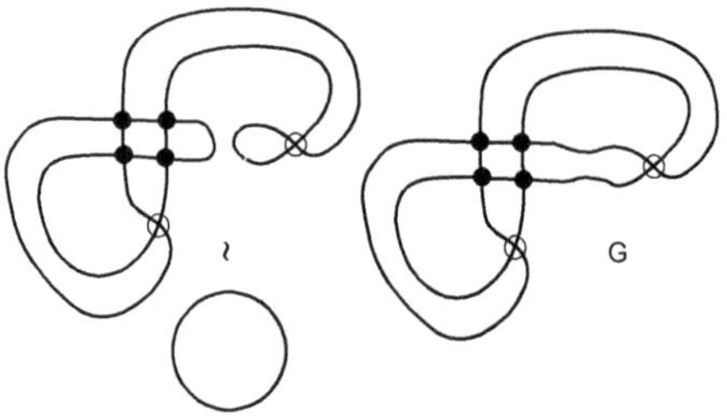

Fig. 13. **Parity Bracket States for the Knot KS**

order to accomplish this, we need to add a corresponding Z-move in the graphical reduction process for the parity bracket. This extra graphical reduction is indicated in Figure 137 where we show a graphical Z-move. The reader will note that graphs that are irreducible without the graphical Z-move can become reducible if we allow graphical Z-moves in the reduction process. For example, the graph associated with the Kishino knot is reducible under graphical Z-moves. However, there are examples of graphs that are not reducible under graphical Z-moves and Reidemister two moves. An example of such a graph occurs in the parity bracket of the knot KS shown in Figure 12. This knot has one even classical crossing and four odd crossings. One smoothing of the even crossing yields a state that reduces to a loop with no graphical nodes, while the other smoothing yields a state that is irreducible even when the Z-move is allowed. The upshot is that this knot KS is not Z-equivalent to any classical knot. Since one can verify that KS has unit Jones polynomial, this example is a counterexample to a conjecture of Fenn, Kauffman and Maturov [24] that suggested that a knot with unit Jones polynomial should be Z-equivalent to a classical knot.

Parity is clearly an important theme in virtual knot theory and will f gure in many future investigations of this subject. The type of construction that we have indicated for the bracket polynomial in this section can be varied and applied to other invariants. Furthermore the notion of describing a parity for crossings in a diagram is also susceptible to generalization. For more on this theme the reader should consult [30, 37] and [27] for our original use of parity for another variant of the bracket polynomial.

8. The Arrow Polynomial for Virtual and Flat Virtual Knots and Links

This section describes an invariant for oriented virtual knots and links, and for f at oriented virtual knots and links that we call the *arrow polynomial* [4, 29]. This invariant is considerably stronger than the Jones polynomial for virtual knots and links, and it is a very natural extension of the Jones polynomial, using the oriented diagram structure of the state summation. The construction of the arrow polynomial invariant begins with the oriented state summation of the bracket polynomial. This means that each local smoothing is either an oriented smoothing or a *disoriented smoothing* as illustrated in Figures 14 and 15. In [4] we show how the arrow polynomial can be used to estimate virtual crossing numbers. In [29] we discuss the arrow polynomial and also a generalization, the *extended bracket polynomial.* The extended bracket polynomial is harder to compute than the arrow polynomial, and we shall not discuss its properties in this introduction. We recommend that the interested reader consult our paper [29] on this subject. We describe a new extension of the arrow polynomial to long virtual knots at the end of the present section.

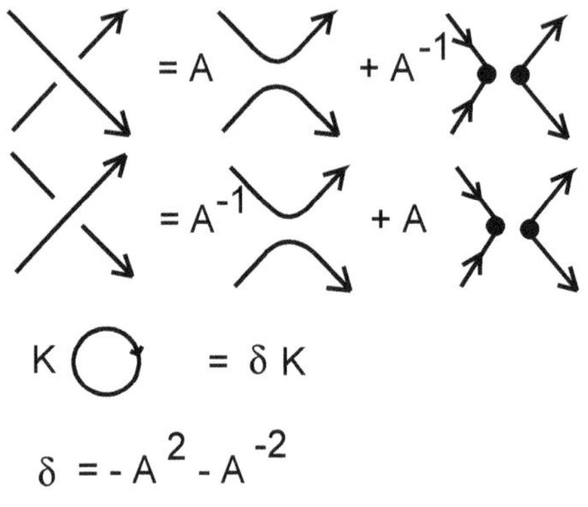

Fig. 14. **Oriented Bracket Expansion**

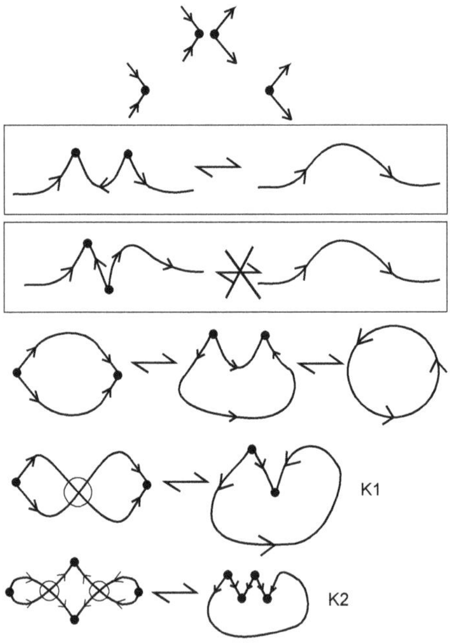

Fig. 15. **Reduction Relation for the Arrow Polynomial.**

In Figure 14 we illustrate the oriented bracket expansion for both positive and negative crossings in a link diagram. An oriented crossing can be smoothed in the oriented fashion or the disoriented fashion as shown in Figure 14. We refer to these smoothings as *oriented* and *disoriented* smoothings. To each smoothing we

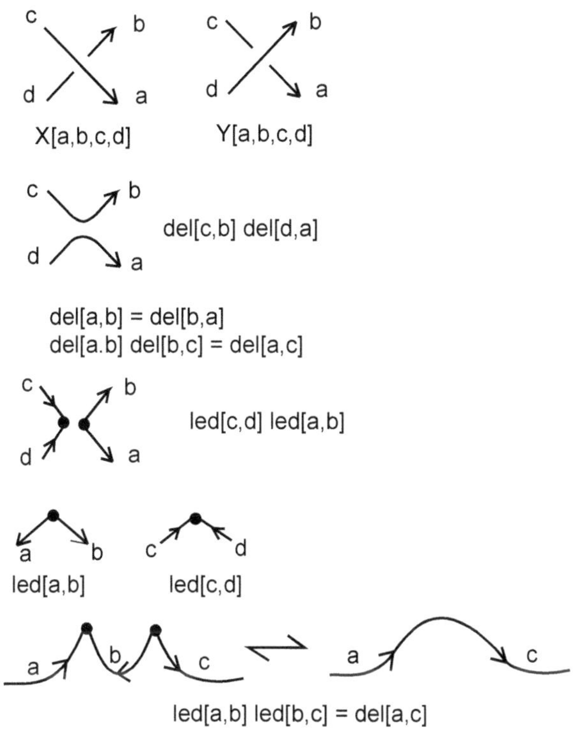

Fig. 16. **Formal Kronecker Deltas.**

make an associated configuration that will be part of the arrow polynomial state summation. The configuration associated to a state with oriented and disoriented smoothings is obtained by applying the reduction rules described below. See Figures 15 and 16. The arrow polynomial state summation is defined by the formula:

$$\langle\langle K \rangle\rangle = \mathcal{A}[K] = \Sigma_S \langle K|S \rangle d^{||S||-1}[S]$$

where S runs over the oriented bracket states of the diagram, $\langle K|S \rangle$ is the usual product of vertex weights as in the standard bracket polynomial, and $[S]$ is a product of extra variables $K_1, K_2, \cdots$ associated with the state S. These variables are explained below. Note that we use the designation $\langle\langle K \rangle\rangle$ to indicate the arrow polynomial in this paper and in the figures in this paper. In [29] we have used this notation to refer to the extended bracket generalization of the arrow polynomial. Since we do not use the extended bracket in this paper, it is convenient to use the double bracket notation here for the arrow polynomial.

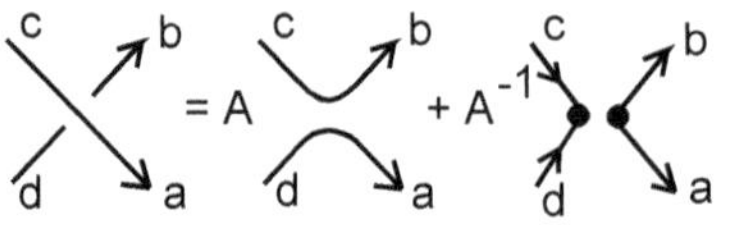

X[a,b,c,d] = A del[c,b] del[d,a] + (1/A) led[a,b] led[c,d]

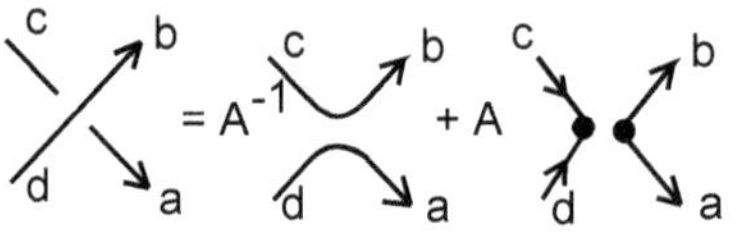

Y[a,b,c,d] = (1/A) del[c,b] del[d,a] + A led[a,b] led[c,d]

Fig. 17. **Simple arrow polynomial Expansion via Formal Kronecker Deltas.**

Due to the oriented state expansion, the loops in the resulting states have extra combinatorial structure in the form of paired *cusps* as shown in Figure 14. Each disoriented smoothing gives rise to a cusp pair where each cusp has either two oriented lines going into the cusp or two oriented lines leaving the cusp. We reduce this structure according to a set of rules that yields invariance of the state summation under the Reideimeister moves. The basic conventions for this simplif cation are shown in Figures 15 and 16. Each cusp is denoted by an angle with arrows either both entering the vertex or both leaving the vertex. Furthermore, the angle locally divides the plane into two parts: One part is the span of an acute angle; the other part is the span of an obtuse angle. We refer to the span of the acute angle as the *inside* of the cusp.

Remark on State Reduction. View Figures 15 and 16. The basic reduction move in these f gures corresponds to the elimination of two consecutive cusps on a single loop. Note that we allow cancellation of consecutive cusps along a loop where the cusps both point to the same local region of the two local regions delineated by the loop. We do not allow cancellation of a "zig-zag" where two consecutive cusps point to opposite (local) sides of the loop. We shall note that in a classical knot or link diagram, all state loops reduce to loops that are free from cusps.

Figure 15 illustates the basic reduction rule for the *arrow polynomial.* The reduction rule allows the cancellation of two adjacent cusps when they have *insides on the same side* of the segment that connects them. When the insides of the cusps are on opposite sides of the connecting segment, then no cancellation is allowed.

Each state circle is seen as a circle graph with extra nodes corresponding to the cusps. All graphs are taken up to virtual equivalence, as explained earlier in this paper. Figure 15 illustrates the simplif cation of two circle graphs. In one case the graph reduces to a circle with no vertices. In the other case there is no further cancellation, but the graph is equivalent to one without a virtual crossing. The state expansion for $\mathcal{A}[K] =<< K >>$ is exactly as shown in Figure 14, but we use the reduction rule of Figure 15 so that each state is a disjoint union of reduced circle graphs. Since such graphs are planar, each is equivalent to an embedded graph (no virtual crossings) via the detour move, and the reduced forms of such graphs have $2n$ vertices that alternate in type around the circle so that n are pointing inward and n are pointing outward. The circle with no vertices is evaluated as $d = -A^2 - A^{-2}$ as is usual for these expansions, and the circle is removed from the graphical expansion. We let K_n denote the circle graph with $2n$ alternating vertex types as shown in Figure 15 for $n = 1$ and $n = 2$. Each circle graph contributes $d = -A^2 - A^{-2}$ to the state sum and the graphs K_n for $n \geq 1$ remain in the graphical expansion. Each K_n is an extra variable in the polynomial. Thus a product of the K_n's corresponds to a state that is a disjoint union of copies of these circle graphs. By evaluating each circle graph as $d = -A^2 - A^{-2}$ (as well as taking its arrow variable K_n) we guarantee that the resulting polynomial will reduce to the original bracket polynomial when each of the new variables K_n is set equal to unity. Note that we continue to use the caveat that an isolated circle or circle graph (i.e. a state consisting in a single circle or single circle graph) is assigned a loop value of unity in the state sum. This assures that $\mathcal{A}[K]$ is normalized so that the unknot receives the value one.

We have the following Proposition, showing that the phenomenon of cusped states and extra variables K_n only occurs for virtual knots and links.

Proposition 2. In a classical knot or link diagram, all state loops reduce to loops that are free from cusps.

Proof. The result follows from the Jordan Curve Theorem. Each state loop for a classical knot is a Jordan curve on the surface of the 2-sphere, dividing the surface of the sphere into an inside and an outside. If a state loop has a non-trivial arrow reduction, then there will be non-empty collections of inward-pointing cusps and a collection of outward-pointing cusps in the reduced loop. Each cusp must be paired with another cusp in the state (since cusps are originally paired). No two inward-pointing cusps on a loop can be paired with one another, since they will have incompatible orientations (similarly for the outward-pointing cusps). Therefore, for a given non-trivial reduced loop λ in a classical diagram there must be another non-trivially reduced loop λ' inside λ and another such loop λ'' outside λ to handle the necessary pairings. This means that the given state of the diagram would have inf nitely many loops. Since we work with knot and link diagrams with f nitely many crossings, this is not possible. Hence there are no non-trivially reduced loops in the states of a classical diagram. This completes the proof. //

We now have the following state summation for the arrow polynomial

$$\mathcal{A}[K] = \Sigma_S < K|S > d^{||S||-1}\mathcal{V}[S]$$

where S runs over the oriented bracket states of the diagram, $< K|S >$ is the usual product of vertex weights as in the standard bracket polynomial, $||S||$ is the number of circle graphs in the state S, and $\mathcal{V}[S]$ is a product of the variables K_n associated with the non-trivial circle graphs in the state S. Note that each circle graph (trivial or not) contributes to the power of d in the state summation, but only non-trivial circle graphs contribute to $\mathcal{V}[S]$.

Theorem 6. With the above conventions, the arrow polynomial $\mathcal{A}[K]$ is a polynomial in A, A^{-1} and the graphical variables K_n (of which f nitely many will appear for any given virtual knot or link). $\mathcal{A}[K]$ is a regular isotopy invariant of virtual knots and links. The normalized version

$$\mathcal{W}[K] = (-A^3)^{-wr(K)}\mathcal{A}[K]$$

is an invariant virtual isotopy. If we set $A = 1$ and $d = -A^2 - A^{-2} = -2$, then the resulting specialization

$$\mathcal{F}[K] = \mathcal{A}[K](A = 1)$$

is an invariant of f at virtual knots and links.

Proof Sketch. We have already discussed the f rst Reidemeister move. Invariance under virtual detours is implicit in the def nition of the state sum. The arrow polynomial has framing behaviour under the f rst Reidemeister moves as shown in Figure 18. In Figures 19 and 20 we show how the removal of a disoriented loop gives rise to invariance under the directly oriented second Reidemeister move. In the state expansion, the disoriented loop is removed but multiplies its term by $d = -A^2 - A^{-2}$. The other two disoriented local conf gurations receive vertex weights of A^2 and A^{-2}, and *each of these conf gurations has the same set of reduced states*. Thus these three conf gurations cancel each other from the state sum. This leaves the remaining local state with parallel arcs, and gives invariance under the directly oriented second Reidemeister move. Invariance under the reverse oriented second Reidemeister move is shown in Figure 20. Invariance under the third Reidemeister move follows from an analysis of Figure 21.

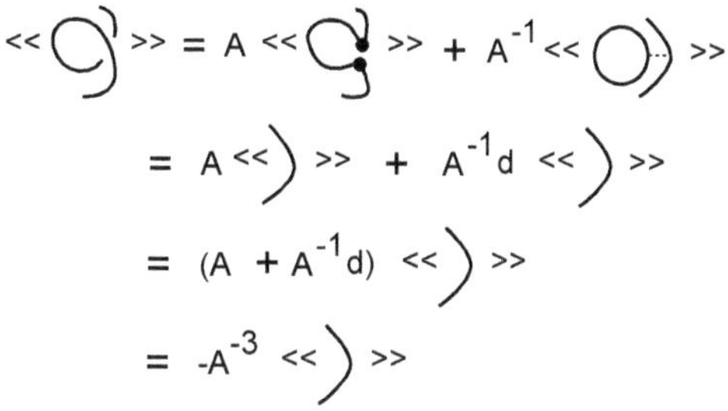

Fig. 18. **The Type One Move**

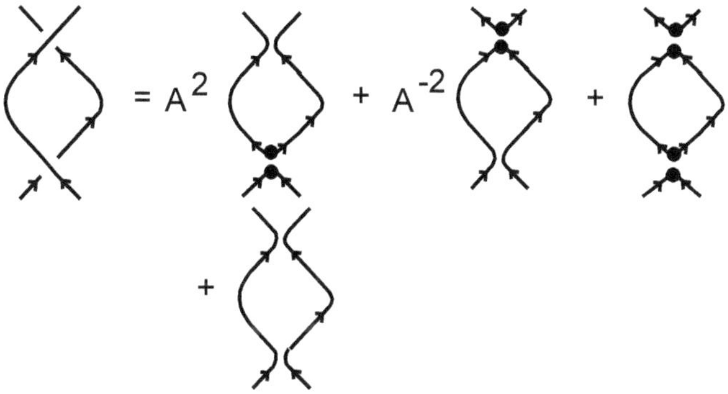

Fig. 19. **Oriented Second Reidemeister Move**

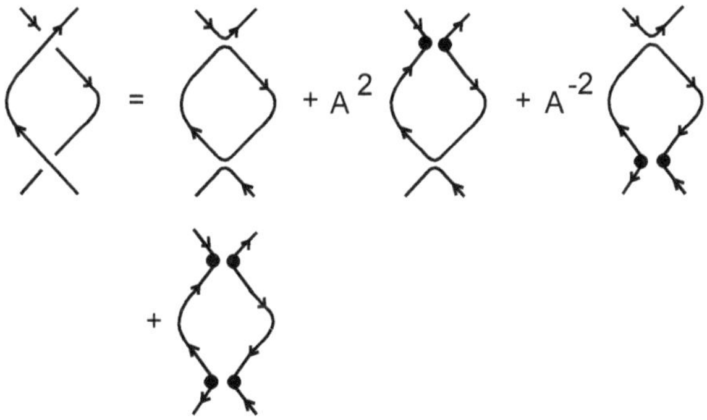

Fig. 20. **Reverse Oriented Second Reidemeister Move**

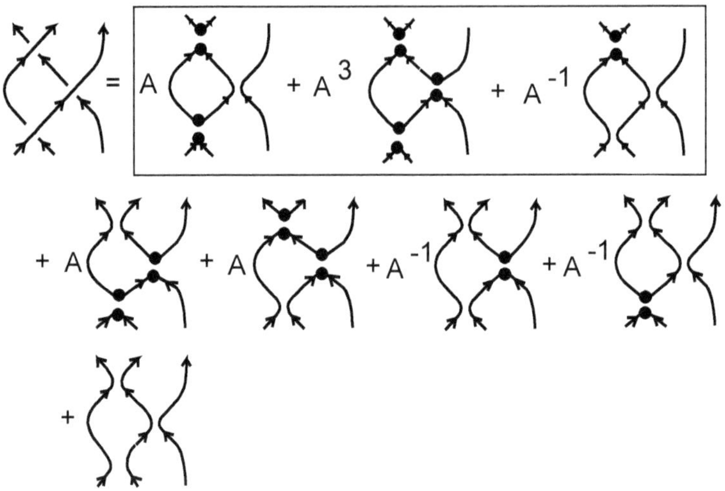

Fig. 21. **Third Reidemeister Move**

Here is a f rst example of a calculation of the arrow polynomial invariant. View Figure 22. The virtual knot K in this f gure has two crossings. One can see that this knot is a non-trival virtual knot by simply calculating the odd writhe $J(K)$ (def ned in section 5). We have that $J(K) = 2$, proving that K is non-trivial and non-classical. This is the simplest virtual knot, the analog of the trefoil knot for virtual knot theory. The arrow polynomial polynomial gives an independent verif cation that K is non-trivial and non-classical.

In the next example, shown in Figure 23, we have a long virtual diagram L with two crossings. The calculation of the arrow polynomial for L is given in Figure 23 and shows that it is a non-trivial and non-classical. In fact, this same formalism proves that $Flat(L)$ is a non-trivial f at link (as we have also by using the odd writhe.). *Note that the arrow polynomial is an invariant of f at diagrams when we take* $A = 1$. In the Figure 23 we have illustrated for this example how non-cancelling cusps can occur on a long state in the expasion of the invariant. This means that, effectively, there are more variables for the arrow polynomial as an invariant of long knots, and this gives it extra powers of discrimination for in the long category. Note the closure of L of Figure 23 is a trivial knot. Thus we see that the arrow polynomial can sometimes discriminate between a long knot and its closure. Another example of this type is shown in Figure 24. In both of these cases the cusps on the long state are in an order that would be differerent if we were to turn the long knot by 180 degrees. Thus the invariant distinguishes between the long knot and its reverse.

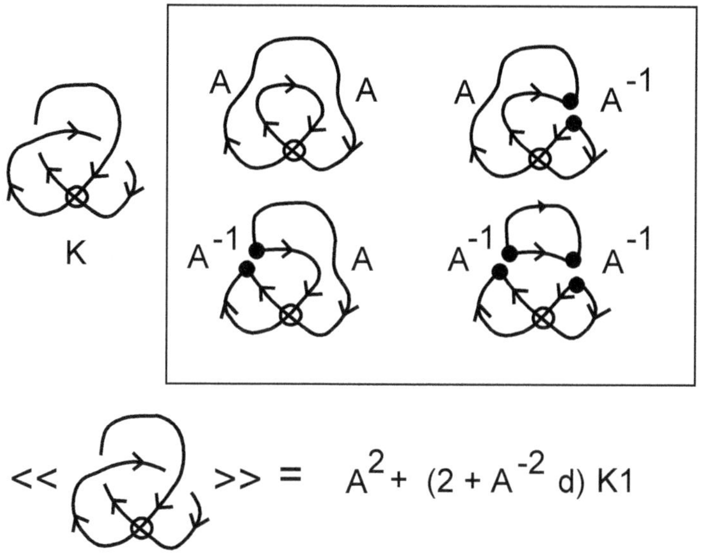

Fig. 22. **Example1**

The next example is given in Figure 25. Here we calculate the arrow polynomial for a non-trivial virtual knot with unit Jones polynomial. Specialization of the calculation to $A = 1$ shows that the corresponding f at knot is non-trivial as well.

Figures 26 exhibits the calculation of the arrow polynomial for the Kishino diagram, showing once again that the Kishino knot is non-trivial and that its underlying f at diagram is also non-trivial.

The example shown in Figures 27 and Figure 28 shows the result of expanding a virtualized classical crossing using the arrow polynomial state sum. Virtualization of a crossing was described in Section 6. In a virtualized crossing, one sees a classical crossing that is f anked by two virtual crossings. In Section 6 we showed that the standard bracket state sum does not see virtualization in the sense that it has the same value as the result of smoothing both f anking virtual crossings that have been added to the diagram. The result is that the value of of the bracket polynomial of the knot with a virtualized classical crossing is the same as the value of the bracket polynomial of the original knot after the same crossing has been *switched* (exchanging over and undercrossing segments).

As one can see from the formula in Figure 27, this smoothing property of the bracket polynomial will not generally be the case for the arrow polynomial

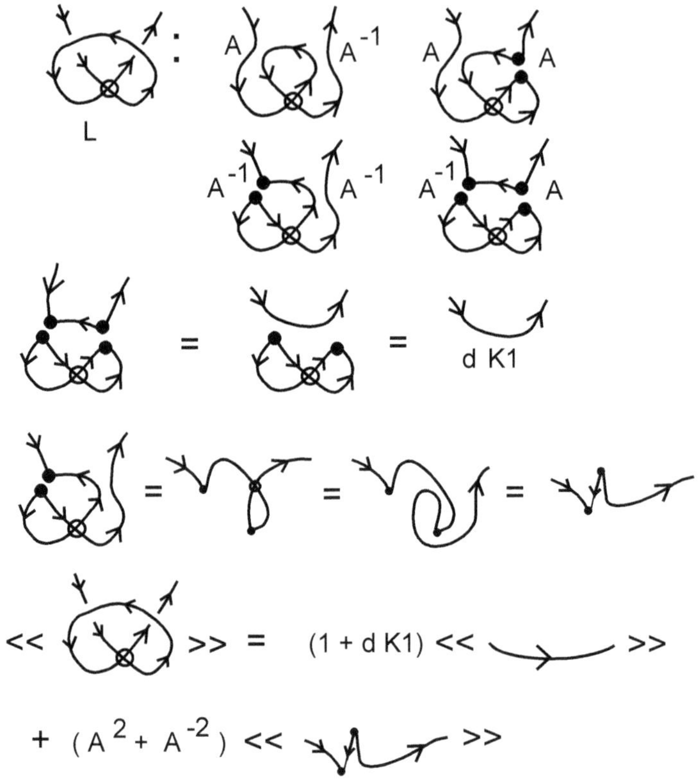

Fig. 23. **Example2**

state sum. In Figure 28 we show that this difference is indeed the case for an inf nite collection of examples. In that f gure we use a tangle T that is assumed to be a classical tangle. arrow polynomial expansison of this tangle is necessarily of the form shown in that f gure: a linear combination of an oriented smoothing and a reverse oriented smoothing with respective coeff cients $a(T)$ and $b(T)$ in the Laurent polynomial ring $Q[A, A^{-1}]$. We leave the verif cation of this fact to the reader. In Figure 28 we show a generic diagram that is obtained by a *single* virtualization from a classical diagram, and we illustrate the calculation of its arrow polynomial invariant. As the reader can see from this Figure, there is a non-trivial graphical term whenever $b(T)$ is non-zero. Thus we conclude that the single virtualization of any classical link diagram (in the form shown in this f gure) will be non-trivial and non-classical whenever $b(T)$ in non-zero. This is an inf nite class of examples, and the result can be used to recover the results about single

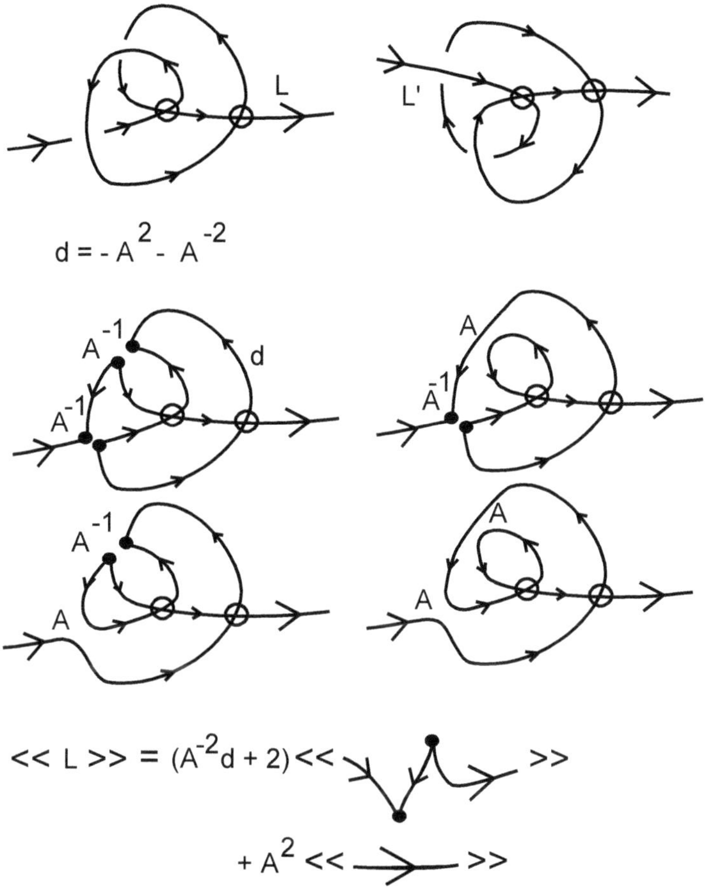

Fig. 24. **Example3**

virtualization that we obtained in a previous paper with Heather Dye [15] using the surface bracket polynomial.

For more information about the arrow polynomial, we refer the reader to our paper [4] where we prove that the maximal monomial degree in the K_n variables is a lower bound for the virtual crossing number of the virtual knot or link. There are many open problems associated with this estimate for the virtual crossing number. Also the reader of [4] will encounter examples of virtual knots and links that are undetectable by the arrow polynomial.

$$(A^3 + Ad + 2A^{-1} + A^{-3}d) \;+\; (Ad + A^{-1}d^2 + Ad)\,K1^2$$

Fig. 25. **Arrow Polynomial for the Virtualized Trefoil**

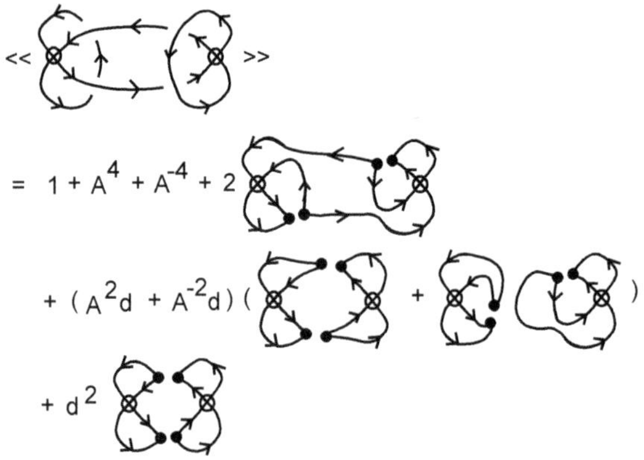

$$= \; 1 + A^4 + A^{-4} + 2\,K2 \;-\; (A^4 + A^{-4} + 2)\,K1^2$$

Fig. 26. **Arrow Polynomial for the Kishino Diagram**

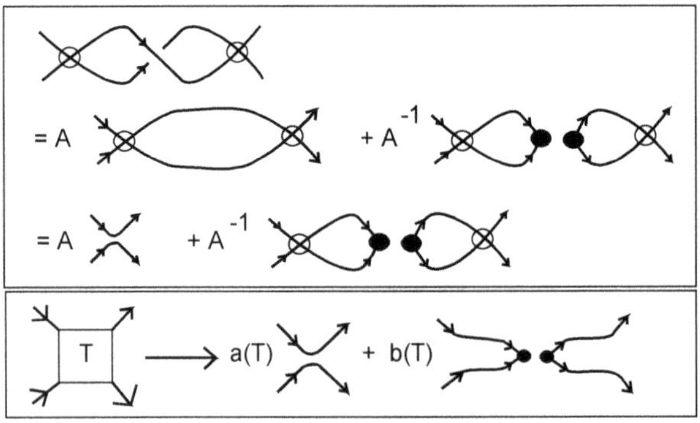

Fig. 27. **The 1-Virtualization of a Classical Diagram (A)**

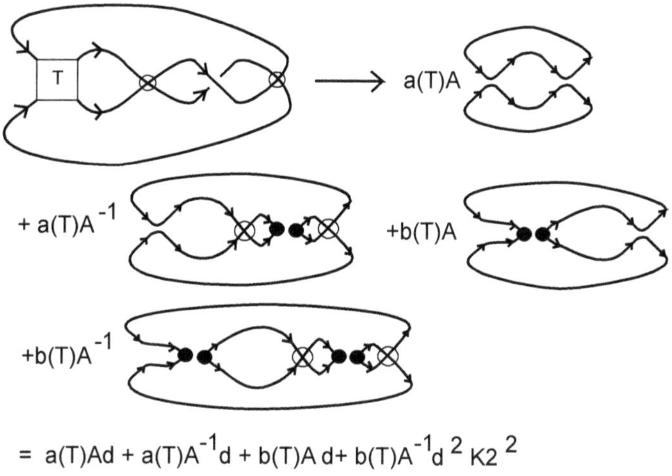

Fig. 28. **The 1-Virtualization of a Classical Diagram (B)**

8.1. Generalizing Parity to the Arrow Polynomial

The purpose of this section is to indicate how our construction of the Parity Bracket in Section 7 generalizes to the arrow polynomial. The reader will f nd it easy to verify that at the site of a Reidemeister Two move either both crossings are odd, or both are even. At the site of a Reidemeister Three move either all three crossings are even, or two of the three are odd. It follows from this that if we decide (as in the Parity Bracket) to make graphical vertices from all the odd crossings and expand the oriented bracket (for the arrow polynomial) then we must factor the graphs that result from the state expansion on the even crossings of the diagram by an equivalence relation generated by patterns of graph reduction corresponding to the second Reidemeister moves on graphical vertices. These reductions come in a number of possible orientations, and it is possible to have a single cusp involved as illustrated in Figure 29. In this f gure we do not indicate the orientations. Because the cusps in the graphs are involved with the graphical reductions one takes the resulting graphs up to virtual graphical equivalence, these reductions and the cusp cancellation is restricted to single edges of the graphs. The result is a parity version of the arrow polynomial. We will not calculate examples of this invariant in this survey. The Parity Arrow Polynomial will be taken up in a subsequent paper.

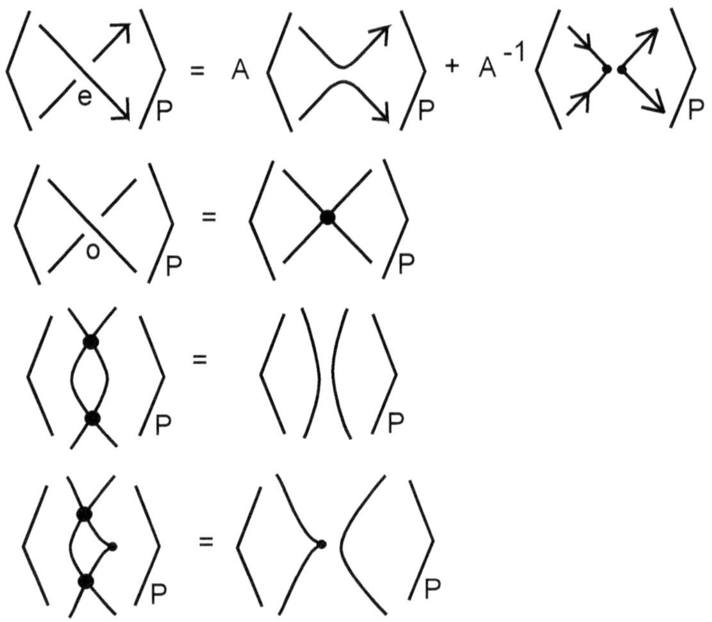

Fig. 29. **Relations for the Parity Arrow Polynomial**

9. Khovanov Homology

In this section, we describe Khovanov homology along the lines of [1, 31], and we tell the story so that the gradings and the structure of the differential emerge in a natural way. This approach to motivating the Khovanov homology uses elements of Khovanov's original approach, Viro's use of enhanced states for the bracket polynomial [39], and Bar-Natan's emphasis on tangle cobordisms [2]. We use similar considerations in our paper [13]. The purpose of this section is to provide background for the next section where we describe a categorifcation of the arrow polynomial and give examples of detection that can be accomplished by this homological invariant.

Two key motivating ideas are involved in f nding the Khovanov invariant. First of all, one would like to *categorify* a link polynomial such as $\langle K \rangle$. There are many meanings to the term categorify, but here the quest is to f nd a way to express the link polynomial as a *graded Euler characteristic* $\langle K \rangle = \chi_q \langle H(K) \rangle$ for some homology theory associated with $\langle K \rangle$.

The bracket polynomial [17, 18] model for the Jones polynomial [9–11, 51] is usually described by the expansion

$$\langle \times \rangle = A\langle \asymp \rangle + A^{-1}\langle)\,(\rangle \tag{1}$$

and we have

$$\langle K \bigcirc \rangle = (-A^{2} - A^{-2})\langle K \rangle \tag{2}$$

$$\langle \gamma \rangle = (-A^{3})\langle \smile \rangle \tag{3}$$

$$\langle \gamma \rangle = (-A^{-3})\langle \smile \rangle \tag{4}$$

Letting $c(K)$ denote the number of crossings in the diagram K, if we replace $\langle K \rangle$ by $A^{-c(K)}\langle K \rangle$, and then replace A by $-q^{-1}$, the bracket will be rewritten in the following form:

$$\langle \times \rangle = \langle \asymp \rangle - q\langle)\,(\rangle \tag{5}$$

with $\langle \bigcirc \rangle = (q + q^{-1})$. It is useful to use this form of the bracket state sum for the sake of the grading in the Khovanov homology (to be described below). We shall continue to refer to the smoothings labeled q (or A^{-1} in the original bracket formulation) as *B-smoothings*. We should further note that we use the well-known convention of *enhanced states* where an enhanced state has a label of 1 or X on each of its component loops. We then regard the value of the loop $q + q^{-1}$ as the sum of the value of a circle labeled with a 1 (the value is q) added to the value of a circle labeled with an X (the value is q^{-1}). We could have chosen the more neutral labels of $+1$ and -1 so that

$$q^{+1} \Longleftrightarrow +1 \Longleftrightarrow 1$$

and

$$q^{-1} \Longleftrightarrow -1 \Longleftrightarrow X,$$

but, since an algebra involving 1 and X naturally appears later, we take this form of labeling from the beginning.

To see how the Khovanov grading arises, consider the form of the expansion of this version of the bracket polynonmial in enhanced states. We have the formula as a sum over enhanced states s :

$$\langle K \rangle = \sum_{s} (-1)^{n_B(s)} q^{j(s)}$$

where $n_B(s)$ is the number of B-type smoothings in s, $\lambda(s)$ is the number of loops in s labeled 1 minus the number of loops labeled X, and $j(s) = n_B(s) + \lambda(s)$. This can be rewritten in the following form:

$$\langle K \rangle = \sum_{i,j} (-1)^i q^j dim(\mathcal{C}^{ij})$$

where we def ne $\mathcal{C}^{ij}$ to be the linear span (over $k = Z/2Z$ as we will work with mod 2 coeff cients) of the set of enhanced states with $n_B(s) = i$ and $j(s) = j$. Then the number of such states is the dimension $dim(\mathcal{C}^{ij})$.

We would like to have a bigraded complex composed of the $\mathcal{C}^{ij}$ with a differential

$$\partial : \mathcal{C}^{ij} \longrightarrow \mathcal{C}^{i+1\,j}.$$

The differential should increase the *homological grading* i by 1 and preserve the *quantum grading* j. Then we could write

$$\langle K \rangle = \sum_j q^j \sum_i (-1)^i dim(\mathcal{C}^{ij}) = \sum_j q^j \chi(\mathcal{C}^{\bullet j}),$$

where $\chi(\mathcal{C}^{\bullet j})$ is the Euler characteristic of the subcomplex $\mathcal{C}^{\bullet j}$ for a f xed value of j.

This formula would constitute a categorif cation of the bracket polynomial. Below, we shall see how *the original Khovanov differential ∂ is uniquely determined by the restriction that $j(\partial s) = j(s)$ for each enhanced state s.* Since j is preserved by the differential, these subcomplexes $\mathcal{C}^{\bullet j}$ have their own Euler characteristics and homology. We have

$$\chi(H(\mathcal{C}^{\bullet j})) = \chi(\mathcal{C}^{\bullet j})$$

where $H(\mathcal{C}^{\bullet j})$ denotes the homology of the complex $\mathcal{C}^{\bullet j}$. We can write

$$\langle K \rangle = \sum_j q^j \chi(H(\mathcal{C}^{\bullet j})).$$

The last formula expresses the bracket polynomial as a *graded Euler characteristic* of a homology theory associated with the enhanced states of the bracket state summation. This is the categorif cation of the bracket polynomial. Khovanov proves that this homology theory is an invariant of knots and links (via the Reidemeister moves) , creating a new and stronger invariant than the original Jones polynomial.

The differential is based on regarding two states as *adjacent* if one differs from the other by a single smoothing at some site. Thus if (s, τ) denotes a pair consisting in an enhanced state s and site τ of that state with τ of type A, then we consider all enhanced states s' obtained from s by smoothing at τ and relabeling only those loops that are affected by the resmoothing. Call this set of enhanced states $S'[s, \tau]$. Then we shall define the *partial differential* $\partial_\tau(s)$ as a sum over certain elements in $S'[s, \tau]$, and the differential by the formula

$$\partial(s) = \sum_\tau \partial_\tau(s)$$

with the sum over all type A sites τ in s. It then remains to see what are the possibilities for $\partial_\tau(s)$ so that $j(s)$ is preserved.

Proposition. The partial differentials $\partial_\tau(s)$ are uniquely determined by the condition that $j(s') = j(s)$ for all s' involved in the action of the partial differential on the enhanced state s. This unique form of the partial differential can be described by the following structures of multiplication and comultiplication on the algebra A = $k[X]/(X^2)$ where $k = Z/2Z$ for mod-2 coefficients, or $k = Z$ for integral coefficients.

(1) The element 1 is a multiplicative unit and $X^2 = 0$.
(2) $\Delta(1) = 1 \otimes X + X \otimes 1$ and $\Delta(X) = X \otimes X$.

These rules describe the local relabeling process for loops in a state. Multiplication corresponds to the case where two loops merge to a single loop, while comultiplication corresponds to the case where one loop bifurcates into two loops. (The proof is omitted.)

There is more to say about the nature of this construction with respect to Frobenius algebras and tangle cobordisms. The partial boundaries can be conceptualized in terms of surface cobordisms. The equality of mixed partials corresponds to topological equivalence of the corresponding surface cobordisms, and to the relationships between Frobenius algebras and the surface cobordism category. The proof of invariance of Khovanov homology with respect to the Reidemeister moves (respecting grading changes) will not be given here. See [1, 2, 31]. It is remarkable that this version of Khovanov homology is uniquely specified by natural ideas about adjacency of states in the bracket polynomial.

10. Categorifying the Arrow Polynomial $\langle\langle K \rangle\rangle = \mathcal{A}[K]$

It is not hard to see that Khovanov homology modulo two extends to a homology theory for virtual links. In [13] we extend this to a new homology theory that categorif es the arrow polynomial.

In order to consider gradings for Khovanov homology in relation to the arrow polynomial $\mathcal{A}[K]$ we have to examine how the arrow number of state loops change under a replacement of an A-smoothing by a B-smoothing. Such replacement, when we use oriented diagrams involves the replacement of a cusp pair by an oriented smoothing or vice versa. Furthermore, we may be combining or splitting two loops. Refer to Figure 30 for a depiction of the different cases. This f gure shows the three basic cases. These are all the ways that loops can interact and change their respective arrow numbers. We apply these results to the grading in Khovanov homology.

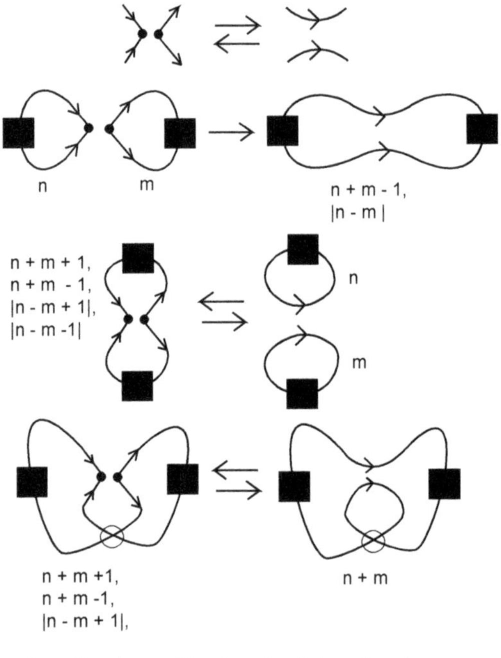

Fig. 30. **Arrow Numbers for Interacting Loops**

10.1. Dotted gradings and the dotted categorif cation

We call an enhanced state *dotted* if it has an odd arrow number, and otherwise it is undotted. The number of dotted X's and the number of dotted 1's in the state correspond to the number of enhanced state loops (labeled with X or with 1) that are dotted. We def ne a grading

$$g(s) = \sharp(\dot{X}) - \sharp(\dot{1})$$

Theorem. Take $[[K]]_g$ to be the space of enhanced states for K endowed with new grading as above. Def ne ∂' to be the composition of ∂ with the new grading projection and set $\partial'' = \partial - \partial'$. Then the homology of $[[K]]_g$ (with respect to ∂') is invariant (up to a degree shift and a height shift).

With this result we have a version of Khovanov homology that uses the arrow information. In [13] we have investigated this categorif cation and other related ones for the arrow polynomial. So far this categorifcation is effective in that *we [12] have many pairs of virtual knots that are distinguished by the arrow categorif cation, but not distinguished by either Khovanov homology or the arrow polynomial.* An example of such a pair (due to Aaron Kaestner) is shown in Figure 31 and 32. In Figure 31 we illustrate the knot $VK5[129]$ and in Figure 32 we illustrate the knot $VK5[267]$ from Jeremy Green's tables (See the website of Dror Bar Natan for these tables.). The examples shown in these f gures do not seem to be related by any form of mutation and the difference betweem them is not dectected by the arrow polynomial, but it is detected by the categorif cation of the arrow polynomial.

Note however, that one can compute the Parity Bracket (See Section 7) for these examples. In the case of $VK5[129]$ there are f ve crossings and four odd crossings. It is easy to see that the Parity Bracket expansion contains an irreducible graph with four nodes. On the other hand, the diagram for $VK5[267]$ has two odd crossings and hence its Parity Bracket (which one can compute to be non-trivial with an irreducible graphs with two nodes) cannot be equal up to normalization with the Parity Bracket for $VK5[129]$. Thus this pair of knots is distinguished by using parity. This example shows the power of the parity method.

This approach through examples and combining categorif cation and parity is very fruitful for producing examples and for sharpening concepts. We are programming an integral version of Khovanov homology for virtual knots that is structured along the lines proposed by Vassily Manturov [35], and we expect all this work to ref ect back on better understanding of Khovanov homology for classical knots. This is the f rst instance of non-trivial use of categorif ed homology in virtual knot theory.

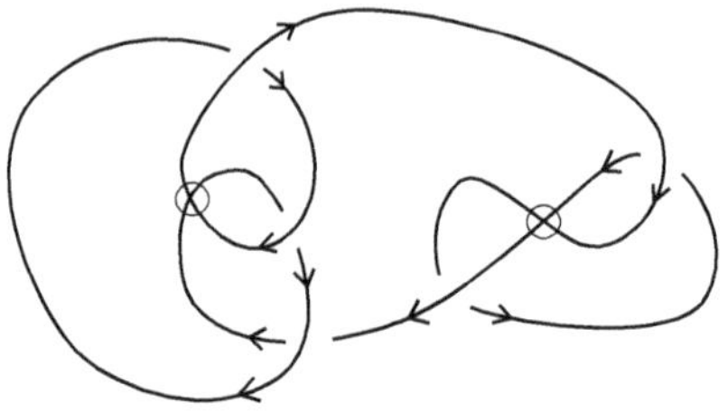

Fig. 31. **VK5[129]**

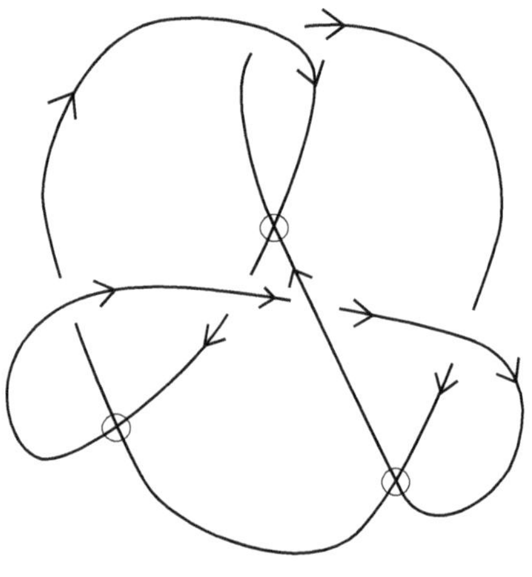

Fig. 32. **VK5[267]**

11. Discussion

In this paper we have given an introduction to the subject of virtual knot theory and we have concentrated on the easy extension of the Jones polynomial to virtual knots and links and a more subtle generalization that we have called the *arrow polynomial*. The arrow polynomial uses the oriented structure of link diagrams in a way that is particularly suited to virtual knot theory and obtains an invariiant of virtual knots and links that has inf nitely many variables. The ideas behind the arrow polynomial have been used to create an extended bracket polynomial [29] that uses more of the oriented structure and has graphs representing its extra variables. The arrow polynomial itself can be seen to be a reformulation or generalization of some of the constructions of Miyazawa and Kamada [40–44]. After discussing the arrow polynomial, we gave a quick summary of Khovanov homology and a brief description of our categorif cation of the arrow polynomial [4, 13] and an indication of the fact that this categorif cation gives a yet stronger invariant and generates even more new problems.

We also designed this paper to present examples of the use of parity in virtual knot theory. We began with a discussion of the odd writhe in Section 5. In the Gauss code of the diagram each crossing appears twice with a sequence of symbols in between the two appearances. When this sequence is odd we say that the given crossing is odd. When this sequence is even, we say that the given crossing is even. Many possibilites for new invariants are inherent in this parity distinction between even and odd crossings. We have given one example of this kind of application in Section 7 where we discuss the Parity Bracket of Vassily Manturov and apply it to solving a conjecture that we had made about virtual knot diagrams with unit Jones polynomial. A second example is given at the end of the paper to distinguish the knots in Figures 31 and 32. There is much potential for the use of parity in virtual knot theory.

References

1. D. Bar-Natan, On Khovanov's categorif cation of the Jones polynomial. *Algebraic and Geometric Topology* **2** (2002), 337-370.
2. D. Bar-Natan, Khovanov's homology for tangles and cobordisms, *Geom. Topol.* **9** (2005), 1443-1499.
3. J. S. Carter, S. Kamada and M. Saito, Stable equivalences of knots on surfaces and virtual knot cobordisms. Knots 2000 Korea, Vol. 1, (Yongpyong), *JKTR* **11** (2002) No. 3, 311-322.
4. H. A. Dye and L. H. Kauffman, Virtual crossing number and the arrow polynomial, *JKTR*, Vol. 18, No. 10 (October 2009), 1335-1357. arXiv:0810.3858.
5. S. Eliahou, L. Kauffman and M. Thistlethwaite, Inf nite families of links with trivial Jones polynomial, *Topology* **42**, 155-169.
6. R. Fenn and V. Turaev. Weyl Algebras and Knots, *Journal of Geometry and Physics*.57 (2007) 1313-1324, arXiv:math/0610481.
7. A. Bartholomew, R. Fenn, N. Kamada, S. Kamada, New invariants of long virtual knots, arXiv:0705.4639.

8. Mikhail Goussarov, Michael Polyak and Oleg Viro, Finite type invariants of classical and virtual knots, math.GT/9810073.
9. V. F. R. Jones, A polynomial invariant of links via Von Neumann algebras, *Bull. Amer. Math. Soc.*, 1985, No. 129, 103-112.
10. V.F.R.Jones, Hecke algebra representations of braid groups and link polynomials. Ann. of Math. 126 (1987), pp. 335-338.
11. V.F.R.Jones, On knot invariants related to some statistical mechanics models. Pacif c J. Math., vol. 137, no. 2 (1989), pp. 311-334.
12. Aaaron Kaestner and Louis H. Kauffman, Arrow Polynomial Categorif cation (to appear).
13. Heather Ann Dye, Louis Hirsch Kauffman, Vassily Olegovich Manturov, On two categorif cations of the arrow polynomial for virtual knots. (to appear in JKTR) arXiv:0906.3408
14. H. Dye and L. H. Kauffman, Virtual knot diagrams and the Witten-Reshetikhin-Turaev invariant. math.GT/0407407, *JKTR* **14** (2005), no. 8, 1045–1075.
15. H. Dye and L. H. Kauffman, Minimal surface representations of virtual knots and links. *Algebr. Geom. Topol.* **5** (2005), 509–535. math. GT/0401035.
16. L. H. Kauffman, R. Fenn and M. Jordan-Santana, Biquandles and virtual links, *Topology and its Applications* **145** (2004), 157-175.
17. L.H. Kauffman, State Models and the Jones Polynomial, *Topology* **26** (1987), 395–407.
18. L. H. Kauffman, "Knots and Physics", World Scientif c, Singapore/New Jersey/London/Hong Kong, 1991, 1994, 2001.
19. L. H. Kauffman and S. Lambropoulou, Virtual braids. Fund. Math. 184 (2004), 159–186.
20. L. H. Kauffman and S. Lambropoulou, Virtual braids and the L-move, *JKTR* **15**, No. 6, August 2006, 773-810.
21. L. H. Kauffman, Virtual Knot Theory , *European J. Comb.* **20** (1999), 663-690.
22. L. H. Kauffman, A Survey of Virtual Knot Theory in *Proceedings of Knots in Hellas '98*, World Sci. Pub. 2000 , 143-202.
23. L. H. Kauffman, Detecting Virtual Knots, in *Atti. Sem. Mat. Fis. Univ. Modena Supplemento al Vol. IL*, 241-282 (2001).
24. L. H. Kauffman, R. Fenn and V. O. Manturov, Virtual Knot Theory – Unsolved Problems, *Fund. Math.* **188** (2005), 293–323, math.GT/0405428.
25. L. H. Kauffman and V. O. Manturov, Virtual knots and links, (Russian) Tr. Mat. Inst. Steklova 252 (2006), Geom. Topol., Diskret. Geom. i Teor. Mnozh., 114–133; translation in Proc. Steklov Inst. Math. 2006, no. 1 (252), 104–121, math.GT/0502014.
26. L. H. Kauffman and Vassily Manturov, Virtual Biquandles, *Fundamenta Mathematicae* **188** (2005), 103-146.
27. L. H. Kauffman, A self-linking invariant of virtual knots. *Fund. Math.* **184** (2004), 135–158, math.GT/0405049.
28. L. H. Kauffman, Knot diagrammatics. "Handbook of Knot Theory", edited by Menasco and Thistlethwaite, 233–318, Elsevier B. V., Amsterdam, 2005. math.GN/0410329.
29. L. H. Kauffman, An Extended Bracket Polynomial for Virtual Knots and Links. JKTR, Vol. 18, No. 10 (October 2009). Page: 1369 - 1422, arXiv:0712.2546.

30. L. H. Kauffman and V. O. Manturov, Parity in Virtual Knot Theory, (in preparation).
31. M. Khovanov, A categorif cation of the Jones polynomial, *Duke Math. J.* **101** (2000), no. 3, 359–426.
32. T. Kishino, 6 kouten ika no kasou musubime no bunrui ni tsiuti (On classif catio of virtual links whose crossing number is less than or equal to 6), Master Thesis, Osaka City University, 2000.
33. G. Kuperberg, What is a virtual link? *Algebr. Geom. Topol.* **3** (2003), 587–591 (electronic). arXiv:math.GT/0208039 v1 5 Aug 2002.
34. W. B. R. Lickorish, "An Introduction to Knot Theory", Springer-Verlag (1997).
35. V. O. Manturov, Khovanov homology for virtual links with arbitrary coeff cients, math.GT/0601152.
36. V. O. Manturov, Parity in knot theory. *Sbornik: Mathematics*, (2010),201(5), 65-110.
37. V. O. Manturov, Parity and Cobordisms of Free Knots (to appear in Sbornik), arXiv:1001.2827
38. M. Thistlethwaite, Links with trivial Jones polynomial, *JKTR* **10**, No. 4 (2001), 641-643.
39. O. Viro, Khovanov homology, its def nitions and ramif cations. Fund. Math. 184 (2004), 317342.
40. Y. Miyazawa, Magnetic graphs and an invariant for virtual links *JKTR* **15** (2006), no. 10, 1319–1334.
41. Y. Miyazawa, A Multivariable polynomial invariant for unoriented virtual knots and links, *JKTR* **17**, No. 11, Nov. 2008, 1311-1326.
42. N. Kamada, An index of an enhanced state of a virtual link diagram and Miyazawa polynomials. *Hiroshima Math. J.* **37** (2007), no. 3, 409–429.
43. N. Kamada, Miyazawa polynomials of virtual knots and virtual crossing numbers, in "Intelligence of Low Dimensional Topology" , Knots and Everything Series – Vol. 40, edited by S. Carter, S. Kamada, L. H. Kauffman, A. Kawauchi and T. Khono, World Sci. Pub. Co. (2006), 93-100.
44. N. Kamada and Y. Miyazawa, A two-variable invariant for a virtual link derived from magnetic graphs. *Hiroshima Math. J.* **35** (2005), 309-326.
45. N. Kamada and S. Kamada, Abstract link diagrams and virtual knots. *JKTR* **9** (2000), No. 1, 93-106.
46. G. Sawollek, On Alexander-Conway polynomials for virtual knots and links. math.GT/9912173 v2, January 2001.
47. D.S. Silver and S.G. Williams, An invariant for open virtual strings, *JKTR* **15**, No. 2 (2006), 143 - 152.
48. L. H. Kauffman and S. Jablan (in preparation).
49. V. Turaev, Virtual strings. *Ann. Inst. Fourier (Grenoble)* **54** (2004), no. 7, 2455–2525 (2005).
50. S. Winker. PhD. Thesis, University of Illinois at Chicago (1984).
51. E. Witten. Quantum Field Theory and the Jones Polynomial. *Comm. in Math. Phys.* **121** (1989), 351-399.

Name	Country	E-mail
Directors		
JABLAN Slavik	Republic of Serbia	sjablan@gmail.com
KAUFFMAN Louis Hirsch	United States of America	kauffman@uic.edu
LAMBROPOULOU Sofia	Greece	sofia@math.ntua.gr
PRZYTYCKI Jozef Henryk	United States of America	przytyck@gwu.edu
Local Organizer		
LI Jiayu	People's Republic of China	jyli@ictp.it, lijia@amss.ac.cn
School Lecturers		
ADAMS Colin Conrad	United States of America	colin.c.adams@williams.edu
BAR-NATAN Dror	United States of America	drorbn@math.toronto.edu
BUCK Dorothy	United States of America	d.buck@imperial.ac.uk
CANTARELLA Jason	United States of America	cantarel@math.uga.edu, jason.cantarella@gmail.com
CHMUTOV Sergei	Russian Federation	chmutov@math.ohio-state.edu
FENN Roger	United Kingdom	rogerf@sussex.ac.uk
FLAPAN Erica	United States of America	eflapan@pomona.edu
GORDON Cameron McAllan	United States of America	gordon@math.utexas.edu
KHOVANOV Mikhail	United States of America	khovanov@cpw.math.columbia.edu
LOMONACO Samuel J.	United States of America	lomonaco@umbc.edu
MANTUROV Vassily Olegovich	Russian Federation	vomanturov@yahoo.es, vomanturov@yandex.ru
MILLETT Kenneth Cary	United States of America	millett@math.ucsb.edu
MORTON Hugh	United Kingdom	morton@liv.ac.uk
OZSVATH Peter	United States of America	petero@math.columbia.edu
RAWDON Eric	United States of America	ejrawdon@stthomas.edu
RICCA Renzo Luigi	Italy	renzo.ricca@unimib.it
Speakers		
BELIAKOVA Anna	Belarus	anna@math.uzh.ch
CARTER J. Scott	United States of America	carter@jaguar1.usouthal.edu
DEHORNOY Patrick	France	dehornoy@math.unicaen.fr
LAWRENCE Naimark Ruth Elke	Israel	ruthel@math.huji.ac.il
LE Thang Tu Quoc	United States of America	letu@math.gatech.edu
THISTLETHWAITE Morwen Bernard	United Kingdom	morwen@math.utk.edu
TURAEV Vladimir	France	vturaev@yahoo.com
Participants		
ABROSIMOV Nikolay	Russian Federation	abrosimov@math.nsc.ru
AFANASYEV Denis Mikhailovich	Russian Federation	
ALI Usman	Pakistan	usman76swat@yahoo.com
ANGELESKA Angela	the Former Yugoslav Republic of Macedonia	aangeles@mail.usf.edu
ARCHIBALD Jana Frances	Canada	jfa@math.utoronto.ca
ASHRAF Rehana	Pakistan	rehana.sms@gmail.com
ASTORINO Marco	Italy	
AYIVOR Audry Fafa	Ghana	audry.ayivor@uct.ac.za
BABIERRA Ariel Lagdameo	Philippines	ariellbabierra@yahoo.com
BANDIERI Paola	Italy	paola.bandieri@unimore.it
BELLINGERI Paolo	Italy	paolo.bellingeri@math.unicaen.fr
BENVEGNU Alberto	Italy	

BERTUCCO Abramo	Italy	abramo.b@gmail.com
BOBTCHEVA Ivelina Julianova	Italy	
BOUZIANI Abdelfatah Amara	Algeria	aefbouziani@yahoo.fr
BUEHLER Irmgard Adelheid	Switzerland	
CANFORA Fabrizio	Italy	canfora@cecs.cl
CATTABRIGA Alessia	Italy	cattabri@dm.unibo.it
CHAMBERS Dwayne Vincent	Jamaica	dwayne.chambers@cgu.edu
CHAN PALOMO Luis Celso	Mexico	celso@matem.unam.mx
CHU Karene Ka Yin	Canada	karenechu@gmail.com
CRISTOFORI Paola	Italy	paola.cristofori@unimore.it
DALVIT Ester	Italy	ester.dalvit@gmail.com, dalvit@science.unitn.it
DANCSO Zsuzsanna	Hungary	zsuzsi@math.toronto.edu
DIAMANTIS Ioannis Athanasios	Greece	
DINAR Yassir Ibrahim	Sudan	yassird@gmail.com
DUZHIN Fedor	Russian Federation	fduzhin@gmail.com, fduzhin@ntu.edu.sg
ERTEM Umit	Turkey	uertem@science.ankara.edu.tr
FELSHTYN Alexander	Germany	felshtyn@diamond.boisestate.edu
GABROVSEK Bostjan	Slovenia	bostjan.gabrovsek@gmail.com
GIRIBET Gaston Enrique	Argentina	gaston@df.uba.ar
GOLLA Marco	Italy	
GOUNDAROULIS Dimoklis	Greece	dgound@gmail.com, dgound@mail.ntua.gr
GUNDYREV Ivan	Russian Federation	
GUTIERREZ GUZMAN Sergio Cesar Alejandro	Mexico	
GUZMAN TRISTAN Araceli	Mexico	guzman@matem.unam.mx
HALACHEVA Iva Hristova	Bulgaria	ihalacheva@gmail.com
HUERFANO Stella	Colombia	rshuerfanob@unal.edu.co
HUYNH Vu Quang	Viet Nam	hqvu@hcmuns.edu.vn
ILYUTKO Denis Petrovich	Russian Federation	dilyutko@yahoo.com
IQBAL Zaffar	Pakistan	zaffarsms@yahoo.com
JABLONSKA Barbara Emilia	Poland	jablonsk@math.tu-berlin.de
JASSO HERNANDEZ Edna Fanny	Mexico	fjasso@matem.unam.mx
JUYUMAYA Jesus Rojas	Chile	
KANIA-BARTOSZYNSKA Joanna	United States of America	joanna.kaniabartoszynska@gmail.com, jkaniaba@nsf.gov
KAREV Maxim	Russian Federation	max.karev@gmail.com
KASHANI Seyyed Mohammad Bagher	Islamic Republic of Iran	kashanism@yahoo.com, kashanim@modares.ac.ir
KHIMSHIASHVILI Giorgi Nikolaevich	Georgia	gogikhim@yahoo.com
KHURSHUDYAN Martiros	Armenia	
KIM Soojeong	Republic of Korea	soojkim@math.uiowa.edu
KOLPAKOV Alexander Aleksandrovich	Russian Federation	
KOSCHORKE Ulrich Matthias	Germany	koschorke@mathematik.uni-siegen.de
KRICKER Andrew James	Australia	andrew.kricker@gmail.com
KULIKOV Vadim	Finland	
KULKARNI Dheeraj Dattatray	India	

LEE Stephen Peter	Canada	petervdovjak@yahoo.ca
LICATA Joan Elizabeth	United States of America	
LOPES Pedro Miguel Marques Francisco	Portugal	pelopes@math.ist.utl.pt
LUBAWSKI Wojciech	Poland	w.lubawski@impan.gov.pl
MAGGIONI Francesca	Italy	francesca.maggioni@unibg.it
MALESIC Joze	Slovenia	joze.malesic@guest.arnes.si, joze.malesic@fmf.uni-lj.si
MASTIN John Matthew	United States of America	mmastin@math.uga.edu
MAXIMO MAURICIO Mauro Alexandre	Portugal	
MECCHIA Mattia	Italy	mecchia@dmi.units.it
MOON Hyeyoung	Republic of Korea	hmoon@math.uiowa.edu
MROCZKOWSKI Maciej	Poland	mmroczko@math.univ.gda.pl
MULAZZANI Michele	Italy	mulazza@dm.unibo.it
MULLNER Daniel	Germany	
MUMBA Nephtale Bvalamanja	Malawi	registar@mzuni.ac.mw
NAZIR Shaheen	Pakistan	shaheem.nazeer@gmail.com
NIKONOV Igor	Russian Federation	nikonov@mech.math.msu.su
NIZAMI Abdul Rauf	Pakistan	arnizami@sms.edu.pk
O'DONNOL Danielle Summer	United States of America	
O'HARA Jun - Imai Jun	Japan	
PAGLIANTINI Cristina	Italy	pagliantini@mail.dm.unipi.it
PAN Weiwei	United States of America	
PANAGIOTOU Eleni	Greece	
PARSLEY Robert Jason	United States of America	
PASCALI Maria Antonietta	Italy	pascali@mat.uniroma1.it
PLACHTA Leonid - Plakhta	Ukraine	dept25@gmail.com
PRICE Candice Renee	United States of America	cprice@math.uiowa.edu, candice-price@uiowa.edu
PROTOGENOV Alexander Pavlovich	Russian Federation	alprot@appl.sci-nnov.ru
PUTYRA Krzysztof Karol	Poland	
RAMADEVI Pichai	India	
RAVNIK Miha	Slovenia	miha.ravnik@fmf.uni-lj.si
RESTREPO Alfredo	Colombia	arestrep@uniandes.edu.co
ROSICKI Witold	Poland	wrosicki@math.univ.gda.pl
ROUKEMA Fionntan Wouter Murray	Ireland	
SAITO Masahiko	Japan	
SANTIAGO GONZALEZ Grissel	Mexico	gissel@matem.unam.mx
SAZDANOVIC Radmila	Republic of Serbia	
SHABBIR Khurram	Pakistan	khurramsms@gmail.com
SHEIKHALISHAHI Akram	Islamic Republic of Iran	akram.alishahi@gmail.com
SINGER Berit Kristina	Germany	berit.singer@hispeed.ch
SORBALO Eugen	Moldova	
SPERA Mauro	Italy	mauro.spera@univr.it
STUKOW Michal	Poland	trojkat@math.univ.gda.pl
SULLIVAN John Matthew	United States of America	
TELLONI Agnese Ilaria	Italy	agneseilaria.telloni@unimore.it
THIEL Anne-Laure	France	thiel@math.u-strasbg.fr

TOMASSETTI Giovanna	Italy	tomassetti@dmmm.uniroma1.it
TRONCOSO CONA Roberto Enrique	Chile	r.troncoso.c@gmail.com
VENDRAMIN Leandro Claudio	Argentina	lvendram@dm.uba.ar
VERSHININ Vladimir	France	
WANG Yue	People's Republic of China	kellywong@cjlu.edu.cn
WIDMER Tamara	Switzerland	
WONG Helen	United States of America	
YANG Zhiqing	People's Republic of China	yangzhq@dlut.edu.cn
ZHANG Mingxing	People's Republic of China	zhangmx@dlut.edu.cn

printed on:20th Apr 2011

The Abdus Salam
International Centre for Theoretical Physics

IAEA

Advanced School and Conference on Knot Theory and its Applications to Physics and Biology

Organizer(s): Directors: S. Jablan, L.H. Kauffman, S. Lambropoulou, J. Przytycki. Local organizer: Li Jiayu
Trieste - Italy, 11 - 29 May 2009

Venue: Leonardo da Vinci Building Main Lecture Hall

Final programme

Monday, 11 May 2009 (Room:Leonardo da Vinci Building Main Lecture Hall)

08:30 - 10:30	(Room: Leonardo da Vinci Building, Lobby) --- REGISTRATION AND ADMINISTRATIVE FORMALITIES ---
10:30 - 11:00	**Le Dung Trang** / *ICTP, Trieste, Italy* **Opening**
11:00 - 12:00	**S. Lambropoulou** / *National Technical University, Athens, Greece* **Introductory talk on Knot Theory**
12:00 - 14:00	--- Lunch break ---
14:00 - 15:00	**L.H. Kauffman** / *University of Illinois at Chicago, Chicago, USA* **Combinatorial knot theory, bracket polynomial, Khovanov homology**
15:00 - 15:15	--- Coffee break ---
15:15 - 16:15	**S. Lomonaco** / *University of Maryland Baltimore County, Baltimore, USA* **Quantum computing and quantum knots**
16:15 - 16:20	--- Break ---
16:20 - 17:20	**C.K. Millett** / *University of California at Santa Barbara, Santa Barbara, USA* **Physical knot theory**

Tuesday, 12 May 2009 (Room:Leonardo da Vinci Building Main Lecture Hall)

08:30 - 09:30	**H. Morton** / *The University of Liverpool, Liverpool, UK* **Basic knot theory, quantum invariants, cabling**
09:30 - 09:45	--- Break ---
09:45 - 10:45	**J.H. Przytycki** / *The George Washington University, Washington DC, USA* **Fox coloring, Burnside groups, skein modules, Khovanov categorification of skein modules**
10:45 - 11:00	--- Break ---
11:00 - 12:00	**S. Lambropoulou** / *National Technical University, Athens, Greece* **Tangles, braids, knots in 3-manifolds**
12:00 - 14:00	--- Lunch break ---
14:00 - 15:00	**L.H. Kauffman** / *University of Illinois at Chicago, Chicago, USA* **Combinatorial knot theory, bracket polynomial, Khovanov homology**
15:00 - 15:10	--- Coffee break ---
15:10 - 16:10	**S. Lomonaco** / *University of Maryland Baltimore County, Baltimore, USA* **Quantum computing and quantum knots**
16:10 - 16:15	--- Break ---
16:15 - 17:15	**C.K. Millett** / *University of California at Santa Barbara, Santa Barbara, USA* **Physical knot theory**

Wednesday, 13 May 2009 (Room:Leonardo da Vinci Building Main Lecture Hall)

08:30 - 09:30	**H. Morton** / *The University of Liverpool, Liverpool, UK* **Basic knot theory, quantum invariants, cabling**
09:30 - 09:45	--- Coffee break ---
09:45 - 10:45	**J.H. Przytycki** / *The George Washington University, Washington DC, USA* **Fox coloring, Burnside groups, skein modules, Khovanov categorification of skein modules**
10:45 - 11:00	--- Break ---
11:00 - 12:00	**S. Lambropoulou** / *National Technical University, Athens, Greece* **Tangles, braids, knots in 3-manifolds**
12:00 - 14:00	--- Lunch break ---
14:00 - 15:00	**L.H. Kauffman** / *University of Illinois at Chicago, Chicago, USA* **Combinatorial knot theory, bracket polynomial, Khovanov homology**
15:00 - 15:10	--- Break ---

15:10 - 16:10 **S. Lomonaco** / *University of Maryland Baltimore County, Baltimore, USA*
Quantum computing and quantum knots

16:10 - 16:15 --- Break ---

16:15 - 17:15 **C.K. Millett** / *University of California at Santa Barbara, Santa Barbara, USA*
Physical knot theory

Thursday, 14 May 2009 (Room:Leonardo da Vinci Building Main Lecture Hall)

08:30 - 09:30 **H. Morton** / *The University of Liverpool, Liverpool, UK*
Basic knot theory, quantum invariants, cabling

09:30 - 09:45 --- Coffee break ---

09:45 - 10:45 **J.H. Przytycki** / *The George Washington University, Washington DC, USA*
Fox coloring, Burnside groups, skein modules, Khovanov categorification of skein modules

10:45 - 11:00 --- Break ---

11:00 - 12:00 **S. Jablan** / *Mathematical Institute Sanu, Belgrade, Republic of Serbia*
Knots, computers, conjectures

12:00 - 14:00 --- Lunch break ---

14:00 - 15:00 **S. V. Chmutov** / *Ohio State University, Mansfield, USA*
Vassiliev invariants

15:00 - 15:10 --- Break ---

15:10 - 16:10 **S. Jablan** / *Mathematical Institute Sanu, Belgrade, Republic of Serbia*
Knots, computers, conjectures

16:10 - 16:15 --- Break ---

16:15 - 17:15 **V.O. Manturov** / *Moscow State University, Moscow, Russian Federation*
Virtual knots, knot combinatorics, free links and graph links

Friday, 15 May 2009 (Room:Leonardo da Vinci Building Main Lecture Hall)

08:30 - 09:30 **S. V. Chmutov** / *Ohio State University, Mansfield, USA*
Vassiliev invariants

09:30 - 09:45 --- Coffee break ---

09:45 - 10:45 **V.O. Manturov** / *Moscow State University, Moscow, Russian Federation*
Free links and 4-graphs

10:45 - 11:00 --- Break ---

11:00 - 12:00 **S. Jablan** / *Mathematical Institute Sanu, Belgrade, Republic of Serbia*
Knots, computers, conjectures

12:00 - 14:00 --- Lunch break ---

14:00 - 15:00 **S. V. Chmutov** / *Ohio State University, Mansfield, USA*
Vassiliev invariants

15:00 - 15:10 --- Break ---

15:10 - 16:10 **V.O. Manturov** / *Moscow State University, Moscow, Russian Federation*
Graph-links (joint work with Dr. P.D. Ilyutko)

16:10 - 17:15 **OPEN DISCUSSION**

Saturday, 16 May 2009 (Room:Adriatico Guest House Kastler Lecture Hall) (**Saturday**)

15:00 - 18:00 **V.O. Manturov and D.P. Ilyutko** / *Moscow State University, Moscow, Russian Federation*
WORKSHOP ON "COMBINATORICS OF KNOTS AND VIRTUAL KNOT THEORY"

Monday, 18 May 2009 (Room:Leonardo da Vinci Building Main Lecture Hall)

08:30 - 09:30 **E. Rawdon** / *University St. Thomas, Saint Paul MN, USA*
Physical knots

09:30 - 09:45 --- Coffee break ---

09:45 - 10:45 **L.H. Kauffman** / *University of Illinois at Chicago, Chicago, USA*
Combinatorial knot theory, bracket polynomial, Khovanov homology

10:45 - 11:00 --- Break ---

11:00 - 12:00 **D. Bar-Natan** / *University of Toronto, Toronto, Canada*
(u, v, and w knots) x (topology, combinatorics, low algebra, and high algebra)

12:00 - 14:00 --- Lunch break ---

14:00 - 15:00 **M. Khovanov** / *Columbia University, New York, USA*
Link homology

15:00 - 15:10 --- Break ---

15:10 - 16:10 **P. Ozsvath** / *Columbia University, New York, USA*
Knot Floer homology

16:10 - 16:15 --- Break ---

16:15 - 17:15 **D. Buck** / *Imperial College London, London, UK*
DNA and knots

Tuesday, 19 May 2009 (Room:Leonardo da Vinci Building Main Lecture Hall)

08:30 - 09:30 **E. Rawdon** / *University St. Thomas, Saint Paul MN, USA*
Physical knots

09:30 - 09:45 --- Coffee break ---

09:45 - 10:45 **C. Gordon** / *The University of Texas at Austin, Austin, USA*
3-manifolds and knots

10:45 - 11:00 --- Break ---

11:00 - 12:00 **D. Bar-Natan** / *University of Toronto, Toronto, Canada*
(u, v, and w knots) x (topology, combinatorics, low algebra, and high algebra)

12:00 - 14:00 --- Lunch break ---

14:00 - 15:00 **M. Khovanov** / *Columbia University, New York, USA*
Link homology

15:00 - 15:10 --- Break ---

15:10 - 16:10 **P. Ozsvath** / *Columbia University, New York, USA*
Knot Floer homology

16:10 - 16:15 --- Break ---

16:15 - 17:15 **D. Buck** / *Imperial College London, London, UK*
DNA and knots

19:00 - 22:00 --- RECEPTION AT ADRIATICO GUEST HOUSE - TERRACE (weather permitting) or AT ADRIATICO CAFETERIA ---

Wednesday, 20 May 2009 (Room:Leonardo da Vinci Building Main Lecture Hall)

08:30 - 09:30 **R. Fenn** / *University of Sussex, Falmer, Brighton, UK*
Basic knot theory, fundamental group, rack, birack

09:30 - 09:45 --- Coffee break ---

09:45 - 10:45 **C. Gordon** / *The University of Texas at Austin, Austin, USA*
3-manifolds and knots

10:45 - 11:00 --- Break ---

11:00 - 12:00 **D. Bar-Natan** / *University of Toronto, Toronto, Canada*
(u, v, and w knots) x (topology, combinatorics, low algebra, and high algebra)

12:00 - 14:00 --- Lunch break ---

14:00 - 15:00 **M. Khovanov** / *Columbia University, New York, USA*
Link homology

15:00 - 15:10 --- Break ---

15:10 - 16:10 **P. Ozsvath** / *Columbia University, New York, USA*
Knot Floer homology

16:10 - 16:15 --- Break ---

16:15 - 17:15 **S.V. Chmutov** / *Ohio State University, Mansfield, USA*
Kontsevich integral and Goussarov's theorem

Thursday, 21 May 2009 (Room:Leonardo da Vinci Building Main Lecture Hall)

08:30 - 09:30 **R. Fenn** / *University of Sussex, Falmer, Brighton, UK*
Basic knot theory, fundamental group, rack, birack

09:30 - 09:45 --- Coffee break ---

09:45 - 10:45 **C. Gordon** / *The University of Texas at Austin, Austin, USA*
3-manifolds and knots

10:45 - 11:00 --- Break ---

11:00 - 12:00 **D. Bar-Natan** / *University of Toronto, Toronto, Canada*
(u, v, and w knots) x (topology, combinatorics, low algebra, and high algebra)

12:00 - 14:00 --- Lunch break ---

14:00 - 15:00 **M. Khovanov** / *Columbia University, New York, USA*
Link homology

15:00 - 15:10 --- Break ---

15:10 - 16:10 **P. Ozsvath** / *Columbia University, New York, USA*
Knot Floer homology

16:10 - 16:15 --- Break ---

16:15 - 17:15 **L.H. Kauffman** / *University of Illinois at Chicago, Chicago, USA*
Topological quantum information theory

Friday, 22 May 2009 (Room:Leonardo da Vinci Building Main Lecture Hall)

08:30 - 09:30 **R. Fenn** / *University of Sussex, Falmer, Brighton, UK*
Basic knot theory, fundamental group, rack, birack

09:30 - 09:45 --- Coffee break ---

09:45 - 10:45 **E. Flapan** / *Pomona College, Claremont CA, USA*
Knots, graphs and chemistry

10:45 - 11:00 --- Break ---

11:00 - 12:00 **J. Cantarella** / *University of Georgia, Athens GA, USA*
Physical knot theory

12:00 - 14:00 --- Lunch break ---

14:00 - 15:00 **J. Cantarella** / *University of Georgia, Athens GA, USA*
Physical knot theory

15:00 - 15:10	--- Break ---
15:10 - 16:10	**E. Flapan** / *Pomona College, Claremont CA, USA* **Knots, graphs and chemistry**
16:10 - 16:15	--- Break ---
16:15 - 16:45	**G.N. Khimshiashvili** / *Georgian Academy of Sciences, Tbilisi, Georgia* **Holomorphic dynamics of knots**
16:45 - 17:30	**OPEN DISCUSSION**

CONFERENCE (Room:Leonardo da Vinci Building Main Lecture Hall)
25 May 2009

08:30 - 09:00	--- Registration of the Participants ---
09:00 - 09:50	**D. Bar-Natan** / *University of Toronto, Toronto, Canada* **Convolutions on Lie Groups and Lie Algebras and Ribbon 2-Knots**
09:50 - 10:10	--- Coffee break ---
10:10 - 11:00	**V. Turaev** / *University of Zurich, Zurich, Switzerland* **Sections of fiber bundles over surfaces and TQFTs**
11:00 - 11:10	--- Break ---
11:10 - 12:00	**M. Khovanov** / *Columbia University, New York, USA* **Categorification of the Hecke algebra**
12:00 - 14:00	--- Lunch break ---
14:00 - 14:50	**R. Lawrence** / *Einstein Institute of Mathematics, Jerusalem, Israel* **On quantum knot and 3-manifold invariants**
14:50 - 15:10	--- Break ---
15:10 - 16:00	**C. Adams** / *Bronfman Science Center, Williams College, Williamstown MA, USA* **Three Knotty Tales: complementary regions, spiral index and spanning surfaces for alternating knots**
16:00 - 16:10	--- Break ---
16:10 - 16:40	**S. Jablan** / *Mathematical Institute Sanu, Belgrade, Republic of Serbia* **Tutte and Jones polynomials of link families**
16:40 - 16:45	--- Break ---
16:45 - 17:05	**R. Sazdanovic** / *The George Washington University, Washington DC, USA* **Categorification of orthogonal polynomials**
17:05 - 17:10	--- Break ---
17:10 - 17:30	**H. Wong** / *Bowdoin College, Brunswick, USA* **Quantum invariants of random Heegaard splittings**

17:30 - 18:30 **POSTER SESSION**

Tuesday, 26 May 2009 (Room:Leonardo da Vinci Building Main Lecture Hall)

09:00 - 09:50 **R.L. Ricca** / *University of Milano-Bicocca, Milano, Italy*
Topology bounds the energy of knots and links

09:50 - 10:10 --- Coffee break ---

10:10 - 11:00 **K.C. Millett** / *University of California at Santa Barbara, Santa Barbara, USA*
Knots and slipknots in random walks and equilateral polygons

11:00 - 11:10 --- Break ---

11:10 - 12:00 **A. Beliakova** / *University of Zurich, Zurich, Switzerland*
A unified quantum SO(3) invariant for rational homology 3-spheres

12:00 - 14:00 --- CONFERENCE PHOTO - Lunch break ---

SESSION 1 (Room:Leonardo da Vinci Building Main Lecture Hall)
26 May 2009

14:00 - 14:40 **J. Kania-Bartoszynska** / *National Science Foundation, Arlington, USA*
Topological applications of quantum invariants

14:40 - 14:50 --- Break ---

14:50 - 15:30 **M. Mulazzani** / *University of Bologna, Bologna, Italy*
Extending homeomorphisms from punctured surfaces to handlebodies

15:30 - 15:40 --- Break ---

15:40 - 16:00 **J. F. Archibald** / *University of Toronto, Toronto, Canada*
An extension of the multivariable Alexander polynomial to tangles

16:00 - 16:10 --- Break ---

16:10 - 16:30 **Z. Dancso** / *University of Toronto, Toronto, Canada*
An extension of the Kontsevich integral to knotted trivalent graphs

16:30 - 16:40 --- Break ---

16:40 - 17:00 **I. Buehler** / *University of Zurich, Zurich, Switzerland*
Integrality property of the colored Jones Polynomial

17:00 - 17:10 --- Break ---

17:10 - 17:30 **N. Abrosimov** / *Sobolev Institute of Mathematics, Novosibirsk, Russian Federation*
On the volume of a spherical octahedron with symmetries

17:30 - 18:30 **POSTER SESSION**

SESSION 2 (Room:Leonardo da Vinci Building Euler Lecture Hall)
26 May 2009

14:00 - 14:40	**J. O'Hara** / *Tokyo Metropolitan University, Tokyo, Japan* **Conformally invariant energies of knots and links**
14:40 - 14:50	--- Break ---
14:50 - 15:30	**A. Felshtyn** / *Boise State University, Boise, USA* **How to categorify dynamical zeta functions**
15:30 - 15:40	--- Break ---
15:40 - 16:00	**D. O'Donnol** / *Rice University, Houston TX, USA* **Intrinsically n-linked Spatial Graphs**
16:00 - 16:10	--- Break ---
16:10 - 16:30	**J. Parsley** / *Wake Forest University, Winston-Salem, USA* **Cohomology reveals when helicity is a diffeomorphism invariant**
16:30 - 16:40	--- Break ---
16:40 - 17:00	**E. Panagiotou** / *National Technical University, Athens, Greece* **The linking number and the writhe of uniform random walks and polygons in confined spaces**
17:00 - 17:10	--- Break ---
17:10 - 17:30	**B. Jablonska** / *Technical University of Berlin, Berlin, Germany* **Surfaces associated to a knotted space curve**
17:30 - 18:30	**POSTER SESSION**

Wednesday, 27 May 2009 (Room:Leonardo da Vinci Building Main Lecture Hall)

09:00 - 09:50	**C. Gordon** / *The University of Texas at Austin, Austin, USA* **Unsolvable problems about higher-dimensional knot groups**
09:50 - 10:10	--- Coffee break ---
10:10 - 11:00	**J.H. Przytycki** / *The George Washington University, Washington DC, USA* **Skein module motivation for Gram determinants of curves on surfaces**
11:00 - 11:10	--- Break ---
11:10 - 12:00	**J. S. Carter** / *University of South Alabama, Mobile AL, USA* **Symmetric extensions of dihedral quandles and triple points of non-orientable surfaces**
12:00 - 14:00	--- Lunch break ---
14:00 - 14:40	**S. Lambropoulou** / *National Technical University, Athens, Greece* **Framed knots, singular knots and Yokonuma-Hecke algebras**

14:40 - 14:50	--- Break ---
14:50 - 15:30	**S. V. Chmutov** / *Ohio State University, Mansfield, USA* **Tutte polynomial of graphs and Jones polynomial of virtual links**
15:30 - 15:40	--- Break ---

SESSION 1 (Room:Leonardo da Vinci Building Main Lecture Hall)
27 May 2009

15:40 - 16:20	**J. M. Sullivan** / *Technical University of Berlin, Berlin, Germany* **Distortion of Knots**
16:20 - 16:30	--- Break ---
16:30 - 16:50	**D. Goundaroulis** / *National Technical University, Athens, Greece* **The Yokonuma-Temperley-Lieb algebras**
16:50 - 17:00	--- Break ---
17:00 - 17:20	**M. Mroczkowski** / *University of Gdansk, Gdansk, Poland* **Kauffman Bracket Skein Module of the connected sum of two real projective spaces**
17:20 - 18:20	**POSTER SESSION**

SESSION 2 (Room:Leonardo da Vinci Building Euler Lecture Hall)
27 May 2009

15:40 - 16:20	**V. Vershinin** / *University of Montpellier II, Montpellier, France* **Pure virtual braid groups and Lie algebras**
16:20 - 16:30	--- Break ---
16:30 - 16:50	**F. Maggioni** / *University of Bergamo, Bergamo, Italy* **On vortex knots and unknots**
16:50 - 17:00	--- Break ---
17:00 - 17:20	**F. Canfora** / *Centro de Estudios Cientificos, Valdivia, Chile* **A phenomenological approach to 2D Potts and 3D Ising models**
17:20 - 18:20	**POSTER SESSION**

Thursday, 28 May 2009 (Room:Leonardo da Vinci Building Main Lecture Hall)

09:00 - 09:50	**L.H. Kauffman** / *University of Illinois at Chicago, Chicago, USA* **Virtual knot theory and the Jones polynomial**
09:50 - 10:10	--- Coffee break ---

10:10 - 11:00	**S.J. Lomonaco** / *University of Maryland Baltimore County, Baltimore, USA* **An intuitive overview of the theory of quantum knots**
11:00 - 11:10	--- Break ---
11:10 - 11:30	**P. Ramadevi** / *Indian Institute of Technology, Mumbai, India* **Detection of Chirality and mutation of knots and links**
11:30 - 11:40	--- Break ---
11:40 - 12:00	**P. Bellingeri** / *University of Basse Normandie, Caen, France* **Exact sequences of pure braid groups: applications and generalizations**
12:00 - 14:00	--- Lunch break ---
14:00 - 14:50	**P. Dehornoy** / *University of Caen, Caen, France* **Braid Order: History and Connection with Knots**
14:50 - 15:00	--- Break ---

SESSION 1 (Room:Leonardo da Vinci Building Main Lecture Hall)
28 May 2009

15:00 - 15:20	**A. Protogenov** / *Institute of Applied Physics of the RAS, Nizhny Novgorod, Russian Federation* **Topologically Ordered States and their Hamiltonians**
15:20 - 15:25	--- Break ---
15:25 - 15:45	**Pedro Lopes** / *Instituto Superior Tecnico, Lisbon, Portugal* **Ribbon 2-knots, graded forests and rational knots**
15:40 - 15:50	--- Break ---
15:50 - 16:10	**J. Malesic** / *University of Ljubljana, Ljubljana, Slovenia* **Meridional number of a link in the solid torus and skein modules**
16:10 - 16:20	--- Break ---
16:20 - 16:40	**Usman Ali** / *Abdus Salam School of Mathematical Sciences, Lahore, Pakistan* **Groebner bases of braid monoids and normal forms of braids**
16:40 - 16:50	--- Break ---
16:50 - 17:10	**Zaffar Iqbal** / *Abdus Salam School of Mathematical Sciences, Lahore, Pakistan* **Growth of elements of Artin monoids**
17:10 - 17:15	--- Break ---
17:15 - 17:35	**Rehana Ashraf** / *Abdus Salam School of Mathematical Sciences, Lahore, Pakistan* **Recurrence Relations for Alexander Conway Polynomial**
17:35 - 18:30	**POSTER SESSION**

SESSION 2 (Room:Leonardo da Vinci Building Euler Lecture Hall)
28 May 2009

15:00 - 15:40	**M. Spera** / *University of Verona, Verona, Italy* **On some differential and symplectic geometric aspects of linking numbers**
15:40 - 15:50	--- Break ---
15:50 - 16:10	**C. Price** / *University of Iowa, Iowa City, USA* **Solving Tangle Equations: An overview of the tangle model associated with site specific recombination and topoisomerase action**
16:10 - 16:20	--- Break ---
16:20 - 16:40	**A. Kolpakov** / *Novosibirsk State University, Novosibirsk, Russian Federation* **Spherical structures on torus knots and links**
16:40 - 16:50	--- Break ---
16:50 - 17:10	**J. Licata** / *Stanford University, Stanford CA, USA* **Legendrian knots and contact lens spaces**
17:10 - 17:15	--- Break ---
17:15 - 17:35	**A. Angeleska** / *University of South Florida, Tampa FL, USA* **Gene Rearrangements via Assembly Graphs**
17:35 - 18:30	**POSTER SESSION**

Friday, 29 May 2009 (Room:Leonardo da Vinci Building Main Lecture Hall)

09:00 - 09:50	**T. Le** / *Georgia Institute of Technology, Atlanta GA, USA* **Hyperbolic volume, L^2-torsion and homology growth**
09:50 - 10:10	--- Coffee break ---
10:10 - 11:00	**H. Morton** / *The University of Liverpool, Liverpool, UK* **Doubly periodic textile patterns**
11:00 - 11:10	--- Break ---
11:10 - 12:00	**M. Thistlethwaite** / *The University of Tennessee, Knoxville TN, USA* **A survey of methods for detecting non-reversibility of knots**
12:00 - 14:00	--- Lunch break ---
14:00 - 14:40	**M. Saito** / *University of South Florida, Tampa FL, USA* **Frobenius Modules and Essential Surface Cobordisms**
14:40 - 14:50	--- Break ---
14:50 - 15:40	**V.O. Manturov** / *Moscow State University, Moscow, Russian Federation* **Free Knots, Gauss Words, and Their Cobordisms**

15:40 - 15:50 --- Break ---

SESSION 1 (Room:Leonardo da Vinci Building Main Lecture Hall)
29 May 2009

15:50 - 16:10 **D.P. Ilyutko** / *Moscow State University, Moscow, Russian Federation*
Graph-Links: the Kauffman bracket, the Jones polynomial, and theorems of minimality

16:10 - 16:20 --- Break ---

16:20 - 16:40 **D. Afanasyev** / *Moscow State University, Moscow, Russian Federation*
On Virtual Crossing Number Estimates For Virtual Links and Long Virtual Knots

16:40 - 16:50 --- Break ---

16:50 - 17:10 **I. Nikonov** / *Moscow State University, Moscow, Russian Federation*
Circle systems of graph-links

17:10 - 18:00 **POSTER SESSION**

SESSION 2 (Room:Leonardo da Vinci Building Euler Lecture Hall)
29 May 2009

15:50 - 16:10 **Vu Huynh** / *National University of Ho Chi Minh, Ho Chi Minh City, Viet Nam*
Twisted Alexander polynomial of links in the projective space

16:10 - 16:20 --- Break ---

16:20 - 16:40 **L. Vendramin** / *University of Buenos Aires, Buenos Aires, Argentina*
RiG: Racks in GAP

16:40 - 16:50 --- Break ---

16:50 - 17:10 **Zhiqing Yang** / *Dalian University of Technology, Dalian, People's Republic of China*
An invariant from knot group and the unknotting number

17:10 - 17:15 --- Break ---

17:15 - 17:35 **L. Plachta** / *Academy of Sciences of Ukraine, Lviv, Ukraine*
Reductions of diagrams, S-graphs and braid index of links

17:35 - 18:00 **POSTER SESSION**

The Abdus Salam
International Centre for Theoretical Physics

ADVANCED SCHOOL AND CONFERENCE ON KNOT THEORY AND ITS APPLICATIONS TO PHYSICS AND BIOLOGY

11 May - 29 May 2009

ICTP, Miramare - Trieste, Italy

The Abdus Salam International Centre for Theoretical Physics (ICTP) is organizing an **Advanced School and Conference on Knot Theory and its Applications to Physics and Biology,** to be held from 11 May to 29 May 2009, in Trieste, Italy. The activity will be directed by Slavik Jablan (Mathematical Institute, Belgrade, Serbia), Louis H. Kauffman (University of Illinois at Chicago, U.S.A), Sofia Lambropoulou (National Technical University, Athens, Greece) and Jozef H. Przytycki (The George Washington University, Washington DC, U.S.A.).

The aim of the activity is to introduce young mathematicians to knot theory, low dimensional topology and its applications in physics and in the life sciences. Some familiarity with basic modern algebra and the rudiments of topology is assumed.

The first and second week (11 May - 22 May) will be an advanced school; eight mini-courses of high level on knot theory, low dimensional topology, and applications of knot theory in biology and physics will be offered. The mini-courses will be presented by leading and famous scholars in these fields (see the list of Lecturers on the right side of this poster).

The third week (25 May - 29 May) will be an international conference on knot theory and its applications. In addition to plenary lectures given by leading and famous scholars, there will also be sessions where participants can present their work.

Main Topics

General Knot Theory and Invariants of Three-Manifolds,
DNA and Chemical Knotting,
Physical Knotting and Computational Knot Theory,
Knots, Physics and Quantum Information Theory,
Categorification (Khovanov and Knot Floer Homology).

The list of Plenary Speakers of the Knot Theory Conference, 25 - 29 May 2009, includes:

Colin Adams (USA), Dorit Aharonov (Israel), Dror Bar-Natan (Canada)*, Anna Beliakova (Switzerrland), Joan Birman (USA), Scott Carter (USA), Patrick Dehornoy (France), Slavik Jablan (Serbia), Vaughan F.R. Jones (USA), Louis H. Kauffman (USA), Michael Khovanov (USA), Sofia Lambropoulou (Greece), Ruth Lawrence (Israel), Thang T.Q. Le (USA), Vassily Manturov (Russia), Ken Millett (USA), Hugh Morton (UK), Peter Ozsvath (USA)*, Jozef H. Przytycki (USA), Lev Rozansky (USA)*, De Witt Sumners (USA), Morwen Thistlethwaite (USA), Vladimir Turaev (USA)*, Oleg Viro (USA).

Participation

Mathematicians from all countries that are members of the UN, UNESCO or IAEA may attend the School. Although the main purpose of ICTP is to help researchers from developing nations through a programme of training activities within the framework of international cooperation, students and scientists from developed countries are also welcome to attend. As the activity will be conducted in English, participants must have an adequate working knowledge of that language.
As a rule, travel and subsistence expenses of the participants are borne by their home institutions. However, limited funds are available for some participants from, and working in, developing countries, not more than 45 years of age, to be selected by the Organizers. **Such financial support is available only for those who attend the entire activity.** Every effort should be made by candidates to secure support for their fare (or at least half-fare) from their home country. **There is no registration fee to attend the school.**

The closing date for receipt of requests for participation is 1 December 2008.

Interested Participants should fill in and submit the **Online Application Form** that can be found at: http://agenda.ictp.it/smr.php?2034. Kindly also attach a Curriculum Vitae (for junior applicants a recommendation letter from their supervisor is also requested). Please upload all file attachments in PDF format.

Secretariat:

SMR 2034 - Advanced School and Conference on Knot Theory and its Applications to Physics and Biology, c/o Ms. K. Mabilo
the Abdus Salam International Centre for Theoretical Physics
Strada Costiera 11
I-34014 Trieste, Italy
Telephone: +39-040-2240455 Telefax: +39-040-2240490 E-mail: smr2034@ictp.it
ICTP Home Page: http://www.ictp.it

Trieste, September 2008

Directors

S. Jablan (Serbia)

L.H. Kauffman (USA)

S. Lambropoulou (Greece)

J.H. Przytycki (USA)

Local Organizer

Li Jiayu
ICTP, Trieste, Italy

Lecturers

Dror Bar-Natan (USA)*
Dorothy Buck (USA)
Jason Cantarella (USA)
Roger Fenn (UK)
Erica Flapan (USA)
Slavik Jablan (Serbia)
Vaughan F.R. Jones (USA)
Louis H. Kauffman (USA)
Michael Khovanov (USA)
Sofia Lambropoulou (Greece)
Samuel J. Lomonaco Jr. (USA)
Vassily Manturov (Russia)
Ken Millett (USA)
Hugh Morton (UK)
Peter Ozsvath (USA)*
Jozef H. Przytycki (USA)
Eric Rawdon (USA)
Dale Rolfsen (Canada)
Jon Simon (USA)
Andrzej Stasiak (Switzerland)
De Witt Sumners (USA)

(*) to be confirmed

DEADLINE

for receipt of applications

1 December 2008

School web page:

http://agenda.ictp.it/smr.php?2034

SERIES ON KNOTS AND EVERYTHING

Editor-in-charge: Louis H. Kauffman *(Univ. of Illinois, Chicago)*

The Series on Knots and Everything: is a book series polarized around the theory of knots. Volume 1 in the series is Louis H Kauffman's Knots and Physics.

One purpose of this series is to continue the exploration of many of the themes indicated in Volume 1. These themes reach out beyond knot theory into physics, mathematics, logic, linguistics, philosophy, biology and practical experience. All of these outreaches have relations with knot theory when knot theory is regarded as a pivot or meeting place for apparently separate ideas. Knots act as such a pivotal place. We do not fully understand why this is so. The series represents stages in the exploration of this nexus.

Details of the titles in this series to date give a picture of the enterprise.

Published:*

Vol. 1: Knots and Physics (3rd Edition)
by L. H. Kauffman

Vol. 2: How Surfaces Intersect in Space — An Introduction to Topology (2nd Edition)
by J. S. Carter

Vol. 3: Quantum Topology
edited by L. H. Kauffman & R. A. Baadhio

Vol. 4: Gauge Fields, Knots and Gravity
by J. Baez & J. P. Muniain

Vol. 5: Gems, Computers and Attractors for 3-Manifolds
by S. Lins

Vol. 6: Knots and Applications
edited by L. H. Kauffman

Vol. 7: Random Knotting and Linking
edited by K. C. Millett & D. W. Sumners

Vol. 8: Symmetric Bends: How to Join Two Lengths of Cord
by R. E. Miles

Vol. 9: Combinatorial Physics
by T. Bastin & C. W. Kilmister

Vol. 10: Nonstandard Logics and Nonstandard Metrics in Physics
by W. M. Honig

Vol. 11: History and Science of Knots
edited by J. C. Turner & P. van de Griend

*The complete list of the published volumes in the series, can also be found at
http://www.worldscibooks.com/series/skae_series.shtml

Vol. 12: Relativistic Reality: A Modern View
edited by J. D. Edmonds, Jr.

Vol. 13: Entropic Spacetime Theory
by J. Armel

Vol. 14: Diamond — A Paradox Logic
by N. S. Hellerstein

Vol. 15: Lectures at KNOTS '96
by S. Suzuki

Vol. 16: Delta — A Paradox Logic
by N. S. Hellerstein

Vol. 17: Hypercomplex Iterations — Distance Estimation and Higher Dimensional Fractals
by Y. Dang, L. H. Kauffman & D. Sandin

Vol. 18: The Self-Evolving Cosmos: A Phenomenological Approach to Nature's Unity-in-Diversity
by S. M. Rosen

Vol. 19: Ideal Knots
by A. Stasiak, V. Katritch & L. H. Kauffman

Vol. 20: The Mystery of Knots — Computer Programming for Knot Tabulation
by C. N. Aneziris

Vol. 21: LINKNOT: Knot Theory by Computer
by S. Jablan & R. Sazdanovic

Vol. 22: The Mathematics of Harmony — From Euclid to Contemporary Mathematics and Computer Science
by A. Stakhov (assisted by S. Olsen)

Vol. 23: Diamond: A Paradox Logic (2nd Edition)
by N. S. Hellerstein

Vol. 24: Knots in HELLAS '98 — Proceedings of the International Conference on Knot Theory and Its Ramifications
edited by C. McA Gordon, V. F. R. Jones, L. Kauffman, S. Lambropoulou & J. H. Przytycki

Vol. 25: Connections — The Geometric Bridge between Art and Science (2nd Edition)
by J. Kappraff

Vol. 26: Functorial Knot Theory — Categories of Tangles, Coherence, Categorical Deformations, and Topological Invariants
by David N. Yetter

Vol. 27: Bit-String Physics: A Finite and Discrete Approach to Natural Philosophy
by H. Pierre Noyes; edited by J. C. van den Berg

Vol. 28: Beyond Measure: A Guided Tour Through Nature, Myth, and Number
by J. Kappraff

Vol. 29: Quantum Invariants — A Study of Knots, 3-Manifolds, and Their Sets
by T. Ohtsuki

Vol. 30: Symmetry, Ornament and Modularity
by S. V. Jablan

Vol. 31: Mindsteps to the Cosmos
by G. S. Hawkins

Vol. 32: Algebraic Invariants of Links
by J. A. Hillman

Vol. 33: Energy of Knots and Conformal Geometry
by J. O'Hara

Vol. 34: Woods Hole Mathematics — Perspectives in Mathematics and Physics
edited by N. Tongring & R. C. Penner

Vol. 35: BIOS — A Study of Creation
by H. Sabelli

Vol. 36: Physical and Numerical Models in Knot Theory
edited by J. A. Calvo et al.

Vol. 37: Geometry, Language, and Strategy
by G. H. Thomas

Vol. 38: Current Developments in Mathematical Biology
edited by K. Mahdavi, R. Culshaw & J. Boucher

Vol. 39: Topological Library
Part 1: Cobordisms and Their Applications
edited by S. P. Novikov & I. A. Taimanov

Vol. 40: Intelligence of Low Dimensional Topology 2006
edited by J. Scott Carter et al.

Vol. 41: Zero to Infinity: The Fountations of Physics
by P. Rowlands

Vol. 42: The Origin of Discrete Particles
by T. Bastin & C. Kilmister

Vol. 43: The Holographic Anthropic Multiverse
by R. L. Amoroso & E. A. Ranscher

Vol. 44: Topological Library
Part 2: Characteristic Classes and Smooth Structures on Manifolds
edited by S. P. Novikov & I. A. Taimanov

Vol. 45: Orbiting the Moons of Pluto
Complex Solutions to the Einstein, Maxwell, Schrödinger and Dirac Equations
by E. A. Rauscher & R. L. Amoroso

Vol. 46: Introductory Lectures on Knot Theory
edited by L. H. Kauffman, S. Lambropoulou, S. Jablan & J. H. Przytycki

Vol. 47: Introduction to the Anisotropic Geometrodynamics
by S. Siparov